AF478744

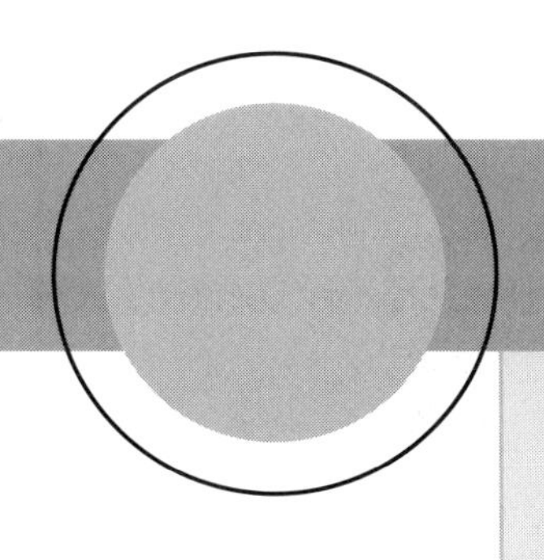

# Advanced Digital Logic Design

## *Using Verilog, State Machines, and Synthesis for FPGAs*

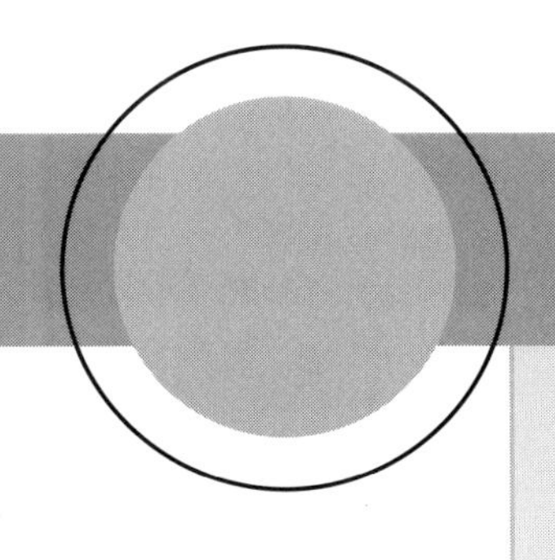

# Advanced Digital Logic Design

## *Using Verilog, State Machines, and Synthesis for FPGAs*

## Sunggu Lee

Australia   Canada   Mexico   Singapore   Spain   United Kingdom   United States

**Advanced Digital Logic Design: Using Verilog, State Machines, and Synthesis for FPGAs**
by Sunggu Lee

**Associate Vice-President
and Editorial Director:**
Evelyn Veitch

**Publisher:**
Bill Stenquist

**Sales and Marketing Manager:**
John More

**Developmental Editor:**
Kamilah Reid Burrell

**Permissions Coordinator:**
Vicki Gould

**Production Services:**
RPK Editorial Services

**Copy Editor:**
Shelly Gerger-Knechtl

**Proofreader:**
Patricia Sklenarik

**Indexer:**
Sunggu Lee/RPK Editorial Services

**Production Manager:**
Renate McCloy

**Creative Director:**
Angela Cluer

**Interior Design:**
Carmela Pereira

**Cover Design:**
Andrew Adams

**Compositor:**
PreTEX, Inc.

**Printer:**
Transcontinental Printing

COPYRIGHT © 2006 by Nelson, a division of Thomson Canada Limited.

Printed and bound in Canada

1   2   3   4   07   06   05   04

For more information contact Nelson, 1120 Birchmount Road, Toronto, Ontario, Canada, M1K 5G4.
Or you can visit our Internet site at http://www.nelson.com

Library of Congress Control Number: 2005921008

ISBN: 0-534-55161-0

ALL RIGHTS RESERVED. No part of this work covered by the copyright herein may be reproduced, transcribed, or used in any form or by any means—graphic, electronic, or mechanical, including photocopying, recording, taping, Web distribution, or information storage and retrieval systems—without the written permission of the publisher.

For permission to use material from this text or product, submit a request online at www.thomsonrights.com

Every effort has been made to trace ownership of all copyright material and to secure permission from copyright holders. In the event of any question arising as to the use of any material, we will be pleased to make the necessary corrections in future printings.

**North America**
Nelson
1120 Birchmount Road
Toronto, Ontario M1K 5G4
Canada

**Asia**
Thomson Learning
5 Shenton Way #01-01
UIC Building
Singapore 068808

**Australia/New Zealand**
Thomson Learning
102 Dodds Street
Southbank, Victoria
Australia 3006

**Europe/Middle East/Africa**
Thomson Learning
High Holborn House
50/51 Bedford Row
London WC1R 4LR
United Kingdom

**Latin America**
Thomson Learning
Seneca, 53
Colonia Polanco
11560 Mexico D.F.
Mexico

**Spain**
Paraninfo
Calle/Magallanes, 25
28015 Madrid, Spain

*To Hoonhee and Chanhee*

# Contents

## 5 State Machine Design 155



## 6 FPGA and Other Programmable Logic Devices 203

## 9 Design of a Pipelined RISC Microprocessor  338



## A THUMB Instruction Set Listing  387

# B Answers to Selected Problems 408



# Index 457

# Preface

This textbook is intended to serve as a practical guide for the design of complex digital logic circuits such as digital control circuits, network interface circuits, pipelined arithmetic units, and RISC microprocessors. As such, the book assumes that the reader has some background in basic digital logic design. However, Chapter 1 provides a condensed overview of basic digital logic design for those readers that may require such an overview.

There are various types of complex digital logic circuits. Some require extremely high performance (with clock speeds in excess of 200 MHz), others require extremely low power consumption (requiring the use of techniques such as self-timed design and variable-speed clocks), and still others require mixed analog/digital logic design. The design of such specialized circuits are not covered in this textbook, but could be considered as a follow-on to the contents of this book.

## Book Focus

The main focus of this book is on the design of digital logic circuits that are complex because they are extremely large, involve complicated calculations, or require complicated state machines. As an example, suppose we wish to design a Universal Serial Bus (USB, a common I/O interface used in modern PCs) interface for a new personal digital assistant device. Then we must design a specialized digital logic circuit that complies with the USB protocol, which basically consists of a set of actions and expected responses using packets defined in a specific format. The USB protocol is typical of the protocols used in modern computer I/O or network interfaces. Although some of these protocols can be implemented in software, some must be implemented in hardware in order to attain the required level of performance, in order to lower systems costs, or so that they can be integrated with other circuits.

In order to design complex digital logic circuits, such as the USB interface described above, several approaches can be adopted. One method of dealing with design complexity is to use a hierarchical design approach, wherein the entire circuit is designed as an interconnection of subcircuits, which may themselves be composed of smaller subcircuits, and so on, until each lowest level subcircuit can be described in a fairly simple manner. Another method is to use a high-level algorithmic description of the operation of the circuit to be designed. The proposed approach, however, is to use the above methods in conjunction with a systematic design approach involving the use of hardware description languages, state machines, and software and hardware tools for logic design.

Hardware description languages are used for *design specification, design simulation*, and as a means for describing a design in a manner that enables it to be *synthesized* into a working hardware circuit. Commonly used hardware description languages include *Verilog* and *VHDL*. This book will show how to use Verilog for the design of highly complex digital logic systems. A companion textbook, entitled *Advanced Digital Logic Design: Using State Machines, VHDL, and Synthesis for FPGAs*, includes the same material using the VHDL language.

State machines can be used to design complex sequential logic circuits in a systematic manner. State machines can be considered as the basic building blocks for complex digital circuits. This book will present state machines and the state machine-based design approach as a step-by-step systematic design procedure. Then, several complex design problems will be specified completely and solved using state machines and hardware description languages.

For the design of complex digital logic systems, what is most important, however, is the availability of software and hardware tools that enable design to proceed in an "automatic" manner as much as possible. A *synthesis* tool enables the automatic generation of a detailed logic design from a high-level circuit description using, for example, a state machine or code written in a hardware description language. This book will show how to write Verilog code so that it can be easily synthesized into a working hardware circuit. These methods will be adopted in the complex design examples presented in the last three chapters of this book.

For the hardware implementation of complex digital logic circuits, the most commonly used solutions are application-specific integrated circuits (ASICs) and field-programmable gate arrays (FPGAs). FPGAs especially are useful for hardware prototyping and the manufacture of hardware devices to be produced in small to medium quantities (and sometimes even fairly large quantities). This book will introduce the reader to FPGA technology and show how it can be used in conjunction with synthesis and hardware description language tools for the "mostly automatic" design and implementation of complex digital logic circuits. All of the code examples presented in this book have been verified using FPGA implementation.

# Book Audience and Organization

This book can be used for a junior- or senior-level undergraduate course on advanced digital logic design or Verilog-based digital logic design. For digital logic design using VHDL, the reader is referred to the companion textbook mentioned earlier.

Since hardware design is best taught by example, the latter half of this textbook is devoted to the design of various types of digital logic circuits, while the first half teaches the basic techniques and methods to be used. This book starts off with a study of hardware description languages, which have become essential tools for any serious engineer involved with digital logic hardware design. Verilog and VHDL (Very high-speed integrated circuit Hardware Description Language) have become the two most predominant hardware description languages in use today, with one or the other language favored in different parts of the world and different commercial/academic/government segments. Due to the many intended uses for these languages and the desire to be able to express numerous types of complex hardware behaviors, these languages have become increasingly complex, and several books expressly devoted to Verilog and VHDL have been written. However, it is possible to use only a subset of the features of these languages to describe the intended behavior and/or structure of highly complex digital logic circuits.

Concepts and techniques for Verilog are discussed in Chapter 3. Verilog code that is meant to model a hardware circuit is presented in a manner that facilitates hardware implementation through synthesis. Following this introduction to Verilog, all examples used throughout the rest of the textbook are presented using complete Verilog code that has been tested thoroughly for functional correctness and synthesizability.

The method of digital logic design used during the past thirty or so years and still taught in introductory digital logic courses builds up digital circuits in a hierarchical manner starting basic logic gates, such as NAND or NOR gates. However, this type of design method becomes extremely tedious and error-prone as the target circuits become large and complicated. Thus, in modern digital logic design, circuit designs typically are created "automatically" using computer-aided design tools referred to as *synthesis* tools. As the name implies, a synthesis tool can create (or synthesize) a gate-level circuit from a high-level description written in a hardware description language, a block diagram drawn using a schematic drawing tool, or a state diagram drawn using a state machine entry tool. Chapter 4 describes how synthesis tools can be used to automate the digital logic design process.

Chapter 5 presents systematic approaches to the design of large and complex digital logic circuits. By using register-transfer language and synthesis, large circuits described using computer program-like algorithms can be implemented in a mostly "automatic" manner. However, it is the opinion of the author that engineers should learn first how something works by manual calculation before using automated tools to simplify and speed up the calculation (or design) involved. For example, one should learn how to perform multiplication by pencil-and-paper before using a calculator. In this way, one can tell when simple mistakes are made by the calculator (or the person entering numbers into the calculator). Also, one has a deeper understanding of the process involved. In a like manner, this book presents "manual" digital logic design (based on the use of register-transfer language and state machines) in the first part of Chapter 5 before progressing to "automatic" digital logic design (based on the use of synthesis tools) methods in the second half of Chapter 5 and the later design example chapters. Then, in the last part of Chapter 5, a design is presented for the control of a small graphics-mode LCD panel. Such a design could be used for numerous interesting projects, including a digital "pet," a simple video game, or the display for a cell phone with internet capabilities.

Another important trend in modern digital logic design is the widespread use of PLD (programmable logic device), FPGA (field-programmable gate array), and ASIC (application-specific integrated circuit) chips. In particular, FPGAs have become popular not only for fast hardware prototyping, but also for small-to medium-volume hardware products requiring fast time-to-market. ASICs, which require the fabrication of customer-designed chips by semiconductor companies, have much higher initial costs (and fabrication delay times) but lower per-unit costs. FPGAs will be used to implement the various design projects discussed in this textbook. PLD and FPGA technology is discussed in Chapter 6.

The rest of this book presents a series of complete design examples for several representative complex digital logic design problems. Chapter 7 presents a very practical design for connecting to and monitoring the USB port of a modern PC. The USB interface is an example of a popular modern I/O interface that also can be used for networking. Chapter 8 presents designs for fast arithmetic circuits, including a pipelined multiply-and-accumulate (MAC) unit, which is a common arithmetic unit for all types of digital signal processing applications. Lastly, Chapter 9 presents a complete design for a 32-bit pipelined RISC microprocessor that supports most of the instruction set for the THUMB$^{TM}$ architecture, which is a standard subset of the ARM$^{TM}$ architecture, a commercial microprocessor architecture used in the highly successful ARM series of microprocessors. By studying these detailed design examples, it is hoped that the reader will gain an in-depth understanding of the processes involved in the design of (and gain an ability to readily design) complex digital logic circuits.

# Chapter Organization

Each chapter is organized as follows. Each chapter begins with "Important Concepts," a brief list of the main concepts to be covered in that chapter. Of course, all of the contents of each chapter are important. However, the "Important Concepts" are listed in order to remind the student of what is particularly important in that chapter—these are the main concepts that the student should attempt to learn by studying that chapter. At the end of the chapter, there is a brief summary of the chapter in the form of a "Chapter Review." Following the chapter review is a brief guide to other "Resources" on the topics covered in that chapter. These resources, which are listed in a "Bibliography," include books, conference or journal papers, and web pages. Finally, each chapter concludes with a set of "Problems" that review how well the student has learned the concepts introduced in that chapter. Difficult or time-consuming problems are marked with an asterix ($*$) in front of the problem number.

Instructional resources can be obtained by contacting the publisher or by referring to the web page for this textbook

http://engineering.thomsonlearning.com/0534551610

These resources include a solutions manual for instructors, PowerPoint presentation slides for all chapters, all of the code and figures used in this textbook, as well as other relevant material.

# Acknowledgments

The assistance of many students, colleagues, and other people were instrumental in producing this textbook. In particular, I would like to thank the students who took my "Computer Design" class for suffering through lectures taught with draft versions of this book, helping me to identify typographical and organizational errors, and offering suggestions for the improvement of this book. I would also like to thank my graduate students for helping me to verify some of the code presented in this book, working as teaching assistants in my Computer Design class, and helping with the laboratory exercises associated with this book. Finally, I would like to thank the POSTECH Teaching Innovation Center for helping to fund the development of some of the web material and the PowerPoint slides created in conjunction with this textbook.

SUNGGU LEE
Pohang University of Science and Technology
slee@postech.ac.kr

# CHAPTER 1

# Condensed Overview of Introductory Digital Logic Design

## Important Concepts

- The terminology associated with the design of digital logic circuits.
- Basic number formats used in digital logic design.
- The basic combinational logic components and how to produce combinational logic circuit designs using those components.
- The basic sequential logic components and how to produce sequential logic circuit designs using those components.

Digital logic design is a complex subject matter with many theoretical as well as practical aspects. This subject matter cannot possibly be covered in its entirety within a short chapter. Thus, for a comprehensive study of digital logic design, the reader is referred to one of the many textbooks devoted to introductory digital logic design, [Wakerly 2000] [Hayes 1993]. This chapter is only intended as a condensed review for readers who already have some familiarity with digital logic design, perhaps by having taken a prior introductory digital logic design course. It is particularly important for the reader to review the *terminology* and *primary methods* associated with digital logic circuits and their design.

Electronic circuits can be categorized into *analog* and *digital* circuits, which process *analog* and *digital* signals, respectively. An *analog* signal is a continuous valued signal, such as the pitch of a musical instrument, while a *digital* signal is a discrete-valued signal, such as the output of a traffic signal light. However, a digital circuit can also be designed to work with analog signals if each analog signal value is quantized (rounded) into one of a set of discrete signal values before being processed by the digital circuit.

In order to work effectively with digital signals, each digital signal value typically is represented using a binary number notation, consisting of only '0' and '1' digits, as opposed to a decimal number notation, which consists of digits in the range 0 to 9. In digital logic, each '0' or '1' digit is referred to as a *bit*. A sequence of eight bits is referred to as a *byte*. Since long bit sequences can become tedious to write or read, it is also common practice to represent such bit sequences using *hexadecimal* notation, in which each set of 4 bits is represented using a hexadecimal digit, which is a value in the range 0 to 15 (the numbers 10 through 15 are represented using the symbols A through F).

Since a bit can attain only one of two values, 0 or 1, a bit also can be used as a representation for a *logic* value, which can be either *false* (represented by 0) or *true* (represented by 1). Just like it is possible to perform *logical operations* (AND, OR, NOT, etc.) on logic values, it also is possible to perform the equivalent logic operations on bit values. More interestingly, by using different combinations of logic operations on sequences of bit values, it is also possible to perform arithmetic operations such as addition, subtraction, multiplication, or division. Even more interestingly, by storing bit values in memory devices and then combining those stored values with new sequences of bit inputs, it is possible to perform many interesting new operations, including the control of many different types of complex circuits. The former types of circuits are referred to as *combinational logic* circuits while the latter types of circuits are referred to as *sequential logic* circuits.

The manipulation of binary (bit) values using logic operations to perform useful circuit functions is based on solid mathematical theory. *Boolean algebra* is a mathematical system, developed by an English mathematician named George Boole in 1854, for manipulating combinations of Boolean (or true/false) statements to produce other Boolean statements. *Switching algebra*, developed by Claude E. Shannon in 1938, is an adaptation of Boolean algebra that can be used to manipulate logic expressions formed from binary-valued signals. Claude E. Shannon also showed how to implement such logic expressions in a physically realizable form by building circuits made from relays (electro-mechanical switches), in which the state of each relay (open or closed) is represented by a binary variable (with a value of 0 or 1, respectively).  Logic expressions, also referred to as *switching expressions* or *logic equations*, can be manipulated in many different ways in order to form simpler circuits: circuits with only certain types of logic functions, circuits with a certain structure, etc.  Circuit devices used to implement the basic logic functions (AND, OR, etc.) are referred to as *logic gates*.

There are several possible methods for representing '0' and '1' binary values in electronic circuits.  Such values, which are carried by wires and changed by going through logic devices, can be represented by voltage levels or transitions, current levels or transitions, signal frequencies, or signal amplitudes.  Frequencies and amplitudes are typically used in communication systems such as cellular phone systems.  In digital logic circuits, the most commonly used method is based on voltage levels.  The term *positive logic* refers to the convention where a high-voltage value represents a logic 1 and a low-voltage value represents a logic 0.  The inverse convention (low voltage for logic 1 and high voltage for logic 1) is referred to as *negative logic*.  Typically, a certain range of voltage levels (e.g., 0 volts to 2.0 volts) is designated as logic 0 while another range of voltage levels (e.g., 2.7 volts to 5.0 volts) is designated as logic 1.  Since variations in voltage levels can occur due to imperfect logic devices, noise in wires, and imperfect external inputs, ranges of voltage levels are used for both logic 0 and logic 1.  Any voltage level in between the two ranges is an undefined value, and can be interpreted as logic 0 or logic 1 by the circuit.

Digital logic circuits typically are implemented using *integrated circuit (IC)* chips or *printed circuit board (PCB)* boards.  An IC chip, which typically is packaged in a rectangular box about the size of a human nail, contains several to several tens of millions of transistors integrated together using semiconductor technology.  IC chips can be categorized on the basis of the number of logic gates that they contain.  A *small scale integration (SSI)* chip consists of around one to 20 logic gates, a *medium scale integration (MSI)* chip consists of around 20 to 200 logic gates, a *large scale integration (LSI)* chip consists of around 200 to 200,000 logic gates, and even more highly integrated chips are referred to as *very large scale integration (VLSI)* chips. A PCB (printed circuit board) is a board with chips soldered onto its surface and wires (referred to as "traces") printed (just like ink on a sheet of paper) onto its surface or an inner layer.  A computer typically consists of several of these PCB boards inserted into sockets in other PCB boards.

● ● ● ● ● ● ● ● ● ● ● ● ● ● ● ● ●

## 1.1  Number Formats

A sequence of binary digits (bits) can have different meanings depending on how the binary digit sequence (bit sequence) is interpreted.  For instance, the sequence of bits "10111000" can be interpreted as the number 184, the number $-72$, part of the Intel x86 microprocessor assembly instruction "MOV *reg, immediate*," part of a floating-point number, part of an MPEG audio stream, etc.  For use in digital logic designs, bit sequences are typically interpreted as sets of logic values, as integers, or as real numbers.

In digital logic, an integer number can be represented as an *unsigned* or *signed number*. In an unsigned number, there is no concept of a *sign* — thus, only positive numbers can be represented. In the decimal number system used in everyday activities, a sequence of decimal digits is interpreted as a number formed from the sum of $10^{i-1} \times a_i$ for all $i$, where $a_i$ is the $i$th rightmost decimal digit and 10 is the *radix* of the decimal number system. The binary number system uses a radix of two. Thus, analogous with decimal numbers, an $n$-bit unsigned binary number $b_{n-1}b_{b-2}\ldots b_1b_0$ can be interpreted as the value

$$b_{n-1}b_{n-2}\ldots b_1b_0 = \sum_{i=0}^{n-1} b_i \times 2^i \qquad (1.1)$$

A special unsigned-number representation that is used sometimes is the *binary coded decimal (BCD)* format. In this number format, each group of four consecutive bits, beginning with the rightmost bits, is converted to a decimal digit using the interpretation of Equation 1.1. Then the entire number is interpreted as a decimal number with these recently converted decimal digits. Clearly, using this format, each group of four consecutive bits representing a decimal digit cannot exceed nine when interpreted using Equation 1.1. This results in a large number of disallowed bit sequences. The main reason for using a BCD format is that bit sequences easily can be converted to decimal numbers, and vice versa.

In order to be able to represent both positive and negative integers, a *signed number* representation must be used. In such a number representation, one bit of the bit sequence typically is used to represent the *sign*, either positive or negative, of the number. Three commonly used signed-number formats that use such a *sign bit* are the *sign-magnitude*, *1's complement*, and *2's complement* number formats. In the sign-magnitude format, the *most significant bit (msb)*, the bit in the leftmost position of the bit sequence, is interpreted as the sign of the number, with '1' used for negative numbers and '0' used for positive numbers. The rest of the bits are interpreted as the *magnitude* (the absolute value of the number), which can be obtained using Equation 1.1.

Also commonly used in digital logic are the 1's complement and 2's complement formats. In both 1's complement and 2's complement, the sign of the number is interpreted using the msb (with '1' for negative and '0' for positive numbers). The two methods differ in how the magnitude of the number is computed. In 1's complement, if the number is negative, the magnitude is computed by inverting every bit in the bit sequence and then applying Equation 1.1. In 2's complement, if the number is negative, the magnitude is computed by inverting every bit in the bit sequence, adding one, and then applying Equation 1.1. If the number is positive, then the magnitude simply can be computed using Equation 1.1 directly in both 1's complement and 2's complement notation. 2's complement notation typically is used to represent integer numbers in computer systems. The reason for this choice is that, when numbers are represented in 2's complement notation, subtraction can be implemented using a regular adder. The number to be subtracted simply needs to have its sign inverted by "taking the 2's complement" of the number, which means that every bit is inverted and a one is added.

One interesting alternative format for representing signed numbers is the excess-k format. In this method, the binary bit sequence is first interpreted as an unsigned binary number $M$. Then the value $k$ is subtract from $M$. The resulting value is the desired signed-number interpretation. For example, if $k = 127$ and the binary sequence 10110111 is observed, then the number is $183 - 127 = 56$, while 00110111 is interpreted as $55 - 127 = -72$. One reason for using this type of format is to make the all-0 sequence the smallest possible value and the all-1 sequence the largest possible value.

<table><tr><td>

**Example 1.1**

</td><td>

### Unsigned and Signed Number Formats

</td></tr></table>

Consider the 8-bit bit sequence 10010111. When interpreted as an unsigned binary number, Equation 1.1 can be used to compute the decimal equivalent of the number as

$$\text{Number} = \sum_{i=0}^{n-1} b_i \times 2^i$$

$$= 1 \times 2^0 + 1 \times 2^1 + 1 \times 2^2 + 0 \times 2^3 + 1 \times 2^4 + 0 \times 2^5 + 0 \times 2^6 + 1 \times 2^7$$

$$= 1 + 2 + 4 + 0 + 16 + 0 + 0 + 128$$

$$= 151 \ .$$

When interpreted as a binary coded decimal (BCD) number, the number is equal to 97 since $1001 = 9$ and $0111 = 7$ (each set of consecutive bits is interpreted using Equation 1.1. When interpreted as a sign-magnitude signed number, this number is a negative number (since the most significant bit, denoted as *msb*, is '1') whose magnitude is $1 + 2 + 4 + 0 + 16 + 0 + 0 = 23$; thus, the entire bit sequence 10010111 is interpreted as $-23$. However, if this bit sequence is interpreted as a 1's complement number, it is equal to $-104$, since all of the bits must be inverted (resulting in 01101000) before applying Equation 1.1 to obtain the magnitude. Finally, if 10010111 is interpreted as a 2's complement number, then the sign is negative and the magnitude is equal to $01101000 + 1 = 01101001 = 64 + 32 + 8 + 1 = 105$.

Now let us consider subtraction in the signed-number formats. Let $B = 10010111$ (the same number as above) and $A = 11101110$. When interpreted as a 2's complement number, $A$ is equal to $-18$, since $00010010 = 18$ (the 2's complement of $A$ is equal to $00010001 + 1$). Then, taking the 2's complement of $B$ results in $-B = 01101001$. Thus, computing $A - B$ in 2's complement results in $A - B = A + (-B) = 11101110 + 01101001 = 101010111$. The result is a 9-bit number, since there is a carry-out. If we truncate this number to an 8-bit by ignoring the carry-out, then the result is $01010111 = 64 + 16 + 4 + 2 + 1 = 87$ in decimal. This clearly is correct, since $-18 + 105 = 87$. However, if this same computation is performed using 1's complement notation, then the result is $A + (-B) = 11101110 + 01101000\ 101010110$. If the carry-out again is discarded, the result is $01010110 = 64 + 16 + 4 + 2 = 86$, which clearly is wrong, since $-17 + 104 = 87$ ($A = -17$ and $B = -104$ in 1's complement notation). Finally, performing this computation in an analogous manner using sign-magnitude format is difficult, since forming $-B$ is more complicated than simple bit inversion (or bit inversion plus one). Thus, in conclusion, 2's complement notation is advantageous because it permits the implementation of subtraction using addition of the 2's complement of the subtrahend in a simple manner.

## 1.2 Combinational Logic

Many types of useful circuits can be created by forming *combinations* of logic operations on input signals to produce the desired output signals. When such circuits are created so that the outputs are *only* dependent on the values of the current input signals, the circuits are referred to as *combinational logic* circuits. These circuits consist of a set of basic combinational logic devices and wire connections between these devices.

### 1.2.1 Combinational Logic Devices

The basic combinational logic devices are based on basic logical operations. Thus, there is an *AND gate* that implements the AND logic function, an *OR gate* that implements the OR logic function, and a *NOT gate* (also referred to as an *inverter*) that implements the logical NOT function. Given a single input, the only possible logic functions are logical NOT and "pass through," which simply transfers the input value to the output. In a logic-gate implementation, the former is implemented by an *inverter* and the latter is implemented by a *buffer*, which is a device that not only transfers the input signal to the output but also "strengthens" the signal. The flow of electrical signals in wires can be likened to the flow of water in pipes. A device producing an electrical signal is like a water source. If multiple water taps are turned on at the same time, the water pressure drops. Likewise, if there are several destinations for an electrical signal, the signal is weakened when it arrives at its destinations. In a digital circuit, a weak electrical signal results in a slow-changing logic signal. If the signal becomes too weak, then it no longer may be interpreted at the correct digital logic value at its destination. A buffer can be used to ensure that this does not happen.

Given two inputs, there are sixteen possible logic functions. Besides the AND and OR gates, the *NAND gate* implements the NOT of the AND of the two inputs, while the *NOR gate* implements the NOT of the OR of the two inputs. An *exclusive-OR gate* outputs a logic 1 only when the two inputs are different. Alternatively, an *exclusive-NOR gate* outputs a logic 1 only when the two inputs are the same. Figure 1.1 shows the symbols used for the most common basic combinational logic devices and their truth tables (input-to-output logic function tables). Note that the bubble notation is used to denote a signal which is asserted when it is logic 0 (instead of logic 1). In terms of functionality, a bubble behaves like an inverter. Bubbles can be placed at the inputs or outputs of any logic device. For instance, a 2-input OR

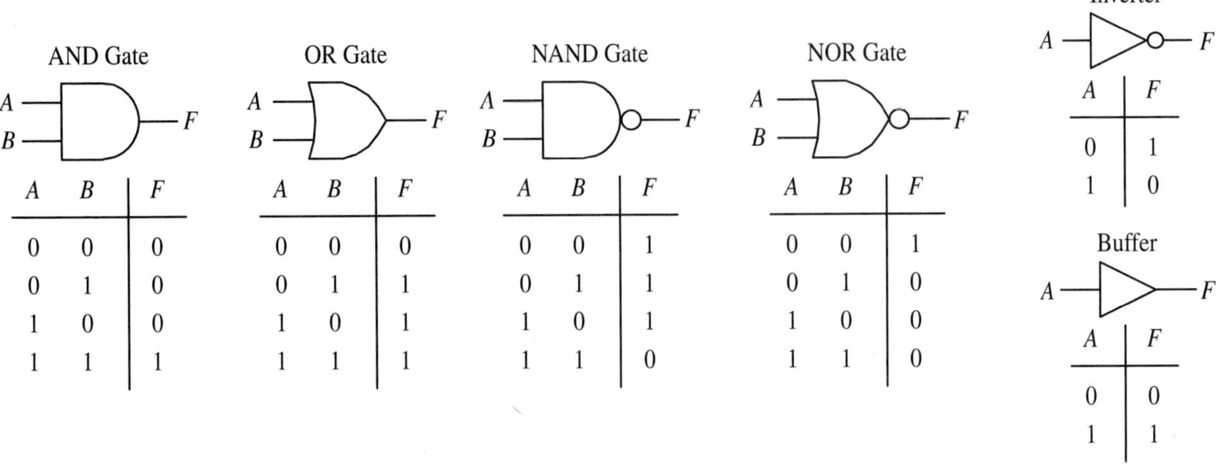

**Figure 1.1**  Symbols and truth tables for several common basic logic gates.

gate with bubbles at its two inputs behaves like an OR gate with inverters at its two inputs, which is in effect a NAND gate due to DeMorgan's theorem — a Boolean algebra theorem that states that $(A + B)' = A'B'$ and $(AB)' = A' + B'$ (the "$+$" symbol is used to denote logical OR while juxtaposition or the "$\cdot$" symbol is used to denote logical AND).

An important concept in combinational logic is the *universal gate*, which is a combinational logic device that can be used to implement *any* switching expression. The NAND and NOR gates are examples of universal gates. Figure 1.2 shows how a NAND gate can be used to implement the AND, OR, and NOT logic functions. Since any switching expression can be written as a combination of AND, OR, and NOT operations, it is clear that any switching expression can be implemented using only NAND gates. This proves that a NAND gate is a universal gate. On a related

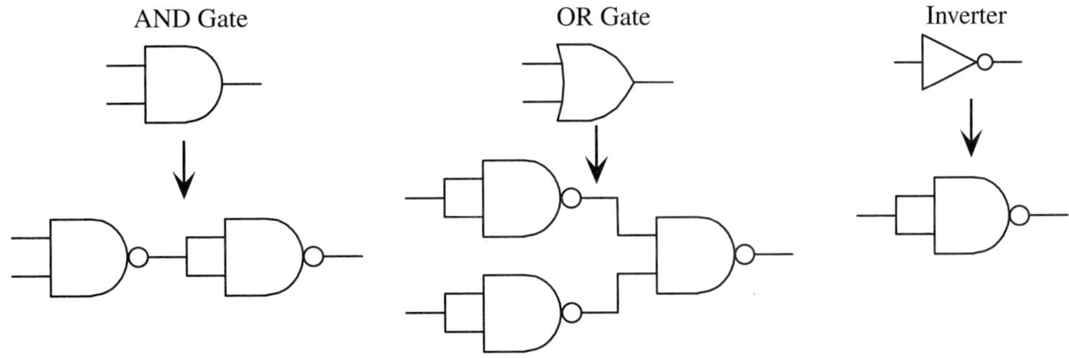

**Figure 1.2**  NAND-gate implementations of an inverter, AND gate, and OR gate.

note, a *functionally complete* set of gates is a set of logic devices that can be used to implement *any* switching expression. Thus, $\{AND, OR, NOT\}$ is a functionally complete set of gates. $\{NAND\}$ is also functionally complete, as is any universal gate expressed as a singleton set.

There are many other useful combinational logic devices that can be built up from the basic logic gates. A *multiplexer (MUX)* is a circuit that is used to route one of a set of inputs to an output line. Thus, if there are $n$ data inputs, one of those data inputs will be routed to the single data-output line and all other data inputs will be ignored. In order to select the data input to be routed to the output, there are a set of $\lceil log_2 n \rceil$ select input lines. If the decimal equivalent of the unsigned binary interpretation of the select lines is $k$, then the $k$th input is routed to the output line. For example, with an 8:1 MUX (8 data inputs, 1 data output), there are $\lceil log_2 8 \rceil = 3$ select lines (the $\lceil x \rceil$ notation denotes the smallest integer greater or equal to $x$). If the binary value of the three select lines is 110, then the data input I6 (given that the eight data inputs are named I0 through I7) is selected and routed to the data-output line.

Another logic device that is sometimes used is the *demultiplexer*, which performs the inverse function of the multiplexer. A $1:2^k$ demultiplexer routes the single data input to one of $2^k$ output lines, with the particular output line chosen dependent on the values of $k$ select inputs. The selected output line attains the same value as that of the data-input line. The other output lines output a default value, typically logic 0, or an arbitrary value that is intended to be ignored.

Two other useful combinational logic devices are *decoders* and *encoders*. A $k:2^k$ decoder decodes a $k$-bit binary number. The binary encoding for the number $i$, transmitted through a set of wires, causes the $i$th output of the decoder to become logic 1, while all other outputs of the decoder remain at logic 0. Thus, if the outputs of the decoder are considered as distinct decimal numbers, then the binary value for the number $i$ has been converted to the corresponding decimal value. A $2^k:k$ encoder performs the inverse function of the decoder. Thus, if the $i$th input of the encoder is logic 1 while all other inputs are logic 1, then the values of the $k$ output lines will be the binary encoding for $i$. If more than one input of the encoder is logic 1 (or none of the inputs are logic 1), then the behavior of the encoder depends on the particular logic design used for the encoder. A *priority encoder* assigns a priority to the inputs of the encoder in order to solve the problem of multiple input lines with logic 1 values (only the highest priority input is encoded). Figure 1.3 shows examples of symbols for decoders, encoders, and priority encoders. Note how an enable signal can be used to disable or enable the function being implemented (the bubble on the enable input to the 3:8 decoder indicates an active-low enable).

Addition is one of the most common operations performed in digital logic circuits. An *adder* can be built up in a hierarchical manner from the basic logic gates. Consider an $n$-bit adder: an adder that adds two $n$-bit numbers to produce an $n$-bit sum and possibly a carry-out bit. If we consider an arbitrary bit position $i$, the value of the $i$th sum bit $sum_i$ depends on the $i$th bit values of the two input numbers and the carry-over (referred as the *carry-in*) from the computation for the $(i-1)$th bit position. This is analogous to the addition of numbers in the decimal number system. However, in binary addition, the sum at a single bit position (digit) can only be 0 or 1. Any larger value must result in a carry-out to the next bit position.

A single-bit adder, referred to as a *full adder*, can be designed based on the following equations:

$$\text{sum}_i = a_i b_i' \text{cin}_i' + a_i' b_i \text{cin}_i' + a_i' b_i' \text{cin}_i + a_i b_i \text{cin}_i$$

and

$$\text{cout}_i = a_i b_i + a_i \text{cin}_i + b_i \text{cin}_i$$

where $a_i$ and $b_i$ are the $i$th bits of the two numbers to be added, $\text{cin}_i$ is the $i$th carry-in bit, $\text{cout}_i$ is the $i$th carry-out bit, and $\text{sum}_i$ is the $i$th sum bit. A simple $n$-bit adder structure (referred to as a *ripple-carry adder*) can be designed by simply connecting the $\text{cout}_i$ output of the $i$th full adder to the $\text{cin}_{i+1}$ output of the $(i+1)$th full adder for all possible $i$ values.

As discussed in Section 1.1, subtraction of a number $B$ can be implemented as the addition of the 2's complement of $B$. Thus, an $n$-bit *subtracter*, which is a device used to perform the subtraction of one $n$-bit number from another, can be designed using an adder. If $B$ is to be subtracted from $A$, then the 2's complement of $B$ has to be added to $A$. The 2's complement of a $B$ is obtained by inverting every bit in $B$ and then adding the number 1. Bit inversion can be implemented with an inverter. It can also be implemented with an exclusive-OR gate in which one of the two inputs is tied to logic 1 and the other input is connected to the logic signal to be inverted (since a logic 1 exclusive-ORed with any binary number $x$ produces logic 1 if $x$ is logic 0 and logic 0 if $x$ is logic 1). Thus, an *adder/subtracter* circuit (a circuit capable of addition or subtraction) can be created by simply using an adder with exclusive-OR gates attached to the second data input of the adder and a control signal attached to the second inputs of the exclusive-OR gates used and the carry-in to the adder (addition of a carry-in bit with value 1 accomplishes the final addition of the number 1 required for the 2's complement operation). Figure 1.4 shows the design for $n$-bit adder/subtracter circuit.

An *arithmetic logic unit (ALU)* is a device that can perform several useful logic and arithmetic operations in addition to simple addition and subtraction. Such a

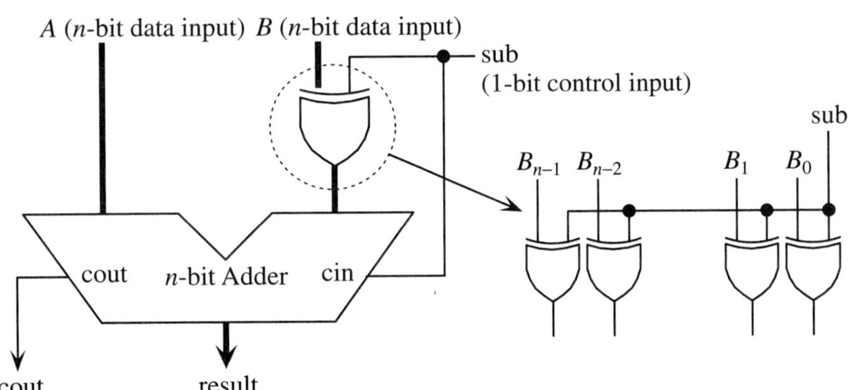


**Figure 1.4**
A design for an $n$-bit 2's complement adder/subtracter.

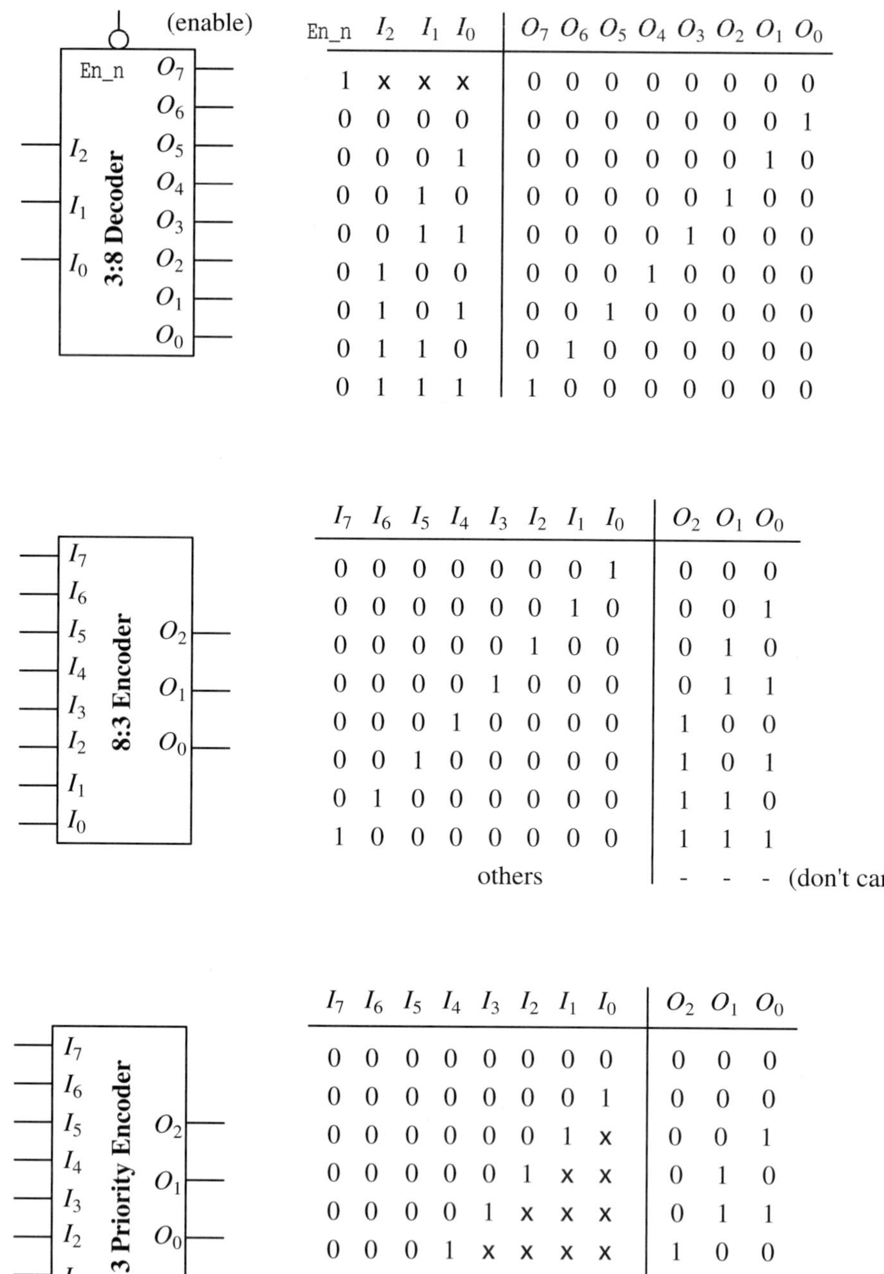

| En_n | $I_2$ | $I_1$ | $I_0$ | $O_7$ | $O_6$ | $O_5$ | $O_4$ | $O_3$ | $O_2$ | $O_1$ | $O_0$ |
|---|---|---|---|---|---|---|---|---|---|---|---|
| 1 | x | x | x | 0 | 0 | 0 | 0 | 0 | 0 | 0 | 0 |
| 0 | 0 | 0 | 0 | 0 | 0 | 0 | 0 | 0 | 0 | 0 | 1 |
| 0 | 0 | 0 | 1 | 0 | 0 | 0 | 0 | 0 | 0 | 1 | 0 |
| 0 | 0 | 1 | 0 | 0 | 0 | 0 | 0 | 0 | 1 | 0 | 0 |
| 0 | 0 | 1 | 1 | 0 | 0 | 0 | 0 | 1 | 0 | 0 | 0 |
| 0 | 1 | 0 | 0 | 0 | 0 | 0 | 1 | 0 | 0 | 0 | 0 |
| 0 | 1 | 0 | 1 | 0 | 0 | 1 | 0 | 0 | 0 | 0 | 0 |
| 0 | 1 | 1 | 0 | 0 | 1 | 0 | 0 | 0 | 0 | 0 | 0 |
| 0 | 1 | 1 | 1 | 1 | 0 | 0 | 0 | 0 | 0 | 0 | 0 |

| $I_7$ | $I_6$ | $I_5$ | $I_4$ | $I_3$ | $I_2$ | $I_1$ | $I_0$ | $O_2$ | $O_1$ | $O_0$ |
|---|---|---|---|---|---|---|---|---|---|---|
| 0 | 0 | 0 | 0 | 0 | 0 | 0 | 1 | 0 | 0 | 0 |
| 0 | 0 | 0 | 0 | 0 | 0 | 1 | 0 | 0 | 0 | 1 |
| 0 | 0 | 0 | 0 | 0 | 1 | 0 | 0 | 0 | 1 | 0 |
| 0 | 0 | 0 | 0 | 1 | 0 | 0 | 0 | 0 | 1 | 1 |
| 0 | 0 | 0 | 1 | 0 | 0 | 0 | 0 | 1 | 0 | 0 |
| 0 | 0 | 1 | 0 | 0 | 0 | 0 | 0 | 1 | 0 | 1 |
| 0 | 1 | 0 | 0 | 0 | 0 | 0 | 0 | 1 | 1 | 0 |
| 1 | 0 | 0 | 0 | 0 | 0 | 0 | 0 | 1 | 1 | 1 |
| others | | | | | | | | - | - | - (don't care) |

| $I_7$ | $I_6$ | $I_5$ | $I_4$ | $I_3$ | $I_2$ | $I_1$ | $I_0$ | $O_2$ | $O_1$ | $O_0$ |
|---|---|---|---|---|---|---|---|---|---|---|
| 0 | 0 | 0 | 0 | 0 | 0 | 0 | 0 | 0 | 0 | 0 |
| 0 | 0 | 0 | 0 | 0 | 0 | 0 | 1 | 0 | 0 | 0 |
| 0 | 0 | 0 | 0 | 0 | 0 | 1 | x | 0 | 0 | 1 |
| 0 | 0 | 0 | 0 | 0 | 1 | x | x | 0 | 1 | 0 |
| 0 | 0 | 0 | 0 | 1 | x | x | x | 0 | 1 | 1 |
| 0 | 0 | 0 | 1 | x | x | x | x | 1 | 0 | 0 |
| 0 | 0 | 1 | x | x | x | x | x | 1 | 0 | 1 |
| 0 | 1 | x | x | x | x | x | x | 1 | 1 | 0 |
| 1 | x | x | x | x | x | x | x | 1 | 1 | 1 |

**Figure 1.3** Examples of a 3:8 decoder (with active-low enable), 8:3 encoder, and 8:3 priority encoder.

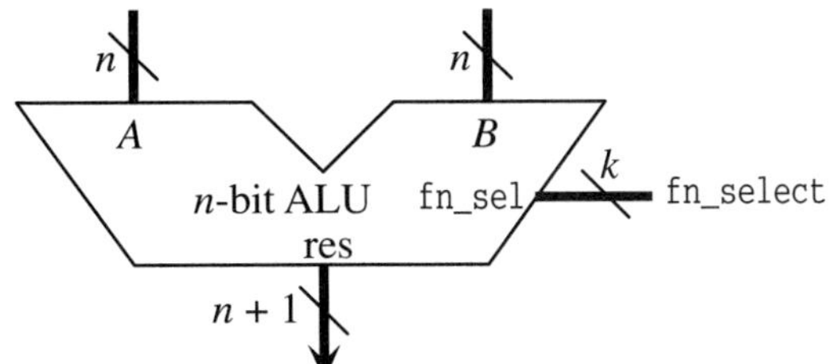


**Figure 1.5**
An $n$-bit arithmetic logic unit (ALU) with $k$-function select signals.

device has a set of inputs (referred to as the *function select* signals) used to select the particular logic or arithmetic function being performed. Figure 1.5 shows an $n$-bit ALU with $k$-function select signals. In this figure, as in Figure 1.4, note that a multi-bit signal (a bus) can be denoted using a thick line and/or a slash notation.

### 1.2.2   Combinational Logic Circuit Design

In order to be able to design combinational logic circuits to perform specific desired tasks, we must be able to create and manipulate switching expressions derived from the task specifications. General methods for this type of combinational logic design are based on graphical maps and tabular lists. Graphical-map methods are, in general, intuitively easier for humans to use, while tabular-list methods, such as the *Quine-McCluskey* method, are simpler to use for the construction of computer-aided design programs. The Quine-McCluskey method is a lengthy procedure that is described in detail in digital logic textbooks [Hayes 1993].

The most commonly used graphical-map method is the *Karnaugh map (K-map)* method. To use this method, the task specification first must be reduced to a truth table. Such a truth table shows the output values expected for each combination of input values. Given this type of truth table, it is possible to derive simplified switching expressions for each output value.

### Minimization of Switching Expressions

Let us first consider what it means to derive a simplified switching expression. Suppose that $f = abcd + bc'$ and $g = (a + bc)d + c'$. Which expression is simpler? Is $f$ simpler because it involves the OR of two AND terms? Or is $g$ simpler because it involves only ORs and ANDs with two inputs each? What is required is a formal definition of what it means for one switching expression to be simpler than another. Towards this end, let us define a *literal* as an instance of a variable, which may be in complemented or uncomplemented form, within a switching expression. Then, a commonly used criterion of "simplicity" is the number of literals used within a switching expression. Thus, $f$ uses six literals and $g$ uses five literals, which implies that $g$ is simpler than $f$. Given a combinational logic function to be implemented, the *minimal* switching expression is one that uses the fewest possible number of literals.

Besides the number of literals used, the number of *levels* used within a switching expression is also an important factor. The number of levels used refers to the number

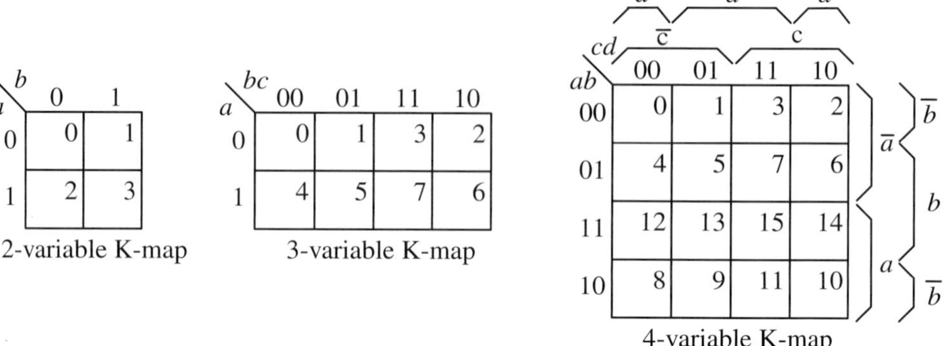

**Figure 1.6**
K-maps of two to
four variables.



of time steps required to evaluate a switching expression, given that all operations that can be performed concurrently are performed within the same time step. Then, if we refer to $f$ and $g$ above, $f$ requires one level consisting of two AND gates followed by one level consisting of one OR gate, while $g$ requires one AND-gate level, followed by an OR gate, followed by another AND gate, followed by a final OR gate. Thus, $f$ is a two-level expression and $g$ is a four-level expression. Here, $f$ also has a special form in which a level of AND gates (referred to as *product* terms) is followed by a level of OR gates (referred to as *sum* terms). This type of switching expression is said to have a *sum-of-products* form and can be implemented with an AND-OR two-level circuit (a level of AND gates followed by a level of OR gates). Alternatively, it is also possible to have an expression like $h = (a + b)(a' + c + d)$, which is referred to as having a *product-of-sums* form, implementable as an OR-AND two-level circuit.

A *Karnaugh map (K-map)* is a tool used to systematically derive a minimized sum-of-products or a minimized product-of-sums switching expression. K-maps can be drawn for functions with two to five input variables. If a circuit requires more than five input variables, it must either be broken down into smaller subcircuits hierarchically or another logic minimization method, such as the Quine-McCluskey method, must be used. Figure 1.6 shows K-maps of two to four variables (K-maps with five variables are difficult to draw and use). The row and column headings of the K-maps are labeled with binary numbers corresponding to the values of the corresponding input variables shown in the left-topmost corner. The K-map entries (boxes) are labeled with decimal numbers corresponding to the decimal interpretation of the binary number formed by the concatenation of the corresponding row and column heading labels. Thus, in the four-variable K-map, the entry with the row labeled 10 and the column labeled 01 is labeled with 9, since the decimal interpretation of 1001 is 9.

Let us now investigate *how* the K-map can be used for logic minimization. First, the truth table for the desired circuit must be entered into a K-map of the appropriate size. Figure 1.7 shows an example of a truth table and the corresponding K-map with the entries grouped to produce a minimal sum-of-products expression. A three-variable K-map was chosen, since the truth table was for a function with

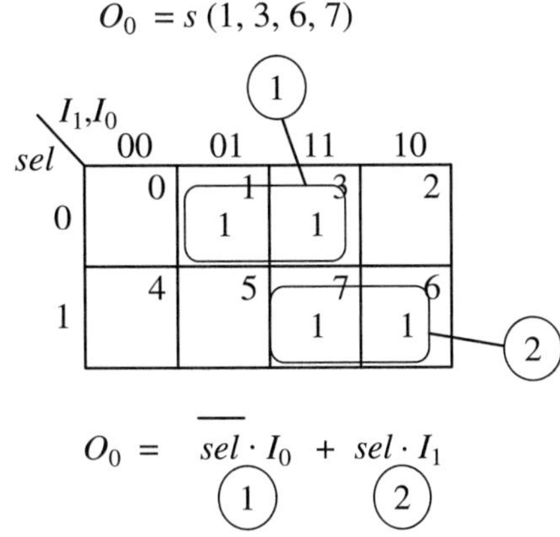



**Figure 1.7**
An example of a truth table and the corresponding K-map with the minterms grouped together to form a minimal sum-of-products expression.

three inputs. Note that a function with three inputs can also be denoted using the form $O_0(sel, I_1, I_0) = \sum(1, 3, 6, 7)$, as shown in Figure 1.7 (the numbers within the parentheses correspond to the *minterms* for this function). If the $i$ th row of the truth table (counting from row 0) produces a logic 1 output, then a 1 is entered into the entry labeled $i$ in the corresponding K-map. Rows producing an output of logic 0 do not need to be entered into the K-map, as this information is redundant (since all entries must either be 0 or 1, any blank entry can be inferred to be 0).

The K-map can be used to produce simplified switching expressions by *grouping* together adjacent 1 entries. Each K-map entry corresponds to a *minterm*, which is a product term in which every input variable is included in complemented or uncomplemented form. The K-map is drawn so that adjacent entries correspond to minterms with differences in only *one* input variable. Thus, in the K-map of Figure 1.7, the entry labeled 3 corresponds to the minterm $I_0' \cdot I_1 \cdot s$ while the adjacent entry labeled 7 corresponds to the minterm $I_0 \cdot I_1 \cdot s$. Clearly, these entries differ in the value of only one input variable ($I_0$). It also follows that, if adjacent 1-entries (such as 3 and 7 in the K-map of Figure 1.7) are grouped together, then a product term containing fewer literals than the original two minterms can be used to "cover" both minterm entries. This type of grouping can be indicated by circling those entries using a single encompassing circle. If four adjacent 1-entries can be circled together, then an even simpler product term can be used to cover all four minterms. In this manner, groups of $2^i$ ($i$ is a non-negative integer) adjacent 1-entries can be circled together to produce simpler product terms.

For this K-map simplification method to work, the entries must be ordered in *Gray code* (a code in which only one bit changes when going from one entry to the next), ordering in both the column and row orderings, as shown in Figure 1.6. Then, even entries in the boundary columns can be considered to be adjacent to one another, since only one bit changes when going from one entry to the adjacent entry on the other boundary. Likewise, entries in boundary rows are adjacent to one another, and a four-variable K-map can be viewed as being wrapped on a toroidal surface. Once a labeled K-map has been circled using the largest groups of 2, 4, 8,

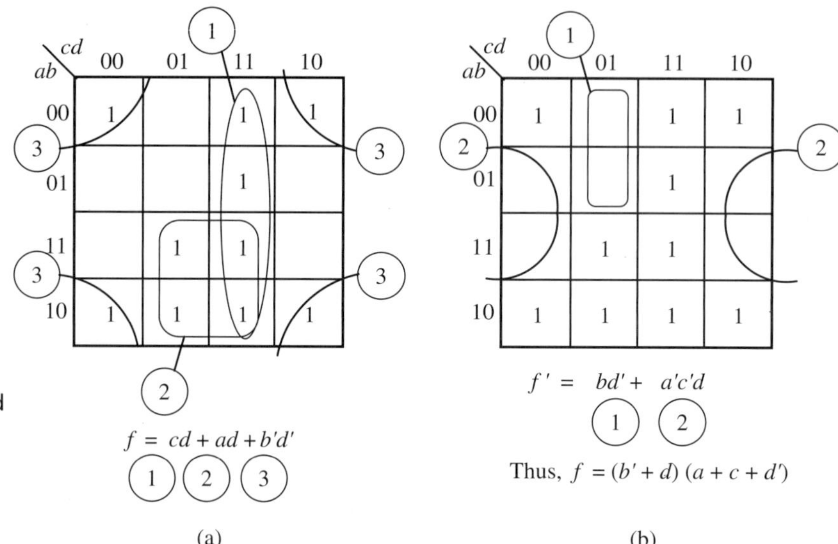



**Figure 1.8**
Using K-maps to produce (a) a minimized sum-of-products expression and (b) a minimized product-of-sums expression.

or 16 adjacent entries possible, the minimized sum-of-products expression can be obtained as the sum of the product terms corresponding to each circle. Note that *every* 1-entry must be included in at least one circle, and redundant circles do not need to be included in the final sum-of-products expression. By labeling the circles, it is easy to check the correspondence between labeled circles and product terms and, thereby, reduce the possibility of human errors in simplification. Examples of this technique are shown for a three-variable K-map in Figure 1.7 and a four-variable K-map in Figure 1.8(a).

As a variation on the above method, it is possible to work with truth tables with "*don't care*" output entries. If an output value is a don't care (denoted by $\times$) in a truth table, then the corresponding entry in a K-map is also a don't care and can be denoted as such using $\times$. In a K-map with don't-care terms, the don't-care terms can be included into adjacent 1-entry circles if desired. However, since they are don't-care terms, they don't necessarily have to be included into any circle. The use of don't-care terms can produce larger 1-entry circles, thereby resulting in simpler product terms, which in turn results in simpler sum-of-products expressions.

Another variation on the above method is a "circle-the-0's" K-map approach. Suppose that the 1-entries in a K-map correspond to the minterms for a function $f$. Then, instead of circling adjacent 1-entries, we can circle adjacent 0-entries using the largest possible groups of size $2^i$, where $i$ is any non-negative integer value. If all 0-entries are covered while excluding all 1-entries, this corresponds to the logical inverse of the function that we are trying to implement. Thus, the original function $f$ can be produced by first writing down the function $f'$ corresponding to the 0-circles and then forming the complement of $f'$ using DeMorgan's theorem. Since $f'$ is a sum-of-products form, $f$ will have a product-of-sums form. $f$ will be a *minimized* product-of-sums expression if all 0-entries in the K-map are covered, while excluding all 1-entries, and using the largest possible $2^i$ sized groupings. An example of this method is shown in Figure 1.8(b). By examining both the minimized sum-of-products and product-of-sums expressions, it is possible to produce a more minimized hardware implementation than when only one type of K-map minimization is used.

### Random Logic Circuit Implementations

Combinational-logic circuit design is based on the manipulation of switching expressions and the hierarchical interconnection of previously designed components. The straightforward logic implementation of a sum-of-products switching expression can be accomplished using AND, OR, and NOT (inverter) gates. Circuit design using the basic logic gates is referred to as *random logic* design. Interestingly, any circuit consisting of AND, OR, and NOT gates easily can be converted into a circuit consisting of only NAND gates. The significance of being able to use only one type of gate, such as a NAND, is that this can result in a lower-cost implementation (for example, if only that type of gate is available in a chip package or design library).

Figure 1.9 shows an example of a sum-of-products switching expression and different circuit implementations for that switching expression. Note that a 2-level NAND-NAND circuit can be implemented by replacing every OR or AND gate in a 2-level AND-OR circuit with NAND gates. However, note that, if an input variable is connected directly to an input of an OR gate in a AND-OR circuit, then the complemented value of that input or an inverter (which can be implemented using a 2-input NAND gate with its two inputs tied together) must be used in the corresponding NAND-NAND circuit, as was done in Figure 1.9(b). The straightforward conversion from an AND-OR circuit to a NAND-NAND circuit is possible because of DeMorgan's theorem. Also, as shown in Figure 1.9(b), the bubble at the output of a NAND in the first level can be cancelled with a bubble at the input of the NAND gate in the second level. An analogous procedure can be used to convert any 2-level

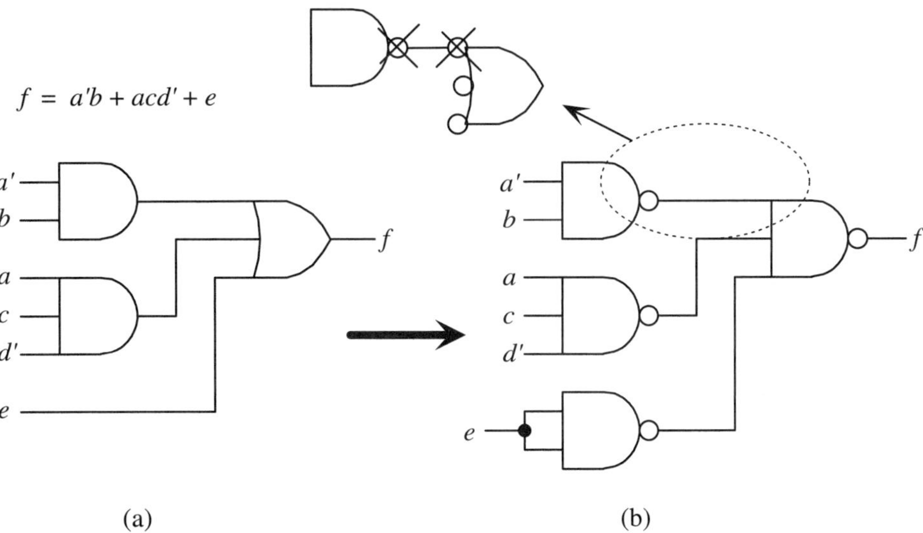


**Figure 1.9**
An example of a sum-of-products expression implemented using (a) AND and OR gates and (b) only NAND gates.

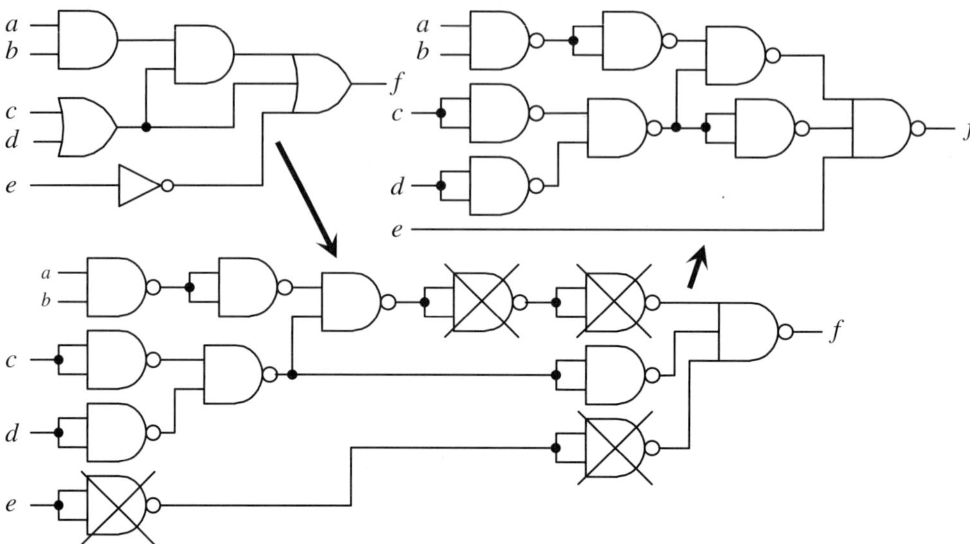

**Figure 1.10**
A general method for converting a circuit with AND, OR, and NOT gates into a circuit with only NAND gates.

OR-AND circuit to a 2-level NOR-NOR circuit. The explanation of this procedure is left as a problem exercise for the reader.

An arbitrary switching expression can be implemented using only NAND gates by first converting the expression to a sum-of-products form and then using a 2-level NAND-NAND circuit. It is also possible to perform the conversion directly from an arbitrary switching expression or an arbitrary multi-level circuit. This can be done by substituting all AND, OR, and NOT gates with their respective NAND-gate equivalent circuits, as shown in Figure 1.2. After this substitution step, any sequence of two NOT gates can be cancelled out. Figure 1.10 shows an example of this conversion process for an arbitrary multi-level circuit. Figure 1.10 also shows how a 3-input NAND gate can be converted to use only 2-input NAND gates (note that the conversion is not as simple as using one 2-input NAND followed by another 2-input NAND). In a similar manner, NAND gates with larger numbers of inputs can also be converted to circuits with only 2-input NAND gates.

### Design using Multiplexers and Decoders

Although its primary purpose is to multiplex several data-input lines to one output line, an MUX can also be used as a *universal logic device*. A $2^k$:1 MUX, which has $2^k$ data inputs and $k$ select inputs, can be used to implement any combinational logic function with up to $k + 1$ inputs. Given a $(k + 1)$-input logic function (a switching expression with $(k + 1)$ variables), $k$ of the inputs can be connected to the $k$ select lines. Then, the $(k + 1)$th input, in complemented and uncomplemented form, can be connected to several of the input-data lines. Suppose $v_{k+1}$ is the $(k + 1)$th input. Each input-data line of the $2^k$:1 MUX is connected to logic 0, logic 1, $v_{k+1}$, or $v'_{k+1}$, depending on the combination of the prior $k$-input values that will select that input-data line. Thus, a truth table (similar to those shown in Figure 1.1) can be used to

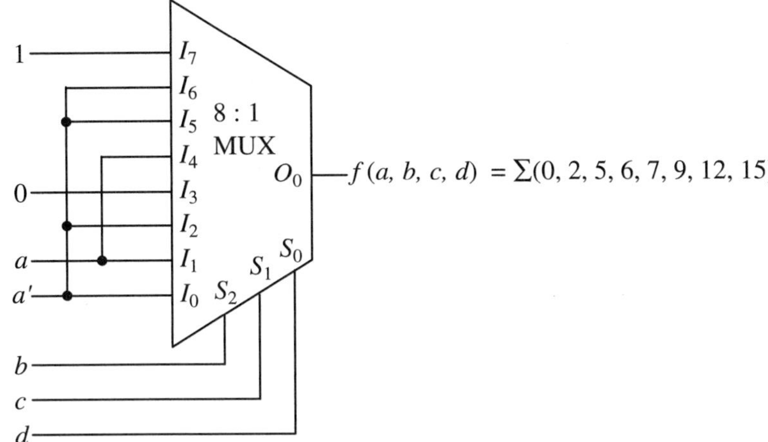



**Figure 1.11**
An example of an 8:1 MUX connected to implement an arbitrary four-variable combinational logic function.

derive the necessary connections for the $2^k : 1$ MUX. Figure 1.11 shows an example of an 8:1 MUX and the connections used to implement a four-variable combinational logic function (note that $k = 3$, since $2^3 = 8$).

It is also possible to design arbitrary combinational logic circuits using decoders and a few basic logic gates. A $k : 2^k$ decoder converts a $k$-bit binary signal into one of $2^k$ signals, depending on the $k$-bit binary-value input. Thus, the $2^k$ output signals can also be considered as the possible minterms of the $k$-bit binary signal, where a minterm is the logical AND of all of the input signals, with each input appearing in complemented or uncomplemented form. Any switching expression can be written as a sum-of-products. Then, since each product is either a minterm or can be converted into a sum of the minterms, any switching expression can be written as a sum-of-minterms. Then it follows that a $k : 2^k$ decoder followed by a large OR gate can be used to implement any switching expression of $k$ variables. If there are bubbles at the outputs of the $k : 2^k$ decoder (the decoder outputs are active-low), then a large NAND gate can follow the $k : 2^k$ decoder since a NAND gate is equivalent to an OR gate with inverted inputs.

Figure 1.12 shows an example implementation of an arbitrary three-variable switching expression using a 3:8 ($k = 3$) decoder and a NAND gate. Note that the NAND gate has been drawn as an OR-like symbol with inverted inputs. This is



**Figure 1.12**
Implementation of an arbitrary three-variable switching expression using a 3:8 decoder and a NAND gate.

perfectly acceptable due to DeMorgan's theorem. In this figure, the NAND gate has been drawn specifically in this manner to show the cancellation of the "bubbles" at the outputs of the decoder and the inputs of the NAND gate.

### Design of Large Combinational Logic Circuits

The techniques presented in the previous subsections can be used for combinational logic circuits with only a few inputs (less than five or six). For larger combinational logic circuits, information about the intended behavior of the circuit and/or hierarchical design must be used. Of course, automated techniques based on the use of hardware description languages and computer-aided design tools can also be used, as described in Chapters 2 through 5. Memory devices and programmable-logic devices can also be used to implement large combinational logic circuits, as detailed in Chapter 6. Many arithmetic circuits can be built up in a hierarchical manner using the adder as a basic component. The design of fast adders, multipliers, and other arithmetic units from simple adder blocks are covered in Chapter 8.

## 1.3  Sequential Logic

A *sequential circuit* consists of a combinational logic subcircuit and memory elements. The distinguishing characteristic of a sequential circuit is that its outputs are dependent on the values of the past inputs as well as the current circuit inputs. Figure 1.13 shows a block diagram for a general sequential circuit. There are two sets of outputs emanating from the combinational logic subcircuit. One set of outputs feeds the inputs of the memory devices, which in turn feed back into the combinational logic subcircuit. The other set of outputs produces the primary outputs for the sequential circuit. The memory devices are used to store the effect of the past values of primary inputs — this information is referred to as the *state*. The outputs of the memory devices constitute the *current state* and the inputs to the memory devices constitute the *next state*.

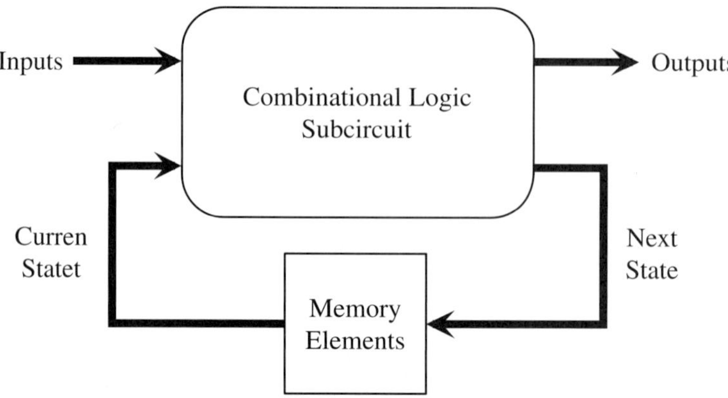


**Figure 1.13**
A general sequential circuit.

Sequential logic circuits can be categorized into *asynchronous sequential* and *synchronous sequential* circuits. The state of an asynchronous sequential circuit can change at any time, while the state of a synchronous sequential circuit only can change at specific times dictated by a special input signal referred to as the *clock*. An asynchronous sequential circuit can be constructed by using a *delay element*— a device which simply delays the transmission of a signal value by a certain time amount—or a *wire*—which has a natural delay dependent on the length of the wire and the properties of the material used in the wire. Thus, a circuit with a feedback loop constitutes an asynchronous sequential circuit. The design and analysis of an asynchronous sequential circuit is, in general, an extremely difficult problem because all parts of the circuit can be changing continuously. However, as shown in the next section, there are a few well-known asynchronous sequential circuits that are used as standard components in higher-level sequential circuits.

A synchronous sequential circuit is a sequential circuit in which the memory elements can only change state at specific times dictated by a *clock* signal. During other times, inputs and internal signals can undergo numerous changes, perhaps as part of a long series of logic operations or a long arithmetic computation, without adversely affecting the circuit (since the state does not change). A synchronous sequential circuit can be designed and analyzed in a methodological fashion by abstracting out the detailed circuit timing behavior and simply designing it as a set of states and combinational operations used to transition from one state to the next.

### 1.3.1  Sequential Logic Devices

The main commonly used sequential logic devices are different types of memory devices. The memory devices typically used with sequential logic circuits are *latches* and *flip-flops*. As its name implies, a *latch* is a device which can "latch up" into a certain state under particular input conditions. A *flip-flop* is a device which can transition between logic 1 and logic 0 states only at clock transition times.

### Latch Designs

A commonly used latch is the *R-S latch*, whose design is shown in Figure 1.14. As shown in this figure, an R-S latch is designed by connecting two NOR gates in a cross-coupled manner. The steady-state behavior of this circuit can be analyzed by

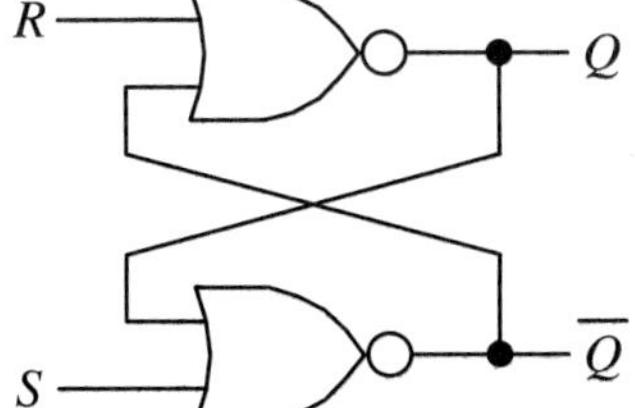


**Figure 1.14**
A design for an R-S latch.

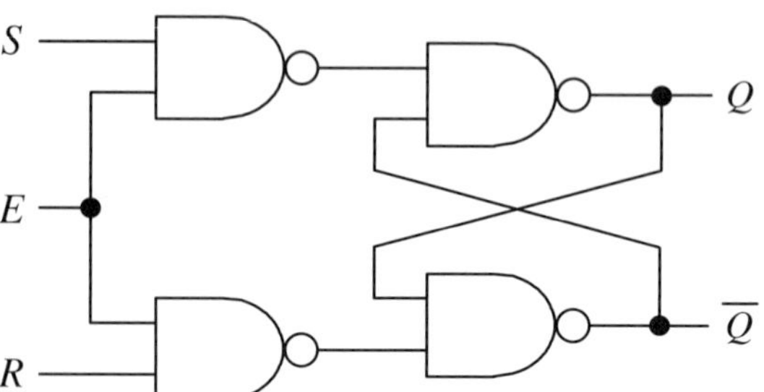

**Figure 1.15**
A design for an
enabled R-S latch.

starting from known logic values. For example, when the $R$ input is 1 and the $S$ input is 0, the output of the first NOR gate ($Q$) is 0, since the NOR of any value with 1 is 0. Then it follows that the output of the second NOR gate ($\overline{Q}$) is 1, since the NOR of the $S$ input (0) with $Q$ (0) is 1. Next, if $R$ is 0 and $S$ is 1, then an analogous derivation can be used to show that $Q = 1$ and $\overline{Q} = 0$. Thus, if $Q$ is considered as the *state* of this latch, $S$ sets the state, and $R$ resets the state. If both $R$ and $S$ are 1, then the result will be $Q = 0$ and $\overline{Q} = 0$. However, this typically is regarded as a *disallowed input* combination, because we would like $\overline{Q}$ to always be the opposite value of $Q$.

If both $R$ and $S$ are 0, the state of the latch depends on the previous $Q$ and $\overline{Q}$ values. If the previous $Q$ value was 1, then 1 NORed with the $S$ input 0 results in $\overline{Q} = 0$, which in turn results in $Q = 1$, since $\overline{Q} = 0$ NORed with $R = 0$ is 1. Since this final value of $Q$ is identical to the initial value of $Q$, the circuit is in steady-state. Likewise, it can be shown that, if the previous $Q$ value was 0, then $R = 0$ and $S = 0$ will result in a final $Q$ value of 0. Thus, $R = 0$ and $S = 0$ causes the latch to store the previous state value.

A circuit similar to the R-S latch can be constructed by using NAND gates instead of the NOR gates in Figure 1.14. This latter circuit is sometimes referred to as an $\overline{S}\text{-}\overline{R}$ *latch* since the top input, labeled $\overline{S}$, causes the $Q$ output to be set when $\overline{S} = 0$ while the bottom input, labeled $\overline{R}$, causes the $Q$ output to be reset when $\overline{R} = 0$. The 00 input combination is disallowed while the 11 input combination causes the latch to retain its previous state.

The R-S latch (and also the $\overline{S}\text{-}\overline{R}$ latch) can be made into a more useful device by adding an *enable* input. Such a circuit, referred to as an *enabled R-S latch*, is shown in Figure 1.15. In this circuit, two NAND gates are followed by a cross-coupled NAND structure. The cross-coupled NAND circuit is an $\overline{S} - \overline{R}$ latch. The inputs to these NAND gates are the outputs of two additional NAND gates. Since the enable ($E$) input is connected to both of these NAND gates, these NAND gates act as inverters when $E = 1$. When $E = 0$, $\overline{S} = 1$ and $\overline{R} = 1$, regardless of the primary input values ($S$ and $R$).

The enabled R-S latch can be converted into an *enabled D latch* by inserting an inverter between the $S$ and $R$ inputs, as shown in Figure 1.16. In an enabled D latch, the state follows the $D$ input value when $E = 1$ and retains the previous value when $E = 0$. Also, since the $R$ input is always the opposite value of the $S$ input, there is no danger of a disallowed input combination.

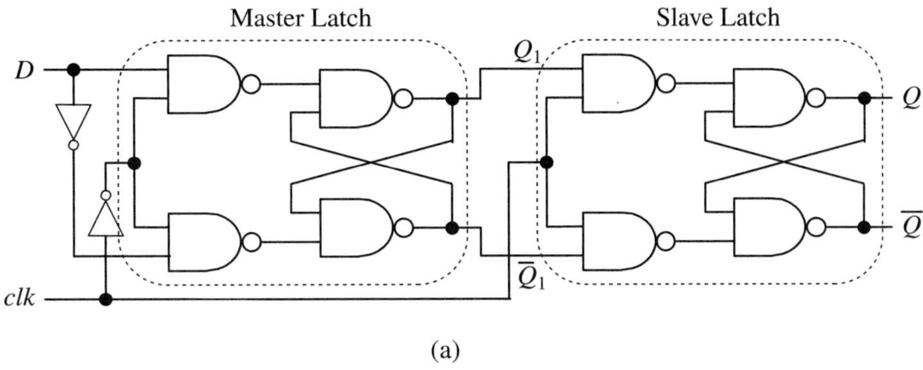


(a)

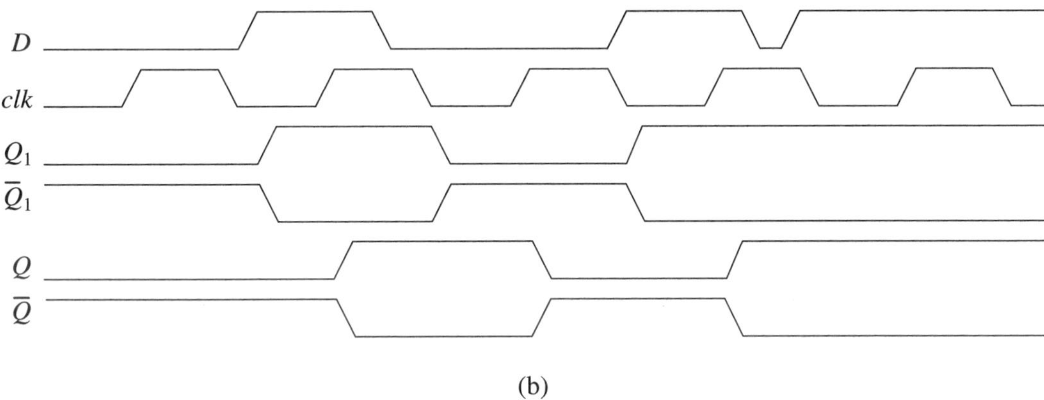


(b)

**Figure 1.16**   A master-slave latch (a) design and (b) timing diagram.

Enabled R-S latches and enabled D latches are *pulse-triggered* devices. The enable input $E$ can be considered as a *clock* signal. Then, it can be seen that the state of these latches only can change during specific values of the clock signal (in this case 1). It is also possible to have an active-low enable input, in which case the state can only change when the clock signal is at the 0 level.

Enabled latches can be connected in a *master–slave* configuration. A *master–slave latch* is a latch structure consisting of two latches, one of which feeds data into the other. Figure 1.16(a) shows the design for a master–slave latch. The first latch, labeled the *master*, is configured as an enabled D latch with $\overline{clk}$ acting as the enable input. The second latch, labeled the *slave*, is an enabled R-S latch which receives its $S$ and $R$ inputs from the $Q$ and $\overline{Q}$ outputs of the master latch and its enable input from $clk$. Since the master latch is enabled by $\overline{clk}$ and the slave latch is enabled by $clk$, the state of the master latch changes when $clk = 0$ and the state of the slave latch changes when $clk = 1$. Thus, the slave latch "follows" the master latch, as shown in the timing diagram of Figure 1.16(b).

### Flip-Flop Designs

Flip-flops are memory devices in which the state only can change during transitions of the clock signal. If the design and timing diagram of the master–slave latch shown in Figure 1.16 is studied closely, it can be seen that this design implements such a device, since the value of the data $D$ at the time of the rising edge of *clk* is the final value stored (and displayed as the output $Q$) in this device. This is referred to as a *D flip-flop*. Flip-flops are triggered off of the rising or falling edges of the clock signal, but not both. Based on its triggering edge, flip-flops can be categorized into *positive edge-triggered* and *negative edge-triggered* devices, respectively. The symbols for positive and negative edge-triggered D flip-flops are shown in Figure 1.17. As can be seen, a negative edge-triggered D flip-flop has a bubble at the clock input, which is denoted using the notation shown.

A *J-K flip-flop* is another type of edge-triggered memory device. The operation of a J-K flip-flop is similar to an R-S latch, with the $J$ input corresponding $S$ (set) and the $K$ input correspond to $R$ (reset). However, the J-K flip-flop also allows the 11 input combination ($J = 1$ and $K = 1$), in which case the state of the flip-flop is *toggled* (i.e., the state is inverted from its previous stored value). As with the D flip-flop, the state of a J-K flip-flop only can change during the positive transition (for a *positive edge-triggered* device) or the negative transition (for a *negative edge-triggered* device) of the clock input. The symbol and truth table for a J-K flip-flop are both shown in Figure 1.17. In the truth table, note that an up-arrow symbol is used to denote the upward transition of the clock signal.

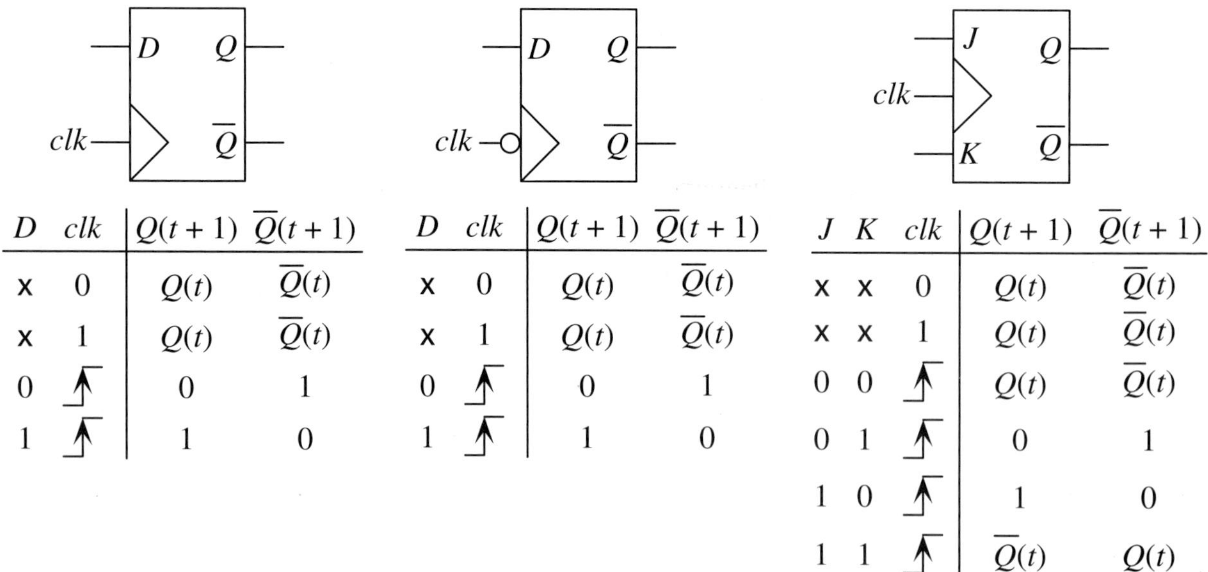

| $D$ | *clk* | $Q(t+1)$ | $\overline{Q}(t+1)$ |
|---|---|---|---|
| x | 0 | $Q(t)$ | $\overline{Q}(t)$ |
| x | 1 | $Q(t)$ | $\overline{Q}(t)$ |
| 0 | ↑ | 0 | 1 |
| 1 | ↑ | 1 | 0 |

| $D$ | *clk* | $Q(t+1)$ | $\overline{Q}(t+1)$ |
|---|---|---|---|
| x | 0 | $Q(t)$ | $\overline{Q}(t)$ |
| x | 1 | $Q(t)$ | $\overline{Q}(t)$ |
| 0 | ↑ | 0 | 1 |
| 1 | ↑ | 1 | 0 |

| $J$ | $K$ | *clk* | $Q(t+1)$ | $\overline{Q}(t+1)$ |
|---|---|---|---|---|
| x | x | 0 | $Q(t)$ | $\overline{Q}(t)$ |
| x | x | 1 | $Q(t)$ | $\overline{Q}(t)$ |
| 0 | 0 | ↑ | $Q(t)$ | $\overline{Q}(t)$ |
| 0 | 1 | ↑ | 0 | 1 |
| 1 | 0 | ↑ | 1 | 0 |
| 1 | 1 | ↑ | $\overline{Q}(t)$ | $Q(t)$ |

**Figure 1.17**   A positive edge-triggered D flip-flop, a negative edge-triggered D flip-flop, and a positive edge-triggered J-K flip-flop.

## 1.3.2   Synchronous Sequential Circuit Design

A synchronous sequential circuit can be designed by deriving the next-state transition logic and the output logic. A systematic method for this type of design is based on the use of *finite state machines (FSMs)*. A finite state machine is a model for a synchronous sequential circuit that uses a finite set of states and conditions for transitions between those states. There are two general types of finite state machines: *Moore machines* and *Mealy machines*. In a Moore machine, the outputs of the circuit are dependent on the current state only. In a Mealy machine, the outputs are dependent on both the current state and the current inputs.

Both Moore and Mealy machines can be designed in a systematic manner using the *finite state machine design (FSM design)* method. This method consists of the following sequence of steps:

1. Form a *state diagram* for the target circuit.

2. Create a set of state variables and encode the states in the state diagram using these state variables.

3. Form a *state transition table* using the state diagram from Step 1 and the state encoding from Step 2.

4. Derive simplified logic equations for the next state variables and the outputs from the state transition table.

5. Form the design for the target circuit by using one flip-flop for each state variable and combinational logic corresponding to the simplified logic equations (derived in Step 4) for the next state and output variables.

**Example 1.2**   **Sequence Detector**

The finite state machine method can be illustrated with a design example. Let us design a circuit for a 101 sequence detector, which is a circuit that accepts a serial data-input stream and outputs an output of 1 when it observes the 101 input pattern and an output of 0 at all other times. Note that overlapping sequences of 101 patterns must also be detected. The first step in the finite state machine method requires us to form a state diagram. A state diagram consists of a set of states and (possibly) labeled transitions between those states. Figure 1.18 shows a state diagram for a 101 sequence detector. Note that each state of this diagram "remembers" a portion of the desired 101 sequence. The state transitions are labeled in the manner $x/y$, where $a$ corresponds to values of the input variables resulting in that transition, and $y$ corresponds to the values of the output variables given the state from which the transition starts and the input values $x$. Thus, since the outputs are dependent on both the current-state and the input variables, this type of diagram corresponds to a Mealy machine. In this type of diagram, it is important to draw a transition for each possible set of input values that can be encountered in each state.

The second step of the finite state machine requires a state assignment. Since there are four states, two state variables are sufficient to encode these four states. For simplicity of exposition, let us simply use a binary encoding for the states. In general, however, some state encodings may result in simpler circuits than other state encodings, although the differences typically are not that large.

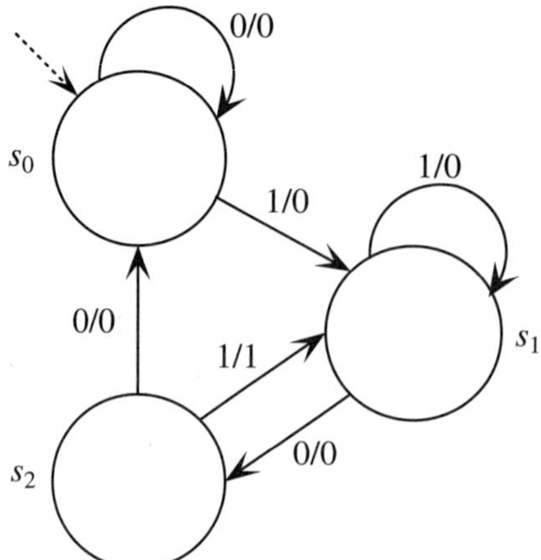

**Figure 1.18**
A state diagram for a sequence detector for the "101" sequence.

Given the binary state encoding and the state diagram of Figure 1.18, we can derive the state transition table shown in Figure 1.19. Note that the state transition table shows the next-state variables and outputs for each combination of current-state variable values and input values.

The state transition table can be viewed as a combined set of truth tables for each of the next-state and output variables. Thus, in Step 4, we can derive simplified combinational logic equations by using K-maps, a logic minimization computer program, or other methods. In this case, simple observation of the state transition table (as a set of truth tables) can be used to derive the necessary switching expressions due to the simplicity of the state diagram used. Thus, the following switching expressions can be derived.

$$NA = A'Bd$$

$$NB = A'B + d$$

$$detect = Ad$$

| | Current State | | Inputs | Next State | | Outputs |
|---|---|---|---|---|---|---|
| | $A$ | $B$ | $d$ | $NA$ | $NB$ | $detect$ |
| $s_0$: | 0 | 0 | 0 | 0 | 0 | 0 |
| $s_0$: | 0 | 0 | 1 | 0 | 1 | 0 |
| $s_1$: | 0 | 1 | 0 | 1 | 1 | 0 |
| $s_1$: | 0 | 1 | 1 | 0 | 1 | 0 |
| $s_2$: | 1 | 1 | 0 | 0 | 0 | 0 |
| $s_2$: | 1 | 1 | 1 | 0 | 1 | 1 |

**Figure 1.19**
The state transition table for the "101" sequence detector.

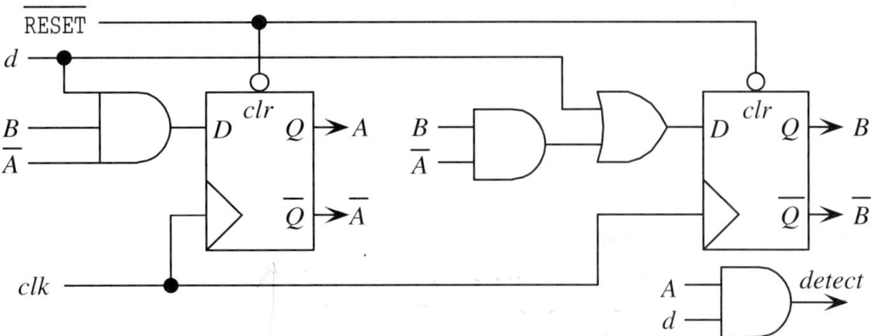

**Figure 1.20**
The complete circuit for the "101" sequence detector.

Finally, in Step 5, we need to derive the required circuit design based on the simplified logic equations for the next-state and output variables. The synchronous sequential circuit for this target circuit can be designed using D flip-flops or J-K flip-flops. If J-K flip-flops are used, then equations must be derived for the J and K inputs of all flip-flops. However, if D flip-flops are used, then the D inputs to the flip-flops, which are used to encode the current state, correspond to the simplified logic equations for the next-state variables. Based on this observation, we can derive the final circuit design of Figure 1.20. Note that an asynchronous reset input is used to initialize the flip-flops. The use of this type of flip-flop initialization mechanism is recommended in order to be able to start the circuit from a known state every time.

### 1.3.3  Hazards and Glitches

*Glitches*, unintended temporary spikes of voltage, can occur in combinational logic circuits. Depending on the manner in which such circuits are designed, such glitches can be ignored or cause the circuits to fail completely. If the combinational logic circuit is a part of a larger *synchronous sequential* logic circuit, then the presence of glitches normally can be ignored. However, if the combinational logic feeds a clock input (which is *not* recommended practice) or is part of an *asynchronous sequential* logic circuit, then glitches may cause the circuit to "latch up" into undesired states, thereby causing circuit failure. For example, in the R-S latch circuit of Figure 1.14, if the *S* input is stable at 0 but the *R* input has a glitch that produces a temporary voltage spike resulting in a momentary 1 value, then the latch may be cleared unintentionally.

Glitches can be caused by *hazards*, which are situations in which a signal oscillates one or more times in response to a change in a single input value. There are two types of hazards. A *static hazard* is a situation in which a signal value which should stay at one value (either 0 or 1) oscillates one or more times before settling down to its original value. A static hazard can occur when there are two logic paths from a single input to a single output. A *dynamic hazard* is a situation in which a signal value (which should make one clean transition from 0 to 1 or from 1 to 0) oscillates one or more times before settling into its final value. Such a situation can occur if there are three or more logic paths from a single input to a single output.

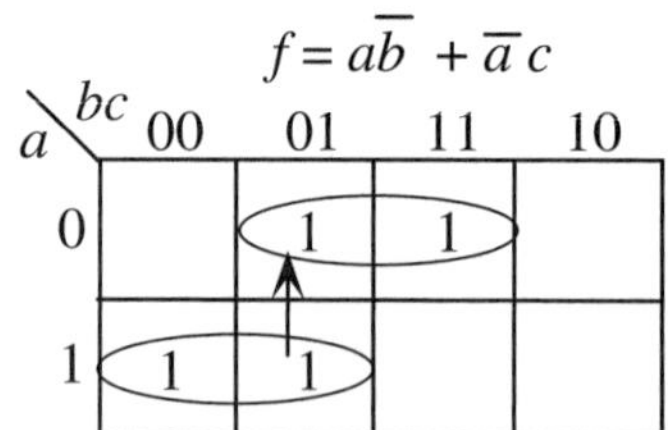



**Figure 1.21**
An example of a
K-map and switching
expression with a
static hazard.

Transition from $(a, b, c) = (1, 0, 1)$ to $(a, b, c) = (0, 0, 1)$

In a general two-level AND-OR circuit, a static hazard can be prevented by adding a *consensus* term to the original switching expression. Consider the switching expression and K-map shown in Figure 1.21. In this figure, if the input $a$ changes from 1 to 0 while the other two inputs remain unchanged, then there will be a momentary transition in the output $f$ (from 1 to 0) before settling back down to 1. This effect occurs because we need to cross over from one K-map circle to another when transitioning from one minterm to another. In order to avoid this static hazard, we can add the product term $\overline{b}c$, referred to as a *consensus* term. With the addition of this consensus term, which is logically redundant, we can avoid the static hazard situation, because both minterms involved in the problematic transition are covered by the consensus product term.

### 1.3.4  Metastability

For the design of reliable synchronous sequential circuits, it is sometimes necessary to deal with the problem of *metastability*. Metastability refers to a situation in which a logic signal, which normally has a high- or low-voltage value (for logic 1 and 0, respectively), pauses at an intermediate-voltage value (referred to as a *metastable value*) for an indefinite amount of time. Such a situation can occur if a data input to a flip-flop changes at the same time as the clock input to that flip-flop. A flip-flop normally requires a *setup time* (for the data input) before the arrival of the active clock transition. If the setup time requirement is not met, then the flip-flop output can become unpredictable.

Metastability typically is not a problem within a properly designed synchronous sequential circuit, because all signal changes can be synchronized to the clock. Any signal traveling from one part of the circuit to a data input of a flip-flop in another part of the circuit will change its value immediately after a clock edge. Thus, provided that the clock period is sufficiently long enough, this signal will have settled down into its new value before it is clocked into a flip-flop.

Metastability problems can occur at the input interface to a circuit. Some external input signals (generated by a human or another circuit) *cannot* be synchronized to the clock signal for a circuit. In such a situation, we cannot predict or control *when* such an external input signal will change its value. Thus, in order to synchronize an external input signal, this signal needs to be input into a flip-flop; the output of this flip-flop, which is the "buffered" signal value, will be synchronized to the rest of the circuit. However, during this process, since we cannot control or predict when the

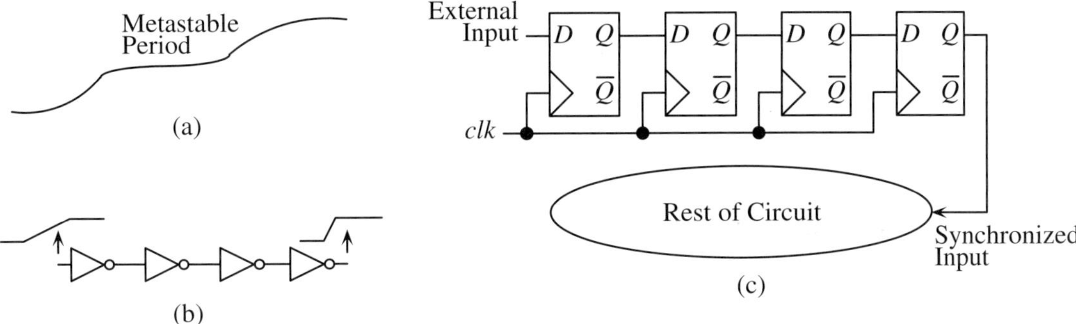

**Figure 1.22**   (a) A metastable signal, (b) a buffer chain, used to produce an input with a *sharp* transition, and (c) a flip-flop chain, used to significantly reduce the probability of metastable signals.

external input signal will change, it may change at approximately the same time as the clock signal; thereby, possibly resulting in a metastable output.

Although *metastability*, as described above, is an important problem that can cause circuits to fail, metastability can never be avoided completely. The best that can be done is to reduce the probability of metastable signals to an acceptably low value, such as once in 100 years. The probability of a metastable signal can be reduced by using a signal with a sharp transition as an external input signal. The shorter the duration of the transition, the smaller is the probability that the signal will still be transitioning when the clock signal arrives at a flip-flop. Thus, by passing the external input signal through a series of buffers, before using it in the rest of the circuit, an external input signal can be made to have very sharp transitions, as shown in Figure 1.22(b).

Another, more effective solution to the metastability problem uses a series of flip-flops connected in a chain, with the external input connected to the first flip-flop of this chain. There is a small probability that the output of the first flip-flop will be a metastable signal. However, since this signal is itself clocked into a second flip-flop, there is a low probability that the metastable signal will survive beyond this second flip-flop. The probability of a metastable signal becomes lower with each successive flip-flop in the chain. For applications requiring a "normal" level of protection against metastability, a two-flip-flop sequence typically is sufficient. Of course, when this method is used, there is a delay of several clock cycles before the external input signal can be used in the rest of the circuit. Figure 1.22(c) shows the design of a flip-flop chain used to reduce the probability of metastable signals.

# 1.4  Resources

Introductory digital logic design is a required course for most electrical engineering and computer engineering-related college degree programs. Thus, there are numerous textbooks that cover this topic. Representative digital logic design textbooks include [Hayes 1993] and [Wakerly 2002].

### 1.4.1   Bibliography

[1.1] HAYES, J. P., *Introduction to Digital Logic Design*, Addison-Wesley, Reading, MA, 1993.

[1.2] WAKERLY, J. F., *Digital Design: Principles and Practices, 3rd Ed.*, Prentice Hall, Upper Saddle River, NJ, 2002.

● ● ● ● ● ● ● ● ● ● ● ● ● ● ● ● ●

# 1.5   Problems

**P1.1.** As described in this chapter, logic values can be represented using voltage values or transitions, current values or transitions, different signal frequencies, or different signal amplitudes. Besides these methods, list at least two other methods that can be used for representing logic 0 and 1 values.

**P1.2.** Show how to construct two-input NAND and NOR gates using relays (electromechanical switches: mechanical switches whose states, closed or open, are controlled by voltage values).

**P1.3.** Even though *positive logic* may appear to be a much more natural method for representing logic 0 and 1, *negative logic* still is sometimes used by digital logic designers. Why?

**P1.4.** Given the binary bit sequence 100110010110, what is the value being represented if (a) unsigned binary, (b) 1's complement, (c) 2's complement, (d) sign-magnitude, and (e) excess-2047 notation is being used?

**P1.5.** Derive a general method for converting an $n$-bit BCD number (assuming that $n$ is a multiple of four) into its corresponding binary number.

**P1.6.** Derive a general method for converting an $n$-bit binary number into its corresponding binary-coded decimal (BCD) number.

**P1.7.** Prove that a 2-input NOR gate is a *universal gate*.

**P1.8.** Prove that a 2:1 MUX is a *universal gate*.

**P1.9.** Use a 16:1 MUX to implement the five-variable function

$$f(a, b, c, d, e) = \sum(1, 5, 6, 8, 9, 16, 22, 27, 30, 31)$$

**P1.10.** Use a 3:8 decoder (with complemented outputs) and an 8-input NAND gate to implement a 1-bit full-adder device.

**P1.11.** SOP and POS expressions:

$$g(A, B, C, D) = \sum(0, 1, 4, 5, 7, 9, 13, 14, 15)$$

Is the SOP or POS form simpler for this function?

**P1.12.** Next, show how the same simplification can be performed using a three-variable K-map.

$$h(w, x, y, z) = \sum(1, 2, 5, 8, 9, 11, 13, 14)$$

**P1.13.** Implement the $g$ function of Problem P1.11 using NAND gates only. Next, implement the same function using 2-input NAND gates only.

**P1.14.** Implement the $h$ function of Problem P1.12 using NAND gates only. Next, implement the same function using NOR gates only.

**P1.15.** Modify Figure 1.13 so that it becomes a general block diagram for a Moore machine.

**P1.16.** Show the design for an R-S latch using only 2-input NAND gates. Design it as a latch with $R$ and $S$ inputs and not $\overline{S}$ and $\overline{R}$ inputs.

**P1.17.** Design a sequence detector for the serial data-input sequence 01110111 using the complete five-step finite state machine method.

**P1.18.** Design a 3-bit up-down binary counter using the five-step finite state machine method. Use J-K flip-flops in the circuit implementation.

**P1.19.** Discuss the *metastability* problem and methods to avoid or minimize the probability of metastable values. Try to include issues and methods not covered in this chapter.

# CHAPTER 2

# Digital Logic Design Using Hardware Description Languages

## Important Concepts

- The concept of a *hardware description language (HDL)*, and the advantages of using such a language to describe a digital logic circuit.

- The concept of *synthesis*, and how synthesis can be used in conjunction with a hardware description language to enable fast design of large and complex digital logic circuits.

- The *register transfer level (RTL)* of description for a digital logic circuit.

- How digital logic circuits are modeled and simulated.

# 2.1 Hardware Description Languages

Nowadays, almost all serious engineers designing digital logic circuits of any reasonable complexity use an *hardware description language (HDL)* in their design. Using an HDL to describe and simulate a digital logic circuit has the following advantages over schematic-based design entry.

1. Since an HDL description is simply a text file, it is more portable than a schematic design, which must be viewed and edited using a drawing tool specific to the *computer-aided design (CAD)* environment being used.

2. A schematic only can describe a design in a structural manner, showing the modules and the connections between those modules required to implement a specific design (each module used must be included in a library describing its behavior, or it must be described using another schematic). On the other hand, an HDL description can be written in a behavioral or structural manner. An HDL can describe the intended behavior of a circuit with no reference whatsoever to structural detail.

3. When combined with an effective *synthesis* tool, the use of behavioral HDL permits fast prototyping of large and complex digital logic circuits. Synthesis refers to the automatic generation of a logic circuit design from a high-level description of the desired operation of the target circuit.

4. The *test vectors* (lists of test input values for all primary inputs) for a circuit and the results expected when those test vectors are applied to the circuit can be described using the same HDL used in the description of the circuit. Thus, a *test bench* (test vectors and the times when those test vectors are applied to the circuit) can be concocted as the circuit is being designed, and the circuit can be simulated using that HDL-based test bench to gain confidence in the design before it is implemented in hardware. By implementing the test bench using an HDL, the test bench becomes highly portable and repeatable. The latter point is important, since it is imperative that a design be re-tested after every change to the design to ensure that it passes all tests that it passed previously (this is referred to as *regression testing*).

Currently, the two HDLs used by most digital logic designers are *Verilog HDL* and *VHDL*. Both Verilog and VHDL are now IEEE standard languages for hardware modeling and simulation. Verilog started off as a proprietary language developed by Philip Moorby in the early 1980's for a company later acquired by Cadence Design Systems. Verilog was then put out in the public domain and later promoted into an IEEE standard known as IEEE 1364. VHDL was developed as a standard hardware description language by the U.S. Department of Defense in 1983 and later become an IEEE standard in 1987 (IEEE 1067-1987). Further improvements were then incorporated, and updated standards were released in 1993 as IEEE 1076-1993 and in 2001 as IEEE 1076-2001.

## 2.2 Design Flow

Figure 2.1 shows the flow of activities (design stages) that occur during the design and implementation of digital logic circuits of any reasonable complexity. First, the

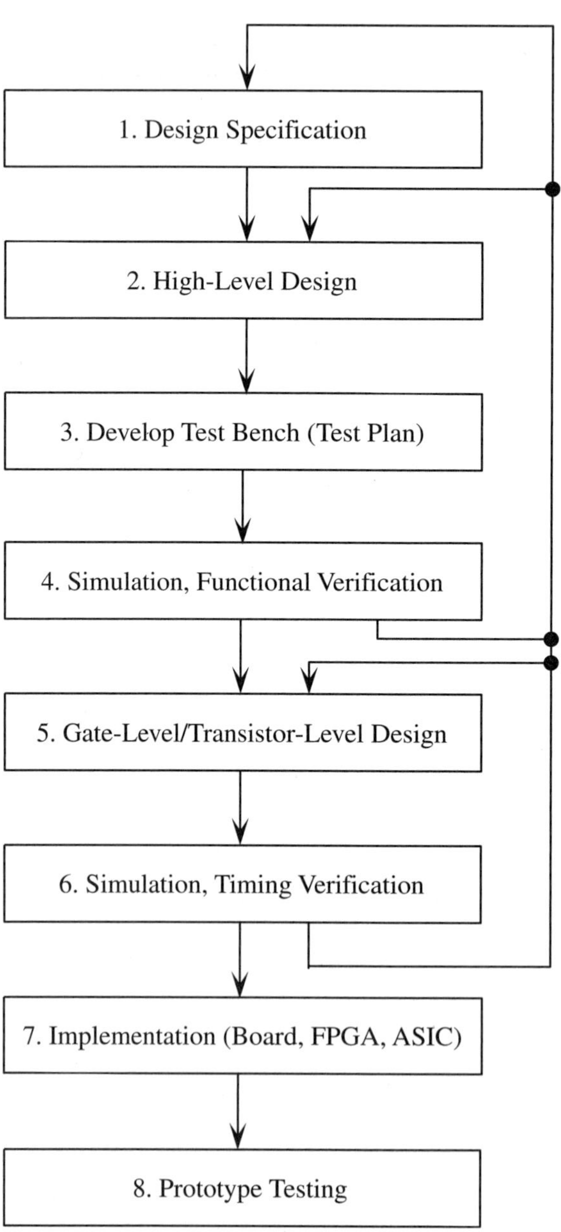


**Figure 2.1**
Design flow for
digital logic design.

designer (or customer) must provide the specifications for the design in terms of *what* the circuit is expected to do and the *constraints* that the circuit must satisfy (clock rate, delays, size, etc.). Next, the designer must create a high-level design for the circuit. Then, he/she must develop a test bench, and if possible, use that test bench to test the high-level design (using a simulation tool). Based on the results of the test, the high-level design can be modified, re-tested, and this sequence repeated as necessary. Once the high-level design functions as desired, it must be refined into a gate-level or transistor-level design (as is appropriate). The gate-level or transistor-level design can be produced using a process referred to as *synthesis*, described in Section 2.3. This gate-level or transistor-level design must then be re-tested, using the same test bench used to test the high-level design (if possible), and the results checked to make sure that it functions correctly *and* satisfies any constraints imposed during design specification. If there is a problem at this level, then the circuit must be re-designed and the entire design sequence repeated as necessary. Note that the design iteration can loop back to Stages 4, 2, or 1 as appropriate. Once Stage 6 is passed, the design can be implemented using a prototype board, *programmable logic device (PLD)*, *field-programmable gate array (FPGA)*, *application-specific integrated circuit (ASIC)*, or custom-designed integrated circuit. After the design has been implemented, it then can be tested with the aid of a test-pattern generator (to generate the test vectors) and a logic analyzer or oscilloscope (to check the outputs). If possible, it is desirable to use an HDL during the design specification phase. However, in many cases, this may not be possible—in which case, there should be at least a formal design document outlining the desired task and the constraints to be imposed on the circuit realization of the desired design. If the design specification itself is written in any HDL, it is possible to "simulate" the design specification in order to determine whether there are errors or essential missing details in the specification itself (e.g., a requirement in the specification may be physically impossible to implement).

## 2.3   Synthesis

A major development in recent years has been the development of effective tools for *synthesis*. Synthesis, a concept that has been around since the early days of digital logic design, refers to the automatic generation of a logic circuit based on a high-level description of the desired operation of the circuit. The output of synthesis is typically a *netlist*, which is a description of all of the connections and components (logic gates or other devices) to be used in a circuit. A subsequent tool can then use the netlist to fabricate (in the case of an ASIC) or program (in the case of an FPGA) an *integrated circuit (IC)* chip. Ideally, the high-level circuit description, which is used as the input to the synthesis tool, can be at an extremely high level (e.g., natural language input (using normal English) describing the circuit in highly general terms). However, the current state-of-the-art has not yet reached this ideal.

Different types of synthesis can be identified depending on the level and type of input used. Synthesis can take place using finite state machine diagrams (*finite state machine synthesis*) or logic schematic diagrams (*logic schematic synthesis*). Some design entry tools permit a design to be described using a finite state machine diagram with labeled circles for the states and labeled arrows for the state transitions. The synthesis tool will then create the necessary state variables, make the necessary state assignments (typically using one-hot[1] or binary encoding), create the necessary state flip-flops and output and next-state transition logic, and then create a netlist for the target implementation platform. If a logic schematic diagram is used as the input to the synthesis tool, then the tool's task is to interpret the input diagram and convert it into the logic devices and connections provided by the target implementation platform. Logic minimization also may be performed during this process.

The most general type of synthesis, however, uses a textual description presented in a hardware description language (HDL), such as Verilog or VHDL, as its input. Hardware description languages typically are used for more than simply describing the structure and behavior of logic circuits. Thus, synthesis tools place restrictions on the types of language constructs and descriptive methods that they will accept. Some synthesis tools permit *behavioral synthesis*, in which circuits can be described using behavioral HDL (using an algorithmic C language-like description method). Most current synthesis tools, however, restrict themselves to behavioral descriptions written in a specific description style referred to as *register transfer level (RTL)* (to be described in Section 2.4). This type of synthesis is referred to as *RTL synthesis*. It is also possible to perform synthesis from structural HDL descriptions. When the structural HDL input describes the specific logic equations to be used to implement the desired circuit, then the resulting synthesis is referred to as *logic synthesis*. In logic synthesis, logic minimization is performed and the resulting logic is mapped onto the devices and connections available in the target implementation platform. For example, if the target implementation platform is an FPGA, then the logic equations may have to be mapped onto generic multiplexers and look-up tables in the programmable cells of the FPGA.

The circuit that results after synthesis may not be exactly what the designer intended. The gate-level design synthesized from a high-level HDL description can differ significantly, depending on the manner in which the HDL code is written and the particular synthesis tool used. If the HDL description contains conditional expressions in which the actions to be taken for all conditions are not stated explicitly, *latches* can be created at unexpected locations. Depending on the manner in which the HDL code is written, chains of logic elements (one output leading to the input of another) can be formed that result in excessive delays. Logic with races, hazards, and other undesirable attributes can be formed. The synthesized logic may even be *incorrect*, in that the synthesized logic does not behave in the same manner as the high-level design.

Thus, it is always a good idea to *examine* the synthesized circuit and perform post-synthesis simulation to verify that the synthesized circuit operates as expected. Many simulation tools permit the user to simulate his/her circuit after it has been

---

[1] One-hot state encoding and the associated one-hot control logic design method is discussed in Chapter 5.

synthesized (or synthesized and implemented onto a target architecture). Also, with most synthesis environments, such as *Synopsys® FPGA Express*, there is a "view schematic" (or similar) command that can be used to examine the synthesized circuit. Post-synthesis schematic examination and/or simulation particularly is important when a new synthesis tool is being used, the designer is unsure about the type of gate-level circuit that will be synthesized for a particular high-level design block, or critical-timing dependencies are involved.

Regardless of the type of design input used, synthesis tools typically provide numberous options on the manner in which synthesis is to be performed. It is possible to specify the amount of effort to be expended for circuit optimization, with higher optimization levels requiring more CPU processing time. Optimization can be directed toward reducing circuit area or toward increasing circuit performance. Some synthesis tools permit us to specify that certain circuit modules should not be optimized (e.g., with "don't touch" directives). Numerous directives and script files can be used to specify *exactly how* the synthesis tool should synthesize the desired logic circuit. Because of these numerous options and the various ways in which synthesis can be performed, there are entire books devoted to synthesis, and the effective use of synthesis can require many man-months of design experience.

In the rest of this book, we attempt to bypass much of the complexity of using synthesis by describing a simple method based on default synthesis settings to synthesize circuits for a specific type of implementation platform: FPGAs. In fact, if several simple syntax rules are followed, it is possible to write high-level HDL code that can be synthesized correctly by most types of synthesis tools. These rules are presented in Chapter 4 for Verilog HDL.

● ● ● ● ● ● ● ● ● ● ● ● ● ● ● ● ●

## 2.4  Register Transfer Level Notation

Current synthesis tools can generate netlists from high-level descriptions written in *register transfer level (RTL)* notation. In RTL notation, *all* operations are described as the transfer of data from one or more registers to another register. During the data transfer, transformations (involving combinational logic operations) can be applied to the data. Thus, variables will correspond to registers, and data transformations will correspond to operations performed on those variables. Operations in RTL are always of the form $REG_x \leftarrow f(REG_1, REG_2, \ldots, REG_n)$, where $REG_i$ corresponds to a register or constant, and $f(\ldots)$ is a function that can be implemented in combinational logic. For instance, operations such as

```
A := B + C;
if (D) E := F; // same as E := (F and D) or (E and (not(D)))
```

are examples of RTL operations. Even the conditional execution statement is an RTL operation, since it is equivalent to the statement shown in the comment. Although it may not be intuitively obvious, it is a fact that the operation of *all* (or almost all) synchronous sequential digital logic circuits, including such complex devices as computers, can be described fully using only RTL operations. RTL can be considered as an intermediate level of description between the gate-level and the processor-memory-switch (PMS) level of description for general digital logic circuits. Table 2.1 shows a list of example RTL statements written in a pseudocode notation.

**Table 2.1:** ▶
Examples of register transfer level (RTL) statements.

| Statement | Intended Operation |
|---|---|
| `X := Y;` | Transfer the contents of $Y$ to $X$ |
| `X := 0;` | Clear the contents of $X$ |
| `X := 0xFF;` | Set lowest 8 bits of $X$ |
| `X := 1;` | Set $X$ to the decimal value 1 |
| `X := X » 1;` | 1-bit right-shift of $X$ |
| `X := X » 4;` | 4-bit right-shift of $X$ |
| `X := M[0xC2];` | Transfer the 194th element of $M$ to $X$ |
| `X := Y | Z;` | $X \leftarrow Y$ OR $Z$ (bitwise operation) |
| `X := Y & Z;` | $X \leftarrow Y$ AND $Z$ (bitwise operation) |
| `X := Y ^ Z;` | $X \leftarrow Y$ EXCLUSIVE-OR $Z$ (bitwise operation) |
| `X := not Y;` | $X \leftarrow$ 1's complement of $Y$ |
| `X := -Y;` | $X \leftarrow$ 2's complement of $Y$ |
| `X := Y + Z;` | $X \leftarrow Y + Z$ |
| `X := Y - Z;` | $X \leftarrow Y - Z$ |
| `X := (c & Y) | ((!c) & X);` | $X \leftarrow (\bar{c}$ AND $X)$ OR $(c$ AND $Y)$ |
| `if (c) X := Y;` | Equivalent to above statement |
| `X := (c & Y) | ((!c) & Z);` | $X \leftarrow (\bar{c}$ AND $Z)$ OR $(c$ AND $Y)$ |
| `X := c ? Y : Z;` | Equivalent to above statement |

● ● ● ● ● ● ● ● ● ● ● ● ● ● ● ●

# 2.5  Logic Simulation

In writing the HDL code for a digital logic design, it can be helpful to know the manner in which that HDL code will be *simulated* by the simulation tool. Most logic simulation tools simulate HDL code in an *event-driven* manner.

Event-driven simulation (also referred to as discrete event simulation) refers to a method of simulation in which there is a global queue of time-ordered events (stored as "record" data structures), from which the event at the head of the queue is executed; thereby resulting in zero or more events being inserted into the global event queue with their new positions dependent on their execution time (the global-system time when they *should* be executed). Once an event is executed, the global-system time is updated to the execution time of that event. The execution of an event may result in new events being generated and inserted into the global time-ordered event queue. Thus, the simulation progresses (forward in time) at a rate dependent on the frequency at which events are executed. If several events are scheduled to execute at the same time, there is no guarantee on the ordering of those events in the global event queue. The following example serves to illustrate this process at work.

**Example 2.1**   **Event-Driven Simulation**

Suppose that we have described the circuit shown in Figure 2.2 using an HDL. The NAND gate has an execution delay of 1 ns, and the OR gate has an execution delay of 2 ns. The primary inputs $a$ and $b$ both have the value '1' at time 0 ns. Then, at time 5 ns, let us assume that $b$ is changed to '0.' This is represented using a global event queue with the single event (named $E_0$) $b \leftarrow 0$ scheduled at time 5 ns.

The simulator removes event $E_0$ from the global event queue and executes it, resulting in the signal $b$ being updated to the value '0' and the global-system time being updated to 5 ns. When this happens, the components affected by this change (the NAND gate and the OR gate) must then execute. This is modeled by the two new events $E_1$ and $E_2$, which are generated and inserted into the global event queue. $E_1$ is the event $c \leftarrow 1$ NAND 0 *with delay* 1 ns scheduled for time 5 ns, while $E_2$ is the event $f \leftarrow 0$ OR *0 with delay* 2 ns scheduled for time 5 ns. Note that the OR gate receives a $c$ input of '0' since $c$ still has the value '0' (the change in $b$ has not yet propagated to $c$ at time 5 ns). Since $E_1$ and $E_2$ have the same execution time, they

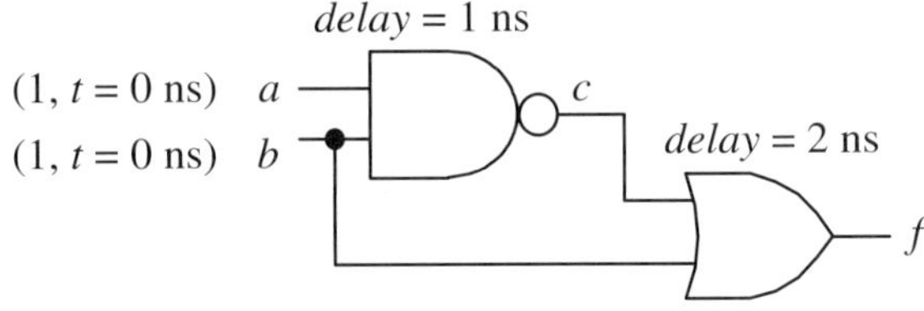


Global Event Queue ($t = 0$ ns)

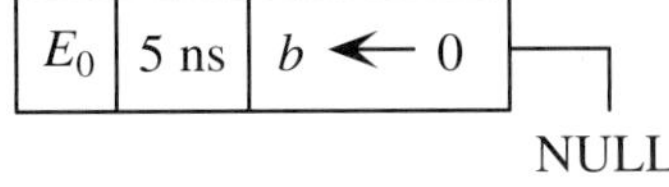


Global Event Queue ($t = 6$ ns)

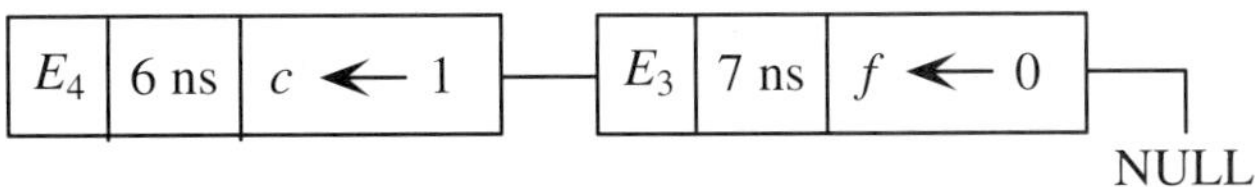


**Figure 2.2**
Sample circuit and global event queue used for Example 2.1.

can be inserted into the global event queue in any order (in this case $E_2$ followed by $E_1$).

Next, the simulator removes $E_2$ from the head of the global event queue and executes it, resulting in the new event $E_3$: $f \leftarrow 0$ scheduled for time 7 ns. Event $E_3$ is inserted after $E_1$, since it has a later execution time than $E_1$.

Since $E_1$ is now the event at the head of the queue, $E_1$ is executed next, resulting in the new event $E_4$: $c \leftarrow 1$ scheduled for time 6 ns. $E_4$ is then itself inserted into the global event queue in front of $E_3$, since $E_4$ has an earlier execution time than $E_3$.

Next, $E_4$ is executed. This results in $c$ being updated to '1' and the global-system time being updated to 6 ns. At this time, the change in $c$ results in a new event $E_5$: $f \leftarrow 1$ OR $0$ *with delay* 2 ns scheduled for time 6 ns, since the OR gate is affected by any change in the signal $c$. $E_5$ is inserted in front of event $E_3$.

The event $E_5$ is then removed from the head of the global event queue and executed. This results in the new event $E_6$: $f \leftarrow 1$ scheduled for time 8 ns. $E_6$ is inserted in the global event queue after $E_3$.

$E_3$ is then removed from the global event queue and executed. This results in the global-system time being updated to 7 ns and the signal $f$ being changed to '0.' No new events are inserted into the global event queue since the change in signal $f$ does not affect any other component or signal.

Next, $E_6$ is executed, since it is the only event in the global event queue. This results in the global system time being updated to 8 ns and the signal $f$ being changed to '1.' Again, no new events are generated by the execution of event $E_6$. Since the global event queue is now empty, no new activity takes place, and the simulator stops execution.

---

When executed using a commercial simulation tool, the *simulation result* (the result of the type of scenario depicted previously) typically is displayed graphically using a *timing diagram*, which is a figure showing how the value of each signal changes with time. Figure 2.3(a) shows the timing diagram for Example 2.1. A timing diagram should display explicitly the sequence of signal changes that occur and the timing relationships between those signals. Note that during the period from 0 ns to 1 ns and 2 ns, the signals $c$ and $f$ have unknown values (denoted by a "hash" marking), since the initial $c$ and $f$ values cannot be determined until the effects of the initial test vector ($a = 1$, $b = 1$) have propagated through their respective logic gates. With an *annotated* timing diagram, shown in Figure 2.3(b), vertical dashed lines are used to indicate important signal transition points (and the time intervals between two or more signal transition points), and arrows may be used to indicate causal effects (i.e., an arrow from one signal to another indicates that a change in the first signal causes a change to occur in the second signal).

The use of test vectors is also illustrated by Example 2.1. The set of test vectors used in this example are $a = 1$, $b = 1$ at time 0 ns and $a = 1$, $b = 0$ at time 5 ns. With the use of this set of test vectors, the output signal $f$ and the intermediate signal $c$ change, as shown in the simulation results of Figure 2.3. Also, we can specify an *expected result* of $f = 1$ after 8 ns (since $f = (a$ NAND $b)$ OR $b = (\overline{a} + \overline{b}) + b = 1$). Thus, the test-bench for this circuit could consist of the two test vectors listed above and the expected results $f = 1$ at time 2 ns and $f = 1$ at time 8 ns. This type of test-bench result could be written in the same HDL used to describe the target circuit itself. The simulation of the test-bench code would then apply the specified test vectors at the specified times and also display output indicating whether the expected results were obtained from the target circuit.

# 2.6 Properties of Actual Circuits

With regards to the physical world, real digital circuits have important characteristics that affect the manner in which they should be modeled and designed. Real digital circuits involve digital logic modules (gates and devices) that execute concurrently and continuously and incur delays in transferring signals from one module to another. Such characteristics also apply to the physical wires (and wire traces within chips and printed circuit boards) interconnecting modules in the circuit. Also, the transfer of signal values in such circuits require the movement of electrical charge. In order to send a logic value from one point to another in a circuit, sufficient electrical charge must be transferred. Thus, the delays incurred depend on the characteristics of the transfer medium and the electrical charge sourcing (or sinking) capability of the device sending the logic value. Due to these characteristics, the following important lessons should be kept in mind when modeling and designing digital logic circuits.

First, all real digital logic modules (gates or devices) have delays, however small they may be. Thus, when a signal arrives at two different modules $M_1$ and $M_2$, and the output of one module ($M_1$) enters the input of another module ($M_2$), module $M_2$ initially will use the *old* value of the output of $M_1$ in computing $M_2$'s output, since there is a delay before the output of $M_1$ changes. Of course, it is possible that the arrival times of the same signal at modules $M_1$ and $M_2$ may differ due to varying wire delays. If such delays are important, then these types of delays must also be modeled

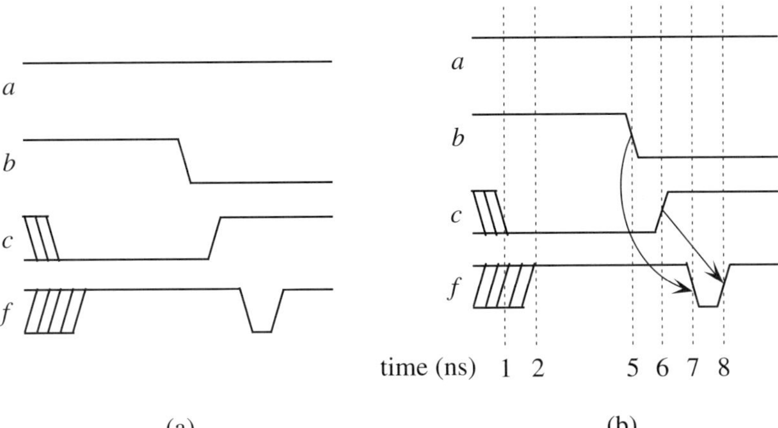


**Figure 2.3**
The (a) timing diagram and (b) annotated timing diagram for Example 2.1.

by modeling *wires* similarly to other logic gates or devices (with propagation delays and perhaps even logic functions).

Second, digital logic modules execute concurrently. If a signal change arrives at two or more discrete modules, all of those modules will execute concurrently. Thus, the HDL must be capable of accurately modeling such concurrent behavior, and the simulator must accurately simulate such concurrent behavior. Modeling of concurrent behavior in the simulator is accomplished with the event-driven model, which results in the global-system time advancing in spurts instead of a continuous manner. Actually, to be more accurate, all digital logic modules in a circuit execute concurrently *and continuously* all of the time. However, the effects of the operation of those modules only show up at certain times (i.e., when input signals change value). Thus, they can be *modeled* as executing only when certain events occur.

Third, the actual delays incurred in a real digital circuit depend on many factors, including process variations during the manufacturing stage. Although manufacturers of chips typically will state the maximum guaranteed delays to be expected for various parts of their chips, chips with different lot numbers can have different delay characteristics (which are all lower, by different amounts, than the maximum guaranteed delays). In fact, even with a single chip, different parts of the chips can have slightly different delay characteristics. Thus, it is *not* a good idea to design a circuit that is dependent on exact or minimum delay values. During circuit simulation, the simulation tool may produce output that *looks* correct because certain signals arrive in a certain order. However, if the arrival order of those signals are dependent on the modeled delays used in the circuit description and *not* guaranteed by the logic design itself, then the actual circuit produced from that design may result in non-working chips or chips that only work *part* of the time.

Finally, the signals in real digital circuits have "signal strengths" that determine how sharp signal transitions are (and thus the delays incurred by those signals) and the number and types of module inputs that can be driven from a particular signal source. Such signal strengths, which also can be modeled approximately in Verilog or VHDL, are determined by the voltage level (i.e., for a logic 1, is the signal voltage very close to the power supply voltage?) and the amount of current that can be sourced or sunk from the signal source. In a typical digital circuit, the transfer of a logic value from an output pin (output of one module) to several input pins (inputs of several modules) requires the transfer of charge to the input pins (for a logic 1) or to the output pin (for a logic 0). The output pin *must* have the capability of *sourcing* or *sinking* all of the required current.

In practical terms, this discussion on signal strengths implies that each module output must drive only a limited number of module inputs. If an excessive number of module inputs must be driven from one module output (for example, with a clock signal generated by an internal module), then a *buffer tree* must be used. A buffer tree is essentially a set of buffers connected in a tree-like manner used to disperse a single signal value in such a manner that the buffers at the leaf nodes of the tree only drive a small number of module inputs. Figure 2.4 shows an example of how a buffer tree could be used to disperse a signal to 16 module inputs, with no single buffer driving more than 4 inputs. Using a buffer tree can have the additional benefit of equalizing the delays to all of the module inputs, provided that the buffer tree is laid out in a wire-length-equalized manner. This type of buffer tree actually is used for clock signals used to drive physically dispersed flip-flops in a synchronous sequential circuit—in such circuits, it is important for the clock signal to arrive at the clock inputs of all flip-flops at approximately the same time instant.

## 2.7 Chapter Review

- Most current digital logic circuits are designed using a hardware description language, of which the most commonly used ones are Verilog and VHDL. Using a hardware description language is more preferable than schematic-based design entry because (1) it is more portable, (2) the design can be described in a behavioral as well as a structural manner, (3) large and complex circuits can be designed in a short period of time by combining behavioral design description with synthesis, and (4) the method to be used to test the target design can be described in the same language and ported along with the target design.

- *Synthesis* refers to the automatic creation of a *netlist* (the logic gate and interconnection information for a physical circuit) from a high-level description of the desired operation of the circuit. If certain rules are followed and a subset of the language constructs of a hardware description language are used, then a behavioral description of the target circuit can be synthesized into a netlist,

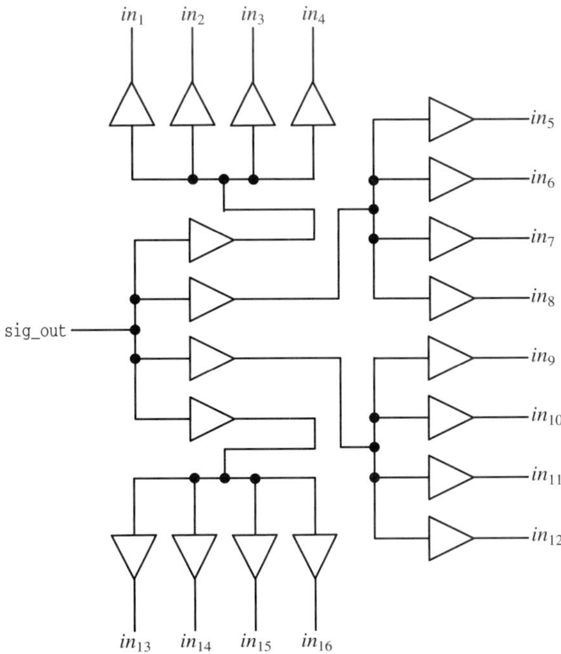


**Figure 2.4**
Using a buffer tree to disperse a signal to 16 module inputs.

which then can be used to produce an integrated circuit chip to realize the target circuit in hardware.

- Digital logic circuits typically are simulated by logic simulation tools using *event-driven simulation*. This type of simulation, as opposed to *time-driven simulation*, simulates only the "events" that occur in the circuit (such as signal changes) and advances the global-system time to the time of occurrence of those events (in spurts). In event-driven simulation, a global time-ordered event queue is maintained, and the execution of an event at the head of this queue can result in more events being inserted into this queue according to the execution time of those events. Thus, there is no guarantee on the execution order of events with the same execution time, and this fact must be considered when a digital logic designer uses a hardware description language to describe concurrently executing events.

# 2.8 Resources

The home page for Aldec Inc. is http://www.aldec.com, a company that supplies an editor and simulator for VHDL and Verilog called Active-HDL. The home page for Synopsys is http://www.synopsys.com, a well-known synthesis tool supplier. [Ashenden 2002], [Sjoholm and Lindh 1997], [Coelho 1989], and [Lipsett *et. al.,* 1989] serve as useful guides to VHDL. In particular, [Sjoholm and Lindh 1997] provides a good discussion of logic synthesis and gives many design examples and tips for writing synthesizable VHDL code. [IEEE 1993] and [IEEE 2001(VHDL)] are the authoritative references on the IEEE VHDL standard. There are also many useful references for the Verilog language, including [Ciletti 2002] and [Zeidmann 1999]. For the authoritative references on the Verilog HDL standard, the interested reader should refer to the standards books [IEEE 1995] and [IEEE 2001(Verilog)].

### 2.8.1 BIBLIOGRAPHY

[2.1] ASHENDEN, P. J., *The Designer's Guide to VHDL, 2nd Ed.,* Academic Press, San Diego, CA, 2002.

[2.2] CILETTI, M. D., *Advanced Digital Design with the Verilog HDL,* Prentice Hall, Upper Saddle River, NJ, 2002.

[2.3] COELHO, D. R., *The VHDL Handbook,* Kluwer Academic, Norwell, 1989.

[2.4] HACHTEL, G. D., and SOMENZI, F., *Logic Synthesis and Verification Algorithms,* Kluwer Academic, Boston, MA, 1996.

[2.5] IEEE, *IEEE Standard VHDL Language Reference Manual,* IEEE Press, New York, 1993. Also available through the IEEE standards web page at http://standards.ieee.org.

[2.6]   IEEE, *IEEE Standard VHDL Language Reference Manual*, IEEE Press, New York, 2001.  Also available through the IEEE standards web page at http://standards.ieee.org.

[2.7]   IEEE, *IEEE Standard Hardware Description Language Based on the Verilog Hardware Description Language*, IEEE Press, New York, 1995.  Also available through the IEEE standards web page at http://standards.ieee.org.

[2.8]   IEEE, *IEEE Standard for Verilog Hardware Description Language 2001*, IEEE Press, New York, 2001.  Also available through the IEEE standards web page at http://standards.ieee.org.

[2.9]   LIPSETT, R., SCHAEFER, C., and USSERY, C., *VHDL: Hardware Description and Design*, Kluwer Academic, Norwell, 1989.

[2.10]  SJOHOLM, S., and LINDH, L., *VHDL for Designers*, Prentice Hall, Upper Saddle River, NJ, 1997.

[2.11]  ZEIDMAN, B., *Verilog Designer's Library*, Prentice Hall, Upper Saddle River, NJ, 1999.

[2.12]  http://www.aldec.com, home page for ALDEC (HDL tool supplier).

[2.13]  http://www.synopsys.com, home page for Synopsys (synthesis tool supplier).

## 2.9  Problems

**P2.1.** List the pros and cons of schematic-based design entry versus HDL-based design entry.

**P2.2.** Identify the various types of synthesis possible (in terms of the circuit-modeling method and the level of description used).  What types are most useful, and what types are most difficult for commercial synthesis tools to support?

**P2.3.** Describe an 8:1 MUX, a 3:8 decoder, and an 8-bit even parity checker using sequences of RTL statements similar to Table 2.1.

**P2.4.** Show how a 2-input exclusive-OR gate (computing $a \oplus b$) can be constructed using four 2-input NAND gates.  Next, in the resulting NAND-gate design, describe the sequence of events that occur when $(a, b) = (0, 0)$ at time $t = 0$, $(a, b) = (0, 1)$ at time $t = 4$, $(a, b) = (1, 1)$ at time $t = 6$, and $(a, b) = (1, 0)$ at time $t = 8$, assuming an event-driven simulator.  Assume that the 2-input NAND gate has a delay of 2 time units, and use a method of description similar to that used in Example 2.1.

**P2.5.** Draw a timing diagram showing all of the events that occur in the solution to Problem P2.4.

**P2.6.** What are *test vectors*?  Identify the test vectors used in Problem P2.4.

**P2.7.** In Example 2.1, why doesn't the unknown period (hash marked area) for signal $f$ in Figure 2.3(b) extend from 0 ns to 3 ns? It seems like it should, since signal $c$ is only determined at 1 ns, and the following OR gate has a 2 ns delay.

**P2.8.** Using the speed of light (denoted by $c$) as a guide, what is the delay required for an electrical signal to traverse a 10 cm length of "wire" embedded in a PCB board (more commonly referred to as a *PCB trace*)? Although the exact speed of travel of an electrical signal on a PCB trace varies depending on the shape and properties of the PCB trace material, it can be assumed that this speed is in between about $0.3c$ and $0.7c$.

**P2.9.** List all of the different ways that you can think of to send a logic signal (for example, a logic 1 followed by a logic 0) from one end of a wire to the other. Identify the pros and cons of each signal-transmission method.

**P2.10.** Assuming that a high-voltage level is used to represent logic 1 and a low-voltage level is used to represent logic 0, discuss how a logic 1 can be sent from the output pin of one chip to ten input pins spread out among ten different chips. What is actually, physically, being transmitted?. Next, discuss what happens when the logic 1 changes to a logic 0 at the output pin. Use an analogy with everyday phenomenon in your discussion (for example, water transmission from one source to several water taps). Include at least one example besides water transmission.

***P2.11.** In this chapter, it is stated that most synthesis tools will accept behavioral descriptions written in RTL (register transfer level) HDL code. Thus, it is important to be able to describe different types of circuits as sequences of RTL operations. Consider a circuit for controlling a *micromouse*, (a small robot that finds its way through a maze) by continuously re-adjusting its direction when it bumps into one of the maze's walls. Describe the behavior of the desired circuit at the RTL level (i.e., define several variables and devise an algorithm (consisting of a sequence of RTL operations similar to those used in Table 2.1) to solve the micromouse control problem).

**P2.12.** As a simpler variation of problem P2.11, create an algorithm to calculate and output the greatest common divisor of two $n$-bit numbers. The algorithm must be an RTL algorithm (i.e., it must consist of a sequence of RTL operations).

**P2.13.** Create an RTL algorithm to implement the digital logic control circuit portion of a vending machine with two item choices: a toothbrush costing one dollar and a tube of toothpaste costing two dollars.

# Introduction to Verilog and Test Benches

**Important Concepts**

- How to use the Verilog *HDL (hardware description language)* for the modeling, simulation, and synthesis of digital logic hardware circuits.

- The main similarities and differences between a hardware description language such as Verilog and a programming language such as C.

- The concept of a test bench, and how to create a Verilog test bench for a Verilog module.

As a full-fledged programming language used for the specification, modeling, synthesis, and simulation of digital logic circuits, the Verilog hardware description language (HDL) is a complex language with numerous special features and library functions. However, for the purposes of describing most types of digital logic circuits so that they can be synthesized or simulated, it is sufficient to use a small subset of the constructs and features available in the Verilog HDL. Following this type of approach, this book attempts to introduce (in a concentrated form) all Verilog concepts required for basic synthesis and simulation. When presenting Verilog code, the emphasis is on *synthesizable* Verilog code (i.e., code that can be used to automatically, through the use of CAD tools, produce working hardware circuits) that easily can be written based on an algorithmic description of the desired behavior of a digital logic hardware circuit. Thus, features of the language that are not essential for this purpose only are covered briefly or not even covered at all. For a complete introduction to the Verilog language, the interested reader should refer to the language specification (IEEE Standard 1364) and the various books specifically devoted to teaching Verilog (some of which are listed at the end of this chapter). The on-line documentation available with most Verilog simulation tools also can be used as a valuable guide for the proper use of the language.

## 3.1   Verilog Basics

### 3.1.1   The Module Definition

A Verilog description of a digital logic circuit begins with **module** and ends with **endmodule**. (Note that all Verilog keywords are written using **bold font**, while all constant and signal names are written using `typewriter font`.) The keyword **module** is followed by the module name and a parenthesized list of the outputs and inputs to the module. Next is a list of the outputs preceded by the keyword **output** and a list of the inputs preceded by the keyword **input**. Bidirectional ports are defined using the keyword **inout**. Following that is a list of any internal signals used in the module description.

The astute reader should note that this sounds suspiciously like a function definition in the C programming language. Indeed, the Verilog syntax largely is based on C, and many C constructs are used unchanged or almost changed in Verilog. If the reader is already proficient in C, this property of Verilog significantly facilitates the learning of Verilog. However, there are significant differences with C, due mainly to the fact that Verilog is describing *hardware*, and not simply describing a sequence of operations to be executed as a computer program. These differences largely can be summarized by the fact that different parts of the module must execute *concurrently* (recall the discussion of concurrent and continuous execution of hardware presented at the beginning of Chapter 2) and the fact that *delays* may need to be specified. There is also the fact that structural connections frequently are used in Verilog; however, this can be likened to the use of subroutines (or subfunctions) in C. In general, the reader can assume the use of C syntax for Verilog, except in the cases described next.

### 3.1.2   Signals and Operators

Information typically is communicated within a digital logic system using *signals*. In digital logic, such a signal represents one of two binary values: 0 or 1. Signals typically are transmitted by sending high- and low-voltage levels, with a high voltage (typically about 3.3 volts in CMOS technology) representing a logic 1 and a low voltage (typically about 0.0 volts in CMOS technology) representing a logic 0 in positive logic systems.

Verilog uses a four-valued logic system to represent signal values. In addition to logic 0 and 1, there are unknown (X) and high-impedance (Z) values. Digital logic hardware only manipulates logic 0's and 1's. However, in certain cases, the exact value of a logic signal is left undetermined (either intentionally or unintentionally).

A *high-impedance* value occurs when a wire value is undetermined because it is *disconnected* from other wires that carry valid signal values.

An *unknown* value occurs when a wire has not yet been initialized to a valid logic value or if it is driven to conflicting logic values by two or more signal sources at the same time.[1]

In Verilog, 0, 1, unknown, and high-impedance logic values are represented as 1'b0, 1'b1, 1'bx, and 1'bz respectively. The 'b represents the fact that it is a binary value and the '1' preceding the single-quote represents the number of bits being represented. In a multiple-bit signal value, the '1' is replaced by the number of bits being represented. Thus, for example, 4'b0101 and 3'b0xz represent the four-bit binary value 0101 and the three-bit binary value consisting of a logic '0' followed by an unknown bit value and a high-impedance bit value, respectively. If the number before the 'b is omitted, then it indicates an arbitrary-sized number (when assigned to a signal, it will become the bit width for that signal). Capital letters also can be used for unknown and high-impedance values. Also, the 'b can be replaced by 'h (for hexadecimal), 'd (for decimal), or 'o (for octal). The default base is decimal. Thus, for example, 19, 5'b10011, 5'h13, 5'd19, and 5'o23 are all equivalent. Finally, an underscore (_) can be used to make long constants easier to read (e.g., 19 could also be written as 5'b1_0011).

In physical hardware systems, signals are carried by wires, both internal and external to chips. Thus, Verilog defines the **wire** type to define such signals. However, some signals need to be manipulated much like C variables. For those cases, Verilog

---

[1] In certain technologies, such as open-collector TTL, driving a common wire to opposing logic values is permitted; the resulting logic value is simply a logical combination (such as NAND) of the logic values being driven (this is referred to as *wired*-NAND logic). However, the situation being described here refers to technologies where this type of behavior is not permitted.

defines the **reg** type.  A signal defined to be of type **reg** may or may not be defined as the output of a register.  Whether or not a register is created for a **reg** signal depends on the usage of that **reg** signal, as determined by a synthesis tool.  Finally, **integer** and **real** signal types can also be defined and used just like C variables.  These latter types typically are implemented like 32-bit **reg** signal types.  Signals that are used without being declared to be of a specific type are assumed to be of type **wire** by default.

Operators that can be used on signals include most of the C language logical, arithmetic, and shift operators, in addition to a few Verilog-specific operators.  Thus, && and || represent the logical AND and OR, respectively, of two binary values.  For multiple-bit values, &, |, and ^ can be used for bit-wise logical AND, OR and exclusive-OR, respectively.  Also, Verilog has added ~ for logical inversion (! can also be used for single-bit values) and { } for concatenation of two or more signal values (e.g., {2'b01,~3'h5} = 5'b0_1101).  The arithmetic and shift operators are identical with C.  One notable exception is that the ++ (increment) and -- (decrement) operators are not permitted in Verilog.  Only single-dimensional arrays (vectors) can be declared in Verilog (using slightly different syntax from C).  However, it is possible to declare an array of multi-bit signals because such a construct is necessary to describe memory arrays.  Thus, for example, **reg** [4:0] sig_a and **reg** [7:0] mem_data [1023:0] declare a 5-bit signal and a 1-Kbyte memory array, respectively.

### 3.1.3  Structural and Behavioral Descriptions

A Verilog module can be defined in a structural or behavioral manner.  Behavioral code follows a C-like syntax to describe how the target circuit should operate or behave.  A structural description describes a module as an interconnection of submodules.  Connections between the submodules are indicated by using identical signal names.  A connection between a module and a submodule is indicated by "mapping" a signal used in the module definition to an input or output port in the definition of the submodule (which may be defined in a separate file).  Such a signal mapping can be accomplished by *positional association* (the place where the signal is entered in the parameter list) or by *explicit assignment* (naming the signal and its associated port in the parameter list in the form **.port**(signal)).  A commonly used convention is to list all outputs before inputs in the parameter list of module definitions.  The following example shows simple Verilog module definitions for a 1-bit full-subtracter module using both behavioral and structural code.  Based on this full-subtracter module description, the reader easily should be able to implement an analogous full-adder circuit.  This is left as an exercise for the reader.

| **Example 3.1** | **Behavioral and Structural Verilog Code** |
|---|---|

The following is behavioral Verilog code for a simple 1-bit full subtracter.  A 1-bit subtracter subtracts two single-bit numbers to produce a single-bit result and a 1-bit "borrow".  For example, $1 - 0 = 1$, $1 - 1 = 0$, and $0 - 1 = 1$.  The first two cases have a borrow value of '0' while the third case has a borrow value of '1.'  A 1-bit full subtracter has a borrow_in input in addition to a borrow_out output.  Several 1-bit full subtracters can be chained together to produce a multiple-bit subtracter.

```
//////////////////////////////////////////////////////////////////
// Comments can be marked with double-slashes (C++ format) or
/* the comment notation used in C. */
module full_sub_behave (res, borrow_out, a, b, borrow_in);
                        // outputs are listed before inputs
  output res;           // result of the subtraction
  output borrow_out;  // the borrow_out value
  input a;              // the value to subtract
  input b;              // the value to be subtracted
  input borrow_in;    // the borrow_out of the "previous" stage

  assign {borrow_out, res} = a - b - borrow_in;
                        // result of subtraction is two bits;   the
                        // msb is "borrow_out" and the lsb is "res".
endmodule
//////////////////////////////////////////////////////////////////
```

There are several notable points in the above code. First, outputs are listed before
inputs in the parameter list. Although this is not required, it is a commonly used
convention that is helpful when several people contribute code for a single project.
Second, all input and output signals are assumed to be of type **wire** since this is the
default type in the absence of explicit type definitions. Third, the **assign** keyword is
used to allow a brief description of the behavior of the two outputs borrow_out and res.
Fourth, the curly brace notation { } is used to indicate the concatenation of two bits
to form a 2-bit field — in general, this notation can be used to concatenate multiple
signals, each consisting of multiple bits. The result of subtracting b and borrow_in from
a is two bits, which are stored in the borrow_out and res signals. Note that subtraction
(the − notation) is one of several predefined operators in the Verilog language.
Most C language operators are supported by Verilog also (note that ++ and -- are not
among the supported operators).

The following is structural Verilog code for the same circuit.

```
//////////////////////////////////////////////////////////////////
module full_sub_struct (res, borrow_out, a, b, borrow_in);
                        // structural full subtracter definition
  output res;           // subtraction result
  output borrow_out;  // the borrow-out value
  input a;              // the input to subtract
  input b;              // the input to be subtracted
  input borrow_in;    // the borrow-out from the previous stage
```

```
xor (c, a, b);       // uses XOR gate library part
xor (res, c, borrow_in);   // res = a XOR b XOR borrow_in
not (not_a, a);
and (d, not_a, borrow_in);
and (e, not_a, b);
and (f, b, borrow_in);
or  (borrow_out, d, e, f); // borrow_out = a'w + a'b + bw,
                           //   where w = borrow_in
endmodule
//////////////////////////////////////////////////////////////
```

This definition uses the **xor** (exclusive-OR), **and**, **or**, and **not** (inverter) built-in library gates. The implied connections easily can be inferred from the structural definition shown (e.g., the output of the first **xor** gate is connected to the first input of the second **xor** gate since the internal wire "c" is connected to the output of the first **xor** gate and the first input of the second **xor** gate). The *schematic* for this circuit, which shows the connections between all gates and devices in a graphical manner, is also shown in Figure 3.1.

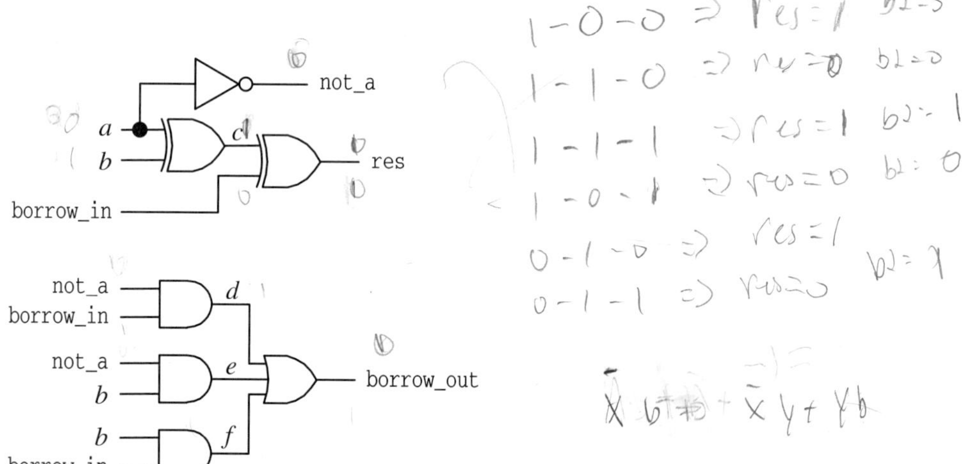


**Figure 3.1**
The schematic diagram corresponding to module `full_sub_struct`.

The derivation of the logic equations that result in this connection of gates is left to the reader.

A list of the primitives (gates and other basic functions) used in Verilog are shown in Table 3.1. Most of these terms will be explained as they are used. The order of the argument parameters for the gates are one or more outputs followed by one or more inputs. The **and**, **nand**, **or**, and **nor** gates are self-explanatory. The **xor** primitive refers to an exclusive-OR gate while the **xnor** primitive refers to an exclusive-NOR gate (a gate that implements the opposite function of an exclusive-OR gate). For all gates listed, the first port element is the output. The **bufif0**, **bufif1**, **notif0**, and

| Combinatorial Logic | Buffers | Operators (besides C language type) |
| --- | --- | --- |
| **and** | **buf** | { } (concatenation) |
| **nand** | **not** | === (case equality) |
| **or** | **bufif0** | !== (case inequality) |
| **nor** | **bufif1** | --- (bit-wise OR) |
| **xor** | **notif0** | ˜ (bit-wise NOT) |
| **xnor** | **notif1** | & (bit-wise AND) |
| | | ^ (bit-wise exclusive-OR) |
| | | ˜^ (bit-wise exclusive-NOR) |

**notif1** primitives refer to tri-state buffers with active-low or active-high enables and uncomplemented or complemented outputs. The **buf** and **not** primitives, which are noninverting and inverting buffers, respectively, can have multiple outputs. The **and**, **nand**, **or**, **nor**, **xor**, and **xnor** gates can have an arbitrary number of inputs.

For simple circuits that can be described in a straightforward manner, the minimal description method described above may be sufficient. However, for large and complex circuits, it is best to use a hierarchical description methodology (analogous to using a hierarchy of function and procedure calls when writing complex computer programs), as shown in Figure 3.2. At each level of the hierarchy, a **module** definition is required. Ideally, for each separate **module** definition, there must be a corresponding *test bench*, described in the next section, which is used to verify the proper operation of that **module**. The top-level module, along with its test bench, is used for the final target circuit. To permit hierarchical decomposition of the target circuit, all modules except the modules at the leaf nodes must use a structural or mixed structural/behavioral description. The lowermost modules in the hierarchy require the use of structural descriptions with built-in logic gates and devices (such as the `full_sub_struct` module), or behavioral descriptions that do not depend on any lower-level modules (such as the `multiplier` module, which can be defined using the Verilog × operator).

It is important to note that the use of a hierarchical Verilog description does *not* mean that we simply can *call* lower-level modules, as in a C program, in order to perform a sub task for the higher-level module.[2] Instead, an instantiation is made to a lower-level module for a separately definable part (or device) of the higher-level module, while other separately definable parts can be described in a behavioral manner.

---

[2]   There are `task` and `function` constructs that can be used for this purpose, as described later in this chapter.

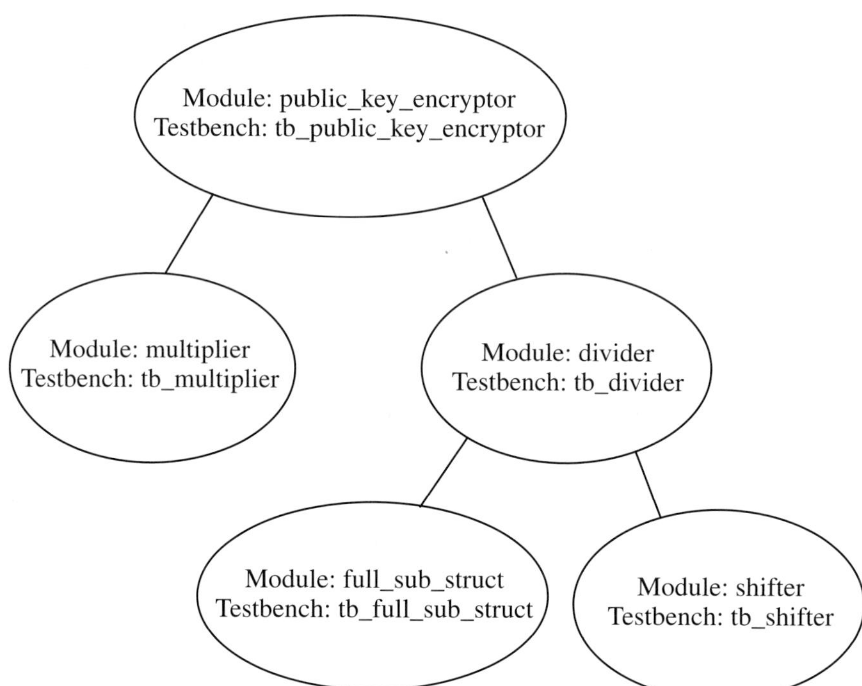


**Figure 3.2**
An example of a hierarchical design description.

## 3.2 Testing and the Test Bench

The importance of *testing* in digital logic design cannot be overemphasized. Testing is used to not only verify the correct operation of the completely designed circuit, but also to aid the designer during the circuit design process. Even when the completely designed circuit has been verified to operate correctly, testing must still be used to weed out hardware circuit parts that may be malfunctioning due to non-design-related problems. This latter type of testing is referred to as *manufacturing testing* while the former type is referred to as *functional testing*.

Regardless of the type of testing being performed, a test essentially consists of asserting values for all of the inputs to the circuit and then observing the resulting behavior at all of the outputs of the circuit. More precisely stated, there must be a sequence of *test vectors* applied to the *unit under test (UUT)*; a test vector is an assignment of values to all of the inputs of the UUT. Such a sequence of test vectors is referred to as a *test bench*, *test suite*, or *test harness*.

### 3.2.1 Manufacturing Testing

In *manufacturing testing*, the main objective is to test each individual hardware circuit (or sample lots of hardware circuits) in order to weed out malfunctioning circuits. Even if the circuit design already has been verified to be correct during the functional testing phase, individual hardware circuits may still malfunction due to *faults* in the fabricated circuit caused by manufacturing defects (such as shorts or opens caused by uneven deposition of a metal layer during the fabrication process for a chip). Manufacturing testing becomes especially important with increasing clock rates, decreasing transistor sizes (within a chip), and tightly packed circuit boards; as all of these factors contribute to smaller manufacturing process tolerances. Manufacturing testing can be used to identify design problems (since some types of designs are more prone to permanent or intermittent hardware defects caused by slight process variations), manufacturing process problems (clustering of defective circuits may point to problems with the fabrication process, e.g., variations in temperature, during the manufacture of those circuits), and other quality-control problems.

Since manufacturing testing has a direct effect on the circuit *yield* (the percentage of hardware circuits produced that are nonfaulty), which in turn has a direct effect on the profitability of a company, much research and development effort has been expended on methods for effective manufacturing testing. In general, manufacturing testing can be performed independently of the function of the target circuit. Since the objective is to weed out circuits that may be malfunctioning due to manufacturing defects, tests should be devised to detect the presence of such manufacturing defects.

As a first step, the types of circuit defects produced during manufacturing must be identified and characterized. Many studies have been conducted for this purpose. In general, the results show that unwanted open connections (opens), unwanted closed connections (shorts), and cross-coupling faults (logic values in one part of the circuit inadvertently changing the logic values in another part of the circuit) are the most common types of manufacturing defects observed in practice. Modeling these different types of circuit defects accurately is extremely difficult.

For the purpose of simply *detecting* malfunctioning circuits, the most commonly used fault model is the *stuck-at* model. In this fault model, every single wire in the target circuit can potentially become *stuck-at*-1 (a permanent logic 1 value due to a manufacturing defect) or *stuck-at*-0 (a permanent logic 0 value due to a manufacturing defect). In a typical circuit, this would imply an extremely large number of combinations of fault patterns. However, many of these fault patterns have the same effect on the behavior of the circuit. For example, a stuck-at-0 fault on one input of a NAND gate has the same effect as a stuck-at-1 fault on the output of that NAND gate. Even when taking into account such logically equivalent faults, the number of possible stuck-at fault patterns is still extremely large. Thus, a commonly used fault model, which is a subset of the stuck-at fault model, is the *single stuck-at* fault model. In this fault model, there is assumed to be at most one stuck-at-1 or stuck-at-0 fault in the entire target circuit. Even though the single stuck-at fault model may be considered a bit restrictive, it is still found to be an effective fault model for testing purposes.

Given a fault model for manufacturing testing, a set of test vectors must be created to detect the faults in that fault model. For instance, with the single stuck-at fault model, an assignment of logic values to all primary inputs of the target circuit will exercise that circuit in some manner. If the output of the circuit with a certain fault is *different* from the output of the circuit without that fault, then that fault can be considered to be *covered* by the given input assignment (or test vector). Of course, there typically will be other faults that will result in the same set of circuit outputs as a nonfaulty circuit for the given test vector—those faults are *not covered*. As more and more test vectors are used, the percentage of potential faults (in the given fault model) that are covered will approach 100%. Given a set of test vectors, the *fault coverage* of that test vector set (given a specific fault model such as the single stuck-at model) will correspond to the percentage of potential faults covered.

With sequential circuits, manufacturing testing becomes a bit more complicated due to the fact that the circuit outputs depend on the current *state* of the circuit. Thus, with such circuits, it is desirable to be able to set the values of all flip-flops to specific desired values. This can be done in one of two ways. First, an *initializing sequence* (i.e., a sequence of input assignments that drives the circuit state into a desired state) can be used. Then, a single test in such a circuit would consist of an initializing sequence followed by another test vector. As can be imagined, such a method can lead to a large number of test vectors, many of which are used simply as initializing sequences. Also, it may not always be possible to set the values of all flip-flops to specific desired values. Thus, in the second method, all flip-flops used in the circuit can be modified to be *scan flip-flops*, which are flip-flops that can be loaded with specific values from the primary inputs (during scan mode) or used in the same manner as unmodified flip-flops (during normal mode). Such scan flip-flops can be created by inserting multiplexers at the $D$ inputs to those flip-flops.

With manufacturing testing, the quality of a test is determined by the *fault coverage* achieved by that test, given a fault model, and the time required to apply all test vectors of that test to the target circuit, which is directly proportional to the number of test vectors in the test. A higher value of fault coverage implies that fewer defective circuits will pass through the fault inspection process. A smaller test vector set implies faster tests, which in turn implies that more circuits can be tested per unit time. However, since there is a limit to the number of circuits that can be produced and consumed per unit time, the size of the test vector set only becomes important if it is so large as to affect the production rate.

Manufacturing testing is an extensive topic area with many interesting research and practical aspects. The interested reader is referred to the many books and journal articles devoted to this topic. This book will concentrate primarily on the second type of testing, functional testing, which is described in the next subsection.

### 3.2.2   Functional Testing

In *functional testing*, the objective is to verify that the target circuit performs all of its operations as intended. Functional testing must be used during all stages of the circuit design process. For large designs, which typically are designed in an hierarchical manner, all lower-level subcircuits must be functionally verified, using separate testing procedures (test benches), before they are included into higher-level subcircuits. Even if all subcircuits have been separately verified, the entire merged circuit must also be verified using a separate set of tests. Then, after the circuit has been realized using a hardware platform (such as an ASIC, FPGA, or a printed circuit board consisting of several separate chips wired together), it must be tested again. If possible, the same set of tests used during the simulation of the completely designed circuit should be used again during the testing of the hardware prototype. Invariably, the first hardware prototype will have errors in it. Comparison of test results with the simulation results for those same tests can help identify and isolate design or manufacturing errors.

The type of testing used throughout the rest of this book is functional testing rather than manufacturing testing. Since functional testing also involves the application of a set of test vectors, there are many similarities with manufacturing testing. Test benches for functional testing will be created for all code presented in the rest of this book. The methodology used for these test benches is explained below.

### 3.2.3   Test Benches

One method for thoroughly testing the functionality of a circuit is to iterate through all possible combinations of the test inputs. However, this may not be possible for some circuits (e.g., the number of possible combinations may be too large, or some combinations of test inputs may not be permitted for a particular circuit). In addition, some functions may require a certain sequence of test inputs. Thus, the particular test bench to be used is circuit dependent. However, in general, we should try to follow the guideline of iterating through all possible test-input combinations whenever possible.

**Test Vector Generation and Verification**

If the number of possible test-input combinations is too large (e.g., it would require several months or years to apply all test-input combinations), there are several heuristic approaches that can be used. First, knowledge of the function of the circuit can be used to eliminate test-input combinations that are meaningless or would never occur in practice. Second, the circuit can be partitioned into several subcircuits, each of which can be tested exhaustively (using all of the test input combinations for each subcircuit). Then the integrated circuit can be tested using a nonexhaustive test set, just to ensure that the circuit has been integrated properly. Third, a representative set of test cases can be chosen to exercise the *unit under test (UUT)* under normal and extreme conditions. Finally, all three of the above methods can be combined in varying degrees.

Let us consider the generation of a "representative" set of test cases for a typical circuit. There should be a representative sample of test inputs for all *corner* cases (extreme conditions, such as when all test inputs have lower-bound values or all test inputs have upper-bound values) as well as the normal cases (non-corner cases). The generation of these corner case and normal case tests should be done in a systematic manner. Thus, for example, we could apply tests with all combinations of lower-bound and upper-bound values for circuit inputs, followed by a large **for** loop iterating through "normal-range" values for all circuit inputs.

When testing a circuit using a large number of test vectors, it is desirable to be able to automatically verify the correct response of the circuit to each test vector. There are several ways in which this can be done. One way is to simply record the sequence of correct responses in a file, and then compare each response in that file to the observed response. Another method is to compute the correct response (for the particular test vector used) and then compare that response to the observed response. However, when using this method, the verification computation must use a completely different method from the method used in the target circuit; otherwise, the same mistake may be repeated in both computations. If such a verification computation is not possible or feasible, then an *acceptability test*—asimple test to determine if the result appears to be "reasonable" or not—should be used (e.g., for a circuit that computes the average of a set of numbers, we could test if the result is smaller than the largest member and larger than the smallest member of the set). Of course, the automatic verification of the circuit should be accompanied by a manual inspection of the outputs of the circuit (the simulation waveforms) to ensure that the automatic verification results are correct.

For a circuit modeled using an HDL, the method typically used to apply the test bench and verify the correct operation of the circuit is computer simulation. Many simulation tools include a menu-driven graphical method of assigning values to the inputs of the target circuit (referred to as *stimulus* inputs). However, using such a menu-driven approach, it can be time consuming to create long test sequences, and such test sequences may not be repeatable the next time the target circuit is modified and the simulation restarted. In addition, a test bench created for one type of CAD software may not be usable when the design is ported to a different type of CAD software.

In light of the above observations, the recommended method for creating a test bench is to create the test bench as simply another HDL module. The test bench module will have the unit under test (UUT) as a submodule, all inputs of the UUT will be generated from within the test bench, and all outputs of the target circuit will be verified from within the test bench using *orthogonal* computations (orthogonal to, i.e., independent of, the computation method used in the target circuit). Thus, in summary, a test bench must include a test-generator subcircuit for test-vector generation, the UUT itself, and a test-verifier subcircuit to check the results of the application of tests on the UUT. A block diagram of such a "complete" test bench is shown in Figure 3.3.

As an HDL module, the test bench will be a special type of module that has no external inputs or outputs. Since it can be written in a C program-like manner, we can create extremely complex patterns of test vectors, and also verify the results of

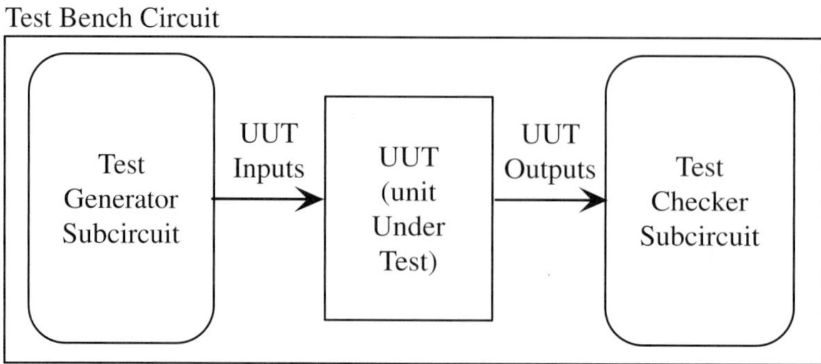


**Figure 3.3**
A block diagram model of a test bench.

applying those test vectors to the target circuit using extremely complex computations. In addition, the test bench normally will *not* be synthesized into a separate circuit. Thus, we will be free to use any available HDL instruction (even those that result in nonsynthesizable circuits). HDL instructions used to control and display the results of the simulation are particularly useful in this regard.

When designing a digital logic circuit, it is important to devise a test bench for the circuit early on. If the circuit is designed hierarchically, then each subcircuit should be tested separately before it is combined with other subcircuits into a higher-level circuit. This type of testing is useful during the design stage. A test bench also should be devised for the anticipated hardware prototype of the circuit. Of course, this test bench should be used in the design model before it is applied to the actual hardware circuit.

## Verilog Test Bench

Once a hardware circuit has been modeled using behavioral or structural Verilog code, a test bench should be devised for that circuit. The test bench normally will *not* be synthesized into a separate circuit. Thus, we are free to use any available Verilog instruction (even those that result in non-synthesizable circuits). Useful instructions of this type include #, **$display**, and **$finish**, which can be used to control the progress of the simulation and display its results. As a Verilog **module**, the test bench is a special type of **module** that has no external inputs or outputs. It has connections to the circuit being tested, typically referred to as the *unit under test (UUT)*. Then there is code to generate the test patterns for all inputs to the UUT. Finally, there should be code to check the outputs of the UUT to verify that the UUT is functioning correctly.

**Example 3.2**   **Simple Test Bench**

Let us create a simple test bench for the circuit of our previous example, Example 3.1. This test bench tests the structural implementation code *full_sub_struct*. The test bench shown in the following results in the simulation output waveform and console display shown in Figure 3.4.

```
///////////////////////////////////////////////////////////////////
`timescale 1ns/100ps     // time unit = 1ns, precision = 100ps

module test_sub();

  reg a, b, borrow_in;   // inputs to full_sub_struct
  wire res, borrow_out;  // outputs of full_sub_struct
  reg expected_res, expected_bo;   // expected results
  integer error_count;   // number of errors

  // make connections to the unit under test (UUT)
  full_sub_struct U1 (res, borrow_out, a, b, borrow_in);
```

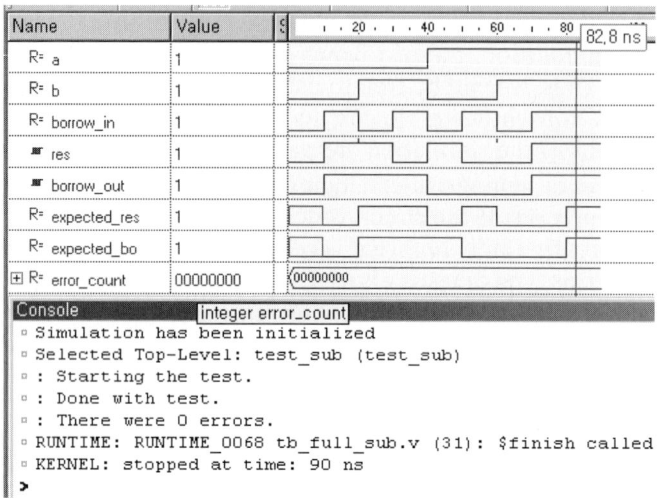

**Figure 3.4**
Waveform and console display for Example 3.2.

```verilog
initial begin // one-time execution block
  $display("Starting the test.");
  error_count = 0;
  {a, b, borrow_in} = 3'b000;
end

// test vector generation and checking process
always begin  // repeated execution block
  #10;        // wait for 10 * 1ns = 10ns
  {expected_bo, expected_res} = a - b - borrow_in;
  if ({expected_bo, expected_res} !== {borrow_out, res}) begin
    $display("Expected (%b, %b) != actual (%b, %b) at time %0t.",
             expected_bo, expected_res, borrow_out, res, $time);
    error_count = error_count + 1;
  end
  if ({a, b, borrow_in} === 3'b111) begin
    #10 $display("Done with test.");
    $display("There were %0d errors.", error_count);
    $finish;
  end
  else          // compute next test vector
    {a, b, borrow_in} = {a, b, borrow_in} + 1;
end // of main always block

endmodule
//////////////////////////////////////////////////////////////////
```

There are several new constructs used in this test-bench code. Here, **'timescale** is a compiler directive used to control the simulator. In this case, it tells the simulator to use a time unit of 1 ns (a delay of 1 time unit = 1 ns) and a time precision of 100 ps (the minimum step size used during simulation is 100 picoseconds). If this directive is omitted, the default values of 1ps/1ps are used—with these values, the simulation waveform may not be viewable until the "zoom" menu button is depressed a number of times. Not shown, but useful, are the **'include** directive (similar to "#include" in C and used to include the contents of another file) and the **'define** directive (similar to "#define" in C and used to define constant values).

There are several commands used to control the display of information on the console and to control the simulation. The **$display** is used to display information in the console window; it has a format similar to the C language *printf* function, except that a newline is automatically appended and there are a few extra formatted print commands such as %t (for printing time information) and %b (for printing binary data). Here, **$time** is a command that returns the current simulator time, while **$finish** is a command to terminate the simulation normally. Another command, **$stop** (not used in this example), can be used to terminate the simulation upon an error condition.

Although the parameter list for this module is empty, there are several signal wires defined. The parameter list is empty since a test bench does not need to have any external inputs or outputs. The list of signal wires should include the signals to be connected to the inputs and outputs of the unit under test (UUT). Signal wires can be defined as **reg**, **wire**, or **integer**. A **reg** variable corresponds to a signal that retains its value once it is assigned; the value is preserved until it is changed by another assignment or the simulation ends. A **wire** variable corresponds to a physical wire; it can only be assigned a value by being driven to that value by a submodule (subcircuit) or an **assign** statement. All undeclared variables are assumed to be of type **wire** by default. Finally, an **integer** variable essentially corresponds to a 32-bit **reg** signal; it can be manipulated like a C language integer value. In this example, the **integer** error_count is used to keep a running count of the simulation errors observed. Multiple-bit signals can be defined by using the bracket [msb:lsb] array notation. Thus, an **integer** essentially is equivalent to **reg**[31:0].

Following the signal declarations is the connection to the unit under test (UUT). Thus, we instantiate the *full_sub_struct* device, assign *U1* as the name for this instantiation (assigning such a name is optional), and then list the connections to the *full_sub_struct* device. The connections can be made by *positional association* (just like a C function call), in which a particular signal is connected to a particular port by listing that signal in the same position as that port in the parameter list of the original module definition for the UUT. An alternate method is to use *named association*, in which a particular signal is connected to a particular port by listing both the port and the signal in the form **.port**(signal). Thus, for example, the connections to the full_sub_struct device shown above could have been written as

```
full_sub_struct U1 (.res(res), .borrow_out(borrow_out),
                    .a(a), .b(b), .borrow_in(borrow_in) );
```

although identical names are used for the signal and the port that it is connected to in this example, different names could be used if desired.

Next is an **initial** block. An **initial** block contains those actions that are taken only once at the beginning of the simulation. It essentially is equivalent to a process that is activated only once. Note how all blocks of code are delimited with **begin** and **end** statements, instead of curly braces { } as in C, since curly braces are used for concatenation in Verilog. Within an **initial** block (and an **always** block, which is discussed next), the designer is free to use a wide range of C language-like constructs to describe the behavior of the circuit being modeled.

The statements within an **initial** block (and an **always** block) execute one after the other (sequentially, just like a normal C program) unless a special type of assignment (labeled <=, to be discussed later) is used within that block. In this case, we first print out a statement announcing the beginning of the simulation, initialize error_count to 0, and then choose an initial test vector of $(0, 0, 0)$ for the three inputs (a, b, borrow_in). Note that the notation 3'b000 is used to denote a 3-bit vector of all zeroes. This type of notation, with the bit length followed by prime ('), followed by the letter b, followed by a bit vector is used to denote a binary constant. A number (such as 0) written without a k'b (k-bit binary) or k'h (k-bit hexadecimal) type of prefix, normally is interpreted as an integer, which essentially is equivalent to 32'b. When listing a long binary constant, a convenient Verilog notation is the use of underscore (_) to make the constant easier to read. Thus, for example, the binary constant consisting of four 0's followed by two 1's could be written as 6'b0000_11.

Finally, the **always** block, which essentially is equivalent to a Unix-type process, contains a series of statements that are executed repeatedly. The **always** is executed forever unless a condition (referred to as a *guard*) is used to indicate *when* the **always** block can be executed or a termination command (such as **$finish** or **$stop**) is encountered. The first statement in this **always** block is #10;, which creates a delay of 10 time units (which equals 10 ns, since one time unit was declared to be equal to 1 ns in the initial **'timescale** statement).

Next, the expected result is computed in a different manner from the target circuit, and the expected result is compared with the output of the target circuit. An error is indicated (and error_count is incremented) if the two results are different. Although a C-like "not equal notation" (!=; with == used for the equals test) could have been used for the comparison, a "case inequality" (!==; with === used for the case equality test) is used in order to test for *exact* equality. This is necessary (especially when used in a test bench) because signals can have the values 'X' (unknown) and 'Z' (high impedance) in addition to '0' and '1.' The case equality (triple equals) or inequality notation tests for exact equality (including 'X' and 'Z') between two sets of signals while the C-like equality (double equals) or inequality notation tests for logical equality (returning the value 'X' if the result cannot be determined exactly).

The final portion of the **always** block updates the test vector and checks for a termination condition. The **always** block is written so that the test vector formed by the bits (a, b, borrow_in) is incremented as a 3-bit vector until the value (1, 1, 1) is reached, at which point a message is printed and the simulation is terminated.

# 3.3 More Advanced Verilog Concepts

### 3.3.1 Concurrent and Sequential Verilog

As stated in Chapter 2, hardware is *concurrent* by nature. For example, if a circuit consists of an AND gate followed by an edge-triggered D flip flop, then both the AND gate and the D flip-flop are *active* all of the time. Whenever there is a change in an input to the AND gate, the effect of the change propagates to the output of the AND gate. Then, when the active clock edge arrives at the D flip-flop, the D flip-flop output Q changes to the value at the D input. At that time instant although the output can only change when the active clock edge arrives, the D flip-flop is active *all of the time* waiting for the proper clock transition to arrive.

Since Verilog is used to model such hardware systems, the top-level statements within a Verilog module definition are simulated as executing concurrently. Thus, if two top-level Verilog statements assign conflicting values to a common signal, then that signal will acquire one of the two conflicting values (in an unpredictable manner) or an unknown (denoted "1'bX" or "1'bx") value.

In order to be able to describe the behavior of a target circuit, it may be necessary to be able to use a sequence of statements that execute sequentially. In order to support such descriptions, the **initial** and **always** statements can be used. A sequence of statements enclosed within an **initial begin** statement and the matching **end** statement (or an **always begin** statement and the matching **end** statement) are simulated as executing in a sequential manner. An **initial** block is executed only once at the beginning of the simulation while an **always** block is executed repetitively as an infinite loop. Also, it is important to note that all **initial** and **always** blocks execute concurrently with respect to each other (only the statements *within* an **initial** or **always** block execute sequentially), since the **initial begin** and **always begin** statements are top-level statements within the **module** definition.

Although both **initial** and **always** blocks can be used in a test bench, **initial** blocks cannot be used in Verilog modules that will be synthesized into circuit modules. Since a test bench only is used to stimulate (and observe the output of) the target circuit, it does not need to be synthesized—thus, all types of Verilog constructs can be used in a test bench. However, for a Verilog module representing an actual hardware module, only Verilog constructs that can model actual physical hardware are permitted. An **initial** block should not be used in such Verilog modules because hardware that executes only once is difficult to construct in general—it is much simpler and more natural to construct hardware circuits that execute repetitively and infinitely. As an example of the use of **always** blocks, the following shows a Verilog module definition for an edge-triggered D flip-flop with an asynchronous active-low reset signal. The active-low signal reset_n is defined with an "_n" suffix; although this type of naming convention is not a requirement of Verilog, a uniform method of naming active-low signals can help prevent mistakes.

```
//////////////////////////////////////////////////////////////////
module Dff (q, q_n, d, clk, reset_n);
  output q, q_n;    // Q and Q' outputs of the D flip-flop
  input d;          // D input to the D flip-flop
  input clk;        // clock signal
  input reset_n;    // active-low reset signal

  reg q, q_n;       // q and q_n need to be defined as type "reg"
                    // instead of the default ("wire") since they
                    // will be assigned values in the always block.
  always @(negedge reset_n or posedge clk) begin
    if (~reset_n) begin  // first, check if the reset is active
      q <= 1'b0;            // if so, assign initial values
      q_n <= 1'b1;
    end
    else begin  // otherwise, the only other case is "posedge clk"
      q <= d;    // store the D input value into the flip-flop
      q_n <= ~d;// at the same time, Q' output is set to (not D)
    end
  end  // of always
endmodule  // of the module definition
//////////////////////////////////////////////////////////////////
```

### 3.3.2   Delay Modeling

All real circuits, devices, gates, transistors, and wires have delays associated with
them. For the accurate simulation of such circuits, we must be able to accurately
model the delays in those circuits. Sometimes, a circuit may not even function properly during simulation if no delays are assigned to any of the circuit parts. One
recommendation in the modeling of a digital circuit is to always assign some delay
(even an arbitrary delay such as 1 ns) to all register–transfer level operations used
in the Verilog code for that circuit. (Note, however, that such exact delays are not
supported in synthesis; some synthesis tools will ignore such delay constructs, while
others will require all such delays to be reset to zero or deleted.)

### Delay Operator #

Delays can be modeled using various methods. The simplest method is to use the
delay operator # followed by the number of time units. If this operator is used by
itself, it causes the **initial** or **always** block (that it is a part of) to pause for that number
of time units (e.g., #10; within an **initial** block causes it to pause for 10 time units).
The # operator also can be used with a submodule instantiation, in which case that
submodule will have the indicated propagation delay (e.g., **not** #10 (a, b); will cause
the **not** gate to have a propagation delay of 10 time units). If the # operator is used
in front of an assignment statement, it causes that assignment to execute after the
specified number of time units (e.g., #10 a = b; or #10 a <= b; causes the assignment
to signal a to occur after 10 time units). On the other hand, if the # operator is

used after the = or = symbol in an assignment statement, then the assignment occurs immediately but does not complete until after the specified number of time units (e.g., a <= #10 b; causes the assignment to a to complete 10 time units later, but using the *current* value of b).  This difference can become important in state machine modeling—in general, it is more preferable to use the latter form (# after the <= symbol) in synchronous sequential machines.

Sometimes, delays associated with hardware devices need to be modeled in a more detailed manner than as simple signal transmission delays.  Verilog permits this to be done by using additional delay parameters.  For example, **nand** #(4, 6) G1 (y, a, b); describes a 2-input NAND gate named G1 (the insertion of a name is optional) with a *rise time* of four time units and a *fall time* of six time units. The rise time indicates the delay incurred when changing the output-signal value from a low value (logic 0) to a high value (logic 1). Analogously, the fall time is the delay incurred in transitioning from high-to-low logic values.  Also, the notation **nand** #(3:4:5, 5:6:7) G1 (y, a, b); is used to specify minimum, typical, and maximum rise and fall times for the G1 NAND gate.  For devices that can be turned off (produce a high impedance output), the *turn-off* time (the delay until the output transitions from a low or high value to a high-impedance value) can be specified.  Thus, **bufif0** (1, 2, 3) G2 (o0, i0, en_n) specifies a tri-state buffer with an active-low enable input (en_n) and a turn-off delay of three time units (minimum, typical, and maximum values can also be specified with these rise, fall, and turn-off times).

### Activation Guard and Wait

Other delay-modeling constructs are @ and **wait**.  The @ construct is an *activation guard* that prevents the associated statement or block from executing until a pre-specified condition is satisfied.  Thus, we saw that **always @(posedge** clk) can be used to model a block that is activated every time there is a positive transition on the clk clock signal. Similarly, **always @**(state) would cause the **always** block to execute whenever there is a change in the state variable.  The @ construct can also be used within an **always** or **initial** block, in which case the corresponding process simply delays until the specified condition becomes true.  Thus, for example, @(**posedge** clk) causes the process to suspend until the next positive transition of the clk signal.

The **wait** construct can be used with **for, until**, or a parenthesized condition check. The **wait for** *delay* statement causes the process to suspend for *delay* time units. The **wait until** *condition* statement causes the process to suspend until *condition* becomes true. The **wait**(*condition*) statement also causes the process to *suspend* until *condition* becomes true.  An example of this can been seen in the final part of the test bench for module power in Section 3.4.3, in which **wait**(done); causes the **always** block that it is a part of to be suspended until the signal done is set to one.

### Inertial Delay and Transport Delay

When a hierarchical circuit description is used, the concepts of *inertial delay* and *transport delay* can become important.  A real digital logic device typically has the property that a glitch in an input port with a duration *less* than the propagation delay of that device does not show up at an output port.  Thus, if a NAND gate has a propagation delay of 1 ns, then a 0.5 ns glitch (a temporary '1' or '0' of duration 0.5 ns) in an input port does not result in a change at the output port. This type of delay behavior is referred to as *inertial delay*, and it is the default delay model for all devices modeled in Verilog. However, real wires in digital logic circuits typically have the property that glitches are transported regardless of the duration of the glitch with respect to the wire delay. This type of delay behavior, in which all glitches in an input port show up at an output port, is referred to as *transport delay*, and it is the default delay model for all wire connections in Verilog. Wires can be declared to have propagation delays by treating the **wire** clause as a sub-module instantiation (e.g., **wire #**10 `wire1`; declares a wire named `wire1` with a propagation delay of 10 time units).

### 3.3.3   Different Types of Assignment Statements

Assignments to signals display different behavior depending on the specific assignment methods used. Assignments can be made using parameter mapping in a sub-module instantiation (similar to a C language function call), using top-level **assign** statements within a **module** definition, or using register assignment statements within **initial** or **always** blocks. These types of assignment methods are referred to, respectively, as *parameter mapping, continuous assignment*, and *procedural assignment*.

As a top-level statement within a **module** definition, assignments to signals are made by instantiating a submodule or by explicitly using an **assign** statement. For both types of assignments, the signal being assigned must be declared as type **wire**. Since type **wire** is the default signal type, the signal does not have to be defined explicitly as type **wire** if it already is defined as an **input** or **output** of the module. The wire can be assigned a value by including it in the parameter list of the submodule instantiation—the corresponding port of the submodule should, of course, be an **output** or **inout** port.

The other wire assignment method requires the use of the **assign** keyword, such as is done in module `full_sub_behave` of Example 3.1. This type of assignment is referred to as *continuous assignment* in Verilog terminology. As the name implies, such a statement implies that the assignment must be made continuously. When such a statement is simulated, however, it only needs to be executed when one of the signals that it is dependent upon changes.

A drawback of continuous assignment is that complex calculations cannot be used in such **assign** statements. For complex assignment calculations, an **initial** or **always** block should be used. At the most, an `if-else` type or `if-elseif-else` type conditional assignment statement can be made. Thus,

```
assign sig1 = cond1 ? value1 : value2;
```

can be interpreted just like a C-language `if-else` assignment, while

```
assign sig1 = cond1 ? value1 :
              cond2 ? value2 :
              cond3 ? value3 :
              value_default ;
```

assigns `value1` to `sig1` if `cond1` is true, `value2` to `sig2` if `cond2` is true, `value3` to `sig3` if `cond3` is true, and `value_default` to `sig1` if all three conditions are false. One major difference with a C-language `if-else` assignment is that the above **assign** statements are "continuously" executed—in practice, this means that the first **assign** statement, for example, is re-executed whenever the `cond1`, `value1`, or `value2` values change.

Assignment to a signal within an **initial** or **always** block is referred to as *procedural assignment*. This type of assignment typically requires the signal to be defined as type **reg**. Since **reg** is not the default signal type, all such signals must be declared explicitly using **reg** declaration statements, even if they are already defined as outputs of the module.

There are three types of procedural assignment statements: *blocking, nonblocking*, and *procedural continuous assignment*. With blocking assignment, which uses the = operator, the actual assignment occurs in a sequential manner, just as would be expected in a series of statements in a C language program. However, with nonblocking assignment, which uses the <= operator (it looks similar to the left arrow used to indicate an assignment in pseudocode), the assignment occurs in parallel (concurrently) with the next statement.

The third type of procedural assignment statement, which normally is not used in code that must be synthesized, is *procedural continuous assignment*, in which the **assign** keyword is used in addition to the = operator within an **initial** or **always** block. With procedural continuous assignment, it becomes possible to *deassign* the value previously assigned. Because of this reason, procedural continuous assignment is not supported by many logic synthesis tools. For this reason, and also because it is not good programming practice, procedural continuous assignment is not used in this textbook.

Nonblocking assignment more closely models the typical behavior of actual hardware than blocking assignment. For example, in a Moore-type finite state machine, several outputs can be assigned values in a single state. However, changes to those outputs do not occur one after the other. Rather, they change concurrently whenever the finite state machine enters that state.

In the following code, examples are shown for each of the four types of assignments discussed in this section. As can be seen, results using nonblocking assignments can be significantly different from results using blocking assignments.

```
/////////////////////////////////////////////////////////////////
module test_assign ();

    reg in1, in2, cond1;         // test inputs of type reg
    wire out1, out2;             // out1, out2 are of type "wire"
    reg out3, out4, out5, out6;  // out3-out6 are of type reg.

    nand (out1, in1, in2);  // nand sub-module defines out1 value
    assign out2 = cond1 ? in1 : in2; // if (cond1 == 1) out2 = in1;
                                     // else out2 = in2;
```

```verilog
   initial begin  // test NAND gate and assign statement
     in1 = 1'b0;
     in2 = 1'b0;
     cond1 = 1'b1;
     #10;          // expect (out1 = 1, out2 = 0) at time 10
     $display("out1 = %0b, out2 = %0b) at time %0t.", out1, out2, $time);
     in2 = 1'b1;
     #10;          // expect (out1 = 1, out2 = 0) at time 20
     $display("out1 = %0b, out2 = %0b) at time %0t.", out1, out2, $time);
     in1 = 1'b1;
     #10;          // expect (out1 = 0, out2 = 1) at time 30
     $display("out1 = %0b, out2 = %0b) at time %0t.", out1, out2, $time);
     cond1 = 1'b0;
     #10;          // expect (out1 = 0, out2 = 1) at time 40
     $display("out1 = %0b, out2 = %0b) at time %0t.", out1, out2, $time);
     in2 = 1'b0;
     #10;          // expect (out1 = 1, out2 = 0) at time 50
     $display("out1 = %0b, out2 = %0b) at time %0t.", out1, out2, $time);
   end

   initial begin  // test blocking assignment
     out3 = 1'b1;
     out4 = 1'b0;
     #60;          // expect (out3 = 1, out4 = 0) at time 60
     $display("out3 = %0b, out4 = %0b) at time %0t.", out3, out4, $time);
     out3 = out4; // out3 should become '0'
     out4 = out3; // out4 should then become '0' also
     #10;          // expect (out3 = 0, out4 = 0) at time 70
     $display("out3 = %0b, out4 = %0b) at time %0t.", out3, out4, $time);
   end

   initial begin  // test nonblocking assignment
     out5 <= 1'b1;
     out6 <= 1'b0;

     #80;          // expect (out5 = 1, out6 = 0) at time 80
     $display("out5 = %0b, out6 = %0b) at time %0t.", out5, out6, $time);
     out5 <= out6;// out5 should become '0'
     out6 <= out5;// out6 should become '1'
     #10;          // expect (out5 = 0, out6 = 1) at time 90
     $display("out5 = %0b, out6 = %0b) at time %0t.", out5, out6, $time);
   end              // note that (out5, out6) values have switched
endmodule  // of module test_assign
/////////////////////////////////////////////////////////////
```

### 3.3.4  Parameters and Modeling a Bidirectional Bus

When designing digital logic circuits, it sometimes becomes necessary to deal with signals driven by multiple sources (e.g., a common bus). This includes bidirectional signals, which can be driven by an external or internal source. However, writing Verilog code that correctly models signals with multiple drivers can be quite tricky. Thus, in this section, a standard method for modeling such signals is shown using, as an example, a bus driven by several sources.

A *bus* is a collection of signal lines that is used as a single unit in a hardware design. Although a bus typically can be driven by multiple sources, it should never be driven by more than one source at any one time instant (unless *wired*-AND logic is being used). A *unidirectional bus* transmits data in only one direction, while a *bidirectional bus* can transmit data in both directions, depending on which source is driving the bus.

The Verilog description for a bidirectional bus is not so simple due to its bidirectional signal behavior and the use of multiple signal sources. These problems can be addressed by using the high-impedance (denoted 1′bZ or 1′bz) signal value and the **inout** port type. Care should be taken to ensure that only one signal source drives the bus with any signal (other than Z) at all times. The following example shows how such a bidirectional bus can be created in Verilog.

| | |
|---|---|
| **Example 3.3** | **Verilog Bidirectional Bus** |

In this example, a 16-bit bus is driven by one bidirectional I/O source and one unidirectional (output only) I/O source. A test bench is used to verify that both the bidirectional and the unidirectional modules work properly. Note the use of the high-impedance signal ('1′bZ' or 1′bz in Verilog) in the following code. Note also that the output enable signal (OE_n) has precedence over the write enable signal (WE_n) in the definition of the bidir_source module.

The concept of a Verilog **parameter** also is illustrated using this example. A **parameter** is a "resettable" constant, typically used to specify time-delay values or bit widths of bus signals. When a Verilog module with a **parameter** is instantiated as a submodule of another module, the **parameter** can be changed to instantiate submodules with different time-delay characteristics or different data widths. This is shown for the instantiations of the bidir_source and unidir_source modules in the tb_bus module definition.

```
//////////////////////////////////////////////////////////////////
// bidir_source.v
module bidir_source (data, OE_n, WE_n);
  parameter WORD_SIZE = 8,     // default data width
            READ_DELAY = 10,   // default read delay
            WRITE_DELAY = 10;  // default write delay
  inout [WORD_SIZE-1:0] data; // bidirectional data signal
  input OE_n;                 // enable output to data bus
  input WE_n;                 // enable input from data bus
```

```verilog
  reg [WORD_SIZE-1:0] data_reg; // connects to external data
  reg [WORD_SIZE-1:0] stored_data; // internal stored data

  assign data = data_reg;
  always @(OE_n or WE_n) begin
    if (~OE_n)
      data_reg <= #READ_DELAY stored_data;
    else if (~WE_n)
      stored_data <= #WRITE_DELAY data;
    else
      data_reg <= #READ_DELAY  'bz;
  end // always
endmodule
//////////////////////////////////////////////////////////////
// unidir_source.v
module unidir_source (data, OE_n);
  parameter WORD_SIZE = 8,      // default data width
            READ_DELAY = 10;    // default read delay
  output [WORD_SIZE-1:0] data; // unidirectional data signal
  input OE_n;                  // enable output to data bus

  assign #READ_DELAY data = OE_n ? 'bz : 1;
                // return an arbitrary constant (1) if OE_n = 0
endmodule
//////////////////////////////////////////////////////////////
// tb_bus.v
`timescale 1ns/1ns
`define WORD_SZ 16
`define PERIOD1 100  // delay for one clock period

module tb_bus();

  // SIGNAL DECLARATIONS FOR EXTERNAL SIGNALS
  wire [`WORD_SZ-1:0] data;  // data being input or output
  reg OE1_n;  // enable read from bidir_source
  reg OE2_n;  // enable read from unidir_source
  reg WE_n;   // enable write to bidir_source

  // internal signal output by this test bench module
  reg [`WORD_SZ-1:0] writeData;  // data to be written
```

```verilog
  // instantiate the modules being tested
  bidir_source #(`WORD_SZ, 30, 40) U1 (data, OE1_n, WE_n);
      // use WORD_SIZE = 16, READ_DELAY = 30, WRITE_DELAY = 40
  unidir_source #(`WORD_SZ, 20) U2 (data, OE2_n);
      // use WORD_SIZE = 16, READ_DELAY = 20

  assign data = WE_n ? 'bz : writeData;
  always begin
    OE1_n = 1'b1;  // initially, disable all control signals
    OE2_n = 1'b1;
    WE_n = 1'b1;
    writeData = `WORD_SZ'b0000_1111_1100_0011;
                    // assign arbitrary 16-bit data to be written
    #`PERIOD1;
    WE_n = 1'b0;   // data should change to 0x0fc3
    #`PERIOD1;
    WE_n = 1'b1;   // now, disable write to bidir_source
    OE2_n = 1'b0;  // and enable read from unidir_source
                    // data should change to 0x0001
    #`PERIOD1;
    OE2_n = 1'b1;
    OE1_n = 1'b0;  // data should change again to 0x0fc3
    #`PERIOD1;
    $finish;
  end // of always
endmodule
//////////////////////////////////////////////////////////////////
```

### 3.3.5   Tasks and Functions

Two constructs that can aid in the description of Verilog modules are *tasks* and *functions*. These must be defined within Verilog **module** definitions. These correspond loosely to subroutines and functions in C, and they can be used to make the Verilog code much more readable, portable, and compact. A Verilog **function** can be used much like a C language macro. It differs from a **task** in that a **function** only can be used to define the code for a combinational logic block (such a **function** could be used to make the code easier to understand and/or to avoid code repetition). The code within a **task** definition simply can be substituted into the position where that **task** is called. Thus, since a **task** can be called from within an **always** or **initial** block, any code construct that can be used within an **always** or **initial** block definition also can be used within a **task** definition.

A Verilog **task** block simply can be considered as a method for permitting designers to manage smaller segments of code. A **task** block can access all variables in the calling block. In addition, it can have **input**, **output**, and **inout** arguments passed to it. A **task** block can also contain locally defined variables that are not accessible from the calling block. Except for the use of locally defined variables, the effect of using a **task** block is identical to having the call to the **task** block replaced by the code within the **task** block. Thus, a **task** block is free to use all constructs that can be used within **initial** and **always** blocks. The following shows an example of a useful "utility" task.

| Example 3.4 | **An Error-Checking Task** |

Within a test bench, it is frequently the case that we wish to compare an expected signal value with the actual signal value, and then, if the two values are found to be different, display an error message and increment the count of the accumulated errors. Since this is a frequently used feature, it is an ideal candidate for implementation as a Verilog **task** block. The following shows a method for implementing this type of task and the associated calling method.

```verilog
///////////////////////////////////////////////////////////////////
`define MAX_NAME 10   // maximum number of characters in a name
module testit();
  integer error_count;
  reg [3:0] result;
  ...
  task error_check;
    input expected, actual;
    input [8*`MAX_NAME:0] name;
    begin
      if (expected !== actual) begin
        $display
           ("ERROR in %s. Expected %0x, got %0x at time %0t.",
               name, expected, actual, $time);
        error_count = error_count + 1; // update global value
      end  // of if (expected !== actual)
    end  // of outermost begin for task error_check
  endtask
  ...
  always @(posedge clk) begin
  ...
     error_check (1'b1, result, "result");   // call error_check
  ...
     $display("Simulation done with %0d errors.", error_count);
     $finish;
  end  // of always
endmodule
///////////////////////////////////////////////////////////////////
```

There are several notable features in this code. First, note that the task definition does not include the task inputs in a parenthesized list after the task name. Second, all constants and signals defined in the enclosing **module** definition can be used in the **task** block. Third, character strings can be represented and used as signals of length $8k$, where $k$ is the maximum number of characters in the string. Fourth, global signal values can be updated within the **task** block. Finally, the **task** block is called just like a C subroutine call from within an **always** block in the enclosing **module** definition.

Verilog functions, unlike tasks, are used to implement combinational behavior, thus, cannot contain timing controls (such as #, @( ), or **wait**), and cannot call tasks. Functions must be declared within a **module** definition, but can be called anywhere that an expression is valid (such as the right-hand side of an assignment statement). A function returns a value by assigning that value to the function name. The task of Example 3.4 can be changed to a function by rewriting it as follows. Note that changes to the previous code are marked with // CCC.

```verilog
/////////////////////////////////////////////////////////////////////
`define MAX_NAME 10   // maximum number of characters in a name
module testit();
  integer error_count;
  reg [3:0] result;
...
  function [31:0] error_check;   // CCC
    input expected, actual;
    input [8*`MAX_NAME:0] name;
    begin
      if (expected !== actual) begin
        $display
          ("ERROR in %s. Expected %0x, got %0x at time %0t.",
            name, expected, actual, $time);
        error_count = error_count + 1; // update global value
      end  // of if (expected !== actual)
      error_check = error_count;  // CCC return error_count
    end  // of outermost begin for function error_check
  endfunction  // CCC
...
  always @(posedge clk) begin
...
    error_count = error_check (1'b1, result, "result"); // CCC
...
    $display("Simulation done with %0d errors.", error_count);
    $finish;
  end  // of always
endmodule
/////////////////////////////////////////////////////////////////////
```

● ● ● ● ● ● ● ● ● ● ● ● ● ● ● ● ● ●

# 3.4 Construction of Complete Verilog Programs

Verilog is used to model actual physical hardware systems. As such, Verilog should follow the actual physical behavior of such hardware systems whenever possible. However, Verilog should also permit design at a high level by permitting highly behavioral descriptions. Synthesis tools then can be used to turn behavioral Verilog code into detailed circuit descriptions (netlists), which in turn can be used to create the actual ASIC or FPGA chips that realize the desired circuit.

A complete Verilog description of a circuit consists of a **module** definition for that circuit and another **module** definition for a test bench for that circuit. The **module** definition of the target circuit should use only synthesizable Verilog constructs. Synthesizable and nonsynthesizable Verilog code is described in detail in Chapter 4. The **module** definition of the test bench can use both synthesizable and nonsynthesizable Verilog constructs. The test bench typically is written in a behavioral manner that resembles a C language program. The target circuit can also be described in a behavioral manner, provided that synthesizable behavioral code is used. However, since Verilog is a language used to describe hardware, there are important differences between the behavior of Verilog code and C code, especially with regards to delay behavior and loops. The following subsections present complete Verilog code examples for several representative combinational and sequential logic circuits.

### 3.4.1 Combinational Logic Circuits

In this section, let us study the Verilog code for combinational logic circuits. Two examples of nontrivial combinational logic circuit design problems are addressed: a code-translation circuit and an *arithmetic logic unit (ALU)* circuit. Other examples of code for combinational logic circuits are presented in Chapter 4.

### Arithmetic Logic Unit

An *arithmetic logic unit (ALU)* is a circuit that, depending on the function select inputs, can perform one of a set of arithmetic or logic operations on one or two input operands. Such a circuit can be described using a single continuous assignment statement or several procedural assignment assignments within an **always** block. Note that a combinational logic circuit can be described using an **always** block as long as all of the circuit inputs are included in the activation guard for that **always** block. The following shows two Verilog code solutions for a generic ALU circuit: one using continuous assignment and the other using procedural assignment statements. These two versions result in equivalent combinational-logic-circuit implementations.

```
//////////////////////////////////////////////////////////////////////////
// MODULE: Arithmetic Logic Unit (ALU) Circuit: alu1.v
// Author: Sunggu Lee
// Created: ...
```

```verilog
// Last Modified: ...
// Description: Implements an arithmetic logic unit (ALU).
//    This circuit simply examines the mode inputs, and then
//    performs a specific arithmetic or logic operation on the
//    data inputs based on the value of the mode inputs using
//    a long continuous assignment statement.

// DEFINITIONS
`define WIDTH 16     // bit-width of data inputs
`define SEL_BITS 3  // number of bits in the mode signal

// MODULE DECLARATION
module alu1 (C, A, B, mode);

  output [`WIDTH-1:0] C;        // data output
  input  [`WIDTH-1:0] A, B;     // data inputs
  input  [`SEL_BITS-1:0] mode; // function select

  // CONTINUOUS ASSIGNMENT STATEMENTS
  assign C = (mode == 'b000) ? A + A : // A * 2 -> shift left
             (mode == 'b001) ? A + B : // add
             (mode == 'b010) ? A - B : // subtract
             (mode == 'b011) ? -A :    // 2's complement of A
             (mode == 'b100) ? A & B : // and
             (mode == 'b101) ? A | B : // or
             (mode == 'b110) ? A ^ B : // exclusive-or
             ~A; // default action - complement all bits of C

endmodule
//////////////////////////////////////////////////////////////////

//////////////////////////////////////////////////////////////////
// MODULE: Arithmetic Logic Unit (ALU) Circuit: alu2.v
// Author: Sunggu Lee
// Created: ...
// Last Modified: ...
// Description: Implements an arithmetic logic unit (ALU).
//    This circuit simply examines the mode inputs, and then
//    performs a specific arithmetic or logic operation on the
//    data inputs based on the value of the mode inputs using
//    procedural assignment statements.

// DEFINITIONS
`define WIDTH 16     // bit-width of data inputs
`define SEL_BITS 3  // number of bits in the mode signal
```

```verilog
// MODULE DECLARATION
module alu2 (C, A, B, mode);

  output [`WIDTH-1:0] C;        // data output
  input  [`WIDTH-1:0] A, B;     // data inputs
  input  [`SEL_BITS-1:0] mode;  // function select

  // SIGNAL DECLARATIONS
  reg [`WIDTH-1:0] C;

  // General format of an always block used for comb. logic
  always @(A or B or mode) begin // activate on any input change
    case (mode)
      'b000: C <= A + A;       // multiply A by 2 = shift left
      'b001: C <= A + B;       // add
      'b010: C <= A - B;       // subtract
      'b011: C <= -A;          // 2's complement of A
      'b100: C <= A & B;       // and
      'b101: C <= A | B;       // or
      'b110: C <= A ^ B;       // exclusive-or
      default: C <= ~A;        // complement all bits of C
    endcase
  end // of main always block

endmodule
/////////////////////////////////////////////////////////////////////
```

As emphasized in Section 3.2, it is always a good idea to test every Verilog module with a test bench, even if it will be included as a submodule in a larger circuit to be tested later on. Thus, in this spirit, the following is a test bench that can be used to test either alu1.v or alu2.v. As described in Section 3.2, a test bench should include the connections to the unit under test (UUT), test-vector generation code, and test-result checking code. Test vectors should be generated for *corner-case* input values (values at the extremes of their domains) and for normal range inputs. The results of all test-vector applications should be checked using code that is different from that used in the UUT.

```verilog
/////////////////////////////////////////////////////////////////////
// MODULE: Test Bench for ALU Circuit: tb_alu.v
// Author: Sunggu Lee
// Date: ...
// Description: Tests the alu1 module.

// Constant Definitions
`define WIDTH 16      // bit-width of data inputs
`define SEL_BITS 3  // number of bits in the mode signal
`define SEL_MAX  7   // maximum mode value = 2^(`SEL_BITS) - 1
```

```verilog
`define MAX_POS_DATA 32767  // = 2^(`WIDTH - 1) - 1
`define MIN_NEG_DATA -32768 // = -2^(`WIDTH - 1)
`define DELAY 10000 // delay value used in testing (10,000 ps)

// Test bench module definition
module tb_alu ();
  // SIGNAL DEFINITIONS
  // Signals used to connect to the UUT (unit under test)
  wire [`WIDTH-1:0] C;        // C output out of UUT
  reg [`WIDTH-1:0] A, B;      // data inputs to UUT
  reg [`SEL_BITS-1:0] mode;   // mode (function select) input
  // Internal signals
  integer error_count = 0;    // total number of errors found
  integer i, j, k, ii, jj;    // test vector generation indices

  // Connect to the UUT (unit under test)
  alu1 uut (C, A, B, mode);

  // Generate the test vectors and check the results
  initial begin
    $display("Start simulation.");
    // first, generate boundary test vectors
    // generate values close to 0
    for (i = -2;  i <= 2;  i = i + 1)
      for (j = -2;  j <= 2;  j = j + 1)
        for (k = 0;  k <= `SEL_MAX;  k = k + 1) begin
          test_and_check_alu();
        end  // of for k loop
    // generate min data input values
    for (i = `MIN_NEG_DATA;  i <= `MIN_NEG_DATA+4;
         i = i + 1)
      for (j = `MIN_NEG_DATA;  j <= `MIN_NEG_DATA+4;
           j = j + 1)
        for (k = 0;  k <= `SEL_MAX;  k = k + 1) begin
          test_and_check_alu();
        end  // of for k loop
```

```verilog
    // generate max data input values
    for (i = `MAX_POS_DATA-4;  i <= `MAX_POS_DATA;
         i = i + 1)
      for (j = `MAX_POS_DATA-4;  j <= `MAX_POS_DATA;
           j = j + 1)
        for (k = 0;  k <= `SEL_MAX;  k = k + 1) begin
          test_and_check_alu();
        end  // of for k loop
    // next, generate a set of normal test vector values
    for (ii = 0;  ii <= 9;  ii = ii + 1)
      for (jj = 0;  jj <= 9;  jj = jj + 1)
        for (k = 0;  k <= `SEL_MAX;  k = k + 1) begin
          i = 41*ii-273;    // generate random set of negative
          j = 89*jj-384;    // and positive numbers
          mode = k;
          test_and_check_alu();
        end  // of for k loop
    #`DELAY;
    if (error_count > 0)
      $display("ERROR: There were %0d errors!", error_count);
    else
      $display("Simulation complete with NO errors.");
    $finish;            // terminate the simulation
  end  // of always block for test vector generation and checking

  // Error checking and reporting task
  task test_and_check_alu;
    reg [`WIDTH-1:0] expected; // used to check test results
    begin
      A = i;
      B = j;
      mode = k;
      case (k)
        0: expected <= i * 2;
        1: expected <= i + j;
        2: expected <= i - j;
        3: expected <= -i;
        4: expected <= i & j;
        5: expected <= i | j;
        6: expected <= i ^ j;
        default: expected <= ~i;
      endcase
```

```
      #`DELAY;
      if (expected !== C) begin
        $display
          ("ERROR: expected (%0b) != actual (%0b) at time %0t ps",
            expected, C, $time);
        error_count = error_count + 1;
      end  // of if
    end  // of outermost begin for task error checking task
  endtask

endmodule
//////////////////////////////////////////////////////////////////
```

## BCD-to-Binary Converter

Various types of binary codes are used in digital logic circuits. Sometimes, two circuits that need to communicate with each other will be using two different types of codes. In such a case, another circuit is required to convert one type of binary code into another. As an example of this type of code-converter circuit, let us consider a BCD-to-binary converter. In binary-coded decimal (BCD), discussed in Chapter 1, each set of four consecutive bits is converted into a decimal digit. The entire number is then interpreted as a decimal number with those decimal digits. Thus, for example, 01011001 is interpreted as 59 since the first set of four bits (0101) corresponds to a decimal five and the second set of four bits (1001) corresponds to a decimal nine. The binary equivalent of this number is 00111011, since this is the binary representation for decimal 59.

There are several different ways to write the Verilog code for a BCD-to-binary converter circuit. It can be written as a sequential circuit or a combinational circuit. When written as a sequential circuit, an algorithmic method can be used to convert a BCD number of arbitrary length into its corresponding binary number. However, since we are considering combinational circuit descriptions in this section, let us consider a fixed-length BCD number and a combinational-logic conversion method.

The simplest solution method for a fixed-length BCD-to-binary conversion problem is to write an extremely long **case** statement. Of course, this type of solution only is feasible for BCD numbers that are not too long. Also, it is not an elegant solution. Even with a 24-bit BCD number, this type of solution would require $2^24 = $ over 16 million **case** clauses.

An alternative method is to use one of the well-known BCD-to-binary conversion algorithms. Since the BCD-to-binary conversion problem is a well-studied problem, there are several standard solution methods. Such solutions typically require a sequence of operations to be performed on the input BCD number, which would seem to imply the need for a sequential circuit. However, by relying on a *synthesis* tool to convert our algorithmic description into a combination of logic gates, it is possible to create a combinational logic circuit from such an algorithmic description. Thus, let us use the following algorithm, described in pseudocode, as the basis of our Verilog code solution, shown below as bcd_to_bin.v.

*Pseudocode for BCD-to-Binary Converter*:

Step 1.　　　$bcd \leftarrow bcd\_data\_input$;
　　　　　　　$bin \leftarrow 0$; (same bit size as $bcd$)
Step 2.　　　For $count \leftarrow 1$ to bcd_bit_size do begin
Step 2a.　　　　$\{bcd, bin\} \leftarrow \{bcd, bin\} >> 1$;
Step 2b.　　　　　For each 4-bit sequence in $bcd$ do
　　　　　　　　　　If the 4-bit sequence $> 7$ then
　　　　　　　　　　　subtract 3 from that 4-bit sequence;
　　　　　　end    / * of for * /
Step 3.　　　$bin$ contains converted number;

```
/////////////////////////////////////////////////////////////////
// MODULE: BCD-to-Binary Converter: bcd_to_bin.v
// Author: Sunggu Lee
// Created: ...
// Last Modified: ...
// Description: Implements a BCD-to-binary code converter based
//   on a standard conversion algorithm.


// DEFINITIONS
`define LENGTH 16   // length of input BCD number

// MODULE DECLARATION
module bcd_to_bin (bin, bcd);

  output [`LENGTH-1:0] bin; // converted binary output
  input  [`LENGTH-1:0] bcd; // BCD number input


  // SIGNAL DECLARATIONS
  reg [`LENGTH-1:0] bin;
  reg [(`LENGTH*2)-1:0] bcd_concat_bin;  // used in conversion
  integer i, j;                          // index variables
  reg [3:0] temp;                        // for temporary use


  // Perform conversion according to conversion algorithm
  always @(bcd) begin  // activated whenever BCD value changes
    bcd_concat_bin = {bcd, `LENGTH'b0};
    for (i = 0;  i < `LENGTH;  i = i + 1) begin
      bcd_concat_bin = bcd_concat_bin >> 1;
      for (j = 0;  j < (`LENGTH/4);  j = j + 1) begin
        temp = {bcd_concat_bin[`LENGTH+j*4+3],
                bcd_concat_bin[`LENGTH+j*4+2],
                bcd_concat_bin[`LENGTH+j*4+1],
                bcd_concat_bin[`LENGTH+j*4]};
```

```verilog
      if (temp > 4'b0111) begin
        temp = temp - 4'b0011;
        bcd_concat_bin[`LENGTH+j*4+3] = temp[3];
        bcd_concat_bin[`LENGTH+j*4+2] = temp[2];
        bcd_concat_bin[`LENGTH+j*4+1] = temp[1];
        bcd_concat_bin[`LENGTH+j*4]   = temp[0];
      end // of if
    end // of for (j ... )
  end // of for (i ... )
  bin = bcd_concat_bin[`LENGTH-1:0];
end // of main always block


endmodule
///////////////////////////////////////////////////////////////////////
```

Upon first inspection, it might seem like the Verilog code shown here is describing a sequential circuit. However, synthesis of this Verilog code will demonstrate that it is indeed describing a combinational logic circuit. Although none of the operations in bcd_to_bin.v are dependent on a clock edge, this does not in itself guarantee the generation of a combinational logic circuit. As will be elaborated on in Chapter 4, care must be taken to ensure that unwanted latches are not created.

Even though an **always** block and **for** loops are used in bcd_to_bin.v, such constructs still result in a combinational logic circuit when used as shown. An **always** block can result in a combinational logic circuit if the activation guard for this block includes all inputs of the circuit. Since such an **always** block is activated *whenever* any of the inputs change (and *only* then), this code properly models the behavior of a combinational logic circuit, which continuously is active but only produces output changes when one or more input values change. Likewise, a **for** loop does not necessarily indicate a sequential circuit. In this case, the **for** loop simply is describing a complex interconnection of combinational logic gates and devices. In a nonoptimized implementation of this circuit, there would be 16 levels of subcircuits, with each subcircuit consisting of four 4-bit subtracters and four 4-bit comparators. In an optimized implementation, such as would be produced by a synthesis tool, it may be possible to combine and merge some or all of the combinational logic in these 16 levels.

The following shows a test bench that can be used to test the proper operation of the circuit described in bcd_to_bin.v. As before, this test-bench code includes test-vector generation and test-result checking code that has been written in a systematic and comprehensive manner.

```verilog
///////////////////////////////////////////////////////////////////////
// MODULE: Test Bench for Code Converter Circuit: tb_bcd_to_bin.v
// Author: Sunggu Lee
// Date: ...
// Description: Tests the tb_bcd_to_bin code converter circuit.
```

```verilog
`timescale 1ns/100ps // 1ns simulation accuracy & 100ps precision
// Constant Definitions
`define LENGTH 16     // length of input BCD number
`define MAX_BCD 9999 // maximum BCD input value possible
`define DELAY  10     // delay value used in testing (10 ns)

// Test bench module definition
module tb_bcd_to_bin ();
  // SIGNAL DEFINITIONS
  // Signals used to connect to the UUT (unit under test)
  wire [`LENGTH-1:0] bin;     // binary number output from the UUT
  reg [`LENGTH-1:0] bcd;      // BCD input to the UUT
  // Internal signals
  integer error_count = 0;    // total number of errors found
  integer i, j;               // test vector generation indices
  reg [`LENGTH-1:0] temp;     // used as a temporary data holder
  reg [3:0] digit;            // used to hold single decimal digit
  reg [`LENGTH-1:0] expected;// used to check test results

  // Connect to the UUT (unit under test)
  bcd_to_bin uut (bin, bcd);

  // Generate the test vectors and check the results
  initial begin
    $display("Start simulation.");
    // generate all possible input values, since max = 9999
    for (i = 0;  i <= `MAX_BCD;  i = i + 1) begin
      temp = i;
      for (j = 0;  j < (`LENGTH/4);  j = j + 1) begin
        digit = temp % 10; // look at least sig. decimal digit
        bcd[j*4 + 3] = digit[3];  // assign test vector input
        bcd[j*4 + 2] = digit[2];
        bcd[j*4 + 1] = digit[1];
        bcd[j*4] = digit[0];
        temp = temp / 10;        // move to next decimal digit
      end // of for (j ... )

      expected = i;     // computation of expected value
      #`DELAY;          // wait for result to be computed

      if (expected !== bin) begin  // test result check
        $display
        ("ERROR: expected (%0b) != actual (%0b) at time %0t ps",
          expected, bin, $time);
```

```
        error_count = error_count + 1;
      end  // of if
    end // of for (i ... )


    #`DELAY;
    if (error_count > 0)
      $display("ERROR: There were %0d errors!", error_count);
    else
      $display("Simulation complete with NO errors.");
    $finish;             // terminate the simulation
  end  // of always block for test vector generation and checking

endmodule
//////////////////////////////////////////////////////////////////
```

### 3.4.2  Sequential Logic Circuits

Since many practical design problems can be solved effectively by using state machines, two state machine examples are examined in this section. Other, longer examples are provided in the later chapters of this book. As will be elaborated on in Chapter 5, a state machine essentially consists of a set of states (encoded using flip-flops) and transitions between those states governed by the values of several status (or condition) bits.

### Moore-Type Finite State Machine

Let us first consider a Verilog implementation of a simple Moore-type finite state machine (or simply "Moore machine"). As stated in Chapter 1, a Moore machine is a state machine in which the outputs of the circuit depend only on the current state and not on the values of the current inputs. On the other hand, the outputs of a Mealy machine depend on both the current state and the values of the current inputs.[3] Figure 3.5 shows an example of a simple Moore machine.

### Behavior Modeling

There are, of course, numerous ways to write the Verilog code for a finite state machine. First, we can choose to use behavioral or structural Verilog code, as initially

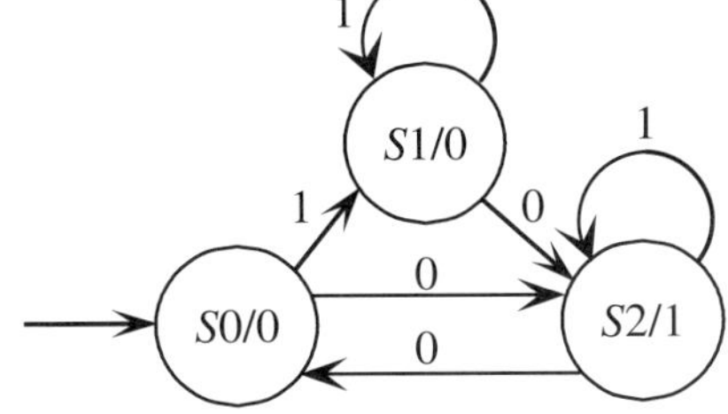


**Figure 3.5**
A simple Moore-type finite state machine.

---

[3]   The reader not familiar with Moore and Mealy machines should review the material in Chapter 1.

discussed in Section 3.1.3. In most cases, behavioral modeling is preferable because such a modeling method results in faster implementations that are less prone to design errors. As long as several rules are followed in the behavioral modeling (to be discussed in Chapter 4), the resulting behavioral code can be *synthesized* into a highly optimized working hardware circuit. The necessary detailed logic design is created automatically by a *synthesis* tool. Since this is an automated process, it is less prone to errors. The following example shows a basic behavioral modeling method that follows a "template" that can be used for finite state machines. Other behavioral modeling methods, using different numbers of **always** blocks, can also be used to model finite state machines. Chapter 5 will also present a more general method for modeling general state machines.

```verilog
//////////////////////////////////////////////////////////////////
// MODULE: Moore Machine Example: simple.v
// Author: Sunggu Lee
// Created: ...
// Last Modified: ...
// Description: Implements a Moore state machine.  This state
//    machine transitions among three states depending on the
//    value of the input variable x.  The output z is set to 1 in
//    state S2 and 0 in all other states.

// DEFINITIONS
`define STATE_BITS 2
`define S0 2'b00
`define S1 2'b01
`define S2 2'b11

// MODULE DECLARATION
module simple (z, state, reset_n, clk, x);

    output z;           // data output
    output [`STATE_BITS-1:0] state; // current state of machine
    input reset_n;      // active-low RESET signal
    input clk;          // clock signal
    input x;            // the single data input

    // SIGNAL DECLARATIONS
    reg z;
    reg [`STATE_BITS-1:0] state;

    // Compute output value based on current state
    always @(state) begin
        case (state)
```

```verilog
      `S0:  begin
              z <= 1'b0;
          end // of case (state == `S0)
      `S1:  begin
              z <= 1'b0;
          end // of case (state == `S1)
      `S2:  begin
              z <= 1'b1;
          end // of case (state == `S2)
      default: begin // desirable for all cases to be covered
              z <= 1'bX;
          end // of case (state == default)
    endcase // of case (state)
end // of always @(state)


// Compute next state based on current state and input values
always @(negedge reset_n or posedge clk) begin
  if (~reset_n)
    state <= `S0;
  else begin
    case (state)
      `S0:  begin
                if (x)
                   state <= `S1;
                else
                   state <= `S2;
          end // of case (state == `S0)
      `S1:  begin
                if (!x)
                   state <= `S2;
          end // of case (state == `S1)
      `S2:  begin
                state <= `S0;
          end // of case (state == `S2)

      default: begin // desirable for all states to be covered
                state <= 2'bXX;
          end // of case (state == default)
    endcase // of case (state)
  end // of else
end // of always loop for computing next state

endmodule
///////////////////////////////////////////////////////////////////////
```

Several new Verilog constructs are introduced in this implementation of a basic state machine. As before, constants used throughout the code are defined using **'define** in order to facilitate global changes and to enhance the readability of the code. Note that constants defined in this manner must be referred to using the opening single-quote notation (as in 'S2). Next, note how a vector of bits (for the state signal) is defined using the bracket array notation.

The main work is divided into two **always** blocks: one to generate the output signals and one to generate the next-state values from the current-state values. In the first **always** block, the @( ) notation specifies an *activation condition* (also referred to as a *guard*) for this process. In this case, the first **always** block is activated whenever there is a change in the state signal. The code within the first **always** block simply uses a **case** statement to specify how the output signal z should change based on the value of the state. A **default** term is added to the end of the **case** block so that all possible cases can be covered—this is a recommended practice, as will be elaborated on in Chapter 4. Note how the value 2'bXX (a bit vector of two unknown or don't-care bits) has been assigned to the state signal (since this case should not occur anyway).

An important construct used in both **always** blocks of the above module definition is the use of the <= notation for assignment. The distinction between this type of nonblocking assignment and blocking assignment (which uses =) was discussed earlier in Section 3.3.3. Although it is possible to use the single equals notation (blocking assignment) in this situation, the <= (nonblocking assignment) construct is used because it more closely models the behavior of actual hardware in this situation. As mentioned previously, all physical hardware circuits execute concurrently all of the time; thus, non-blocking assignments more closely model the situation in which several register transfers take place concurrently. Sequential constructs (which use blocking assignments enclosed within **initial** or **always** blocks) are simply used to *model* the behavior of such circuits.

In the second **always** block, a **case** statement again is used to define the actions to be taken in each state of the state machine—in this case, the action is simply to update the state signal (i.e., compute the next state) based on the current value of the state signal (i.e., the current state) and the primary input values. The guard (the clause within the @( ) notation) specifies that this **always** block is activated if and only if there is a negative transition in the reset_n signal or a positive transition in the clk signal. This is a standard method for defining a synchronous module with an asynchronous reset signal. For a synchronous reset, the negedge reset_n clause could be omitted from the **always** @( ) clause (the clause within the @( ) notation is referred to as a "guard" clause). Note that the first action taken within the **always** block is to check the reset_n signal value—this specifies the actions to take when the reset signal is activated. The tilde notation (~) in front of the reset_n signal value specifies a bit-wise negation operator; in this case, since reset_n is a single-bit value, it is equivalent to **if** (!reset_n). The rest of the **always** block is enclosed within an **else** clause. This part is the synchronous part—it specifies the actions to take upon a positive transition in the clk signal (with reset_n inactive). Note that it is not necessary to use **else if** (clk), since either reset_n must be '0' or clk must be '1' in order for this **always** block to be activated (due to its guard condition).

A Moore machine can be described using other methods besides that demonstrated by the earlier Verilog code. A structural description method will be shown later in this section. It also is possible to combine the **always** block for the output signals with the **always** block for the generation of the next state. This is shown in the following code, where the operations within a single state execute concurrently, since nonblocking assignment (<=) is being used.

```verilog
///////////////////////////////////////////////////////////////
// Possible modification to Moore machine Verilog description.
// Compute output and next state based on current state
always @(negedge reset_n or posedge clk) begin
  if (~reset_n) begin
    state <= `S0;
    z <= 1'b0;
  end
  else begin
    case (state)
      `S0:  begin
              if (x) begin
                state <= `S1;
                z <= 1'b0;
              end
              else begin
                state <= `S2;
                z <= 1'b1;
              end
        end // of case (state == `S0)
      `S1:  begin
              if (!x) begin
                state <= `S2;
                z <= 1'b1;
              end
              else
                z <= 1'b0;
        end // of case (state == `S1)
      `S2:  begin
              state <= `S0;
              z <= 1'b0;
        end // of case (state == `S2)
```

```verilog
          default: begin // desirable for all states to be covered
                   state <= 2'bXX;
                   z <= 1'bX;
              end // of case (state == default)
          endcase // of case (state)
       end // of else
    end // of always loop for computing next state
///////////////////////////////////////////////////////////////////
```

The following shows a test bench that can be used to test the proper operation of this basic state machine. Note that this test bench exercises the state machine thoroughly by making sure that all arcs in the state machine machine are traversed at least once.

```verilog
///////////////////////////////////////////////////////////////////
// MODULE: Test Bench for Moore Machine Example
// Author: Sunggu Lee
// Created: ...
// Last Modified: ...
// Description: Tests the module simple in "simple.v".

// DEFINITIONS
`timescale 1ns/1ns
`define PERIOD1 100  // clock period
`define STATE_BITS 2 // number of bits used for the state signal
`define S0 2'b00
`define S1 2'b01
`define S2 2'b11

// MODULE DEFINITION
module tb_simple();

  // SIGNAL DECLARATIONS
  wire z;              // data output from the state machine
  wire [`STATE_BITS-1:0] state; // current state
  reg reset_n;         // active-low RESET signal
  reg clk;             // clock signal
  reg x;               // input data for the state machine

  integer error_count;// number of errors encountered during test

  // instantiate the unit under test
  simple simple1 (z, state, reset_n, clk, x);

  // initialize clk, error_count, and reset_n values
  initial begin
    clk = 0;           // used to set initial clock signal value
```

```verilog
      error_count = 0; // used to keep track of the errors

   reset_n = 1;       // generate a LOW pulse for reset_n
   #(`PERIOD1/4) reset_n = 0;
   #`PERIOD1 reset_n = 1;
end

// generate a clock with a period of PERIOD1
always #(`PERIOD1/2) clk = ~clk;

// error checking task used in the main process body
task error_check;
   input [`STATE_BITS-1:0] expected1, actual1;
   input expected2, actual2;
   begin
      if (!((actual1 === expected1) && (actual2 === actual2)))
      begin
         error_count = error_count + 1;
         $display("ERROR: state = %0b, output = %0b at time %0t.",
                  actual1, actual2, $time);
      end   // of if
   end   // of outermost begin for task error_check
endtask

// main body - generate and check non-recurring test sequence
initial begin
   $display ("Start simulation.");
   x = 0;                    // initial data value; check reset next
   #`PERIOD1;  error_check(`S0, state, 1'b0, z);
   x = 0;                    // next, go to state `S2
   #`PERIOD1;  error_check(`S2, state, 1'b1, z);
   x = 1;                    // next, go to state `S0
   #`PERIOD1;  error_check(`S0, state, 1'b0, z);
   x = 1;                    // next, go to state `S1
   #`PERIOD1;  error_check(`S1, state, 1'b0, z);
   x = 1;                    // next, stay in state `S1
   #`PERIOD1;  error_check(`S1, state, 1'b0, z);
   x = 0;                    // next, go to state `S2
```

```
  #`PERIOD1;  error_check(`S2, state, 1'b1, z);
  x = 0;                   // next, go to state `S0
  #`PERIOD1;  error_check(`S0, state, 1'b0, z);
  #`PERIOD1;
  if (error_count == 0)
    $display ("Simulation completed with NO errors.");
  else $display ("ERROR: There were %0d errors.", error_count);
  $finish;  // terminate all processes
end // of main initial block to generate and check test vectors

endmodule
///////////////////////////////////////////////////////////////////////
```

There are several notable points in this test-bench code. First, note how the reset pulse is generated as a one-time low pulse using delay statements in an **initial** block. Also, note how a periodic clock signal (with a period of `PERIOD1 = 100 ns) is generated by assigning it an initial value in an **initial** block and then repeatedly complementing its value using an **always** block with a delay of half of the clock period. This is a standard method for generating a periodic clock signal. Finally, a sequence of test vectors are generated (and their results checked) using a sequence of blocking assignment statements within another **initial** block. A Verilog **task** is used for error checking. Note that all **initial** and **always** blocks execute concurrently with respect to one another, as this is the default behavior for all Verilog code that resides outside of an **initial** or **always** block. Although an **always** block normally is used for repetitive test-vector sequences, an **initial** block is used in this case because the test-vector sequence has no repetitive pattern and is not excessively long.

### Structural Modeling

Let us now investigate how to write structural Verilog code for a finite state machine. By using structural Verilog modeling, it is possible to more closely control the actual logic that is created for a design. Since this involves a lot of manual design, there is more possibility for human design error. On the other hand, by more closely controlling the logic implementation, it is possible to create a more highly optimized design than with an automated-design process based on behavioral Verilog coding and synthesis.

In order to use structural Verilog code, we must first decide the lowest-level components that will be used in our design. Let us use the basic SSI-level logic gates (AND, OR, inverter, NAND, NOR, and exclusive-OR) and edge-triggered D flip-flops as our basic components (the edge-triggered D flip-flop definition given as module Dff in Section 3.3.1 is assumed to be part of our design). In order to design at this level, we must derive the logic equations for the next-state and output signals. Table 3.2 shows the state transition table that results from an analysis of the state diagram of Figure 3.5 or the behavioral Verilog code given in simple.v. Note that the variable pair $(A, B)$ is used to encode the current state while the variable pair $(NA, NB)$ is used to encode the next state.

<table>
<tr><td rowspan="2">**Table 3.2:** ▶<br>State transition<br>table for the<br>example Moore<br>machine.</td><td colspan="3">Current State</td><td>Input</td><td colspan="2">Next State</td></tr>
<tr><td></td><td>$A$</td><td>$B$</td><td>$x$</td><td>$NA$</td><td>$NB$</td></tr>
<tr><td></td><td colspan="6"></td></tr>
<tr><td></td><td>S0:</td><td>0</td><td>0</td><td>0</td><td>1</td><td>1</td></tr>
<tr><td></td><td>S0:</td><td>0</td><td>0</td><td>1</td><td>0</td><td>1</td></tr>
<tr><td></td><td>S1:</td><td>0</td><td>1</td><td>0</td><td>1</td><td>1</td></tr>
<tr><td></td><td>S1:</td><td>0</td><td>1</td><td>1</td><td>0</td><td>1</td></tr>
<tr><td></td><td>S2:</td><td>1</td><td>1</td><td>0</td><td>0</td><td>0</td></tr>
<tr><td></td><td>S2:</td><td>1</td><td>1</td><td>1</td><td>0</td><td>0</td></tr>
</table>

From the state transition table of Table 3.2, the following logic equations can be derived:

$$NA = A' \, x'$$

$$NB = A'$$

Next, since the output $z$ is only asserted in state `S2`, which is encoded as $(A, B) = (1, 1)$, the equation for $z$ is clearly

$$z = A\,B + A\,B'$$

$$= A$$

since $(A, B) = (1, 0)$ is a don't-care state. In general, a formal method (such as a K-map) is necessary to produce the equations for the next-state and output signals. However, since this is a simple state machine, the necessary logic equations have been derived directly from the state transition table.

The structural Verilog code for this state machine can be written based on the above logic equations. The following code, which models the Moore machine of the previous section, can be used as a "template" for the structural Verilog modeling of general state machines.

```
//////////////////////////////////////////////////////////////////////
// MODULE: Moore Machine Structural Code Example: simple_struct.v
// Author: Sunggu Lee
// Created: ...
// Last Modified: ...
// Description: Implements a simple state machine in a structural
//    manner.  This state machine transitions among three states
//    depending on the value of the input variable x.  The output
//    z is set to 1 in state S2 and 0 in all other states.  As
//    such, the state machine is a Moore machine.
```

```verilog
// DEFINITIONS
`define STATE_BITS 2
`define S0 2'b00
`define S1 2'b01
`define S2 2'b11

// MODULE DECLARATION
module simple_struct (z, state, reset_n, clk, x);

   output z;            // data output
   output [`STATE_BITS-1:0] state; // current state
   input reset_n;       // active-low RESET signal
   input clk;           // clock signal
   input x;             // the single data input

   // SIGNAL DECLARATIONS
   wire z;
   wire [`STATE_BITS-1:0] state;

   // Signal Declarations for Internal Signals
   wire A, B;     // current state variables
   wire NA, NB;   // next state variables

   // Create alternate names for the state bits.
   assign state[1] = A;
   assign state[0] = B;

   // Create and make connections to state flip-flops.
   // Note that named association is used for the Dff connections
   // since no connection is required for the q_n output.
   Dff dff_A ( .q(A), .d(NA), .clk(clk), .reset_n(reset_n) );
   Dff dff_B ( .q(B), .d(NB), .clk(clk), .reset_n(reset_n) );

   // Implement combinational logic equations using C-like logic
   // operations (can also use connections to basic logic gates).
   assign NA = ~A & ~x;
   assign NB = ~A;
   assign z = A;

endmodule
////////////////////////////////////////////////////////////////
```

Although the structural Verilog code for this state machine is simpler than its equivalent behavioral Verilog code (simple.v), this is not necessarily true in the general case (especially when the state machine becomes large and complex). This structural Verilog code was simplified by the fact that we used the previously defined Dff module and implemented the necessary combinational logic equations using logic operations in C-like notation (except for the tilde, which denotes a logical NOT in Verilog). As noted in the comments, it is also possible to implement the combinational logic equations by creating and making connections to basic Verilog logic gates, such as those shown in Table 3.1. The main difficulty in this type of structural Verilog implementation is the derivation of the necessary logic equations. Care must be taken to ensure that human design errors are not introduced in this stage. A comprehensive test bench also should be used to check for any possible design errors. For this example, the previous test bench tb_simple (with the connections to the simple module replaced by connections to the simple_struct module) can be used to test the simple_struct module.

### Mealy-Type Finite State Machine

As stated in Chapter 1, a Mealy-type finite state machine (or simply Mealy machine) is a state machine in which the outputs of the circuit depend on both the current state and the values of the current inputs. Thus, within a given state, the outputs can change whenever the inputs change. Then, in the general case, when the inputs are not synchronized to the clock signal used in the Mealy machine, any glitches in the inputs may be propagated to the outputs. Thus, care should be used when working with this type of Mealy machine circuit.

As an example of the implementation of a Mealy machine, let us design a Verilog program for a sequence recognizer for the bit sequence 0110. This machine will output a logic 1 whenever the sequence 0110 is observed in the serial-input data stream, which is assumed to be entered synchronously with the system clock signal. The logic 1 will be output one logic gate delay after the last '0' of the sequence is entered into this cicuit. For all other bit sequences, a logic 0 will be output. Overlapping bit sequences will be permitted. As shown in Chapter 1, a solution for this type of problem can be obtained by using the finite state machine method. The first step of this method involves the creation of a state diagram, which in this case will be a Mealy state diagram. Thus, based on the state diagram shown in Figure 3.6, the Verilog code shown next can be written.

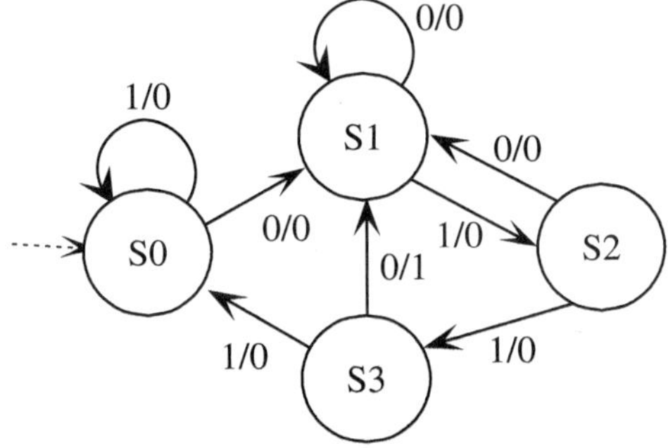


**Figure 3.6**
A state diagram for a Mealy machine for a 0110 sequence detector.

```verilog
///////////////////////////////////////////////////////////////////
// MODULE: Mealy Machine Example (Sequence Detector): seq0110.v
// Author: Sunggu Lee
// Created: ...
// Last Modified: ...
// Description: Implements a Mealy machine that detects the
//   bit-serial input sequence 0110.  This state machine is based
//   on a state diagram solution for this problem.  The use of a
//   synchronous reset is also demonstrated in this code.

// DEFINITIONS
`define STATE_BITS 2
`define S0 2'b00
`define S1 2'b01
`define S2 2'b10
`define S3 2'b11

// MODULE DECLARATION
module seq0110 (z, state, reset_n, clk, x);

  output z;            // data output
  output [`STATE_BITS-1:0] state; // current state of machine
  input reset_n;       // active-low RESET signal
  input clk;           // clock signal
  input x;             // the bit-serial data input

  // SIGNAL DECLARATIONS
  reg z;
  reg [`STATE_BITS-1:0] state;

  // Compute output value based on current state and input x
  always @(state or x) begin // activate upon change in x or state
    if ((state == `S3) && (x == 1'b0))
      z <= 1'b1;
    else
      z <= 1'b0;
  end // of always

  // Compute next state based on current state and input values
  always @(posedge clk) begin  // active on rising clock edge
    if (~reset_n)      // this implements a synchronous reset
      state <= `S0;
    else begin         // reset has priority over normal action
```

```
     case (state)
       `S0:  begin
                if (x == 1'b0)  // first bit of sequence
                  state <= `S1;
                else
                  state <= `S0;
         end // of case (state == `S0)
       `S1:  begin
                if (x == 1'b1)  // second bit of sequence
                  state <= `S2;
                else
                  state <= `S1;
         end // of case (state == `S1)
       `S2:  begin
                if (x == 1'b1)  // third bit of sequence
                  state <= `S3;
                else
                  state <= `S1;
         end // of case (state == `S2)
       default: begin // desirable for all states to be covered
                if (x == 1'b0)  // fourth bit of sequence
                  state <= `S1;
                else
                  state <= `S0;
         end // of case (state == default)
       endcase // of case (state)
     end // of else
  end // of always loop for computing next state

endmodule
//////////////////////////////////////////////////////////////////
```

This Verilog code has been written similarly to the code for simple.v, the Moore-machine Verilog code. The **always** block used to generate the next state (the second **always** block shown) corresponds directly to the state diagram shown in Figure 3.6. Note that, unlike the code shown for simple.v, there is no **negedge** reset_n clause in the activation guard for this **always** block. Thus, the **if** (~reset_n) check within this **always** block only can be invoked at the rising edge of the system clock signal. This will result in a *synchronous reset* as opposed to the *asynchronous reset* mechanism used in simple.v and the Verilog code to be presented later in this textbook.

There are pros and cons involved with the use of synchronous and asynchronous resets. Some digital logic designers prefer the use of synchronous resets, while others prefer asynchronous resets. Both types are supported in most commercial FPGA and ASIC implementations. With a synchronous reset, glitches in the reset signal can be tolerated (of course, such glitches may not matter if the only effect that they may have is to reinitialize the circuit). On the other hand, there is a delay in the implementation of the reset and a heavy dependence on the clock signal. This can become particularly important with low-power circuits that use variable-frequency clock signals. In general, either type of reset mechanism can be used with most types of digital logic circuits.

The **always** block used to create the output signal (the first **always** block shown) can be used as a template for the output generation part of a Mealy machine. In this case, the careful reader will note that this **always** block is describing a simple 3-input AND gate (the output will become a logic 1 only when both state variables are logic 1 and the input signal is a logic 0). The following can serve as a replacement for this **always** block:

```
assign z = ((state == `S3) && (x == 1'b0));
```

This is a *continuous assignment* statement that is activated whenever there are any changes in the signals on the right-hand side. Since this is a combinational logic block, the duration of the logic 1 output signal can become much shorter than a full clock cycle, with the exact length dependent on when the input signal changes to its next value during state `S3. In order to guarantee a logic 1 output pulse of at least one full clock cycle, the output signal needs to be "registered," which means that it needs to be passed through another D flip-flop. This can be done by adding the statement

```
always @(posedge clk) z_long <= z;
```

and then making z_long the circuit output.

The following shows a test bench that can be used to test the proper operation of the Mealy machine seq0110. This test bench exercises the state machine thoroughly by making sure that all arcs in the state machine are traversed at least once.

```
//////////////////////////////////////////////////////////////////////
// MODULE: Test Bench for Mealy Machine Example
// Author: Sunggu Lee
// Created: ...
// Last Modified: ...
// Description: Tests the module "seq0110.v".

// DEFINITIONS
`timescale 1ns/1ns
`define PERIOD1 100  // clock period
`define STATE_BITS 2 // number of bits used for the state signal
`define S0 2'b00
`define S1 2'b01
`define S2 2'b10
`define S3 2'b11

// MODULE DEFINITION
module tb_seq0110();
```

```verilog
// SIGNAL DECLARATIONS
wire z;              // data output from the state machine
wire [`STATE_BITS-1:0] state; // current state
reg reset_n;         // active-low RESET signal
reg clk;             // clock signal
reg x;               // input data for the state machine

integer error_count;// number of errors encountered during test

// instantiate the unit under test
seq0110 uut (z, state, reset_n, clk, x);

// initialize clk, error_count, and reset_n values
initial begin
  clk = 0;           // used to set initial clock signal value
  error_count = 0;   // used to keep track of the errors
  reset_n = 1;       // generate a LOW pulse for reset_n
  #(`PERIOD1/4) reset_n = 0;
  #`PERIOD1 reset_n = 1; // LOW pulse should be at least
end                      // 1 clock period long for synch reset

// generate a clock with a period of PERIOD1
always #(`PERIOD1/2) clk = ~clk;

// error checking task used in main process
task error_check;
  input [`STATE_BITS-1:0] expected1, actual1;
  input expected2, actual2;
  begin
    if (!((actual1 === expected1) && (actual2 === expected2)))
    begin
      error_count = error_count + 1;
      $display("ERROR: state = %0b, output = %0b at time %0t.",
               actual1, actual2, $time);
    end  // of if
  end  // of outermost begin for task error_check
endtask

// main body - generate and check non-recurring test sequence
initial begin
  $display ("Start simulation.");
  x = 0;                    // set initial data value; check reset
  #`PERIOD1;  error_check(`S0, state, 1'b0, z);
  x = 1;                    // next, stay in state `S0
```

```verilog
      #`PERIOD1;  error_check(`S0, state, 1'b0, z);
      x = 0;                   // next, go to state `S1
      #`PERIOD1;  error_check(`S1, state, 1'b0, z);
      x = 0;                   // next, stay in state `S1
      #`PERIOD1;  error_check(`S1, state, 1'b0, z);
      x = 1;                   // next, go to state `S2
      #`PERIOD1;  error_check(`S2, state, 1'b0, z);
      x = 0;                   // next, go to state `S1 again
      #`PERIOD1;  error_check(`S1, state, 1'b0, z);
      x = 1;                   // next, go to state `S2
      #`PERIOD1;  error_check(`S2, state, 1'b0, z);
      x = 1;                   // next, go to state `S3
      #`PERIOD1;  error_check(`S3, state, 1'b0, z);
      x = 1;                   // next, go to state `S0 again
      #`PERIOD1;  error_check(`S0, state, 1'b0, z);
      x = 0;                   // next, go to state `S1
      #`PERIOD1;  error_check(`S1, state, 1'b0, z);
      x = 1;                   // next, go to state `S2
      #`PERIOD1;  error_check(`S2, state, 1'b0, z);
      x = 1;                   // next, go to state `S3
      #`PERIOD1;  error_check(`S3, state, 1'b0, z);
      x = 0;                   // next, check for output 1 in state `S3
      #(`PERIOD1/4);  error_check(`S3, state, 1'b1, z);
                               // next, check for transfer to state `S1
      #(3*`PERIOD1/4);  error_check(`S1, state, 1'b0, z);
      x = 0;                   // next, stay in state `S1
      #`PERIOD1;  error_check(`S1, state, 1'b0, z);
      #`PERIOD1;               // done with all test vectors
      if (error_count == 0)
        $display ("Simulation completed with NO errors.");
      else $display ("ERROR: There were %0d errors.", error_count);
      $finish;                 // terminate all processes
    end // of main initial block to generate and check test vectors

endmodule
/////////////////////////////////////////////////////////////////////
```

This test bench has been written similarly to the test bench for the simple.v Moore machine. The reset pulse is generated as a one-time low pulse using delay statements in an **initial** block. Since a synchronous reset is being used, the low reset pulse should be at least one clock period long in order to ensure that it will be sampled during the active clock edge of the state machine. Also, because of the synchronous reset mechanism, the actual reset will only take place when the next active clock edge occurs *after* reset_n has been set to logic 0. A periodic clock signal is generated by assigning it an initial value in an **initial** block and then repeatedly complementing its value using an **always** block with a delay of half of the clock period. Finally, a sequence of test vectors are generated (and their results checked) using a sequence of blocking assignment statements within another **initial** block. The test-input values for the bit-serial input x are chosen so that all of the states and arcs (state transitions) shown in Figure 3.6 are exercised.

### 3.4.3   Behavioral Modeling of More Complex Circuits

Let us now consider the modeling of more complex types of state machines. Complex state machines can be created through a systematic software-to-hardware conversion process, such as that described in Chapter 5. Such machines can also be modeled as Moore or Mealy finite state machines. However, the resulting model can become too bulky and cumbersome to work with effectively. Thus, a better solution is to simply use behavioral Verilog coding and model the state machine "indirectly" based on a pseudocode algorithmic description.

The specific problem addressed is to compute $z = x^y$ ($x$ raised to the power $y$) given the 8-bit unsigned numbers $x$ and $y$. The result also is assumed to be eight bits and overflow/underflow exceptions are ignored. The following shows a possible pseudocode solution for this problem.

*Pseudocode for Computing* $\{x = x^y\}$:

Step 1.        / * Initialization * /
    result ← 1;
    count ← y;
    multiple ← x;
    If (start)
        if (y == 0)
            go to Step 3;
       else
            go to Step 2;

Step 2.        / * Repeated multiplication * /
    result ← result * multiple;
    count − −;
       if (count == 0)
           go to Step 3;
      else
           go to Step 2;

Step 3.        / * Store result * /
    z ← result;
    go to Step 1;

It is possible to write our Verilog-code solution based closely on this type of pseudocode description. The following code is written assuming that a synthesis tool will produce the proper state machine-based synchronous sequential digital circuit from our behavioral Verilog code. Synthesis issues are discussed in the next chapter.

```verilog
//////////////////////////////////////////////////////////////
// MODULE: State Machine Example: power.v
// Author: Sunggu Lee
// Created: ...
// Last Modified: ...
// Description: Implements an algorithmic state machine.  This
//   circuit performs the task of calculating z = x^y; i.e., z
//   is x multiplied by itself y times.  However, x, y, and z are
//   all constrained to use 8 bits each, and overflow/underflow
//   situations are not considered in this implementation.

// DEFINITIONS
`define MAX_DATA_BITS 8
`define STEP_BITS 2
`define STEP1 2'b00
`define STEP2 2'b01
`define STEP3 2'b10

// MODULE DECLARATION
module power (z, step, reset_n, clk, start, x, y);

  output [`MAX_DATA_BITS-1:0] z; // data output
  output [`STEP_BITS-1:0] step;  // current step number
  input reset_n;                 // active-low RESET signal
  input clk;                     // clock signal
  input start;                   // start signal for computation
  input [`MAX_DATA_BITS-1:0] x;  // data to be multiplied
  input [`MAX_DATA_BITS-1:0] y;  // number of times to multiply

  // SIGNAL DECLARATIONS
  reg [`MAX_DATA_BITS-1:0] z;
  reg [`STEP_BITS-1:0] step;
  reg [`MAX_DATA_BITS-1:0] result; // holds partial result
  reg [`MAX_DATA_BITS-1:0] multiple; // saved "x" value
  reg [`MAX_DATA_BITS-1:0] count;    // decrementing count value

  // Implements the "pseudocode" for this problem
  always @(negedge reset_n or posedge clk) begin
    if (~reset_n)
      step <= `STEP1;          // upon RESET, start with Step 1
    else begin
```

```
    case (step)
      `STEP1: begin         // initial step
          result <= 1;       // initialize variables
          multiple <= x;
          count <= y;
          if (start) begin
            if (y == 0)
              step <= `STEP3;  // go to Step 3
            else
              step <= `STEP2;  // go to Step 2
          end
      end // of case (step == `STEP1)
      `STEP2: begin           // repeated multiplication step
          result <= result * multiple;
          count <= count - 1;
          if (count == 1) // count value BEFORE decrement
            step <= `STEP3;
          else
            step <= `STEP2;       // multiply again
      end // of Step 2
      `STEP3: begin                // final step
          z <= result;    // save the result for output
          step <= `STEP1; // return to Step 1
      end // of final step
      default: begin        // since all cases should be covered
          step <= 2'bXX;
      end // of default
    endcase // of case (step)
  end // of else
end // of always loop to compute the next step

endmodule
//////////////////////////////////////////////////////////////////
```

Since Verilog includes many of the constructs available in a general programming language like C, it is possible to write the Verilog code in the same manner as we would write a C-language solution based on the pseudocode presented. However, such a solution would be extremely difficult to synthesize properly.

The Verilog code for module power shows a compromise solution in which a state machine structure and register–transfer operations are used to emulate the pseudocode solution. The **always** block in module power corresponds to a process that is activated whenever there is a positive transition in the clk clock signal. Upon a low pulse in the active-low reset_n signal, we first transition to `STEP1. The states labeled `STEP1 through `STEP3 correspond to Steps 1 through 3 in our pseudocode solution. Since this code corresponds to digital logic hardware, it is active *all* of the time. To make it behave in a sequential step-by-step manner, we have added a guard condition the @(**posedge** clk) clause) and made each *step* of the pseudocode correspond to a *state* in a state machine synchronized by the positive edge of the clk clock signal. Another change from the pseudocode occurs in `STEP2, where we changed the check for (count == 0) (used in the pseudocode) to (count == 1). This change is necessary because of the use of nonblocking assignments for the count variable. Since all nonblocking assignments occur concurrently, the count value is unchanged at the instant that it's value is checked in the **if** statement in `STEP2—thus, to perform the correct number of multiplications, the transition to `STEP3 must occur when (count == 1). Nonblocking assignments are used because such assignments more closely model the behavior of actual hardware.

A significant difference between the modeling method used in power.v and the modeling method used in simple.v is the use of a single **always** block for both the circuit outputs and the next-state computations. Since this **always** block is of the form **always** @(**negedge** reset_n **or posedge** clk) **begin**, the current state is only checked upon a positive transition of the clock signal (assuming that the initial reset_n pulse has passed). However, since the current state only is determined when a positive clock transition occurs, the check of that current-state value within this **always** block occurs almost one full clock cycle after it is changed. The net effect of this is that the actions within a **case** block for this state machine take effect one clock cycle *after* its corresponding state. For example, the z output is assigned a value in state `STEP3. However, in the simulation waveform for this module, shown in Figure 3.7, it can be seen that the z value is changed in step 0, which corresponds to state `STEP1 (refer to the `**define** statements at the top of power.v).

In the module simple.v, shown earlier in Section 3.4.2, one **always** block is used to generate the output-signal values based on the value of the current state (always @(state) begin), while another **always** block is used to determine the next state values (always @(negedge reset_n or posedge clk) begin). With this type of structure, the first **always** block is activated as soon as the state value is changed in the second **always** block. Thus, the output associated with a particular state will change value in the corresponding state of the first **always** block. The net effect is that, for example, the z output value is changed to 1 as soon as the state is changed to `S2 in the simulation output (instead of being changed one clock cycle later). If this type of effect is necessary in the algorithmic state machine, then it must be written in the same manner as simple.v.

The following code shows a possible test bench for power.v. This test bench uses several new techniques to facilitate the generation of test vectors and the checking of the results of the application of those test vectors. The **module** declaration, signal declarations, and reset and clock waveform generation code are shown next. Since this is a test bench, which will not be synthesized into a circuit with external input and output ports, an empty parameter list is used in the **module** declaration. Constants are defined using `**define** statements.

The code for this test bench module starts off with declarations of the signals to be connected to the inputs and outputs of the UUT, signals used for internal descriptive aids, and subcircuits to be used by this module. For a test bench, the main subcircuit to be declared is the UUT. As shown in Figure 3.3, a complete test bench consists of a test generator subcircuit, the UUT itself, and a test verifier subcircuit. The signals used to connect to the UUT are those signals connecting the test generator to the UUT and the UUT to the test verifier. The other internally used signals, including check_z, check_index and done, are used to aid in the description of the test generator and verifier subcircuits. Note that two sets of index variables {i, j} and {ii, jj} are used in order to avoid conflicts between the test generator and verifier **always** blocks (all top-level statements, include **initial** and **always** blocks, execute continuously and concurrently using the same set of signals defined for that module).

```verilog
/////////////////////////////////////////////////////////////////////
// MODULE: Test Bench for State Machine Example
// Author: Sunggu Lee
// Created: ...
// Last Modified: ...
// Description: Tests the module "power.v".

// DEFINITIONS
`timescale 1ns/1ns
`define PERIOD1 100
`define MAX_DATA 255
`define MAX_DATA_BITS 8
`define STEP_BITS 2

// MODULE DEFINITION
module tb_power();

    // SIGNAL DECLARATIONS
    wire [`MAX_DATA_BITS-1:0] z; // data output
    wire [`STEP_BITS-1:0] step;  // current step
    reg reset_n;                 // active-low RESET signal
    reg clk;                     // clock signal
    reg start;                   // start signal for computation
    reg [`MAX_DATA_BITS-1:0] x;  // data to be multiplied
    reg [`MAX_DATA_BITS-1:0] y;  // number of times to multiply

    reg [`MAX_DATA_BITS-1:0] check_z; // expected output data for z
    reg [`MAX_DATA_BITS-1:0] prev_z;  // previous output data for z
    integer i, j, k, ii, jj; // used to calculate expected results
    integer check_index;      // used to calculate expected results
    integer next_check_index;// used to calculate expected results
    integer cycle1, cycle2;   // used to iterate thru test vectors
    integer error_count;      // number of accumulated errors
    integer done;             // signals that simulation is done
/////////////////////////////////////////////////////////////////////
```

The main body of the test bench consists of three parts: the connection to the UUT, the test-vector generation code, and the test-vector checking code. In this case, the UUT (power) is instantiated with the name power1. Positional association is used to connect the test-bench signals to the UUT's inputs and outputs. Next, test waveforms must be defined for all signals connected to the inputs of the UUT. The waveforms for the reset and clock signals typically are defined separately from the other signals. Using an **initial** block, the reset_n signal is defined to have a low pulse with a duration of `PERIOD1 starting from time unit `PERIOD1/4. Other signals, including clk, are also initialized in this **initial** block. The waveform for the clock signal clk, with a period of `PERIOD1, is defined using a simple **always** statement of the form shown.

Test vectors for general input signals, besides the reset and clock signals, are generated using a computer procedure-like **always** block, such as the one shown next. Since one test vector is applied to the UUT during each clock cycle, a sequence of test vectors can be applied using a sequence of assignment statements separated by # and **wait** statements or a state machine-like structure involving an **always** block triggered by a clock edge and a large **case** block. In this case, both methods are used, with # and **wait** statements and a **case** block conditioned on the cycle1 state variable. The first set of tests (with cycle1 = 0) is used to test the *boundary values* or *corner-cases*, which involve input values at the extremes of their domains. The second set of tests (with cycle1 = 1) involves *intermediate* parameter values. For completeness, it would be best to iterate through all possible combinations of test inputs. However, such a procedure would require an excessively long simulation time. In addition, most of the input combinations would result in overflows of the result register, since an exponentiation function is being performed. Thus, only $x$ and $y$ values in the range 1 through 4 are used in this second set of tests.

```verilog
/////////////////////////////////////////////////////////////////////
// instantiate the unit under test
power power1 (z, step, reset_n, clk, start, x, y);

// initialize inputs
initial begin
  clk = 0;                        // set initial clock value

  cycle1 = 0;            // initialize variables for result checks
  cycle2 = 0;
  error_count = 0;
  done = 0;

  reset_n = 1;                    // generate a LOW pulse for reset_n
  #(`PERIOD1/4) reset_n = 0;
  #`PERIOD1 reset_n = 1;
end

// generate a clock (period = `PERIOD1)
always #(`PERIOD1/2)clk = ~clk;
```

```verilog
// main body consists of two "always" processes
always @(negedge clk) begin  // used to generate test vectors
  case (cycle1)
    0: begin                     // first, test the boundary values
       start = 0;
       check_index = 0;
       #`PERIOD1;
       x = 0;  y = 1;
       #`PERIOD1;
       start = 1;
       wait (z == 0);
       prev_z = z;  check_index = 1;
       x = 1;  y = 0;
       while (z == prev_z)       #`PERIOD1;
       x = 0;  y = `MAX_DATA_BITS;
       prev_z = z;  check_index = 2;
       while (z == prev_z)       #`PERIOD1;
    end
    1: begin     // next, test intermediate test vector values
       for (ii = 1; ii <= (`MAX_DATA_BITS/2); ii = ii + 1)
         for (jj = 1; jj <= (`MAX_DATA_BITS/2); jj = jj + 1)
         begin
           x = jj;  y = ii;
           prev_z = z;  check_index = check_index + 1;
           while (z == prev_z) #`PERIOD1;
         end
       prev_z = z;  check_index = check_index + 1;
    end
    default: begin
       #`PERIOD1;
       $display("Simulation of all test vectors complete.");
       done = 1;
    end
  endcase
  cycle1 = cycle1 + 1;
end // of always loop to generate the tests
//////////////////////////////////////////////////////////////////////
```

The final portion of this test bench involves the test-result checking process implemented using an **always** block. In the main-process body, the output values produced by the UUT are compared with the expected results. The check_index signal is used to facilitate the synchronization of the test-vector generation and checking processes. This check_index signalling mechanism is useful, particularly because the number of clock cycles required to compute each result varies depending on the input parameter values. For the corner cases, the expected results are precomputed and used.

For the intermediate-value cases, the expected results ($x^y$) are computed using an iterative multiplication technique.

This test-bench code is more typical of a general test bench than the previous test benches for a synchronous sequential digital circuit. There is a subcircuit instantiation statement used to connect to the UUT, one **initial** and two **always** blocks used to generate the test vectors, and one **always** block used to check the application of these test vectors to the UUT. There are two "cycle" counters, which are used to control the different phases of test generation and checking (boundary values, normal values, etc.). Numerous **$display** statements are used to indicate the progress of the simulation and the presence of any errors. Several **#** and **wait** statements are used to delay execution for a certain number of time units or until certain boolean events become true. Finally, it is important to note that different index variables are used in **case** (cycle1 == 1) of the first **always** block ($ii$ and $jj$) and **case** (cycle2 == 1) of the second **always** block ($i$ and $j$). This is required because both **always** blocks will execute concurrently. Thus, if the same index variables (e.g., $i$ and $j$) are used in both **always** blocks, then those index variables are incremented by 2 at each step instead of being incremented by 1.

```
//////////////////////////////////////////////////////////////////
  always @(negedge clk) begin // used to check the results
    case (cycle2)
      0: begin // check the boundary value test vector tests
        $display("Starting checks for the power.v module test.");

        wait(check_index == 1);
        check_z = 0;           // first result should be 0^1 = 0
        if (z != check_z) begin
          error_count = error_count + 1;
          $display
            ("ERROR: expected %0d, obtained %0d at time %0t",
                  check_z, z, $time);
        end

        wait(check_index == 2);
        check_z = 1;           // next result should be 1^0 = 1
        if (z != check_z) begin
          error_count = error_count + 1;
          $display
            ("ERROR: expected %0d, obtained %0d at time %0t",
                  check_z, z, $time);
        end

        wait(check_index == 3)
        check_z = 0;           // next result should be 0^8 = 0
```

```verilog
          if (z != check_z) begin
            error_count = error_count + 1;
            $display
              ("ERROR: expected %0d, obtained %0d at time %0t",
                    check_z, z, $time);
          end
      end // of case 0
      1: begin // next, check set of intermediate test results
         next_check_index = check_index + 1;
         for (i = 1;  i <= (`MAX_DATA_BITS/2);  i = i+1)
           for (j = 1;  j <= (`MAX_DATA_BITS/2);  j = j+1) begin
             wait(check_index == next_check_index);
             check_z = j;
             for (k = 1;  k < i;  k = k + 1)
               check_z = check_z * j;
             if (z != check_z) begin
               error_count = error_count + 1;
               $display
                 ("ERROR: expected %0d, obtained %0d at time %0t",
                       check_z, z, $time);
             end
             next_check_index = check_index + 1;
           end
      end // of case 1
      default: begin
         wait(done);
         if (error_count > 0)
           $display("ERROR: There were %d errors.", error_count);
         else
           $display("Simulation completed with NO errors.");
         $finish;
      end // of case default
    endcase
    cycle2 = cycle2 + 1;
  end // of always block to check the test results

endmodule
//////////////////////////////////////////////////////////////////////
```

Since tb_power is a test bench, which normally does not need to be synthesized into
a working circuit, the programmer is free to use *any* construct available in Verilog.  The test-bench code even can be written in the *exact same manner* as an
equivalent programming language program. Thus, constructs such as **for** and **while**
loops, multidimensional arrays, large numbers of integer signals, etc., can be used
liberally.

Figure 3.7 shows the simulation output for this example. It can be seen that this waveform is significantly more complex than the simulation waveform for Example 3.2. Thus, the display output, showing the progress of the simulation and the presence/absence of errors, is extremely helpful.

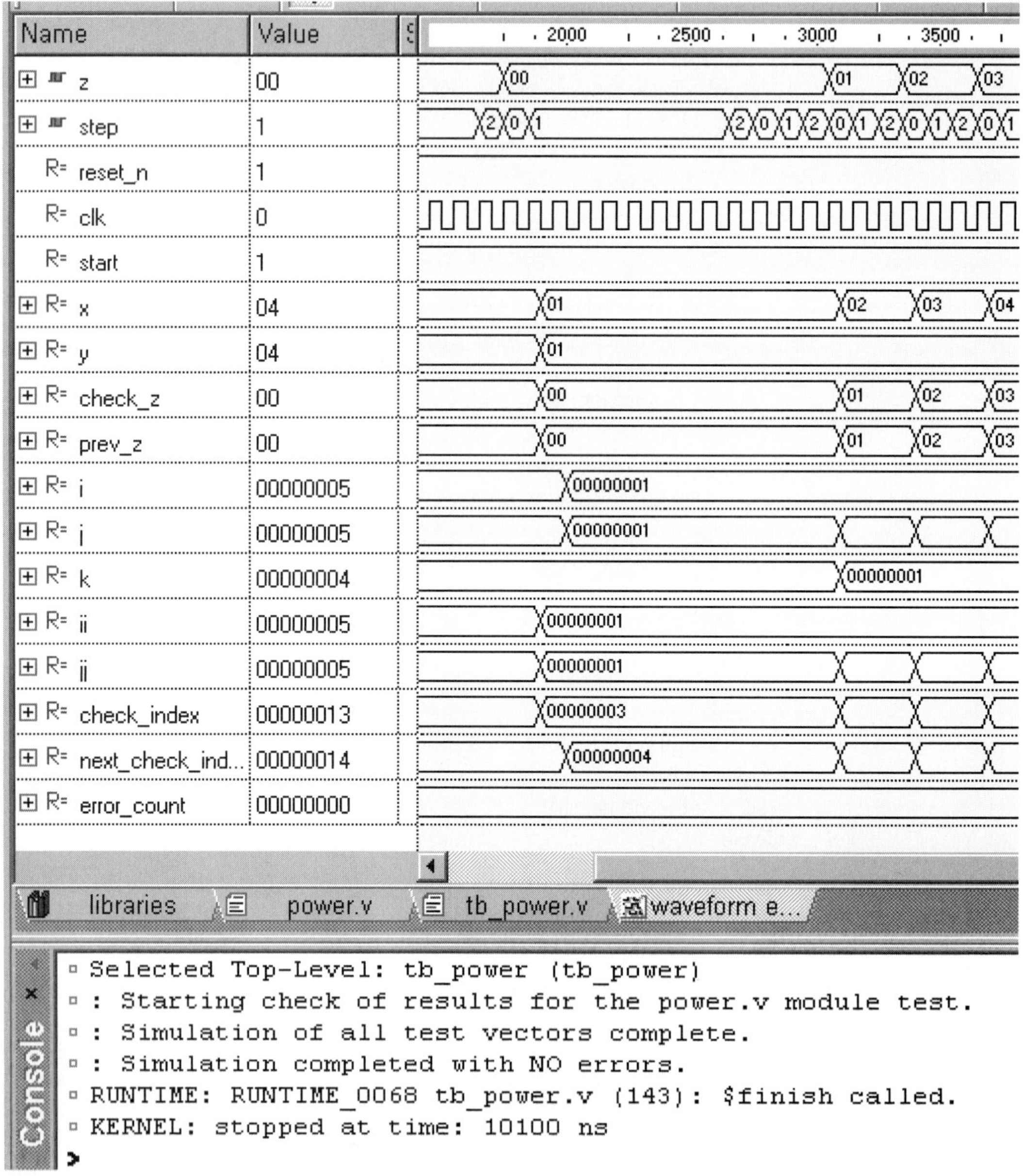


**Figure 3.7**  The simulation waveform and display output for module power.

## 3.5  Chapter Review

- Verilog HDL (Hardware Description Language) is a full-featured language for modeling and simulating digital logic hardware. Although it has many complex features, a subset of those features, mostly based on the C programming language, can be used to model and simulate most types of digital logic circuits. A significant difference between Verilog and C is that top-level statements in Verilog execute continuously (or repetitively) and concurrently because this is the default behavior of hardware, which Verilog is meant to model.

- A Verilog **module** can be written using structural or behavioral code. Structural Verilog code describes a circuit hierarchically, as an interconnection of submodules, which may themselves be described in a structural or behavioral manner. In behavioral Verilog code, **initial** and **always** blocks are used to describe the desired *operation* of a circuit; **initial** blocks execute only once at the beginning of the simulation, while **always** blocks execute in an infinite loop. Guard conditions (of the form `@(condition)`) can be used with an **always** block in order to prevent it from monopolizing CPU time.

- There are several types of assignment statements in Verilog. The **assign** keyword can be used for *continuous assignment*, in which a **wire** is continuously assigned a signal value (if the signal changes, then the wire value will also change). This type of assignment is a top-level Verilog statement. Nonblocking assignment (using `<=`) and blocking assignment (using `=`) are two types of assignments used with **initial** or **always** blocks. As the names imply, a nonblocking assignment statement executes concurrently with the statement that follows, while a blocking assignment statement executes *prior* to the statement that follows. Nonblocking assignment more accurately reflects the actual behavior of hardware.

- In order to simulate a Verilog **module** (for debugging or verification purposes), a *test bench* should be used. Such a test bench can be written as simply another Verilog **module**. A Verilog test-bench **module** has no external inputs or outputs and includes mechanisms for generating test vectors and checking the results of the application of those test vectors to the target circuit. The advantages of using a Verilog test bench for simulation instead of a graphical tool-based test generator are portability (to other platforms and environments), reproducibility, and flexibility (it can be written almost like a regular C program).

- State machines can be modeled using behavioral Verilog or structural Verilog code. For most applications, it is preferable to use behavioral Verilog code as current synthesis tools produce highly optimized designs from behavioral Verilog code, and such code leads to faster and less error-prone design of complex state machines. A sequence of steps in a pseudocode description of an algorithm can be modeled using a long **case** block (with one case for each step of the algorithm) enclosed within an **always** block.

## 3.6 Resources

The home page for Aldec Inc. is `http://www.aldec.com`, which is a company that supplies a Verilog editor and simulator called Active-HDL. Comprehensive guides to Verilog include the references listed below. For the authoritative references on the Verilog HDL standard, the interested reader should refer to standards publications such as [IEEE 1995] and the updated standard [IEEE 2001].

### 3.6.1 Bibliography

[3.1] CILETTI, M. D., *Advanced Digital Design with the Verilog HDL*, Prentice Hall, Upper Saddle River, NJ, 2002.

[3.2] CLARE, C. R., *Designing Logic Systems Using State Machines*, McGraw-Hill, New York, 1973.

[3.3] HACHTEL, G. D., and SOMENZI, F., *Logic Synthesis and Verification Algorithms*, Kluwer Academic, Boston, MA, 1996.

[3.4] IEEE, *IEEE Standard Hardware Description Language Based on the Verilog Hardware Description Language*, IEEE Press, New York, 1995. Also available through the IEEE standards web page at http://standards.ieee.org.

[3.5] IEEE, *IEEE Standard for Verilog Hardware Description Language 2001*, IEEE Press, New York, 2001. Also available through the IEEE standards web page at http://standards.ieee.org.

[3.6] ZEIDMAN, B., *Verilog Designer's Library*, Prentice Hall, Upper Saddle River, NJ, 1999.

[3.7] http://www.aldec.com, home page for ALDEC (HDL tool supplier).

[3.8] http://www.synopsys.com, home page for SYNOPSYS (Synthesis tool supplier).

[3.9] http://www.xilinx.com, home page for XILINX (FPGA H/W and S/W vendor).

## 3.7 Problems

**P3.1.** What is the difference between *structural* and *behavioral* Verilog code? What are the advantages and disadvantages of these two descriptive styles when used for digital logic design?

**P3.2.** What are the main differences between a programming language like C and a hardware description language like Verilog?

***P3.3.** It is mentioned that Verilog uses a four-valued logic system to represent signal values. Although all actual signals in physical circuits are one of two logic values (0 or 1), the unknown (X) and high-impedance (Z) values also are used in Verilog in order to display simulation output and to aid in the description of the desired behavior of a circuit. However, some logic simulation systems are known to use a nine-valued logic system. This type of system uses five other possible signal values in addition to the four values used in Verilog. What do you think these five other signal values are? Suggest five signal values (in addition to the four used in Verilog) that can be useful in describing the behavior of digital logic circuits. Explain your reasoning.

**P3.4.** Why is it the case that Verilog permits the use of single-dimensional arrays but not two-dimensional or higher arrays for logic signals? (*Hint*: Consider the fact that Verilog is used to model *hardware*.)

**P3.5.** Referring to Example 3.1 and using a similar method, write the Verilog **module** definition for a 16-bit BCD (binary-coded decimal) adder using behavioral Verilog code. Now repeat the exercise using structural Verilog code.

**P3.6.** Write a Verilog test bench for the `full_sub_behave` module. Be sure to calculate the expected value using a method different from the method used in the `full_sub_behave` module. Draw the timing diagram for the behavior that should result from the application of your test bench to the `full_sub_behave` module.

**P3.7.** Draw the timing diagram for the test bench `tb_simple()` (which tests the module `simple.v` shown in Section 3.4.2).

**P3.8.** Define the terms *positional association* and *explicit assignment* (equivalent to *named association*). Give examples of each. What are the pros and cons of these two methods?

**P3.9.** Write the Verilog code for a 32-bit ripple-carry subtracter in a structural manner using the 1-bit full-subtracter module (shown in Example 3.1) as a submodule.

**P3.10.** Write the Verilog code for a 32-bit ripple-carry adder in a structural manner using a 1-bit full-adder module as a submodule. (*Note*: Code needs to be written for the submodule as well as the top-level module.

**P3.11.** Write the Verilog code for an 8-bit Gray-code counter. Then write a test bench for this module and draw the expected timing diagram for your test bench.

**P3.12.** How are functions and tasks used in Verilog? Suggest at least two additional uses for functions and two additional uses for tasks besides the usage examples described in Section 3.3.5.

**P3.13.** Describe the differences in writing styles for Verilog code that must be synthesized and Verilog code for a test bench. Be specific and identify as many differences as possible.

*__P3.14.__ Most Verilog simulation tools permit the designer to view the values of logic signals internal to the circuit (and not part of its set of primary outputs). In addition, some tools even permit the designer to force an internal signal to a specific value *while* it is being simulated. These types of capabilities (the ability to view and change internal signals) can be extremely helpful in debugging a digital-logic-circuit design. Clearly, if these capabilities also are present in the actual digital logic circuit after it has been implemented, it also would be extremely helpful during debugging of the physical circuit. Explain how such capabilities (to view and change internal signal values) could be added to physical circuits.

__P3.15.__ Explain the differences between the Verilog keywords **reg**, **wire**, **integer**, **input**, **output**, and **inout**.

__P3.16.__ Devise and explain an example that shows different behavior when using nonblocking assignment (<=) versus blocking assignment (=) within an **always** block.

__P3.17.__ Devise and explain an example that shows how unexpected behavior (errors) can occur when using nonblocking assignments (<=) with no time delays in the assignment statements. (*Hint*: For example, create two **always** processes in which one process changes a register value that the other process reads.)

__P3.18.__ It has been proposed that the C programming language could be used as a hardware description language that describes the desired behavior of a digital logic circuit at an extremely high level. Discuss how this could be done. Include a discussion of the limitations and difficulties of this method as well as the expected benefits. For what aspect of digital logic design would the use of the C programming language be most helpful?

*__P3.19.__ Verilog also can be used to describe switch-level digital logic designs (in which the base components are transistors modeled as switches). What capabilities (constructs, signal values, etc.) must Verilog have, besides the capabilities described in this chapter, for the effective description and simulation of switch-level digital logic designs? It can be helpful to refer to other sources for Verilog syntax (other books, on-line documentation available with Verilog simulation tools, etc.) in order to solve this exercise.

*__P3.20.__ Write a Verilog code description for a traffic-light control system. Use your imagination to create a detailed design specification, high-level behavioral design that simulates properly, and a Verilog test bench. If possible, use a Verilog simulation tool to verify the correctness of your design. (*Note*: This problem may become easier to solve if Chapter 5 is studied first.)

*__P3.21.__ Write a Verilog code description for a digital circuit used to control the circuitry on one floor of a two-elevator system. Use your imagination to create a detailed design specification, high-level behavioral design that simulates properly, and a Verilog test bench. If possible, use a Verilog simulation tool to verify the correctness of your design. (*Note*: This problem may become easier to solve if Chapter 5 is studied first.)

# High-Level Verilog Coding for Synthesis

## Important Concepts

- The concept of synthesis and how to write high-level synthesizable Verilog code.
- The types of logic structures created by different types of Verilog code segments.
- How to check and verify the results of synthesis.
- Heuristics to be used for writing Verilog code intended to be synthesized.

As stated in Section 2.3, *synthesis* refers to the automatic generation of a logic circuit based on a high-level description of the desired operation of the circuit. Most current-synthesis tools support *RTL synthesis*, in which the circuit is described using *register-transfer level (RTL)* notation. *FPGA Express*™ is one such tool, produced by *Synopsys*®, optimized for use with FPGAs. This tool has been used to verify all of the code examples shown in this textbook.

● ● ● ● ● ● ● ● ● ● ● ● ● ● ● ●

## 4.1  Register-Transfer Level Notation

As described in Section 2.4, *register-transfer level (RTL)* notation refers to a notational method in which *all* operations are described as the transfer of data from one or more registers to another register. During the data transfer, transformations (involving combinational logic operations) can be applied to the data. Table 4.1 shows a list of example RTL statements written in Verilog.

Table 4.1: ▶
Examples of register-transfer level (RTL) statements written in Verilog.

| Statement | Intended Operation |
|---|---|
| `X <= Y;` | Transfer the contents of $Y$ to $X$ |
| `X <= 0;` | Clear the contents of $X$ |
| `X <= 'SIZE'b1;` | Set all bits of $X$ (of 'SIZE bits) |
| `X <= 1;` | Set $X$ to the decimal value 1 |
| `X <= X » 1;` | 1-bit right-shift of $X$ |
| `X <= {X[3:0], X['SIZE-1:4]};` | 4-bit end-around right-shift of $X$ |
| `X <= M['SIZE'hC2];` | Transfer the 194th element of $M$ to $X$ |
| `X <= Y \| Z;` | $X \leftarrow Y$ OR $Z$ (bit-wise operation) |
| `X <= Y & Z;` | $X \leftarrow Y$ AND $Z$ (bitwise operation) |
| `X <= Y ^ Z;` | $X \leftarrow Y$ exclusive-OR $Z$ (bit-wise operation) |
| `X <= ~Y;` | $X \leftarrow$ 1's complement of $Y$ |
| `X <= -Y;` | $X \leftarrow$ 2's complement of $Y$ |
| `X <= Y + Z;` | $X \leftarrow Y + Z$ |
| `X <= Y - Z;` | $X \leftarrow Y - Z$ |
| `X <= (c & Y) \| ((~c) & X);` | $X \leftarrow (\bar{c}$ AND $X)$ OR $(c$ AND $Y)$ |
| `if (c) X <= Y;` | Equivalent to above statement |
| `X <= (c & Y) \| ((~c) & Z);` | $X \leftarrow (\bar{c}$ AND $Z)$ OR $(c$ AND $Y)$ |
| `X <= c ? Y : Z;` | Equivalent to above statement |

# 4.2 Combinational-Logic Synthesis

Hardware description languages such as Verilog have facilitated the use of synthesis, as such languages can be used to describe the desired behavior of a circuit in a high-level manner. For the effective use of synthesis, the user should be aware of how to write Verilog code to create specific circuit structures. Let us first consider several general combinational logic structures.

The behavior and output of a combinational logic block is dependent only on the values of the current inputs, and not on the values of past inputs or any "stored state" information. Thus, the Verilog description should not include any history or state-dependent constructs. Timing delays, which are technology and process dependent, should also be avoided.

There are a variety of methods for describing combinational logic structures in Verilog. The same type of structure (e.g., a NAND gate) can be described using a variety of Verilog statements (e.g., structural Verilog using the **nand** primitive gate, a C-like assignment statement such as `c = !(a & b);`, etc.). In the following, a style of Verilog description that is simple to understand, generally applicable, and widely used is presented as the recommended method for the description of most types of combinational logic structures.

## 4.2.1   Using Continuous Assignment for Combinational Logic

Simple combinational logic structures involving a logical combination of basic logic gates or multiplexers should be described using *continuous assignment* statements. As introduced in Section 3.3.3, this type of statement uses the **assign** keyword outside of **always** or **initial** blocks. As such, such statements are "top-level" statements that are executed concurrently (and continuously) with all other top-level statements in a Verilog **module**. This correctly models the behavior of combinational-logic hardware devices, which are active all of the time in a concurrent manner. The left-hand side of a continuous assignment statement must be a **wire** signal, while the variables on the right-hand side can be **reg** or **wire** signals. Table 4.2 shows several examples of combinational logic devices and the continuous assignment statements that can be used to implement them.

The examples in Table 4.2 show several interesting aspects. First, simple logic gates can be synthesized from simple C (the programming language) logic assignment statements. Even arrays of logic gates can be synthesized from the same types of assignment statements, as shown for the array of inverters. Note that for logical negation, ! can only be used for single-bit values (boolean NOT) while ~ (tilde) can be used for single or multiple-bit values. Whether an array of logic gates is created or a single logic gate is created depends on the size of the **wire** signal on the left-hand side. Second, multiplexers and tri-state buffers can be synthesized from simple C conditional assignment statements, with `1'bz` used to denote a single-bit high-impedance value. Third, the curly brace notation can be used to simplify checks of more than one bit at a time, as shown in the decoder example. Fourth, an extended form of the C conditional assignment statement can be used to assign values based

**Table 4.2:** ▶
Examples of continuous assignment statements used to implement combinational logic.

| Device | Continuous Assignment Statement |
|---|---|
| NAND gate | **assign** c = !(a & b); |
| Exclusive-OR gate | **assign** c = a ^ b; |
| Array of inverters | **assign** c_vec = ~a_vec;<br><br>// c_vec and a_vec are vectors |
| 2:1 multiplexer | **assign** o0 = sel ? i1 : i0; |
| 2:4 decoder | **assign** o0 = ({i1, i0} == 2'b00);<br><br>**assign** o1 = ({i1, i0} == 2'b01);<br><br>**assign** o2 = ({i1, i0} == 2'b10);<br><br>**assign** o3 = ({i1, i0} == 2'b11); |
| 4:1 multiplexer | **assign** o0 = (sel == 2'b00) ? i0 :<br><br>(sel == 2'b01) ? i1 :<br><br>(sel == 2'b10) ? i2 :<br><br>i3;  // when (sel == 2'b11) |
| Tri-state buffer | **assign** o0 = enable ? i0 : 1'bz; |
| Tri-state bus | **assign** bus_out = enable ? bus_in : 'bz; |
| 4-bit parity checker | **assign** p_odd = ^data[3:0]; |
| Multiple-bit adder | **assign** result = a_vec + b_vec; |
| Multiple-bit multiplier | **assign** result = a _vec * b_vec; |

on multiple conditions, as shown in the 4-bit multiplexer example. Fifth, operations applied to multiple bits of an array can be denoted by writing the operator preceding the array, as shown for the 4-bit parity checker.

Finally, circuits for performing arithmetic operations can be synthesized by simply writing the corresponding C statements, as shown in the final two examples. However, this type of method only creates common types of arithmetic circuits.[1] To create more complex combinational logic circuits, it may be necessary to use the type of method shown in the next subsection.

---

[1] Depending on the synthesis options used, the arithmetic units created can be slow circuits that utilize only a small number of logic gates (if the "optimize for area" option is chosen) or fast circuits that utilize a large number of logic gates (if the "optimize for speed" option is chosen).

### 4.2.2  Using Always Blocks for Combinational Logic

Although continuous assignment statements are useful for creating simple combinational-logic structures, another method is required to create more complex circuits. Continuous assignment statements are too limited in their expressive capabilities to describe complex combinational logic circuits such as fast adders or multipliers, presented in Chapter 8. Also, many Verilog constructs (such as **if**, **case**, **for**, etc.) can only be used within an **always** (or **initial**) block. Thus, **always** blocks can be used to describe a combinational logic circuit, as long as (1) all inputs to the circuit are included in the *event control* clause of the **always** definition, (2) no other signals are included in the event control clause, and (3) none of the statements within the **always** block are sensitive to rising or falling edges of any signals.

Let us consider a simple example of the use of **always** blocks for combinational logic. An *encoder* is a device that converts a decimal value into its binary equivalent (it performs the inverse function of a decoder). The decimal value is presented to the device by assigning a unique decimal value to each input port and then asserting the signal connected to only one of those input ports—the output ports then form a binary value that is the encoded value of the single asserted input port. An encoder does not function properly if more than one input port is asserted concurrently. If such situations are possible, then it may be necessary to use a *priority encoder*, which is an encoder that assigns a priority order to the input ports and then outputs the binary value corresponding to the *asserted* input port with the highest priority. For both types of encoders, an extra output sometimes is used to indicate the absence of an asserted input port. The device that results from the synthesis of the following code is a 4:2 priority encoder (4 inputs and 2 outputs, excluding the valid output indicator).

```
/////////////////////////////////////////////////////////////////////////
module priority_encoder_4_2 (valid, encoded, i3, i2, i1, i0);
  output valid;        // indicates at least one asserted input
  output [1:0] encoded; // binary encoded output
  input i3, i2, i1, i0; // ix (0 <= x <= 3) indicates decimal x
                        // priority order is i3, i2, i1, i0
  // SIGNAL DECLARATIONS for signals connected to module outputs
  reg valid;
  reg [1:0] encoded;

  always @(i3 or i2 or i1 or i0) begin
    if ((i3 == 1) || (i2 == 1) || (i1 == 1) || (i0 == 1))
      valid <= 1;   // at least one input port is asserted
    else
      valid <= 0;   // no input asserted -> "encoded" is invalid
    casex ({i3, i2, i1, i0}) // case block with don't care values
      4'b1xxx: encoded <= 2'b11;  // i3 asserted
      4'b01xx: encoded <= 2'b10;  // (i3' and i2)
```

```
      4'b001x: encoded <= 2'b01;  // (i3' and i2' and i1)
      default: encoded <= 2'b00;  // only i0 asserted or invalid
    endcase
  end
endmodule
////////////////////////////////////////////////////////////////////
```

There are several noteworthy points in the above Verilog description. First, the module outputs are declared as type **reg** since they are assigned values within an **always** block. The inputs are of type **wire** by default. Second, the **always** block is activated whenever *any* of the input values change—thus, all input signals separated by **or** must be included in the activation guard clause for this **always** block. Third, **if-else** statements and other types of C language-like clauses are present within the **always** block. Fourth, a **casex** block is used to check the values of the four input values. A **casex** block is similar to a Pascal "case" statement or a C-language "switch" statement, except that don't-care values (**x**) can be included in the case clauses; **z** (high impedance) values can also be present and are treated as don't cares. Other possible variations permitted in Verilog are **case** blocks (all bit values must match exactly) and **casez** blocks (0, 1, and **x** values must match exactly while **z** (high impedance) values are treated as don't cares).

A test bench that can be used to test this priority encoder module is shown next.

```
////////////////////////////////////////////////////////////////////
// MODULE: Priority Encoder Test Bench: tb_priority_encoder_4_2
// Author: Sunggu Lee
// Date: ...
// Description: Tests the priority_encoder_4_2 module.

`timescale 1ns/1ns   // 1ns simulation accuracy & 1ns precision
// Constant Definitions
`define MAX_COMB 16  // maximum number of input combinations
`define DELAY    100 // delay value used in testing (100 ns)

// Test bench module definition
module tb_priority_encoder_4_2 ();
  // SIGNAL DEFINITIONS
  // Signals used to connect to the UUT (unit under test)
  wire valid;          // indicates at least one asserted input
  wire [1:0] encoded;  // binary encoded output
  reg i3, i2, i1, i0;  // ix (0 <= x <= 3) indicates decimal x
  // Internal signals
  integer error_count = 0;   // total number of errors found
  integer i;                 // test vector generation index
  reg [3:0] temp;            // used as a temporary data holder
  reg expected_valid;        // used to check test results
  reg [1:0] expected_encoded;
```

```verilog
// Connect to the UUT (unit under test)
priority_encoder_4_2 uut (valid, encoded, i3, i2, i1, i0);

// Generate the test vectors and check the results
initial begin
  $display("Start simulation.");
  // generate all possible input values, since max = 15
  for (i = 0;  i < `MAX_COMB;  i = i + 1) begin
    temp = i;
    i3 = temp[3];  // assign input values
    i2 = temp[2];
    i1 = temp[1];
    i0 = temp[0];

    if (i == 0) begin  // compute expected values
      expected_valid = 0;
      expected_encoded = 0;
    end // of if (i == 0)
    else begin
      expected_valid = 1;
      if (i == 1) expected_encoded = 0;
      else if (i <= 3) expected_encoded = 1;
      else if (i <= 7) expected_encoded = 2;
      else             expected_encoded = 3;
    end // of else (i > 0)

    #`DELAY;           // wait for result to be computed
    if ((expected_valid !== valid) ||
        (expected_encoded !== encoded)) begin  // test result
      $display
      ("ERROR: valid = %0b and encoded = %0b at time %0t ns",
        valid, encoded, $time);
      error_count = error_count + 1;
    end  // of if
  end // of for (i ... )

  #`DELAY;
  if (error_count > 0)
    $display("ERROR: There were %0d errors!", error_count);
  else
    $display("Simulation complete with NO errors.");
  $finish;           // terminate the simulation
end  // of always block for test vector generation and checking

endmodule
//////////////////////////////////////////////////////////////////
```

### 4.2.3  Complex Combinational Logic

As an example of the design of a complex combinational logic circuit, let us consider a *hierarchical carry-lookahead adder*, which is a fast adder structure with a hierarchical tree-like interconnection of "carry lookahead" blocks. This logic circuit requires the use of several advanced techniques for combinational-logic design and detailed analysis before the actual design.

Let us first consider the binary addition of two $n$ bit signals $A$ (consisting of the sequence of 1-bit signals $a_{n-1}a_{n-2}\ldots a_0$) and $B$ (consisting of the sequence of 1-bit signals $b_{n-1}b_{n-2}\ldots b_0$). At the $k$th-bit position ($n-1 \geq k \geq 0$), if both $a_k$ and $b_k$ are 1, then the carry into the $(k+1)$th-bit position will always be a 1. Thus, the term $g_k = (a_k$ and $b_k)$ can be referred to as a *generate* signal, since it results in the "generation" of a carry bit. Alternatively, if only one of $a_k$ or $b_k$ is a 1, then the carry bit acquires the value of the carry from bit position $k-1$. In other words, the term $p_k = (a_k \oplus b_k)$ can be considered as a *propagate* signal, since it "propagates" the carry from bit position $(k-1)$ to bit position $(k+1)$. For a slightly simpler implementation, $p_k = (a_k$ or $b_k)$ also can be used as a propagate signal. The carry into the $(k+1)$th-bit position is a 1 if $(g_k = 1)$ or $(p_k = 1$ and there is a carry into the $k$th-bit position). These generate and propagate signals can be used as the basis for the design of fast adders and subtracters, as detailed in Chapter 8.

In some fast adder designs, the $n$-bit signals are partitioned into $k$ groups. Then, given $m = n/k$ bits in each group, the *group propagate*($P_{from\_index:to\_index}$) and *group generate* ($G_{from\_index:to\_index}$) signals for group $i$ (consisting of bits $from\_index$ to $to\_index$) can be defined as

$$P_{i*m:i*m+m-1} = \prod_{x=i*m}^{i*m+m-1} p_x$$

$$G_{i*m:i*m+m-1} = \sum_{x=i*m}^{x=i*m+m-1} \left(g_x \prod_{y=x+1}^{i*m+m-1} p_y\right).$$

For example, if $i = 1$ and $m = 4$, then $P_{4:7} = p_4 p_5 p_6 p_7$ (there is a propagate out of this group if there are propagates from bit positions 4, 5, 6, and 7) and $G_{4:7} = g_4 p_5 p_6 p_7 + g_5 p_6 p_7 + g_6 p_7 + g_7$ (a carry is *generated* from this group if there is a generate from bit position 7, a generate from bit 6 and a propagate from bit 7, or a generate from bit 5 and propagates from bits 6 and 7, etc.).

The logic implied by the above equations is shown in Figure 4.1 for the case with $n = 16$, $k = 4$, and $m = 4$. Clearly, it would be difficult (or at least tedious, with the same calculations required for different values of $n$) to describe combinational logic implementing the above equations using only continuous assignment statements. Instead, the following Verilog code could be used to produce the group propagate and group generate signals. Note that if the Verilog *parameter* construct is used, then the same code (with a few modifications) can also be used to describe instantiations of this circuit for different values of $n$.[2]

---

[2] The use of Verilog parameter statements is covered in Section 3.3.4.

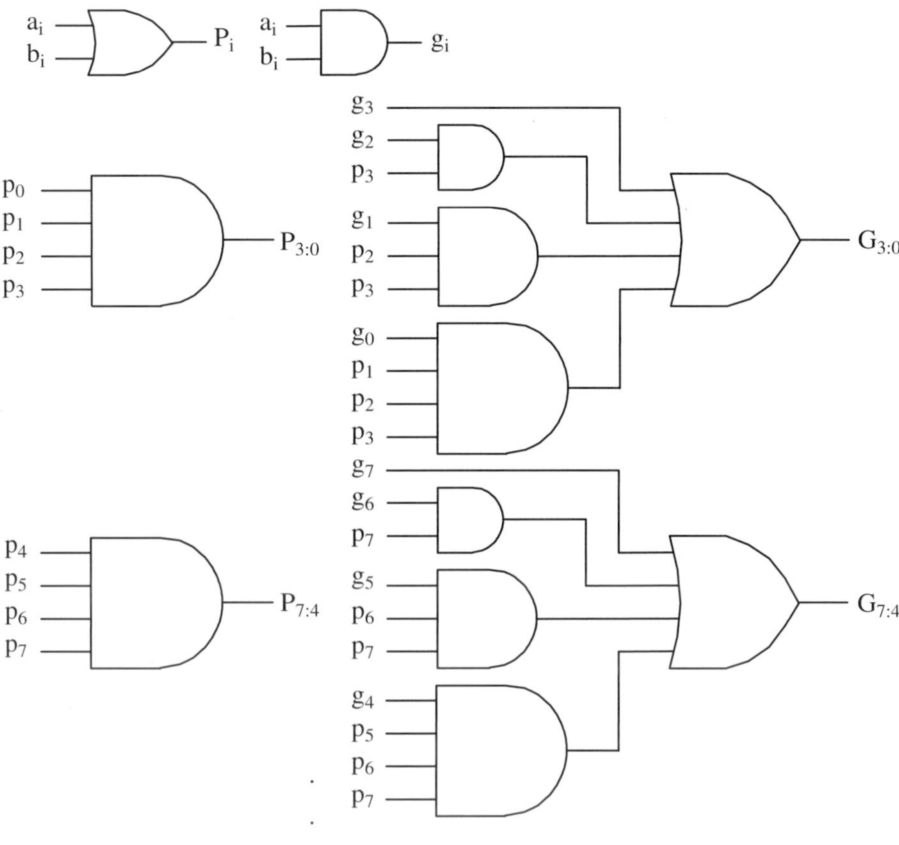


Similarly for  $P_{11:8}$, $G_{11:8}$, $P_{15:12}$, and $G_{15:12}$

**Figure 4.1**
Group-propagate
and group-generate
logic given $n = 16$
bits, $k = 4$ groups,
and $m = 4$ bits-per-
group.

```
//////////////////////////////////////////////////////////////////////////
`define DATA_BITS  16   // number of bits in input and output data
`define NUM_GROUPS 4    // number of stages (or groups) used
`define GROUP_BITS (`DATA_BITS / `NUM_GROUPS)  // bits in 1 stage

// MODULE DECLARATION
module p_g_generator (P, G, a, b);

  output [`NUM_GROUPS-1:0] P;  // group-propagate signals
  output [`NUM_GROUPS-1:0] G;  // group-generate signals
  input [`DATA_BITS-1:0] a, b; // adder inputs
```

```verilog
// SIGNAL DECLARATIONS
reg [`NUM_GROUPS-1:0] P;     // group-propagate signals
reg [`NUM_GROUPS-1:0] G;     // group-generate signals
reg [`DATA_BITS-1:0] p;      // propagate signals
reg [`DATA_BITS-1:0] g;      // generate signals
integer i, j;                // indices used in for loops


// always block used to model combinational logic
always @(a or b) begin
  for (i = 0;  i < `NUM_GROUPS;  i = i + 1) begin
    P[i] = 1;
    G[i] = 0;
    for (j = `GROUP_BITS-1;  j >= 0;  j = j - 1) begin
      p[i*`GROUP_BITS+j] =
               (a[i*`GROUP_BITS + j] | b[i*`GROUP_BITS + j]);
      g[i*`GROUP_BITS+j] =
               (a[i*`GROUP_BITS + j] & b[i*`GROUP_BITS + j]);
      G[i] = G[i] | (g[i*`GROUP_BITS+j] & P[i]);
      P[i] = P[i] & p[i*`GROUP_BITS+j];
    end
  end
end
endmodule
/////////////////////////////////////////////////////////////////////
```

When the p_g_generate architecture shown here is synthesized using Synopsys FPGA Express, it creates two circuits: an unoptimized version using standard logic gates and devices and an optimized version using the components available in the target hardware architecture. Assuming that a Xilinx SpartanXL S10XLPC84-4 FPGA chip is used as the target hardware architecture, Figure 4.2 shows the unoptimized and optimized circuits produced by the synthesis tool.

There are several noteworthy features of this Verilog code and the circuit that is synthesized from it. Firstly, combinational logic is produced even though an **always** block is used. Note that the **always** block is written in the form **always** @(sig1 or ... or sigN)—this implies that the **always** block is activated whenever there is a change in value in any of the signals *sig*1 through *sigN*. The @(sig1 or ... or sigN) construct is referred to as an *event-control* clause. An important rule to follow when using an **always** block to model a combinational block in this manner is that the **always** block should be activated every time there is a change in *any* of the input signals to the combinational block. This correctly models combinational logic, since the outputs of a combinational-logic block only can change when there is a change in one or more of its inputs.

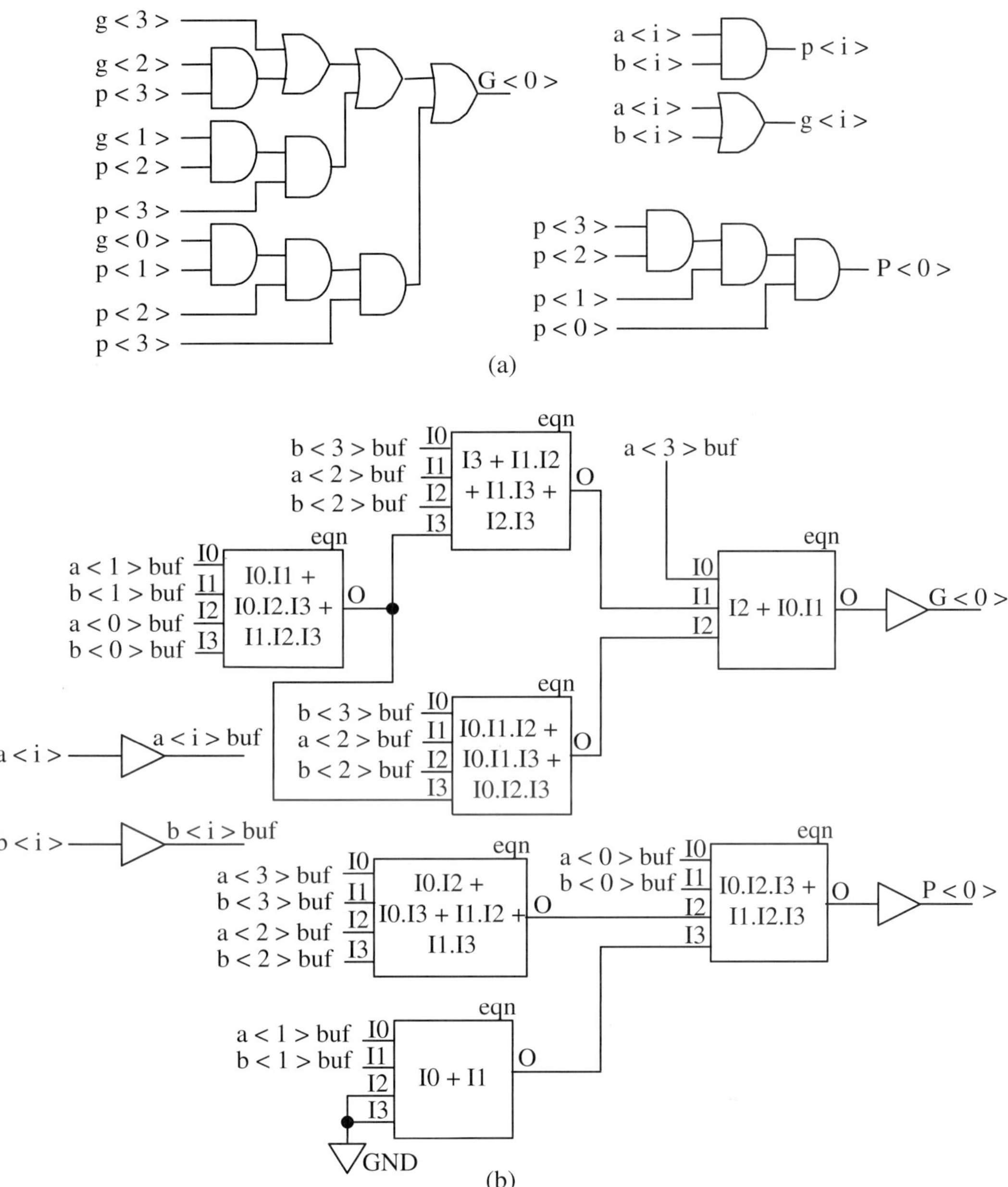

**Figure 4.2**  The logic synthesized for the `p_g_generate` Verilog module. (a) unoptimized and (b) optimized version.

Secondly, some of the **reg** and **integer** (equivalent to a 32-bit **reg**) signals are not present in the logic created by the synthesis software. As stated earlier, the use of a **reg** signal does not mean necessarily that a register should be created. In fact, the synthesis software does not create *any* logic that is not absolutely *required* in order to produce the output signals. Thus, since the **reg** signals $p$ and $g$ and the **integer**s $i$ and $j$ only are used to facilitate the Verilog description (and not necessary in the combinational logic), the synthesis software does not create any logic for them.[3]

Thirdly, the synthesis software may create a logic gate design that appears to be different from that intended by the designer. For instance, the logic equation for the group-propagate signal, which the Verilog code is intended to model, should result clearly in a $k$-input AND gate, with each input consisting of an OR of an $a_i$ and a $b_i$ signal, as shown in Figures 4.1 and 4.2(a). However, the optimized logic created by the Synopsys FPGA Express synthesis software, as shown in Figure 4.2(b), is not of this form. Nevertheless, close inspection of the circuit in Figure 4.2(b) should convince the reader that the two forms are equivalent *logically* (verification of this fact is given as an exercise at the end of the chapter). The synthesis software tends to create logic that is equivalent logically to the intended design, but of a form that results in *faster* logic or logic with *less area* (i.e., lower hardware cost).

A test bench that can be used to test the p_g_generator module is shown next. Although this module will only be used as a subcircuit for the fast-adder circuit to be shown later, it is still a good idea to thoroughly test this p_g_generator module before it is included into a higher-level circuit.

```
////////////////////////////////////////////////////////////////////
// MODULE: P/G Generator Test Bench: tb_p_g_generator
// Author: Sunggu Lee
// Date: ...
// Description: Tests the p_g_generator module.

`timescale 1ns/1ns     // 1ns simulation accuracy & 1ns precision
// Constant Definitions
`define DATA_BITS   16 // number of bits in input and output data
`define NUM_GROUPS 4   // number of stages (or groups) used
`define GROUP_BITS (`DATA_BITS / `NUM_GROUPS) // bits in 1 stage
`define MIN_NEG_DATA -32768 // = -2^(`DATA_BITS - 1)
`define MAX_POS_DATA 32767  // = 2^(`DATA_BITS - 1) - 1
`define INTEGER_SIZE 32     // number of bits in integer
`define DELAY    100 // delay value used in testing (100 ns)
```

---

[3] With some types of synthesis software, it is possible to specify, through the use of options, that certain logic blocks should be left unoptimized, that a hierarchical structure should be maintained, or that certain redundant logic blocks should be created.

```verilog
// Test bench module definition
module tb_p_g_generator ();
  // SIGNAL DEFINITIONS
  // Signals used to connect to the UUT (unit under test)
  wire [`NUM_GROUPS-1:0] P;  // group-propagate signals
  wire [`NUM_GROUPS-1:0] G;  // group-generate signals
  reg [`DATA_BITS-1:0] a, b; // adder inputs
  // Internal signals
  integer error_count = 0;   // total number of errors found
  integer i, j, ii, jj;      // test vector generation indices

  // Connect to the UUT (unit under test)
  p_g_generator uut (P, G, a, b);

  // Generate the test vectors and check the results
  initial begin
    $display("Start simulation.");
    // first, generate boundary test vectors
    // generate values close to 0
    for (i = -2;  i <= 2;  i = i + 1)
      for (j = -2;  j <= 2;  j = j + 1)
        test_and_check_p_g(i, j);
    // generate min data input values
    for (i = `MIN_NEG_DATA;  i <= `MIN_NEG_DATA+4;
         i = i + 1)
      for (j = `MIN_NEG_DATA;  j <= `MIN_NEG_DATA+4;
           j = j + 1)
        test_and_check_p_g(i, j);
    // generate max data input values
    for (i = `MAX_POS_DATA-4;  i <= `MAX_POS_DATA;
         i = i + 1)
      for (j = `MAX_POS_DATA-4;  j <= `MAX_POS_DATA;
           j = j + 1)
        test_and_check_p_g(i, j);
    // next, generate a set of normal test vector values
    for (ii = 0;  ii <= 9;  ii = ii + 1)
      for (jj = 0;  jj <= 9;  jj = jj + 1) begin
        i = 41*ii-273;   // generate random set of negative
        j = 89*jj-384;   // and positive numbers
        test_and_check_p_g(i, j);
      end // of for (jj ...) loop
```

```verilog
    #`DELAY;
   if (error_count > 0)
     $display("ERROR: There were %0d errors!", error_count);
   else
     $display("Simulation complete with NO errors.");
   $finish;            // terminate the simulation
end  // of always block for test vector generation and checking

// Test application and error checking task
task test_and_check_p_g;
  input [`INTEGER_SIZE-1:0] a_in, b_in;  // a and b inputs
  reg [`NUM_GROUPS-1:0] expected_P; // expected P result
  reg [`NUM_GROUPS-1:0] expected_G; // expected G result
  reg [`DATA_BITS-1:0] p, g;        // used for result checking
  integer k;                        // local loop index
begin
  a = a_in;  // apply test vector
  b = b_in;
  for (k = 0;  k < `DATA_BITS;  k = k + 1) begin // compute
    p[k] = a[k] | b[k];                     // expected
    g[k] = a[k] & b[k];                     // results
  end // of for (k ...)
  // for now, simply compute each expected value separately
  expected_P[0] = p[3] & p[2] & p[1] & p[0];
  expected_G[0] = g[3] | (g[2] & p[3]) | (g[1] & p[2] & p[3]) |
                  (g[0] & p[1] & p[2] & p[3]);
  expected_P[1] = p[7] & p[6] & p[5] & p[4];
  expected_G[1] = g[7] | (g[6] & p[7]) | (g[5] & p[6] & p[7]) |
                  (g[4] & p[5] & p[6] & p[7]);
  expected_P[2] = p[11] & p[10] & p[9] & p[8];
  expected_G[2] = g[11] | (g[10] & p[11]) |
                  (g[9] & p[10] & p[11]) |
                  (g[8] & p[9] & p[10] & p[11]);
  expected_P[3] = p[15] & p[14] & p[13] & p[12];
  expected_G[3] = g[15] | (g[14] & p[15]) |
                  (g[13] & p[14] & p[15]) |
                  (g[12] & p[13] & p[14] & p[15]);
```

```
      #`DELAY;  // wait for UUT to compute result
      if (!((expected_P === P) && (expected_G === G)))
      begin
        error_count = error_count + 1;
        $display("ERROR: P = %0b, G = %0b at time %0t.",
              P, G, $time);
      end  // of if
    end  // of outermost begin for error checking task
  endtask

endmodule
////////////////////////////////////////////////////////////////////
```

Now, let us extend this sample to demonstrate the use of a hierarchical Verilog description for combinational logic. A *hierarchical carry-lookahead adder* consists of several "mini-adder" stages (or groups) connected together using a tree of carry-lookahead generator blocks, as shown in Figure 4.3. The "mini-adder" is simply a carry-lookahead adder for the group of bits being handled in that stage. For a simpler (but slightly slower) implementation, a ripple-carry adder can be used as the "mini-adder" instead of a full carry-lookahead adder. A hierarchical carry-lookahead adder of the type shown in Figure 4.3(a) can be modeled in Verilog as shown next.

```
////////////////////////////////////////////////////////////////////
`define DATA_BITS   16  // number of bits in input and output data
`define NUM_GROUPS 4    // number of stages (or groups) used
`define GROUP_BITS (`DATA_BITS / `NUM_GROUPS)  // bits in 1 stage

// MODULE DECLARATION
module hierarchical_cla (sum, cout, x, y, cin);

  output [`DATA_BITS-1:0] sum; // the resulting sum value
  output cout;                 // carry-out bit
  input [`DATA_BITS-1:0] x, y; // the two numbers to be added
  input cin;                   // carry-in to the adder

  // Declaration of "reg" outputs
  reg [`DATA_BITS-1:0] sum;
  reg cout;

  // Declaration of internal signals
  reg [`DATA_BITS:0] carry;     // intermediate carry signals
  reg [`NUM_GROUPS-1:0] temp;   // temporary signal
  wire [`NUM_GROUPS-1:0] GP;    // group propagate signals
  wire [`NUM_GROUPS-1:0] GG;    // group generate signals
  integer i, j;                 // indices used in for loops
```

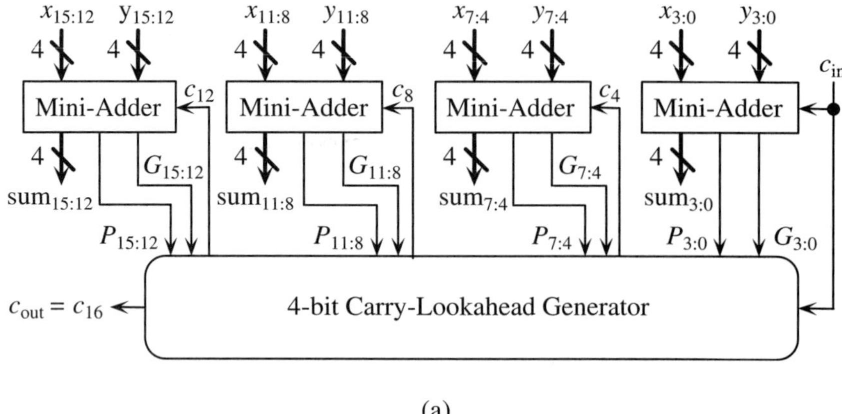



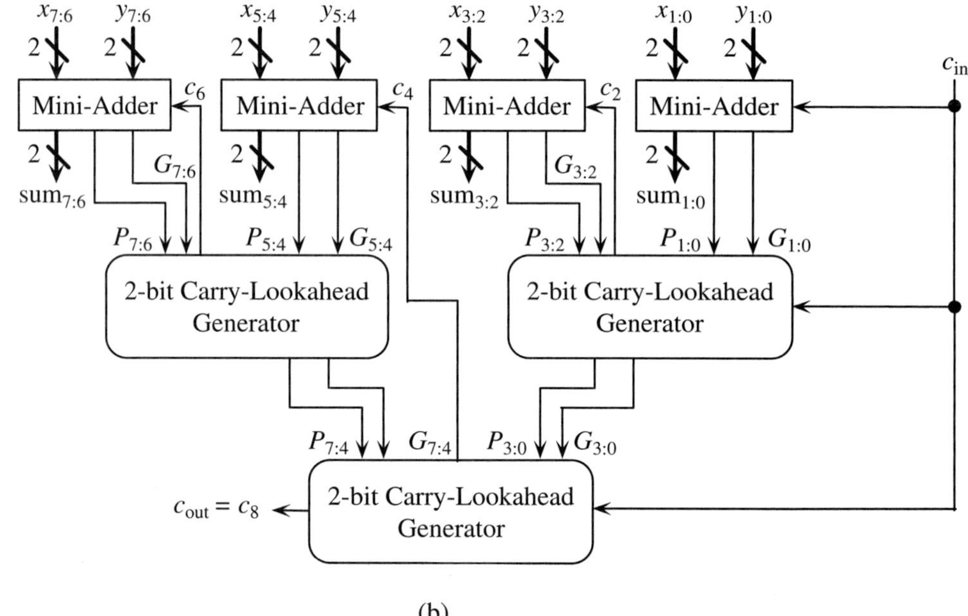



**Figure 4.3**
A block diagram for (a) a 16-bit hierarchical carry-lookahead adder using 4-bit mini-adders and (b) an 8-bit hierarchical carry-lookahead adder using 2-bit mini-adders.

```verilog
// use a lower-level block for GP[i] and GG[i] signals
p_g_generator UNIT0 (GP, GG, x, y);

// create the carry-lookahead logic and ripple-carry adders
always @(x or y or GP or GG or cin) begin
  // first describe carry-lookahead logic block
  carry[0] = cin;
```

```
  for (i = 0;  i < `NUM_GROUPS;  i = i + 1) begin
    carry[(i+1)*`GROUP_BITS] = GG[i];
    carry[(i+1)*`GROUP_BITS] = carry[(i+1)*`GROUP_BITS] |
                (GP[i] & carry[i*`GROUP_BITS]);
  end  // of for (i)
  cout = carry[`NUM_GROUPS * `GROUP_BITS];

  // next, generate the ripple-carry adder stages
  for (i = 0;  i < `NUM_GROUPS;  i = i + 1) begin
    for (j = 0;  j < (`GROUP_BITS-1);  j = j + 1) begin
      {carry[i*`GROUP_BITS + j + 1], sum[i*`GROUP_BITS + j]} =
            x[i*`GROUP_BITS + j] + y[i*`GROUP_BITS + j] +
            carry[i*`GROUP_BITS + j];
    end  // of for (j)
    {temp[i], sum[(i+1)*`GROUP_BITS - 1]} =
          x[(i+1)*`GROUP_BITS - 1] + y[(i+1)*`GROUP_BITS - 1] +
          carry[(i+1)*`GROUP_BITS - 1];
  end  // of for (i)
end  // of always block

endmodule
//////////////////////////////////////////////////////////////////////
```

Figure 4.4 shows the logic created by Synopsys FPGA Express from the previous Verilog code.

There are several interesting aspects of the Verilog description and synthesis of the `hierarchical_cla`. First, the `p_g_generator` module is used as a submodule of this structure by calling the `p_g_generator` module much like you would a C procedure call. This type of hierarchical description method facilitates the description of large or complex circuits. Second, an **always** block again is used in the description of the above combinational block. The carry-lookahead generator logic (which generates the carry inputs to the ripple-carry adder stages) is described in the first half of the **always** block, while the four ripple-carry adder stages are described in the second half of the same **always** block. If desired, these two halves could have been described using separate **always** blocks. In order to use an **always** block to describe combinational logic, *all* input signals used in that **always** block must be included in the *event-control* clause for that **always** block.

In the logic synthesized from the `hierarchical_cla` Verilog code, it again can be seen that the synthesized circuit does not produce the exact logic structures most directly implied by the Verilog code. Optimizations have been performed. It is possible to specify that the lower-level `p_g_generator` module, whose synthesized circuit was shown in Figure 4.2, should retain its structure by using a "retain hierarchy" option, as was done in this case. However, if the "retain hierarchy" option is not used, then a "flattened" logic structure, which may be more efficient, may be produced by the synthesis software.

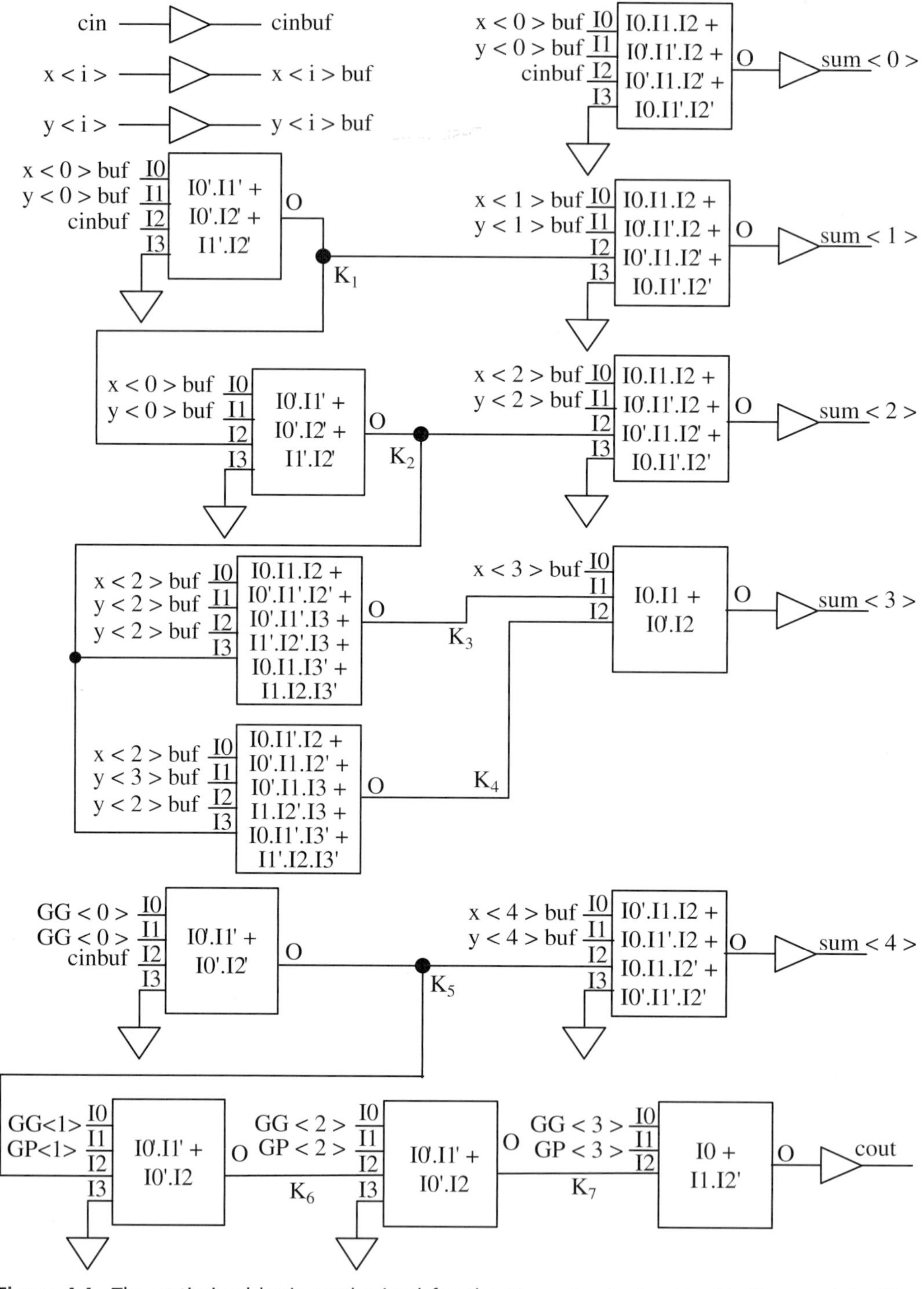

**Figure 4.4** The optimized logic synthesized for the `hierarchical_cla_adder` Verilog module (the `GG<i>` and `GP<i>` signals are produced by the lower-level `p_g_generate` module, and the parts not shown in this figure are exactly analogous to the parts shown).

The following shows a test bench that can be used to test the `hierarchical_cla` module.

```verilog
///////////////////////////////////////////////////////////////////
// MODULE: Hierarchical CLA Test Bench: tb_hierarchical_cla
// Author: Sunggu Lee
// Date: ...
// Description: Tests the hierarchical_cla module.

`timescale 1ns/1ns    // 1ns simulation accuracy & 1ns precision
// Constant Definitions
`define DATA_BITS  16 // number of bits in input and output data
`define NUM_GROUPS 4  // number of stages (or groups) used
`define GROUP_BITS (`DATA_BITS / `NUM_GROUPS) // bits in 1 stage
`define MIN_NEG_DATA -32768 // = -2^(`DATA_BITS - 1)
`define MAX_POS_DATA 32767  // = 2^(`DATA_BITS - 1) - 1
`define INTEGER_SIZE 32     // number of bits in integer
`define DELAY   100  // delay value used in testing (100 ns)

// Test bench module definition
module tb_hierarchical_cla ();
  // SIGNAL DEFINITIONS
  // Signals used to connect to the UUT (unit under test)
  wire [`DATA_BITS-1:0] sum; // addition result
  wire cout;                 // carry out of adder
  reg [`DATA_BITS-1:0] a, b; // adder inputs
  reg cin;                   // carry in into adder
  // Internal signals
  integer error_count = 0;   // total number of errors found
  integer i, j, k, ii, jj;      // test vector generation indices

  // Connect to the UUT (unit under test)
  hierarchical_cla uut (sum, cout, a, b, cin);

  // Generate the test vectors and check the results
  initial begin
    $display("Start simulation.");
    // first, generate boundary test vectors
    // generate values close to 0
    for (i = -2;  i <= 2;  i = i + 1)
      for (j = -2;  j <= 2;  j = j + 1)
        for (k = 0;  k <= 1;  k = k + 1)
          test_and_check(i, j, k);
```

```verilog
        // generate min data input values
        for (i = `MIN_NEG_DATA;  i <= `MIN_NEG_DATA+4;
            i = i + 1)
          for (j = `MIN_NEG_DATA;  j <= `MIN_NEG_DATA+4;
              j = j + 1)
            for (k = 0;  k <= 1;  k = k + 1)
              test_and_check(i, j, k);
        // generate max data input values
        for (i = `MAX_POS_DATA-4;  i <= `MAX_POS_DATA;
            i = i + 1)
          for (j = `MAX_POS_DATA-4;  j <= `MAX_POS_DATA;
              j = j + 1)
            for (k = 0;  k <= 1;  k = k + 1)
              test_and_check(i, j, k);
        // next, generate a set of normal test vector values
        for (ii = 0;  ii <= 9;  ii = ii + 1)
          for (jj = 0;  jj <= 9;  jj = jj + 1) begin
            i = 41*ii-273;    // generate random set of negative
            j = 89*jj-384;    // and positive numbers
            for (k = 0;  k <= 1;  k = k + 1)
              test_and_check(i, j, k);
          end // of for (jj ...) loop
        #`DELAY;
        if (error_count > 0)
          $display("ERROR: There were %0d errors!", error_count);
        else
          $display("Simulation complete with NO errors.");
        $finish;             // terminate the simulation
      end  // of always block for test vector generation and checking

    // Test application and error checking task
    task test_and_check;
      input [`INTEGER_SIZE-1:0] a_in, b_in, c_in;  // a, b, cin
      reg [`DATA_BITS-1:0] expected_sum;// used for result checking
      reg expected_cout;                      // used for result checking
    begin
      a = a_in;  // apply test vector
      b = b_in;
      cin = c_in;

      // compute expected results
      {expected_cout, expected_sum} = {1'b0, a} + {1'b0, b} + cin;
```

```
    #`DELAY;   // wait for UUT to compute result
    if (!((expected_sum === sum) && (expected_cout === cout)))
    begin
      error_count = error_count + 1;
      $display("ERROR: sum = %0b, cout = %0b at time %0t.",
             sum, cout, $time);
    end  // of if
  end  // of outermost begin for error checking task
  endtask

endmodule
////////////////////////////////////////////////////////////////////
```

## 4.3   Sequential Logic Synthesis

The behavior and output of a sequential logic block is dependent on the values of the past inputs as well as the current inputs. The past input information is recorded in the form of state variables stored in memory storage elements, typically latches or flip-flops. A commonly used convention is to refer to level-triggered devices (whose state can change only on a specific level of the enable input) as *latches* and edge-triggered devices (whose state can change only when the clock input transitions from 0 to 1 (for positive edge-triggered devices) or from 1 to 0 (for negative edge-triggered devices)) as *flip-flops*. As in combinational logic, timing delays, which are technology and process dependent, should be avoided in the description of sequential logic that is intended to be synthesized.

There are a variety of methods for describing sequential logic structures in Verilog. The same type of structure (e.g., a D latch) can be described using a variety of Verilog statements (e.g., structural Verilog using the **nand** primitive gate, a conditional assignment statement within an **always** block, etc.). In the following, a style of Verilog description that is simple to understand, generally applicable, and widely used is presented as the recommended method for the description of most types of sequential-logic structures.

### Using Always Blocks for Sequential-Logic Synthesis

Although not impossible, continuous-assignment statements typically are *not* used to describe sequential logic structures. Instead, even the simplest sequential logic structures are described within **always** blocks. Within an **always** block, the signal on the left-hand side of an assignment statement *must* be of type **reg**. Also, within the same Verilog **module**, values must *not* be assigned to a single Verilog signal in two or more different **always** blocks (i.e., assignments to a single signal must all be made within a single **always** block, and not distributed among multiple **always** blocks).  Although such assignments are not illegal in Verilog, they typically are *not* supported by commercial synthesis tools.

Within the **always** block describing a sequential circuit, nonblocking assignment statements (which use the `<=` operator) or blocking assignment statements (which use the `=` operator) can be used to assign values to flip-flop or latch outputs.  As stated in Chapter 3, although nonblocking assignment models the behavior of actual hardware more closely than blocking assignment, the latter may be more usable when describing the desired behavior of a circuit in a high-level manner.

When describing a sequential circuit using an **always** block, an *event control* clause should be used to control *when* the **always** block is activated.  An event control clause is a Verilog `@(sig)`, `@(sig1 or sig2)`, or `@(posedge sig1 or negedge sig2)` type of statement.  When an event control clause is encountered, the simulator does not progress beyond the event control point until there is a change in one of the signals listed in the event control clause.  It is possible to include *event control* clauses in the declaration of the **always** block (as in **always** `@(event)` **begin**) or within the body of the **always** block.

If there are no event control clauses in an **always** block, then that **always** block executes continuously and repetitively.  The following shows the definition of this type of unrecommended **always** block with a blocking assignment (which describes a D latch).

```
always begin        // BAD! -- no event control clause
  if (EN) q = D;  // blocking assignment
end
```

This type of **always** block tends to make the simulation of the module inefficient, as this **always** block (with no event control clause) tends to (unnecessarily) hog all of the CPU resources.  Thus, the use of event control clauses is recommended for all **always** block definitions.  The following shows another definition of a D latch that conforms to this recommendation and uses nonblocking assignment.

```
always @(EN or D) begin  // BETTER -- uses event control
  if (EN) q <= D;        // nonblocking assignment
end
```

In the above, note that a latch is defined by using a conditional assignment statement without an **else** clause.  This is an important point as it shows that incompletely specified conditional assignment statements result in latches being created.  Thus, if a latch is *not* desired in a conditional assignment statement, then an **else** clause *must* be used even if it is not required by the logic being implemented (e.g., `c <= a ? b : 1'bx;` could be used, where `1'bx` denotes a 1-bit don't-care value).  This applies to any type of conditional assignment statement (if-else, case, etc.).  If the **else** clause is not specified, then the default behavior is to assume that the previous value *must* be retained.

Flip-flops should be described using **always** blocks with event control clauses in the **always** block declarations.  Although it is possible to use event control clauses in the body of an **always** block, this is not recommended, since liberal use of this type of method can lead to unsynthesizable or otherwise problematic Verilog code. Instead, all sequential logic elements within a single **always** block should be made sensitive to a single edge of a single clock signal. Thus, a D flip and a J-K flip-flop, both of which trigger off of the positive edge of a common clk clock signal, can be described as follows.

```verilog
always @(negedge reset_n or posedge clk) begin
   if (~reset_n) begin // asynchronous active-low reset
     q1 <= 1'b0;
     q2 <= 1'b0;
   end
   else begin              // posedge clk by default
     q1 <= D;
     case ({J, K})
        2'b01: q2 <= 1'b0;
        2'b10: q2 <= 1'b1;
        2'b11: q2 <= ~q2;
        default: q2 <= q2;
     endcase
   end  // of else
 end  // of always
```

In many cases, it is desirable to use flip-flops with asynchronous reset or preset inputs. Such asynchronous inputs are useful in initializing the flip-flops to known states at the beginning of circuit operation. Asynchronous inputs can be modeled in several ways. However, a recommended method is to include the asynchronous input in the event control clause along with the clock input. The above code shows an example of the use of an active-low asynchronous reset signal. In some implementation technologies, only one type of asynchronous input (either an asynchronous reset or an asynchronous preset) is permitted for a single flip-flop (e.g., this is the case with most Xilinx FPGAs). Also, an event control clause with both a positive edge of a clock signal and a negative edge of the same clock signal (as in @(**posedge** clk or **negedge** clk)) typically is *not* permitted in synthesis. As an example of the recommended method for describing general sequential circuits, let us consider the example Moore machine (module simple) introduced in Section 3.4.2 and reproduced here.

```verilog
//////////////////////////////////////////////////////////////////
// MODULE: State Machine Example: simple.v
// Author: Sunggu Lee
// Date: ...
// Description: Implements a simple state machine.  This state machine
//    transitions among three states depending on the value of the input
//    variable x.  The output z is set to 1 in state S2 and 0 in all
//    other states.  As such, the state machine is a Moore machine.

// DEFINITIONS
`define STATE_BITS 2
```

```verilog
`define S0 2'b00
`define S1 2'b01
`define S2 2'b11

// MODULE DECLARATION
module simple (z, state, reset_n, clk, x);

  output z;            // data output
  output [`STATE_BITS-1:0] state; // current state of state machine
  input reset_n;      // active-low RESET signal
  input clk;          // clock signal
  input x;             // the single data input

  // SIGNAL DECLARATIONS
  reg z;
  reg [`STATE_BITS-1:0] state;

  // Compute output value based on current state
  always @(state) begin
     case (state)
       `S0:  begin
             z <= 1'b0;
         end // of case (state == `S0)
       `S1:  begin
             z <= 1'b0;
         end // of case (state == `S1)
       `S2:  begin
             z <= 1'b1;
         end // of case (state == `S2)
       default: begin // desirable for all states to be covered
             z <= 1'bX;
         end // of case (state == default)
      endcase // of case (state)
    end // of always @(state)

  // Compute next state based on current state and input values
  always @(negedge reset_n or posedge clk) begin
    if (~reset_n)
      state <= `S0;
```

```
    else begin
      case (state)
        'S0:  begin
               if (x)
                  state <= 'S1;
               else
                  state <= 'S2;
          end // of case (state == 'S0)
        'S1:  begin
               if (!x)
                  state <= 'S2;
          end // of case (state == 'S1)
        'S2:  begin
                  state <= 'S0;
          end // of case (state == 'S2)
        default: begin // desirable for all states to be covered
                  state <= 2'bXX;
          end // of case (state == default)
      endcase // of case (state)
    end // of else
  end // of always loop for computing next state


endmodule
//////////////////////////////////////////////////////////////////
```

The Verilog code for the `simple` module follows all of the recommendations for the description of easily synthesizable code. An asynchronous active-low reset signal and a positive edge-triggered clock signal are included in the event control clause used with the second **always** block, which describes the next state transition logic. A default case is used in the case statement within the second **always** block in order to prevent the creation of an unintended latch. The first **always** block is activated whenever there are changes in the `state` value. All assignment statements used are nonblocking assignments. Although either blocking or nonblocking assignments can be used in Verilog, it is recommended that only one type (either all blocking assignments or all nonblocking assignments) be used in a single **always** block. This is because using a mixture of blocking and nonblocking assignments can make that module difficult to debug; in certain cases, such modules also can be rejected by some synthesis tools.

Figure 4.5 shows the circuit that is synthesized for the `simple` module shown above. There are several interesting features of this synthesized circuit. The `state_reg<0>` and `state_reg<1>` blocks correspond to enabled D flip-flops. Although there are two asynchronous clear inputs (`GC` and `CLR`), the `CLR` input is tied to ground, as it is not used. The `reset_n` input is tied to the `GC` active-high asynchronous clear inputs through a buffer and an inverter. Interestingly, the `GC` inputs of both flip-flops also are connected to the `GSR` output of a block labeled `startup`.

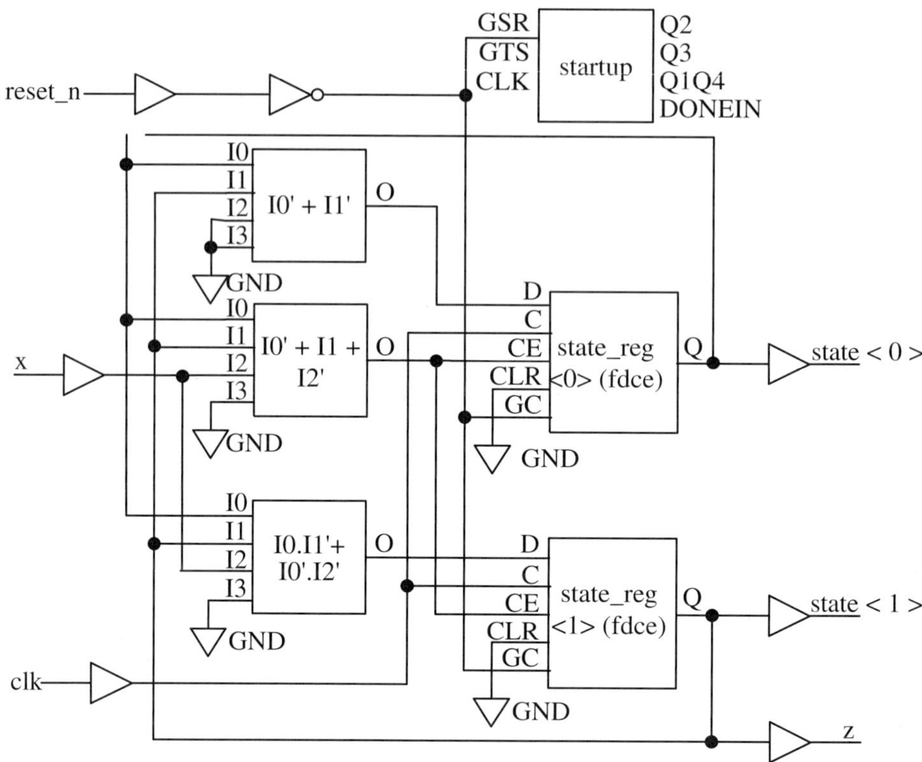

**Figure 4.5**
The logic synthesized for the simple Verilog module (Moore machine).

The startup block, which is created by the synthesis software for the Xilinx target-implementation chip, serves several functions, one of which is to reserve a "wide" globally routed wire for the global-reset signal—this "wide" wire can be used by connecting to the GSR output of the startup block. There are several such "wide" globally routed wires, some with special functions and routing characteristics. For example, the clock inputs of all flip-flops (the C input of the state_reg blocks) are connected typically to a "wide" globally routed wire (referred to as a clock line) in which the routing is performed such that the signal delays from the external clock input to all flip-flop clock inputs are made approximately equal (leading to small clock skews).

## 4.4 Synthesis Heuristics

Synthesis tools, in general, have a certain amount of intelligence and thus can *infer* certain circuit structures from common behavioral description patterns. When using Verilog description with a synthesis tool, the user must take advantage of this capability. By using as high a level of description as is accepted by the synthesis tool, the user can create highly efficient and less error-prone circuits, since most reputable automation tools create highly efficient and error-free circuits for the target architecture specified. Currently, this means that the user should, in general, use RTL Verilog code that conforms to certain guidelines for synthesizable code.

The user must write his/her Verilog circuit description using standard easily synthesizable language constructs. In order to write Verilog code that can be synthesized by most synthesis tools, several heuristic rules should be followed. These rules are based on observations on what language constructs can and cannot be mapped easily to actual hardware components. They are as follows.

1. Delay statements (excluding synchronized delays) cannot be synthesized since it is extremely difficult to construct circuits with exact prespecified delays, due to the fact that manufacturing and process variations result in circuits with varying delay characteristics even if they use the exact same design. Thus, statements using the Verilog # notation cannot be synthesized. Some synthesis tools ignore all such delay statements, while others generate compilation errors. However, it sometimes is necessary to include delay statements in order to be able to simulate a circuit properly. In this case, a general technique that can be used is to define a delay constant at the top of the code (such as **`define** DEL 1) and then use that delay constant throughout the code (as in A <= #`DEL B). The delay constant can be defined as a positive value for the simulation and then redefined as a zero value for synthesis. However, all circuit components, once synthesized, will have *some* delay in the actual circuit (since no signal can travel faster than the speed of light). Note that delay statements in the test bench are inconsequential, as the test bench does not need to be synthesized (we just need to synthesize the circuit that the test bench is testing).

2. **Initial** blocks cannot be synthesized and thus should be avoided. An **initial** block specifies a process that executes only once at the beginning of the simulation. However, it is difficult to design logic circuits that execute only once. It is much more natural for a logic circuit to execute infinitely. For instance, a NAND gate executes infinitely, waiting for changes in its input values. If an operation needs to be performed just once at the beginning of the simulation, then it should be placed in the sequence of actions that take place upon the activation of a reset signal. Thus, an **always** block activated by a transition in the reset signal (where the required operation is executed when the reset is asserted) can be used. An example of this is shown with the asynchronous reset operation in the simple Moore machine circuit of Section 4.3.

3. Full **case** statements, instead of **if** ... **else if** statements, should be used for all situations with more than about two cases. This is because **case** statements synthesize naturally into multiplexers or multiplexer-equivalent structures, which are fast and have a low area overhead. However, because all input values need to be specified for a multiplexer, all cases should be included in the **case** statement used. A good technique is to use the **default** case, and then assign don't-care values to the outputs for that case (as was done in module simple). Given don't-care (1'bx or 1'bX) outputs, the logic can be simplified by the synthesis tool, possibly leading to a more efficient circuit. On a related note, in simulation, even if all cases are specified, a **default** case may be necessary. This is because signal values

can acquire the values 'X' and 'Z,' in addition to '0' and '1.' However, this is not a concern with synthesis, since signals in all actual digital circuits must be interpreted as '0' or '1' (we simply may not know *which* value it will acquire in certain situations).

4. Care should be taken when using conditional assignment statements. This is because a conditional assignment statement with no **else** clause typically implies a *latch* (in which the output signal retains its previous value if the condition being checked is not satisfied). If we don't want a latch to be created, then **else** clauses should be included in all conditional assignment statements, even if we don't care about the value assigned during the *else* clause (e.g., a don't-care value, 1'bx, can be assigned to the target signal in the **else** clause).

5. With some synthesis tools, it is possible to specify *synthesis hints* that can be used by the synthesis tool to produce more efficient circuits. For instance, when using Synopsys FPGA Express, if `// synopsys full case` is included as a comment in a Verilog **case** definition, then the default case (with assignment of don't-care values in the default case) does not need to be specified. The synthesis tool knows that don't-care values should be assigned for any unspecified cases. In a similar manner, if `// synopsys parallel case` is included as a comment in a Verilog **case** definition, then the synthesis tool knows that all cases should be considered to be mutually exclusive (i.e., the overlap between any two cases can be ignored, and optimizations can be performed in the assignment of logic at the inputs to the multiplexer that is created). Since synthesis hints are purely optional, they can be omitted for default (more portable) implementations.

6. For sequential circuits, event control clauses should be used in **always** block declarations, and either all nonblocking assignments or all blocking assignments should be used within the bodies of **always** blocks. Event control clauses should be used to limit when an **always** block is executed. Both positive and negative edges of the same signal (such as a clock signal) should *not* be checked within a single event control clause, as such a structure is not easily synthesized. Given a single **reg** signal, assignments to that signal only should be made within a single **always** block in order to preclude the possibility of conflicts in assigning values to that **reg** signal.

7. Directives, **$display** statements, and comments should be used liberally, as they are useful in debugging and do not adversely affect the synthesis results. In addition to the liberal use of comments, **$display** statements (showing the values of important signals at critical points in the code) can significantly aid the user during code development. All comment and **$display** statements are ignored by the logic synthesis tool. All constants should be defined using **`define** statements at the top of the Verilog code file before being used in the rest of the code. Global changes to constant values are simplified by using this practice. Also, if meaningful names are used for these constants, then it becomes easier to read and debug the Verilog source code. Although it is possible to use **parameter** statements to define constants within Verilog modules, the **`define** directive method is more desirable in most cases (in the author's opinion) because all constant definitions can be placed together at the top of the file, and because constants in the code

can be identified easily (as being constants instead of variables or signals) by the presence of the backquote (') in front of the constant name.

8. Finally, an important rule of thumb is to inspect the synthesized circuit whenever possible. Post-synthesis simulation should also be performed if possible. This type of verification is especially important for first-time users of a given synthesis tool. Synthesis tools, like other automation tools intended to make an engineer's task easier, can produce errors or results that are correct but not what the designer intended. Most synthesis tools provide some method for the designer to be able to view or otherwise analyze the synthesized-logic circuit. If the circuit is too large to inspect completely, the designer can still inspect parts of the circuit in a hierarchical manner. Common mistakes that can be found by this type of visual inspection are the presence of unintended latches, logic gates that have been "optimized out" (when they should not have been), and flip-flops of the incorrect types (e.g., synchronized reset instead of asynchronous reset, etc.).

These techniques should be sufficient for the design of most types of digital logic circuits. For a more detailed study of techniques to create high-quality synthesized circuits, the interested reader is referred to books specifically devoted to Verilog and synthesis (some of which are listed at the end of this chapter).

● ● ● ● ● ● ● ● ● ● ● ● ● ● ● ● ● ● ●

## 4.5  Synthesis using a Commercial Tool

In this section, let us present the synthesis procedure and some examples of simple combinational and sequential circuits produced by a commercial synthesis tool: Synopsys FPGA Express. To use the capabilities of a synthesis tool fully, it is necessary to create several script files and follow a fairly complicated design process described in books such as [Bhatnagar 1999]. However, with the emergence of PC-based CAD tools such as the Xilinx ISE series, it now is possible to use a menu-driven approach to create synthesized circuits using the default settings (changes also can be made to the synthesis settings by using the "synthesis options" menu). With this type of approach, even the beginning user can create a fairly complex circuit (to be programmed into an FPGA) by describing the desired circuit using a behavioral Verilog description, and then using the synthesis tool (with the default settings) to synthesize the desired circuit.

**Example 4.1**   **Combinational Logic**

Let us synthesize the Verilog module `full_sub_behave`, the behavioral Verilog description for the full subtracter of Example 3.1. The main FPGA software used for this example is the Project Navigator in the Xilinx ISE Version 6.2.03i software package. The synthesis tool used is *XST*, which is a part of the Xilinx ISE package. The reader most likely will be using a different set of software tools. Thus, the procedure described here is simply a representative example of the FPGA synthesis process.

An initial series of simple steps is used to start up the Xilinx ISE Project Navigator tool, create a new project, and insert the code to be synthesized. In this example, the new project has been named `full_sub`, and a file with the behavioral Verilog code for

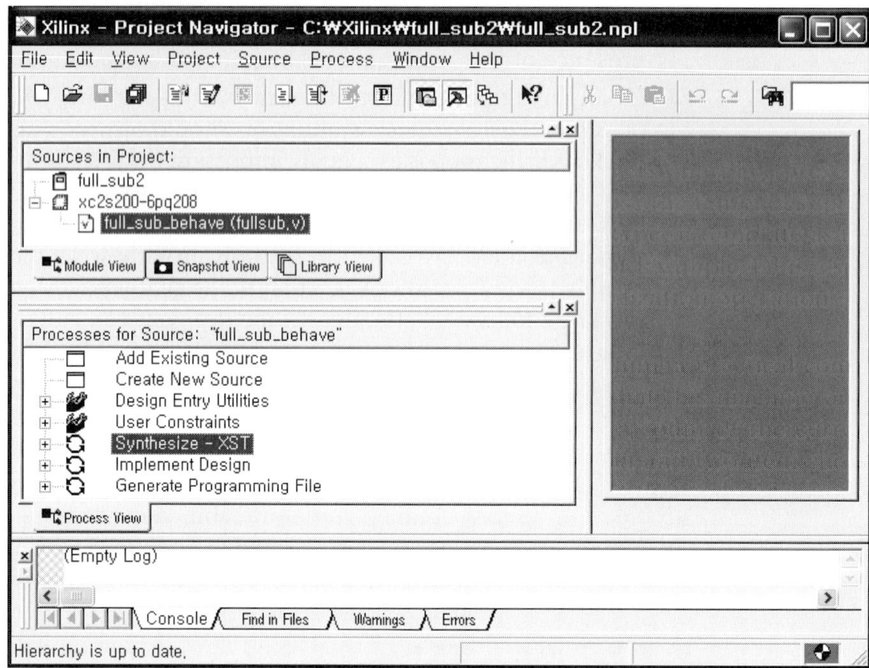

**Figure 4.6**
The "Project Navigator" window for the full-subtracter project.

the full subtracter has been "added" to this project. After these steps, the window that is created is shown in Figure 4.6. When starting up this project, the Xilinx Spartan II series FPGA chip XC2S200-6PQ208 was chosen as the synthesis target. This is shown in the upper-left subwindow of the figure.

Next, the synthesis step must be executed. This can be done by double-clicking the highlighted entry in the middle-left subwindow of Figure 4.6. Prior to this step, it is also to customize the synthesis operation in order to add timing constraints, add area constraints, use fixed-user I/O pad locations, etc., by executing one of the entries under the "User Constraints" tab in the middle-left subwindow. However, simply by executing the highlighted "Synthesize - XST" tab, we can use the default synthesis parameters.

After the synthesis is complete, it is possible to view the post-synthesis report and the synthesized circuit. After opening up the "Synthesis - XST" tab, double-clicking the "View Synthesis Report" tab results in a text form of the synthesis report being displayed in the right subwindow. This report has several interesting pieces of information, most of which can be understood at an intuitive level. In particular, the middle portion of the report displays the information shown in Figure 4.7. From this information, we can infer that this design was implemented using five I/O blocks (three inputs, two outputs) and two LUTs (look-up tables). Of the 2,352 programmable "slices" available in the XC2S200, this design used only one. Of the 144 user I/O pins available in the XC2S200, this design used only five.

A schematic view of the synthesized circuit can be viewed by double-clicking the "View RTL Schematic" tab. The initial schematic view shown is the top-level

```
149  ===============================================================================
150  *                                 Final Report
151  ===============================================================================
152  Final Results
153  RTL Top Level Output File Name      : full_sub_behave.ngr
154  Top Level Output File Name          : full_sub_behave
155  Output Format                       : NGC
156  Optimization Goal                   : Speed
157  Keep Hierarchy                      : NO
158
159  Design Statistics
160  # IOs                                                 : 5
161
162  Macro Statistics :
163  # Xors                                                : 2
164  #           1-bit xor3                                : 2
165
166  Cell Usage :
167  # BELS                                                : 2
168  #           LUT3                                      : 2
169  # IO Buffers                                          : 5
170  #           IBUF                                      : 3
171  #           OBUF                                      : 2
172  ===============================================================================
173
174  Device utilization summary:
175  ---------------------------------
176
177  Selected Device : 2s200pq208-6
178
179  Number of Slices:                         1  out of    2352    0%
180  Number of 4 input LUTs:                   2  out of    4704    0%
181  Number of bonded IOBs:                    5  out of     144    3%
182
183
184  ===============================================================================
185  TIMING REPORT
186
```

full_sub_b...

**Figure 4.7**
Part of the post-synthesis report for the full-subtracter circuit.

input-output symbol diagram for this circuit. By double-clicking on this symbol, it is possible to view the lower-level blocks that this circuit is composed of. As shown in Figure 4.8, this circuit is composed of two Msub blocks connected in a cascaded manner. Lower-level views of these Msub blocks can be observed by double-clicking on them. At the lowest level, logic-gate equivalent circuits can be viewed for each

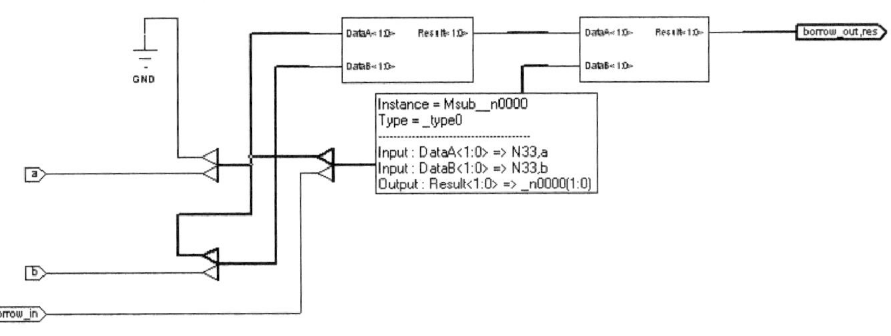


**Figure 4.8**
A schematic view of the circuit synthesized for the full subtracter.

sub-block. Upon analyzing these logic-gate circuits, it can be observed that the following equations are being implemented.

$$borrow\_out = (a + \overline{b}) \oplus ( (a \oplus b) + \overline{borrow\_in}),$$

$$res = a \oplus b \oplus borrow\_in$$

The logic synthesized for the full-subtracter circuit may at first seem to be a bit odd. Although the equation for *res* is equivalent to that derived earlier, the equation for *borrow_out* appears to be quite a bit different. However, it can be verified, using a simple truth table, that this equation is correct and equivalent to the previously derived equation for *borrow_out*. The reason for the use of this convoluted method for deriving *borrow_out* is that this logic circuit attempts to utilize the existing circuitry in the target architecture as efficiently as possible. Although the logic derived by the synthesis software may at times seem odd, it still is acceptable as long as the resulting logic is equivalent to a logic circuit that implements the desired circuit correctly.

Although the synthesis process is complete at this stage, it still is edifying to continue the process a bit further. Referring to Figure 4.6, the "Implement Design" tab (which follows the "Synthesize - XST" tab) is used to map the synthesized circuit to the components available in the target architecture, in this case the configurable blocks present in the XC2S200 FPGA chip. Then the final "Generate Programming File" tab can be used to generate a configuration file (with a ".bit" extension), which is the file that actually is downloaded to the target FPGA chip. After the "Implement Design" phase, it is possible to view exactly *how* the synthesized circuit has been mapped to the target FPGA chip. Figure 4.9 shows this view for the full-subtracter circuit mapped to an XC2S200. This view was generated by double-clicking the

**Figure 4.9**
The "chip layout" view of the full-subtracter circuit mapped onto an XC2S200 FPGA.

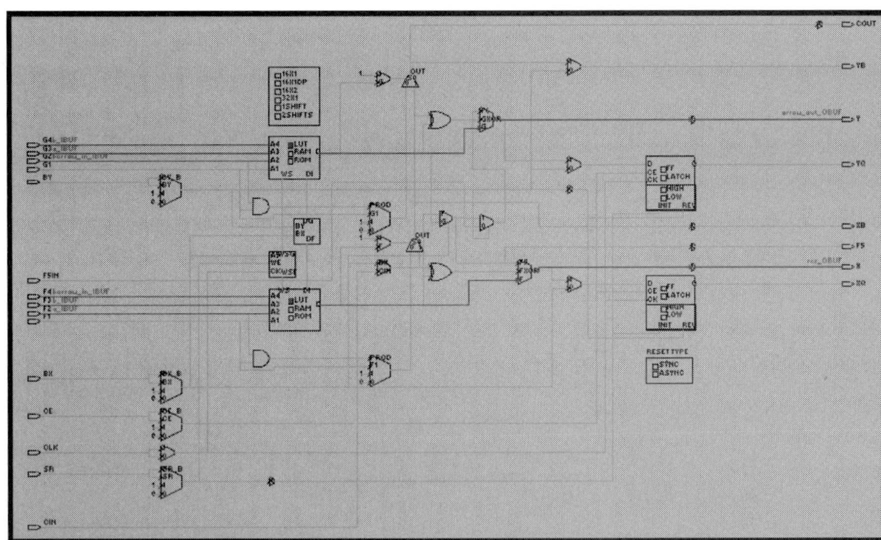

**Figure 4.10**
The configurable logic circuitry used to implement a full-subtracter circuit on an XC2S200 FPGA.

"Implement Design → Place & Route → View/Edit Routed Design (FPGA Editor)" tab. From this figure, it can be seen that there is a very large matrix of programmable logic blocks, of which the full-subtracter circuit has occupied a very small section in the lower-left hand corner of the chip. Note that it is also possible to view lower-level circuit details and to *edit* the mapping from within this tool (FPGA Editor). Figure 4.10 shows the internal circuitry of the highlighted block in Figure 4.9.

The above example shows how a combinational logic circuit can be synthesized from a behavioral Verilog description. Sequential circuits can be synthesized in a similar manner. The following example demonstrates how latches, flip-flops, registers, and simple "common" sequential circuits can be synthesized from a Verilog description.

**Example 4.2**    ### Simple Synchronous Sequential Circuits

Let us use the same sequence of steps discussed in Example 4.1 to synthesize the circuit described by the example Verilog code shown here. This Verilog code illustrates several commonly used constructs and the devices that those constructs typically result in. The circuit synthesized from this code is shown in Figure 4.11.

```
/////////////////////////////////////////////////////////////////////
// MODULE: Sequential Circuit Synthesis Example: seq1.v
// Author: Sunggu Lee
// Date: ...
// Description: This circuit doesn't perform any useful function.
//    It simply includes several constructs commonly used in
//    Verilog code.  By synthesizing this circuit, the user can
//    examine the devices that result from the statements
//    included in this example.
```

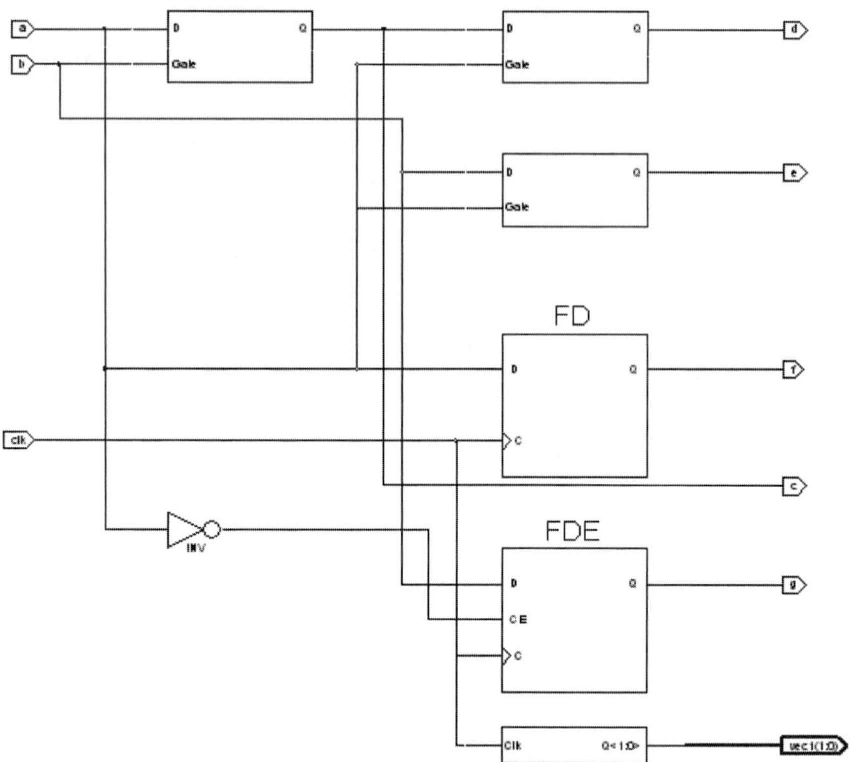


**Figure 4.11**
The optimized synthesized circuit for module seq1 in Example 4.2.

```verilog
// DEFINITIONS
`define VEC_SIZE 2

// MODULE DECLARATION
module seq1 (vec1, c, d, e, f, g, clk, a, b);

   output [`VEC_SIZE-1:0] vec1;      // bit vector output
   output c, d, e, f, g;            // single bit outputs
   input clk;                        // clock signal
   input a, b;                       // data inputs

   // SIGNAL DECLARATIONS
   reg c, d, e, f, g;                // connects to the outputs
   reg [`VEC_SIZE-1:0] vec1;         // bit vector output
   reg [`VEC_SIZE-1:0] vec2;         // other bit vectors
   wire [`VEC_SIZE-1:0] vec3;
```

```verilog
  // Latch example 1
  always @(a or b) begin
    if (b)
      c <= a;
  end

  // Latch example 2
  always @(a or b or c) begin
    if (a)
      d <= c;
    else
      e <= b;
  end

  // Latch example 3
  always @(a or vec3) begin
    if (~a)
      vec2 <= ~vec3;
  end

  // Flip-flop examples
  always @(posedge clk) begin
    f <= a;                   // using nonblocking assignment
    vec1 <= vec1 + 1;         // counter
    case (a)
      1'b0: g <= b;
    endcase                   // note that case 1'b1 is not covered
  end

endmodule
//////////////////////////////////////////////////////////////////////
```

The following observations can be made from the above Verilog description for module `seq1` and the resulting synthesized circuit shown in Figure 4.11.  Latches are inferred from "incomplete" conditional assignment statements.  ("Complete" conditional-assignment statements result in combinational-logic circuits.)  Such conditional-assignment statements can include **if-else** constructs and **case** constructs.  A conditional-assignment statement is considered "incomplete" if not all possibilities are covered in the condition-check statements, or if signals which are assigned values are not assigned values under all possible conditions.  Thus, for example, the sections of code labeled `Latch example  1` and `Latch example 2` result in latches.  In the first latch example, the condition **if** (`!b`) is not covered since there is no **else** clause.  In the second latch example, there is an **else** clause but values are assigned to different signals in the **if** (`a`) and **else** clauses.  Latch example 2 results in two different latches being created, as can be seen in Figure 4.11.

Latch example 3 does not result in the synthesis of any circuit devices. This is because the output vec2 is an internal signal value that is not used to generate any output signal value—thus, this section of code is redundant, and is optimized out by the synthesis tool. In a like manner, any logic that is deemed to be redundant may be optimized out by the synthesis tool. Even a structural Verilog description may not result in the same structural connections (as specified in the Verilog code) once it is synthesized. If a specific structure *must* be retained, then synthesis options must be used to disable "flattening" of the hierarchy, submodules must be created (in order to create an artificial hierarchy), and "Don't Touch" options must be checked for those submodules that must not be optimized.

Next, the **always @(posedge** clk) block is used to create positive edge-triggered flip-flops. The blocks labeled FD and FDE in Figure 4.11 are edge-triggered D flip-flops and enabled edge-triggered D flip-flops, respectively. Since this code only is activated on the positive edge of the clk clock signal, a simple assignment statement (such as f <= a) results in the creation of a D flip-flop. Note that blocking assignments can also be used to create flip-flops (instead of the nonblocking assignments used in this **always** block). This block also shows an example of how a common synchronous sequential circuit can be synthesized by the synthesis tool. Simply by entering the statement vec1 <= vec1 + 1, all of the combinational logic and flip-flops necessary for a `VEC_SIZE bit binary counter can be created via synthesis. The counter circuitry can be examined by delving down into the lower hierarchy of the lowermost block in Figure 4.11. In a like manner, many common functions (e.g., subtraction, multiplication, etc.) and other RTL (register-transfer level) statements can be synthesized into complete circuits by the synthesis tool. Finally, the **case** block within this **always** block results in a D flip-flop with an enable input, as shown in the synthesized circuit of Figure 4.11. If this type of incompletely specified **case** block is included within an **always begin** block (with no activation condition specified), then it results in a latch similar to "Latch example 1".

---

## 4.6  High-Level Verilog Coding

Synthesis tools have matured significantly over the past few years. Although they have not reached the ideal of synthesis from a natural-language circuit description, current synthesis tools can create extremely area-efficient and/or fast logic gate level circuits from high-level descriptions written in RTL (register-transfer level) notation. When targeted for specific architectures (such as Xilinx FPGAs), synthesis tools can make maximum use of the logic blocks and circuit structure of the target architectures. Unless the designer is intimately familiar with the target architecture, the use of synthesis tools typically can result in *better* circuits than manually designed (to the logic gate or transistor level) circuits.

Due to the above facts, in today's design environment, the recommended hardware design method is to simply write a high-level Verilog program for the target circuit, and then use a good synthesis tool to produce the netlist for the target implementation platform. In order to produce circuits that are optimized for the target implementation platform, the designer should use higher rather than lower-level descriptions of the desired circuit behavior—the synthesis tool has more freedom in producing efficient circuits if a higher-level description method is used. For example, if the designer writes a structural Verilog description based on NAND gates or uses a sequence of logic equations to describe the desired circuit, then the synthesis tool can only optimize out redundant logic gates and use simplified logic equations. However, if the designer writes the desired *behavior* of the same circuit, it may be possible for the synthesis tool to produce an entirely different logic design that is more efficient than the logic design originally considered by the designer.

Writing a synthesizable high-level Verilog program is not difficult. The designer can write his/her program in a manner exactly analogous to a C language program for solving the given circuit problem. However, a few changes in the basic program structure may be necessary due to the fact that a Verilog program is describing *hardware*, which has an inherently parallel execution behavior. Also, adherence to the synthesis heuristics listed in Section 4.4 is helpful. In general, though, a synthesizable RTL Verilog program can be based on the same algorithm used in the corresponding C language program. A useful method to use, especially for complex circuits, is to first write and test a C language program for solving the given problem. Then the corresponding Verilog program can be written based on the same algorithm.

**Example 4.3**   **Sequence Detector**

Suppose we wish to design a circuit to detect the serial data-input sequence 01111110 (note that this sequence commonly is used as an end-of-packet marker). Such a sequence detector (or sequence recognizer) circuit could be designed as a Mealy-type finite state machine using the approach presented in Section 3.4.2. However, as shown in this example, it can also be designed in a more direct (behavioral) manner using the following type of intuitive algorithm for this problem.

*Sequence Detector Algorithm (for 01111110 sequence)*:

Step 1.    $prev\_data \leftarrow 11111111$; / * stores 8 previous data bits * /
Step 2.    while (true) do/ * repeat forever * /
$\qquad data\_in \leftarrow next\ data\ input$;
$\qquad prev\_data \leftarrow prev\_data << 1$; / * shift left * /
$\qquad prev\_data[0] \leftarrow data\_in$; / * lsb gets data_in * /
$\qquad$ if ( $prev\_data == 01111110$) then
$\qquad\qquad detected \leftarrow 1$; / * sequence detected * /
$\qquad$ else
$\qquad\qquad detected \leftarrow 0$; / * sequence no longer detected * /
end while

The Verilog code for this sequence detector can be written based on the algorithm. Let us write and analyze two versions: one using blocking assignments and the other using nonblocking assignments. The following Verilog code uses blocking assignments and is almost identical to the sequence-detector algorithm.

```verilog
// Verilog Code Version 1 (using blocking assignments)
module seq_det_1 (detected, reset_n, clk, data_in);
   output detected;   // desired pattern detected?
   input reset_n, clk, data_in;
   reg detected;       // redefined as type reg
   reg [7:0] prev_data;   // used to store previous data bits

   always @(negedge reset_n or posedge clk) begin
      if (~reset_n)
         prev_data = 8'b1111_1111;   // _ is just a separator
      else begin   // on posedge clk
         prev_data = prev_data << 1;
         prev_data[0] = data_in;
         if (prev_data == 8'b0111_1110)
            detected = 1;
         else
            detected = 0;
      end   // of else (posedge clk)
   end   // of main always block
endmodule
```

The only differences in the Version 1 Verilog code (from the algorithm description) are the use of Verilog syntax and the deletion of the assignment of the next input data to data_in. The assignment to data_in is deleted in the Verilog code, since there is an implicit assumption that data bits are entered in synch with the positive edge of the clock signal clk.

Several interesting facts about the Version 1 Verilog code can be observed by synthesizing this code. Figure 4.12 shows the logic that results from the synthesis of the Version 1 Verilog code using Synopsys FPGA Express. There is a flip-flop that is used to hold the detected value. There is a shift register used to hold prev_data values as expected, since a shift-left operation is performed on the 8-bit prev_data **reg** signal. However, notice that there are only seven flip-flops in the shift register! The reason that there are only seven flip-flops (and not eight) is that the synthesis tool has performed some *optimization* on the Version_1 Verilog code. From the code, note that the only output of this circuit is the signal detected. Any logic that does not affect this output signal is redundant and thus can be deleted. In this case, note that the D input to the detected flip-flop is formed from the AND of the data_in', prev_data[0] through prev_data[5], and prev_data[6]. Thus, it can be seen that the prev_data[7] bit, which is shifted out during the shift-left operation, is not needed in computing the value of the detected signal.

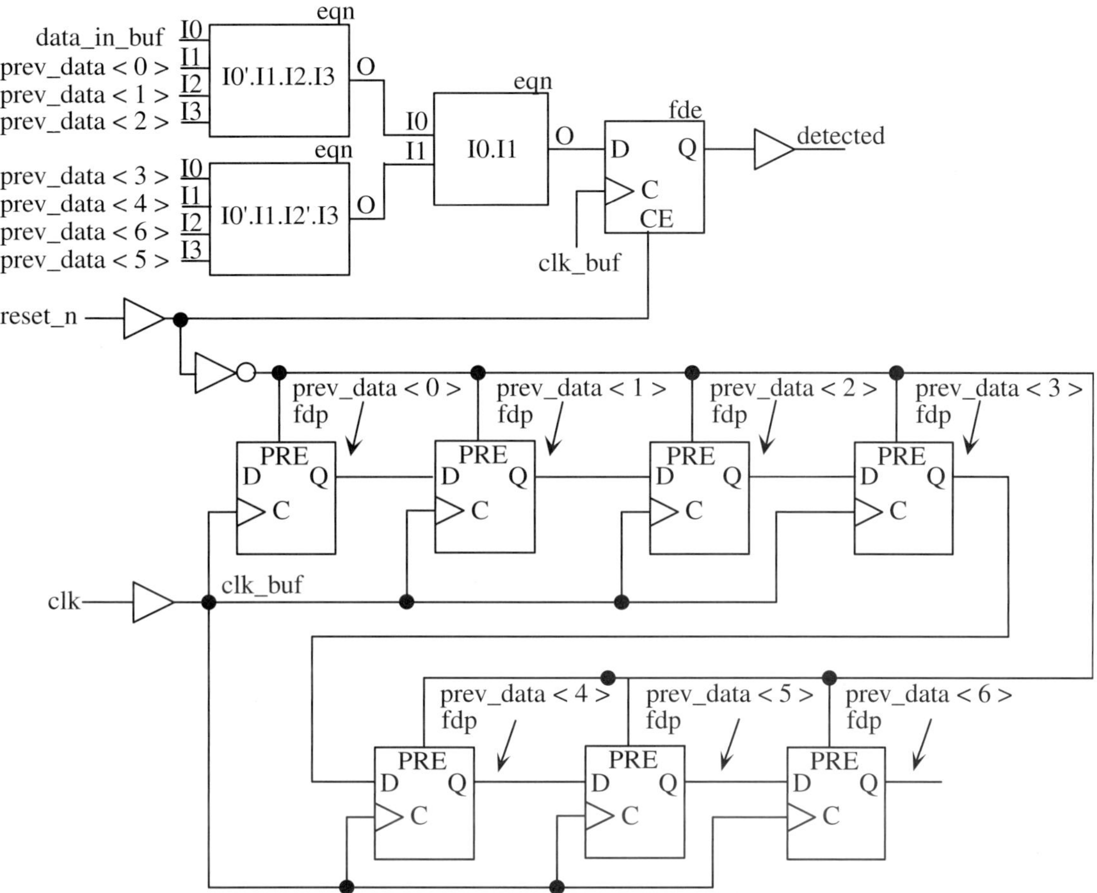



**Figure 4.12**   The optimized synthesized circuit for module seq_det_1 in Example 4.3.

With the above synthesis result in mind, the Verilog code for this sequence detector can also be written using nonblocking-assignment statements, as shown here.

```verilog
// Verilog Code Version 2 (using nonblocking assignments)
module seq_det_2 (detected, reset_n, clk, data_in);
   output detected;  // desired sequence detected?
   input reset_n, clk, data_in;
   reg detected;      // redefined as a type reg
   reg [7:0] prev_data;  // stores previous data bits

   always @(negedge reset_n or posedge clk) begin
      if (~reset_n)
         prev_data <= 8'b1111_1111;  // _ is just a separator
```

```verilog
        else begin   // on posedge clk
            prev_data <= {prev_data[6:0], data_in};
            if ({prev_data[6:0], data_in} == 8'b0111_1110)
                detected <= 1;
            else
                detected <= 0;
        end   // of else (posedge clk)
    end   // of main always block
endmodule
```

At first glance, the Version 2 Verilog code looks significantly different from the algorithm description. However, when the Version 2 Verilog code is synthesized, the resulting logic circuit is exactly the same as that produced for the Version 1 code. This is a natural consequence of the fact that nonblocking assignment models the behavior of sequential circuits *better* (more closely) than blocking assignment (in which a series of operations are performed one after the other).

If the change to nonblocking assignment is taken into account, then it is possible to establish a correspondence between the Version 2 Verilog code and the algorithm description. The first statement after **else begin** denotes a left shift in which the data_in value is shifted into the 1sb position (since the curly brace denotes bit concatenation). Then since this statement is nonblocking, it is executed concurrently with the next statement, which checks for equality with 8'b0111_1110. Since the check is concurrent with the change in the prev_data signal, the prev_data value *before* the left shift must be checked. If we desire the detected signal to be output as soon as the last 0 of the 01111110 sequence is observed, instead of one clock cycle later, then the equality check *must* be conducted as shown.

The Version 1 and Version 2 Verilog code versions show two basic high-level Verilog coding styles. Version 1, which uses blocking assignment, can be considered to be a slightly higher-level description and thus is more desirable. Since hardware is modeled more naturally using nonblocking assignment, the Version 2 code may result in a closer match with the synthesized circuit. The choice of which coding style to use is up to the individual designer. However, it is *not* a good idea to use a mixture of blocking and nonblocking assignment statements in the same Verilog code, as this results in code that is difficult to debug and may not be accepted by some synthesis tools.

- - - - - - - - - - - - - - -

# 4.7  Chapter Review

- Synthesis refers to the automatic generation of a netlist from a high-level description of the desired operation of a circuit. The most general type of synthesis accepts input in the form of RTL Verilog code. RTL (register-transfer level) refers to a level of description in which all digital-logic circuits are described as a set of registers and movements of data between those registers, during which transformations (such as an addition or shift operation) can be applied to the data.

- Combinational logic structures should be described using continuous assignment statements or **always** blocks. Continuous assignment statements, which execute concurrently with all other top-level blocks and statements in a Verilog module, can be used to describe combinational structures involving multiplexers or logical or arithmetic computations. **Always** blocks can be used to describe more complicated combinational logic circuits. The **always** block should include all input signals of the combinational logic circuit in the event control clause for that **always** block.

- Sequential logic structures should be described using **always** blocks with event control clauses in the **always** block declarations (i.e., in the first line, before the first **begin**, of the **always** block). The event control clause should *not* include both the positive and negative edge of a single clock signal. Within a single Verilog **module**, assignments to a single signal must all be made within a single **always** block. Latches are created by **if-else** statements with no **else** clause or incomplete **case** statements.

- The combination of Verilog and a synthesis tool can be used to design and implement complex digital logic circuits in a short amount of time. However, since current synthesis tools do not support all of Verilog and produce varying quality circuits depending on the manner in which the Verilog code is written, certain heuristic rules should be followed when writing Verilog code that will be synthesized. Also, it is always a good idea to check and inspect the synthesized circuit (and if possible to perform post-synthesis simulation) to ensure that the circuit has been synthesized as the designer intended.

- Since most current synthesis tools produce highly efficient and fast circuits, it is recommended that extremely high-level Verilog code be used, thereby providing synthesis tools with the maximum flexibility to produce optimized circuits. Thus, given a hardware design problem, the recommended method is to first devise a high-level algorithm for solving the given problem. The algorithm even can be coded and tested as a regular C language program. Then, the high-level algorithm should be converted to a high-level Verilog description based on one of the **always** block templates shown in Section 4.6.

## 4.8  Resources

The home page for Aldec Inc. is http://www.aldec.com, which is a company that supplies a Verilog editor and simulator called Active-HDL. The home pages for Synopsys and Xilinx, are http://www.synopsys.com and http://www.xilinx.com, are a synthesis tool company and an FPGA company, respectively. [Hachtel and Somenzi 1996] provide a theoretical treatment of synthesis. [Bhatnagar 1999] provides a detailed guide to synthesis using Synopsys Design Compiler and another tool called

PrimeTime. [Cilleti 1999] and [Zeidman 1999] are useful guides to Verilog. For the authoritative references on the Verilog standard, the interested reader should refer to the IEEE standards publications [IEEE 1995] and the updated version [IEEE 2001].

### 4.8.1   Bibliography

[4.1] BHATNAGAR, H., *Advanced ASIC Chip Synthesis: Using Synopsys Design Compiler and Primetime*, Kluwer-Academic, 1999.

[4.2] CILETTI, M. D., *Modeling, Synthesis, and Rapid Prototyping with the Verilog HDL*, Prentice Hall, Upper Saddle River, NJ, 1999.

[4.3] HACHTEL, G. D., and SOMENZI, F., *Logic Synthesis and Verification Algorithms*, Kluwer Academic, Boston, 1996.

[4.4] IEEE, *IEEE Standard Hardware Description Language Based on the Verilog Hardware Description Language*, IEEE Press, New York, 1995. Also available through the IEEE standards web page at http://standards.ieee.org.

[4.5] IEEE, *IEEE Standard for Verilog Hardware Description Language 2001*, IEEE Press, New York, 2001. Also available through the IEEE standards web page at http://standards.ieee.org.

[4.6] ZEIDMAN, B., *Verilog Designer's Library*, Prentice Hall, Upper Saddle River, NJ, 1999.

[4.7] http://www.aldec.com, home page for ALDEC (Verilog tool supplier).

[4.8] http://www.synopsys.com, home page for SYNOPSYS (Synthesis tool supplier).

[4.9] http://www.xilinx.com, home page for XILINX (FPGA H/W and S/W vendor).

## 4.9  Problems

***P4.1.** In this chapter, it is stated that most synthesis tools support RTL synthesis, in which the circuit to be synthesized is written using behavioral descriptions written in RTL (register-transfer level) Verilog code. Thus, it is important to be able to describe different types of circuits as sequences of RTL operations. Consider a *thermostat* circuit, which is a circuit that adjusts the heater and air-conditioner levels so as to maintain the room temperature between a lower-bound and upper-bound temperature set by the user. Describe the behavior of the desired circuit using RTL-level Verilog code (i.e., devise an algorithm (consisting of a sequence of RTL operations) to describe the thermostat circuit, and then convert that algorithm into Verilog code).

**P4.2.** As a simpler variation of the above problem, create an RTL Verilog program to calculate and output all prime numbers up to the largest prime number with *n* bits. Check to make sure that your Verilog code is synthesizable. Also, the algorithm used must *not* simply output a sequence of pre-stored prime numbers. Instead, it must check each integer to determine if it is a prime number by dividing it by all previously found prime numbers.

**P4.3.** Create an RTL Verilog algorithm to implement the digital-logic control-circuit portion of a telephone answering machine.

**P4.4.** Describe an 8-bit comparator circuit (which has two 8-bit inputs and three 1-bit outputs named equal, greater, and less) using sequences of RTL Verilog continuous-assignment statements.

**P4.5.** Write a Verilog **always** block that implements one 8:1 MUX, one 3:8 decoder, and one 8-bit even parity checker.

**P4.6.** Write a Verilog **module** for the circuit of Problem P4.2, a prime number finding circuit.

**P4.7.** Section 4.2.2 shows the Verilog code for a 4:2 priority encoder. In a similar manner, write the Verilog code for a 4:2 encoder. For this circuit, it can be assumed that at most, one input is asserted at any given time instant. (*Hint*: There is no priority on the inputs, and any value can be output for an invalid input condition, such as multiple asserted inputs.)

**P4.8.** Modify the hierarchical_cla module shown in Section 4.2.3 so that it becomes a 64-bit adder with five p_g_generate submodules (arranged in an upside-down tree structure) and uses 4-bit CLAs (carry-lookahead adders) instead of 4-bit RCAs (ripple-carry adders) for each "mini-adder" stage.

***P4.9.** Show that the logic synthesized for the p_g_generate module, as shown in Figure 4.2(b), is logically equivalent to the equations for $P_{i*m+m-1:i*m}$ and $G_{i*m+m-1:i*m}$ shown in Section 4.2.3.

***P4.10.** Show that the logic synthesized for the hierarchical_cla module, as shown in Figure 4.4, is correct (just the parts shown in the figure).

**P4.11.** Recall from Chapter 3 that a test bench is a Verilog module used to test the correct operation of another module by applying test patterns to the module to be tested (the UUT: unit under test) and checking the responses from the UUT. Writing and using comprehensive test benches is very important for writing properly functioning Verilog modules. Write a comprehensive test bench (one that applies and checks all possible test patterns) for a 16-bit binary-coded decimal (BCD) adder circuit.

**P4.12.** Write a comprehensive test bench for an 8-bit Gray code counter circuit.

**P4.13.** What potential problems can occur if two **always** blocks update and check the same Verilog **reg** signal? Create Verilog examples that demonstrate how such problems can occur.

**P4.14.** In Section 4.3, it is mentioned that a D latch can be described using the Verilog **nand** primitive gate. Write the structural Verilog code for a D latch using only 2-input **nand** gates. Write it as a simple sequence of instantiations of the Verilog 2-input **nand** primitive.

**P4.15.** In Section 4.3, it is mentioned that it is not impossible to describe a sequential circuit using continuous-assignment statements. Thus, it must be possible. Show a sequence of continuous-assignment statements that can be used to implement a D latch. (*Hint*: One solution is a slight variation of the solution to the previous problem.)

**P4.16.** Show that the logic created for the FSM module `simple` by Synopsys FPGA Express, as shown in Figure 4.5, is correct.

**P4.17.** In Section 4.4, the first heuristic listed states that delay statements (which use the **#** notation) cannot be synthesized. How then can we insert a specific delay into a circuit that must be synthesized (the synthesized circuit must include the specific delay)? Show at least two methods for accomplishing this task.

**P4.18.** In Section 4.4, the second heuristic states that **initial** blocks cannot be used. How then can we implement an operation that must be performed once at the beginning of circuit operation?

**P4.19.** Show the Verilog implementation for an 8-bit encoder using the `Synopsys parallel case` directive.

* **P4.20.** Synthesize the `power` module described in Chapter 3 "by hand" (i.e., manually). Then synthesize the same circuit using a commercial synthesis tool, draw the synthesized logic in the same manner as Figure 4.5, and show that the synthesized logic matches your hand-synthesized logic diagram and correctly implements the Verilog code shown for the `power` module.

**P4.21.** In the `Version 2` Verilog code shown in Example 4.3, suppose that the check for equality with `8'b0111_1110` is changed to the following:

```
if (prev_data == 8'b0111_1110)
```

What effect would this have on the overall operation of this circuit? Draw a timing diagram that illustrates this change in circuit operation.

# CHAPTER 5

# State Machine Design

**Important Concepts**

- How to design state machines for complex synchronous sequential digital logic circuits.

- How to convert algorithmic descriptions into HDL code that can be synthesized into working circuits.

- The general form of the circuit that will be synthesized from an HDL code description for an algorithm.

There is a large class of design problems that can be solved by *synchronous sequential* digital logic circuits. As will be recalled from introductory digital logic design, a *sequential* circuit is one in which the current outputs depend on the current *and* past inputs. The past inputs are recorded in a summarized form as the *current state* of the circuit. The current state combined with the current values of input signals determine what the outputs will be and the *next state* of the circuit. A *synchronous sequential* circuit is one in which the state only changes at specific times governed by a *clock* signal, which is a periodic signal dispersed throughout the circuit to control flip-flops or latches. Examples of useful synchronous sequential digital logic circuits include microprocessors, network interface circuits (which operate according to a network protocol), automatic teller machines, vending machine control circuits, factory control circuits, robot controllers, and many other types of control circuits.

As shown in the block diagram of Figure 5.1, a synchronous sequential digital logic circuit consists of two main parts: the control logic and the datapath. The datapath consists of components that store, combine, and otherwise manipulate data bits. Such components include registers, adders, multiplexers, etc. The control logic consists of the logic necessary to generate the control signals to control all of the components in the datapath. Of course, some digital logic circuits may not require a datapath part, while others may not require control logic circuitry. The datapath sends status signals (indicating the state of certain data signals) to the control logic, which in turn sends control signals to the datapath based on those status signals and its internal control state.

The operation of a synchronous sequential circuit can be described by an algorithmic description. In fact, some of these circuits can even be replaced by software written in a high-level programming language—the assumption is that there is a general-purpose microprocessor that interfaces to the circuit inputs and outputs, and the microprocessor is controlled by a software program that implements the desired control algorithm. If such a software solution is not acceptable (for example, because of cost, circuit size, or performance reasons), then a dedicated hardware circuit may be necessary.

**Figure 5.1**
The partitioning of a synchronous sequential circuit into the control logic and the datapath.

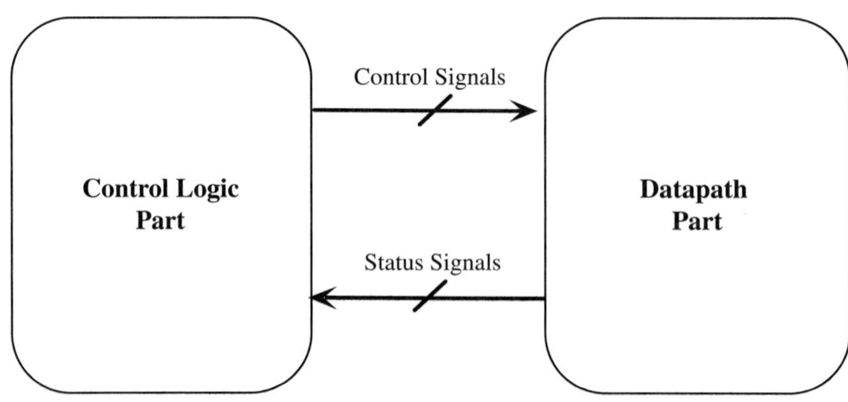

Since the operation of a synchronous sequential circuit can be described by an algorithm, such circuits can be designed from an algorithmic description. Given a problem to be solved using a hardware circuit, we can first devise a high-level algorithm for that problem. Then, the algorithm can be converted into an HDL description using Verilog. Provided that only synthesizable language constructs have been used, the HDL description can then be synthesized for the desired hardware platform. This constitutes the automatic synthesis-based approach to synchronous sequential circuit design. As stated in Chapter 4, digital logic circuits should be described at the *register transfer level (RTL)* (in which all operations are described as the transfer of data between registers) in order to facilitate synthesis.

Before learning the automatic approach to synchronous sequential circuit design, the student should understand the manual approach to synchronous sequential circuit design. In fact, upon examining the circuit created by the synthesis software, the student should discover that it is logically equivalent to the circuit created by the manual method to be discussed in the following section.

# 5.1 Manual State Machine Design

The design of a general synchronous sequential circuit requires the design of the datapath part and the control logic part. The control logic can be designed as a state machine with combinational logic to generate control signals during each of those control states. Once the control logic has been designed, the datapath can then be designed based on an observation of the components necessary to store and manipulate the data as dictated by the control logic.

There are several systematic approaches for the design of the control-logic portion of a synchronous sequential circuit. Introductory digital logic design typically teaches the design and use of *finite state machines (FSMs)*. Although FSMs can be used for the design of circuits with a small number of inputs and outputs, FSM-based approaches are not appropriate for large circuits with large numbers of inputs and outputs. Another type of approach, termed the *algorithmic state machine (ASM)* method, uses a graphical flowchart-like diagram (referred to as an *ASM chart*) to provide a systematic method for converting algorithmic descriptions into hardware. This method has the drawback that it is highly reliant on the use of the graphical ASM chart, which can become cumbersome for circuits with a large number of states. However, the ASM chart method is still a useful tool, preferred by some designers, for the high-level design of synchronous sequential digital logic circuits, and thus will be presented in Section 5.1.6. The main method presented in this section is one based on the use of a *state machine*, which is a generalization of an FSM in which state transitions and outputs are indicated as being dependent on logical combinations of input signals. A state machine can be represented as a graphical diagram, referred to as a *state diagram*, or as pseudocode resembling a high-level language software program.

Suppose that we wish to design a synchronous sequential circuit to solve a given problem. Then the datapath can be created as a set of registers and combinational logic components that manipulate the data. The control logic can be created as a state machine and combinational logic that generates the control signals (to be entered into the datapath) based on the state variables and status signals. Status signals are sent from the datapath to the control logic to aid in the generation of the control signals. A block diagram of this type of general synchronous sequential circuit is shown in Figure 5.2. In this figure, note that a single global clock signal is used for both the control logic and datapath parts. Thus, the general idea is that as the state machine in the control logic goes through its states, data is transferred (and operated on) between registers in the datapath based on the control signals generated in each of those states. Given that this is the general type of circuit desired, it can be designed in a systematic manner by using the following 5-step approach.

### Systematic State Machine Design Method:

1. Describe the desired operation of the target circuit using *pseudocode*.
2. Convert the pseudocode description into an *RTL program*.
3. Design the *datapath* based on the operations indicated in the RTL program.
4. Create a *state diagram* from the RTL program and the signals defined in the datapath.
5. Design the *control logic* based on the state diagram.

Let us now examine each of the steps in the *state machine method* described here. In order to better illustrate the concepts involved, an example design problem will be used. For the example design problem, let us consider the problem of designing

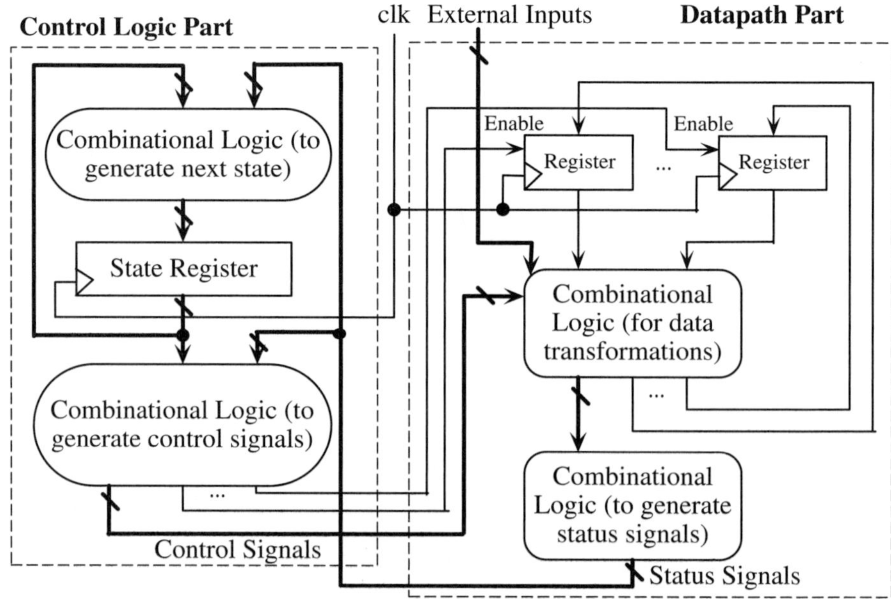


**Figure 5.2**
A general "template" circuit that can be used for the design of a synchronous sequential circuit.

a custom hardware circuit to compute the factorial of an input number. A factorial function commonly is implemented in software. It is a problem commonly used to teach computer programming to undergraduate students. Such a factorial function also can be implemented in hardware by following the systematic 5-step state machine method.

### 5.1.1  Pseudocode

A pseudocode description is a computer program-like procedural description for solving a given problem. It does not have all of the strict syntax requirements of a programming language. Instead, a free-form method, loosely based on the C or Pascal programming languages, can be used. Although pseudocode can be written at various levels of detail, a more detailed description can facilitate the conversion of the pseudocode into a computer program or HDL code.

Let us consider the hardware factorial circuit problem. The factorial of an input number $n$, assumed to be an integer greater than 1, is $n \times (n-1) \times (n-2) \times \ldots \times 1$. Let us assume that $n$ is represented as a 3-bit unsigned number and the factorial result $fact$ is represented as a 16-bit unsigned number. An extra output signal $done$ will be used to indicate when the factorial result is ready. Then the following is the pseudocode for one possible solution to the factorial problem.

*Pseudocode Solution for the Factorial Circuit*:

Step 1.        $fact\langle 15:0 \rangle \leftarrow 1;$
               $done \leftarrow 0;$
Step 2.        For $i = 2$ to $n$ do
                   $fact \leftarrow fact \times i;$
Step 3.        $done \leftarrow 1;$

### 5.1.2  RTL Program

A pseudocode solution for a given problem can be constructed readily as the pseudocode can be written at as high a level as desired. Even when tackling a highly complex design problem, a good starting point is to devise an extremely high-level pseudocode description of the anticipated solution method. Then, by progressively refining this description, the designer can attempt to produce pseudocode that can be converted readily into a working hardware circuit (or a computer program). Thus, recalling the discussion of register-transfer level (RTL) notation in Chapter 4, it clearly would be desirable to refine the pseudocode description into a form that uses only RTL notation. Note, from Table 4.1, that RTL notation includes register transfers, arithmetic operations, logical operations, shifts, and even conditional assignments. This type of pseudocode will be referred to as an *RTL program*.

In order to generate an RTL program from a general pseudocode description, we must refine the pseudocode to the level of a computer program and then convert the computer program constructs used to RTL notation. For the factorial circuit example, it can be seen that the pseudocode description of Section 5.1.1 is already at the level of description of a computer program. Then, in order to convert the pseudocode into an RTL program, we must convert the loop constructs and the variable assignments used. Clearly, assignments of values to variables can correspond to transfers of data (which may undergo some combinational logic transformations) to registers. Loop constructs can be converted into registers for the loop iteration related variables, and conditional branches for jumps out of the loop or jumps back to the top of the loop. Jumps can also be considered as RTL statements since it clearly is possible to maintain a register holding the current program step value, and then change that register value when a jump is required. Then, using these types of conversions, the following shows one possible RTL program for the factorial circuit.

> *RTL Program for the Factorial Circuit*:
>
> Step 1.   $fact\langle 15:0\rangle \leftarrow 1$;
>           $done \leftarrow 0$;
>           $count \leftarrow 2$;
> Step 2.   if ($count \le n$) then begin
>               $fact \leftarrow fact \times count$;
>               $count \leftarrow count + 1$;
>               go to Step 2;
>           end
> Step 3.   $done \leftarrow 1$;

It is possible to write an RTL program in several different ways. If an RTL program can be rewritten to execute in a fewer number of steps, then it will result in a more efficient circuit. All of the RTL statements in a single step of the RTL program will correspond to RTL operations performed during a single state of the control-logic state machine. Several RTL statements can be performed in a single step if all of the corresponding RTL operations can be performed concurrently. Thus, instead of performing the initialization of *count* in a separate step, this operation was combined into Step 1. Also, although not done here, it often is easier to decrement an index value and check for equality with zero (or another fixed constant) than to increment the index value and check if the index is greater or less than some limiting value.

An important difference between an RTL program and pseudocode is that all operations in a single step of the RTL program are assumed to execute concurrently, while all operations in the pseudocode are assumed to execute sequentially (even if those operations occur within a single step). An RTL program is meant to model hardware (which executes in an inherently concurrent manner, as emphasized in the previous chapters), while pseudocode typically models a software process. An implication of this fact is that if a variable is updated and then used within a single step of the RTL program, then the *old* value of the variable will be used instead of the updated value. In the RTL program shown previously, suppose that Step 2 is changed to the following.

> Step 2.   if ($count \le n$) then begin
>               $count \leftarrow count + 1$;
>               $fact \leftarrow fact \times count$;
>               go to Step 2;
>           end

The difference from the original RTL program is that the update of *count* is listed *before* the multiplication of *count* to *fact*. Even in this case, though, the *old* value of *count* will be multiplied to *fact*, since both the update and multiplication operations execute concurrently. Thus, the RTL program will still be correct even with the change shown above. Referring back to the block diagram for a general synchronous sequential-circuit design, shown in Figure 5.2, the *count* register and the *fact* register will be clocked using the same clock edge. Since both registers are assumed to use edge-triggered flip-flops, when a clock edge arrives, the updated value of the *count* register will not be able to propagate back to the data input of the *fact* register until the *next* clock edge.

### 5.1.3  Datapath

Once the desired operation of a circuit has been described using only RTL operations, it is a simple matter to derive the necessary datapath components. First, every unique variable that is assigned a value in the RTL program can be implemented as a register. Depending on the functional operation performed when assigning a value to a variable, the register for that variable may be implemented as a straightforward parallel-in-parallel-out (PIPO) register, a shift register, a counter, or a register preceded by a combinational-logic block. The combinational logic block associated with a register may implement an adder, subtracter, multiplexer, or some other type of combinational-logic function.

Second, if there is a variable that is assigned different values at different points in the RTL program, then a multiplexer or a tri-state bus structure is necessary. However, if the first assignment to the variable is a fixed assignment, then the first assignment can be implemented as an initialization of the corresponding register. Also, for special types of repeated assignments, special types of registers may be usable. For example, if a variable is always incremented by a fixed amount, then a counter can be used. If a shift operation is required, then a shift register can be used.

Third, control-signals inputs are required to control the operation of the various datapath components. Thus, for example, a multiplexer will have select control inputs and an arithmetic logic unit (ALU) will have function select control signal inputs. A register typically will require an enable control signal to enable loading of the register on the next clock edge. An enabled register can be implemented using D flip-flops with 2:1 MUXes at the D inputs. This is shown in Figure 5.3.

Fourth, it may be possible to perform a few optimizations by combining registers, adders, or other components in the datapath. For example, if a variable is used to hold a temporary value, then the register for that variable may be combinable with the register for another variable used in a different time step. Likewise, if two additions occur at different time steps, then it may be possible to combine the adders for those two additions.

Taking all of these points into account, the datapath for the factorial circuit can be designed as shown in Figure 5.4. The *fact* variable is implemented as a 16-bit register. Note that instead of the 16-bit MUX used in front of the *fact* register, it also is possible to simply use a 16-bit register that can be initialized to the value 1 (the lsb is preset and all other bits are reset) or loaded with a new value. The *count* variable

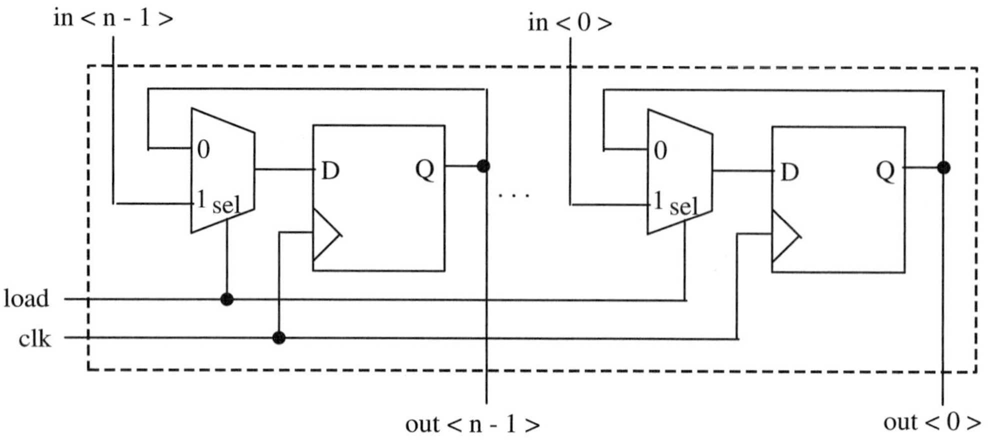



**Figure 5.3**  One possible design for a register with a `load` control signal.

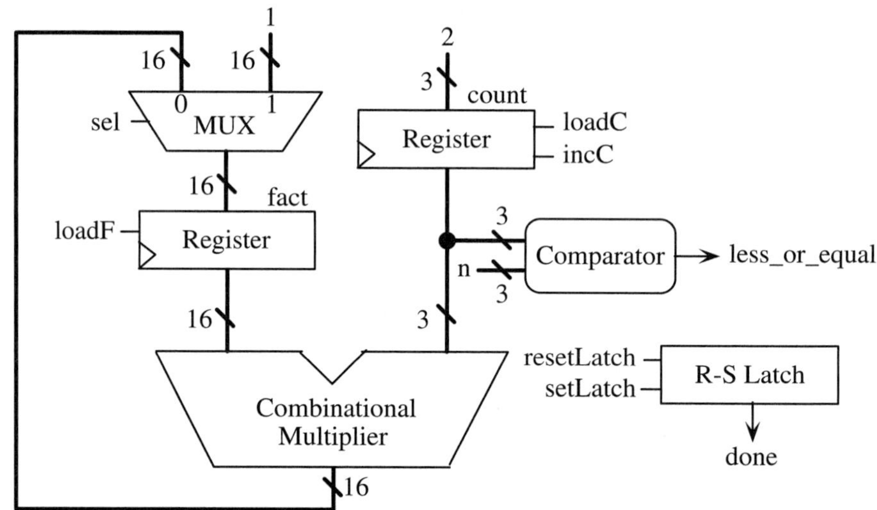



**Figure 5.4**
The datapath for
the factorial circuit.

is implemented as a 3-bit up-counter, since it is either loaded with an initial value or incremented by one. It is assumed that there is a $16 \times 3$ combinational multiplier circuit (such a circuit could be designed using two 16-bit adders and 48 AND gates). Also, a 3-bit combinational comparator circuit is used to produce a status signal named `less_or_equal`. The *done* variable is implemented using a simple R-S latch.

### 5.1.4  State Diagram

The *state diagram* used in this method will be a much more well-defined and restrictive state diagram than that normally used in the finite state machine (FSM) design method. The state diagram will take on a form similar to the RTL program, except that the state diagram normally is shown in a graphical form instead of a program listing. Also, instead of listing a series of RTL operations in each state, a list of control signals will be used. These control signals will correspond to those signals required to control the components of the datapath (in order to get them to perform the RTL operations required). The idea is that during a single state, the control signals listed for that state will be asserted (all other control signals will be de-asserted), thereby resulting in the RTL operations for that state. Another change in the state diagram from the RTL program is that state-change conditions are always shown as simple logical combinations of status signals. If the condition in the RTL program is a local combination of signal values, then no change is necessary. However, if the condition in the RTL program involves a more complex check, such as the arithmetic check "if ($count \leq n$)", then it is replaced by a status signal, such as $less\_or\_equal$, which is in turn created by a combinational-logic circuit (in the datapath) designed to create that status signal. Such condition checks also apply to conditional assignments used in register transfers.

The state diagram for the factorial circuit can then be drawn as shown in Figure 5.5. Arrows with labels refer to conditional-state transfers while unlabeled arrows refer to unconditional-state transfers. The initial state is labeled by showing a dashed arrow into the initial state.

The labels within each state refer to the control signals that must be asserted in order to achieve the RTL operations for that state. All control signals not listed in a state are assumed to be unasserted during that state. Thus, in State s0, control signals sel and loadF are asserted in order to achieve the RTL operation $fact \leftarrow 1$ (select the MUX '1' input and load it into the fact register), loadC is asserted in order initialize the $count$ counter, and resetLatch is used to execute $done \leftarrow 0$.

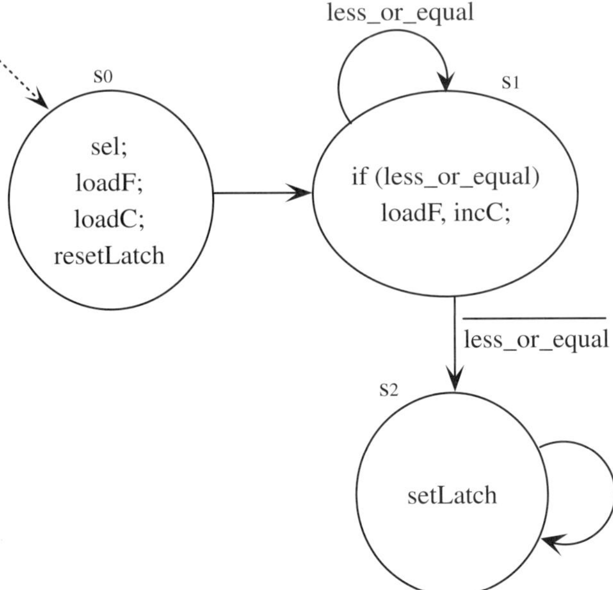


**Figure 5.5**
The state diagram for the factorial circuit.

Then, in State s1, `loadF` is used to execute $fact \leftarrow fact \times count$ and `incC` is used to increment the `count` register (implemented as a counter). The reason that the control signal `loadF` by itself can be used to perform the multiplication is that combinational-logic circuits such as the multiplier are *active* all of the time, and since it is a *combinational* circuit, there are no disable, enable, or select signals required. The transitions from State s1 to s1 or s2 are conditioned on the value of the status variable `less_or_equal`.

Finally, in State s2, the control signal `setLatch` is used to set the *done* signal to '1.' Note that in the final state, an unconditional self-loop has been added to indicate that the circuit will stay in this state until a system-reset signal is encountered—this is necessary in order to prevent the state machine from transitioning to another state *after* this final state.

### 5.1.5   Control Logic

The control logic consists of a state machine and combinational logic to generate the control signals based on the "control state" and status signals. The state machine consists of a set of flip-flops used to store the current state (the control state) and next-state generation logic. The control state can be encoded using one of several possible encoding methods: *one-hot*, *binary*, *Gray code*, etc. The difficulty of the control-logic design problem and the complexity of the resulting control logic varies depending on the specific encoding method used.

**One-Hot Encoding**

In *one-hot* encoding, one flip-flop is used for each state in the state diagram. At any time instant during the operation of the control-logic circuit, exactly one flip-flop (whose $Q$ output corresponds to a state) must be asserted (have its $Q$ output set to 1) and all other flip-flops must be de-asserted (have its $Q$ output reset to 0). This type of encoding, while wasteful for large state diagrams, can result in the *simplest* type of control-logic design method.

There is a one-to-one correspondence from elements of the state machine to elements of the one-hot control logic. Each state will correspond to a flip-flop whose output is the state variable. If there are several states that assert a common control signal, then that control signal is formed from the OR of the $Q$ outputs of the flip-flops for all of the relevant states. A transition between two states in the state diagram will correspond to wires from the $Q$ output of one state flip-flop to the $D$ input of another state flip-flop. However, if there are several inputs into a single state, then the corresponding hardware logic must use an OR gate, since two output signals normally cannot be connected together directly in hardware. Also, a conditional transition is implemented with an AND gate. Likewise, a conditional control-signal assertion is implemented using an AND gate. These one-to-one transformations from state-diagram elements to control-logic elements are summarized in Figure 5.6.

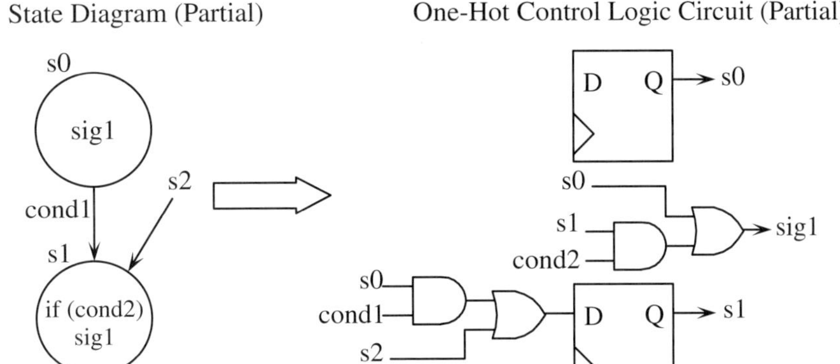



**Figure 5.6**
One-to-one transformations from state-diagram elements to control-logic elements.

**Example 5.1**   **Transformation from a State Diagram to One-Hot Control Logic**

As an example of the use of the transformation method, let us transform the state diagram for the factorial circuit, shown in Figure 5.5, into one-hot control-logic. The transformed one-hot control-logic circuit is shown in Figure 5.7. Each state has been represented using a unique flip-flop, state transitions have been changed into links from flip-flop outputs to flip-flop inputs, multiple inputs into a single state have been changed to OR gates, conditional-state transfers are implemented using AND gates,

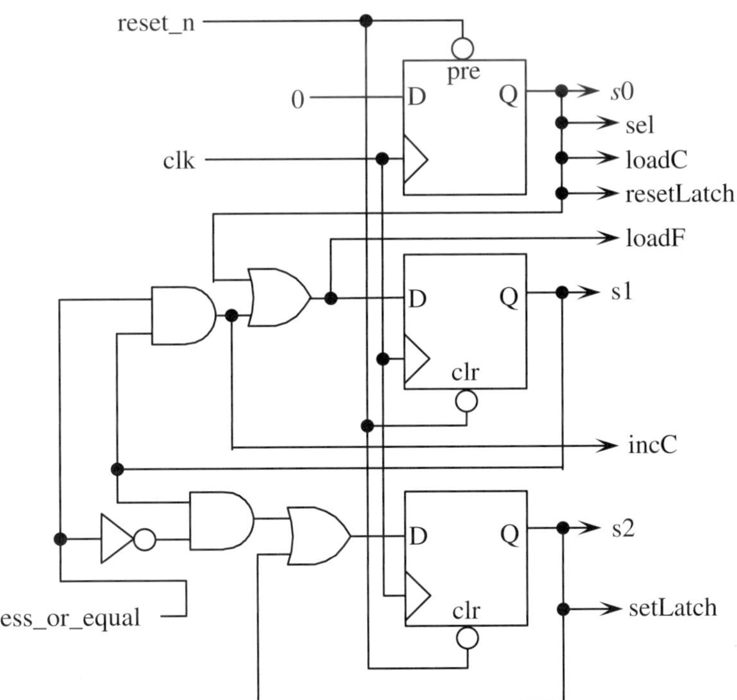



**Figure 5.7**
The one-hot control logic corresponding to the factorial-circuit state diagram.

and control-logic signal outputs are formed from the OR of the states that assert those control-logic signals (naturally, control-logic signals are de-asserted when the corresponding states are not asserted).

For this circuit, however, all control signals except `loadF` are used in only one state—thus, no gates are necessary for those control signals. The control signal `loadF` = s0 OR (s1 AND `less_or_equal`). Also, `incC` = s1 AND `less_or_equal`. Finally, note how all clock inputs are connected to a `clk` signal, a '0' is entered into the D input of the first flip-flop, and the preset input of the first flip-flop and the reset inputs of the other flip-flops are connected to a `reset_n` signal to complete the logic circuit.

## Design using Binary Encoding

For state machines with large numbers of states, the one-hot method becomes extremely inefficient in its use of flip-flops. Given $N$ states, the one-hot method requires $N$ flip-flops, while a method which encodes those $N$ states could use a significantly fewer number of flip-flops. There are numerous ways in which $N$ states can be encoded, corresponding to the various ways in which a decimal number $N$ can be encoded using a sequence of binary bits. The most efficient such encoding method uses $\lceil log_2 N \rceil$ flip-flops. Although there are many different types of encoding methods that require this minimum number of flip-flops, the most commonly used method is a simple binary encoding method.

With binary encoding of states, state transfers must be translated into changes in the encoded states. Thus, for each flip-flop used to store state information, we must determine how the value of that flip-flop will change with changes in the state, which will in turn change according to the state diagram. The method typically used to determine this translation is the *state transition table*, which is an expanded form of the FSM (finite state machine) state transition table. The FSM state transition table is expanded to include the conditions for changes in state. Thus, for example, Table 5.1 is the state transition table, which also demonstrates the use of the binary encoding method, for the factorial-circuit state diagram of Figure 5.5.

Table 5.1: ▶
State transition table for the factorial circuit.

| | Current State | | State Transfer | Next State | | Asserted |
|---|---|---|---|---|---|---|
| | $E_1$ | $E_0$ | Conditions | $NE_1$ | $NE_0$ | Outputs |
| $s0$ | 0 | 0 | | 0 | 1 | sel, loadF, loadC, resetLatch |
| $s1$ | 0 | 1 | less_or_equal | 0 | 1 | loadF, incC |
| $s1$ | 0 | 1 | $\overline{\text{less\_or\_equal}}$ | 1 | 0 | |
| $s2$ | 1 | 0 | | 1 | 0 | setLatch |

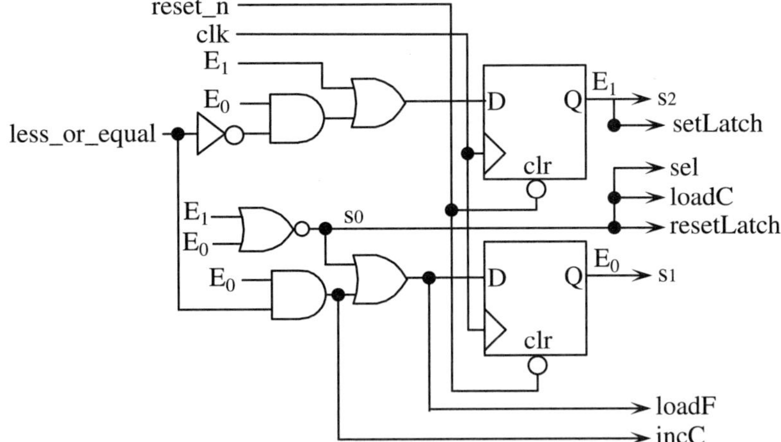

**Figure 5.8**
The binary-encoded control logic for the factorial circuit.

After the state transition table has been created, the logic equations for the next state variables and the outputs can be derived. These logic equations then can be used to derive random combinational logic for the $D$ inputs to the flip-flops encoding the state and the control-signal outputs. This random combinational logic can be implemented using simple SSI gates (AND, OR, inverter, NAND, NOR, XOR), several multiplexers (using the fact that a $2^n - 1$ multiplexer can be used to implement any combinational logic function of up to $n + 1$ variables), or a programmable logic device (such as a PLA or a PROM).

Figure 5.8 shows the random combinational logic and the binary-encoded state flip-flops created based on the state transition table of Table 5.1. The derivation of the relevant logic equations are left as an exercise for the reader. As shown in Chapter 1, K-maps (Karnaugh maps) or the Q-M method (Quine-McCluskey table-based method) can be used to create the simplified logic equations for the next-state and output variables.

### Gray Code Encoding

Besides the binary encoding method, another commonly used state encoding method is the Gray code encoding method. A Gray code is distinguished by the fact that only one bit needs to be changed when changing from one Gray code value to the next. Thus, decimal values are encoded using a binary string in which successive decimal values differ in only bit position of the binary string. Since there are many different types of binary encodings which have this property, there are many different types of Gray codes.

However, the term *Gray code* normally is used to describe a specific type of Gray code (also referred to as a *binary reflective Gray code*) defined recursively in the manner shown in Figure 5.9. This figure shows a systematic progression of binary reflective Gray codes from 1-bit codes, 2-bit codes, 3-bit codes, etc. The 1-bit code consists of 0 and 1. To create a 2-bit code, the 1-bit code is repeated twice, with the second half listed in the inverse order of the first half (i.e., as a reflection of the first half). Then a 0 is inserted into the msb position of the first and second numbers, and

| 1-bit | 2-bit | 3-bit | 4-bit |
|:-----:|:-----:|:-----:|:-----:|
|       |       |       | 0000 |
|       |       |       | 0001 |
|       |       |       | 0011 |
|       |       |       | 0010 |
|       |       | 000   | 0110 |
|       |       | 001   | 0111 |
|       | 00    | 011   | 0101 |
| 0     | 01    | 010   | 0100 |
| 1     | 11    | 110   | 1100 |
|       | 10    | 111   | 1101 |
|       |       | 101   | 1111 |
|       |       | 100   | 1110 |
|       |       |       | 1010 |
|       |       |       | 1011 |
|       |       |       | 1001 |
|       |       |       | 1000 |

**Figure 5.9**
Binary reflective
Gray codes.

a 1 is inserted into the msb position of the third and fourth numbers. The bit 0 position values of the third and fourth numbers are a reflection of the bit 0 position values of the first and second numbers. Next, this code can be extended to a 3-bit code using an analogous procedure. The 2-bit code is listed first, followed by a reflection of the original 2-bit code. Then a 0 is inserted into the msb position of the first half of the sequence, and a 1 is inserted into the msb position of the second half of the sequence. Thus, the 3-bit Gray code sequence is 000, 001, 011, 010, 110, 111, 101, and 100. By continuing in this manner, we can create a Gray code for any number of bits.

The main reasons for using a Gray code state encoding are (1) to simplify the state transition logic and (2) to prevent a spurious unintended state being detected during consecutive state transitions. A Gray code state encoding sometimes results in simpler combinational logic for the next-state transition logic, because only one state variable needs to be changed when transitioning from one state to the next consecutive state. During such state transitions, a straightforward binary encoding may result in two or more state variable changes. If these state variables change at slightly different points in time, then a state decoding circuit will detect the presence of a third state during the time instant when one state variable has already changed while another state variable still has its old value. Of course, even with Gray code encoding, this property may not hold for state transitions between non-consecutive states.

If Gray code encoding is used for the factorial circuit, then the following changes need to be made to the state transition table and its corresponding control logic circuit. In the state transition table of Table 5.1, the state encoding for state s2 will change to 11 (from 10) and all 10 state values will change to 11. Then, with this change, the control logic diagram of Figure 5.8 needs to be changed to use the following set of modified logic equations.

$$s0 = \overline{E_0}$$

$$s1 = \overline{E_1} \cdot E_0$$

$$s2 = E_1$$

$$NE_0 = 1$$

$$NE_1 = s2 + s1 \cdot \overline{\text{less\_or\_equal}}$$

$$= E_1 + \overline{E_1} \cdot E_0 \cdot \overline{\text{less\_or\_equal}}$$

$$= E_1 + E_0 \cdot \overline{\text{less\_or\_equal}}$$

$$\text{sel} = s0 = \overline{E_0}$$

$$\text{loadC} = s0 = \overline{E_0}$$

$$\text{resetLatch} = s0 = \overline{E_0}$$

$$\text{loadF} = s0 + s1 \cdot \text{less\_or\_equal}$$

$$= \overline{E_0} + \overline{E_1} \cdot E_0 \cdot \text{less\_or\_equal}$$

$$= \overline{E_0} + \overline{E_1} \cdot \text{less\_or\_equal}$$

$$\text{incC} = s1 \cdot \text{less\_or\_equal}$$

$$= \overline{E_1} \cdot E_0 \cdot \text{less\_or\_equal}$$

$$\text{setLatch} = s2 = E_1$$

### 5.1.6   State Machine Design using ASM Charts

An algorithmic state machine (ASM) chart can be used instead of a state diagram for manual state machine design. This type of method, referred to as the *ASM chart method*, may be preferable for some designers because of the use of the flowchart-like ASM chart notation, which looks much more like an algorithm than a state diagram.

**ASM Chart**

An ASM chart consists of an interconnection of three types of basic elements: states, condition checks, and conditional outputs. An ASM state, represented as a rectangular block, corresponds to one state of a regular state diagram or finite state machine.

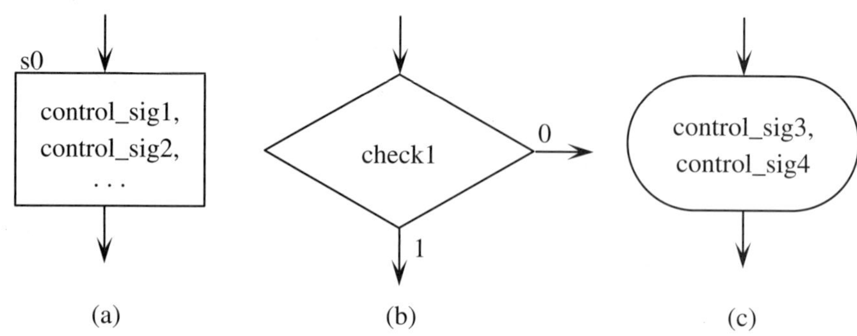



**Figure 5.10**
The three basic
ASM chart elements:
(a) state, (b) condition
check, and
(c) conditional output.

An ASM condition check, represented using a diamond notation with one input and two outputs (for true and false, respectively), is used to conditionally transfer between two states or between a state and a conditional output. An ASM conditional output, represented as an oval shape, corresponds to a conditional output possible with a *Mealy machine*. A *Moore machine*, in which all outputs are dependent on only the current state, does not require the use of an ASM conditional output element. Figure 5.10 shows diagrams of these three ASM chart elements.

To understand the ASM chart, it is useful to study its correspondence with a state diagram. Thus, Figure 5.11 shows a conversion from the state diagram for a sequence detector (for the sequence 101) to its corresponding ASM chart. Note that all circled states in the state diagram have been converted into rectangular ASM states, while all labeled arrows have been converted into ASM condition checks and ASM conditional outputs. Because of the fact that an ASM condition check and an ASM conditional output only are active *during* the state corresponding to the ASM state that they originate from, an ASM state and all ASM condition checks and ASM conditional outputs emanating from it (before reaching another ASM state) can be combined into a block referred to as an *ASM block*. Such ASM blocks are shown in Figure 5.11 using dashed enclosing rectangles. Clearly, an ASM block corresponds to a state and the (possibly labeled) arrows emanating from it in a state diagram.

The main reasons for using ASM charts are to facilitate the creation of hardware algorithms and to enable systematic conversion into working digital-logic circuits. As can be observed from Figure 5.11, an ASM chart closely resembles a *flowchart*, such as that sometimes used to describe a software algorithm before implementing it using a software program. Thus, if a hardware designer can describe the operation of the circuit that he/she wishes to design using a flowchart, then that flowchart (with perhaps a few changes included to enable easier transformation into working hardware) can become the ASM chart required for manual state-machine design. Then, as will be described later in this section, that ASM chart can be converted into digital-logic circuit components in a systematic manner in order to create the desired hardware circuit.

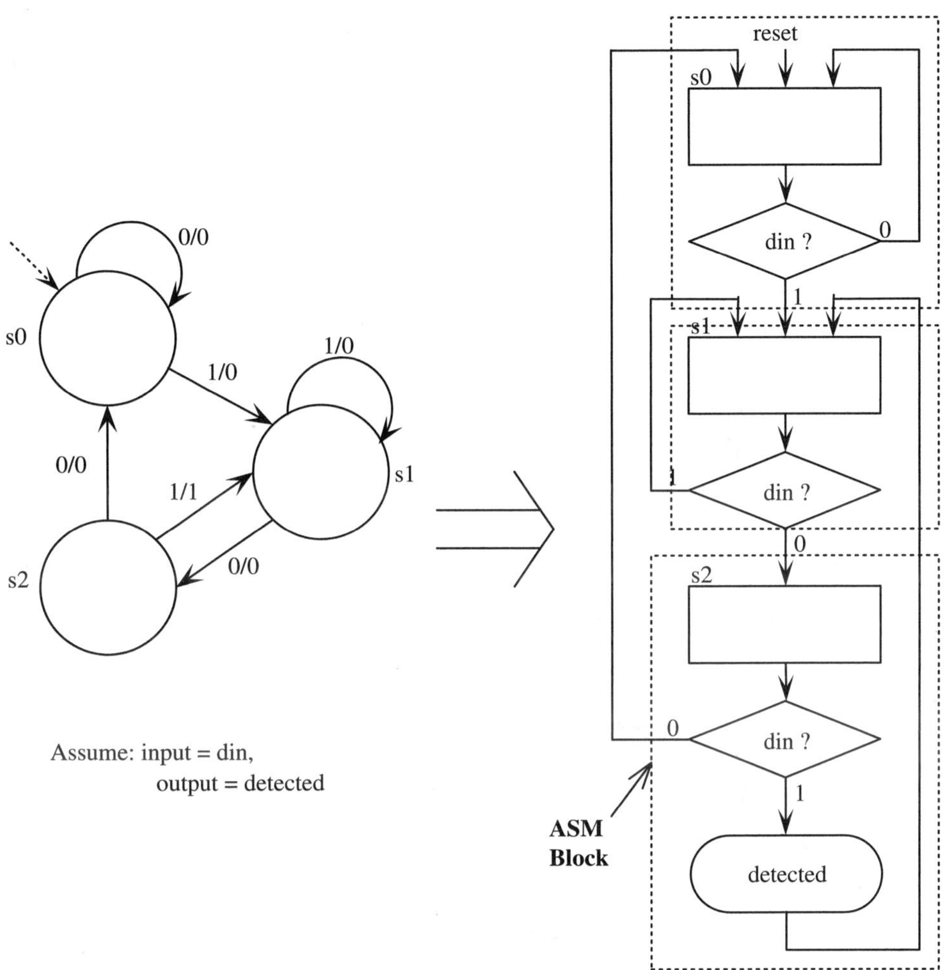



**Figure 5.11**   Conversion from a state diagram to an ASM chart.

## Example 5.2   Four-Phase Handshake Circuit

As an example of a practical hardware design problem that can be effectively handled using the ASM chart method, let us consider the design of the transmitter for a *four-phase handshake* circuit. Whenever two independently operating circuits need to communicate with each other, a *handshaking* protocol needs to be used to enable reliable asynchronous data transfer between the two circuits. A commonly used handshaking protocol is the four-phase handshake shown in Figure 5.12(a).

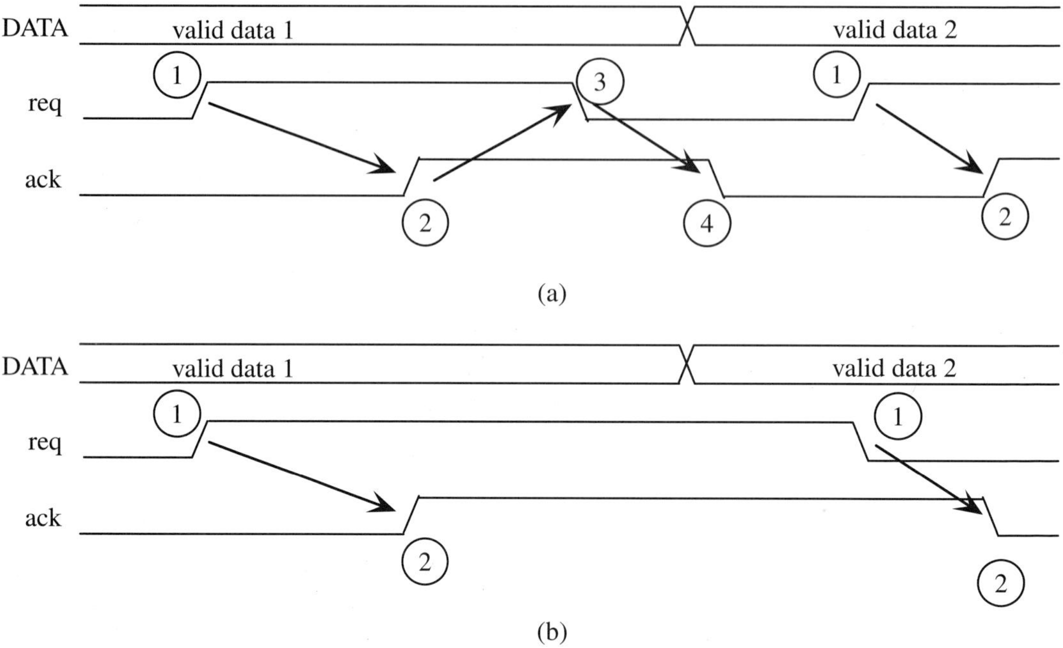

**Figure 5.12** Common handshaking protocols: (a) four-phase and (b) two-phase.

The four-phase handshake protocol works as follows. The sending circuit sends a $req = 1$ signal along with the data. The receiving circuit, upon seeing the $req$ signal, latches in the data, processes it, and then sends an $ack = 1$ signal back. The sending circuit waits until $ack = 1$ is observed, which indicates successful data receipt, and then sets $req = 0$. Upon seeing $req = 0$, the receiving circuit sets $ack = 0$ in preparation for the receipt of the next data item. Finally, the sending circuit has to wait until it observes $ack = 0$ before sending out the next data item and $req = 1$. It is also possible to use a *two-phase handshake*, as shown in Figure 5.12(b). The analysis of this protocol is left as an end-of-chapter problem.

Following the five-step manual ASM chart design method, the pseudocode, RTL program, datapath, ASM chart (instead of the state diagram), and control logic need to be designed. Based on the description of the four-phase handshake protocol given above, the following shows the pseudocode for the target-transmitter circuit.

*Pseudocode for Four-Phase Handshake Transmitter Circuit:*

Step 1.　　$req \leftarrow 0$;
　　　　　　wait until $ready = 1$ (data ready);
Step 2.　　$DATA \leftarrow$ data to be sent;
Step 3.　　$req \leftarrow 1$;
　　　　　　wait until $ack = 1$;
Step 4.　　$req \leftarrow 0$;
　　　　　　wait until $ack = 0$;
Step 5.　　go to Step 1;

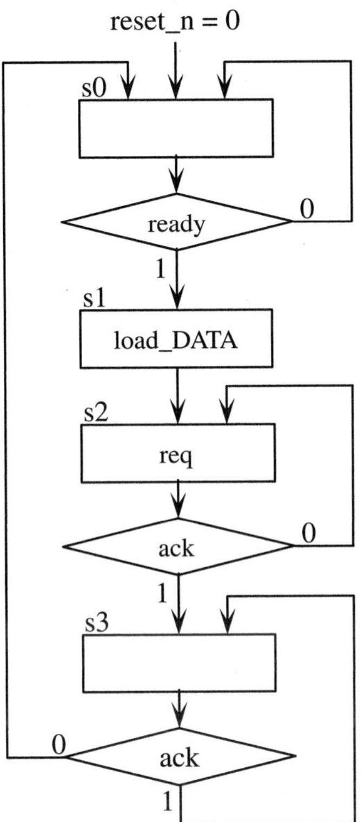

**Figure 5.13**
The ASM chart
for a four-phase
handshake circuit.

Based on this pseudocode description, the ASM chart shown in Figure 5.13 can be designed.  The RTL program for this circuit is equivalent to the ASM chart to be produced, while the datapath simply consists of a single DATA register.

### Conversion from an ASM Chart to Control Logic

One of the main motivations for the use the ASM chart is to enable systematic transformation to digital logic circuits.  Each ASM chart element can be converted into an equivalent digital logic circuit element.  Figure 5.14 shows each ASM chart element and the digital logic component (or components) that it can be converted into.  Then, by interconnecting the converted digital logic components in the same manner as the ASM chart, a working digital logic circuit can be created.  Note that this method uses the *one-hot* state encoding method, which was presented earlier in this section.

As an example of the use of this conversion method, Figure 5.15 shows the control logic (along with the datapath and the circuit interface) derived from the ASM chart for the four-phase handshake transmitter circuit.  The reader should compare this figure with the ASM chart shown in Figure 5.13.

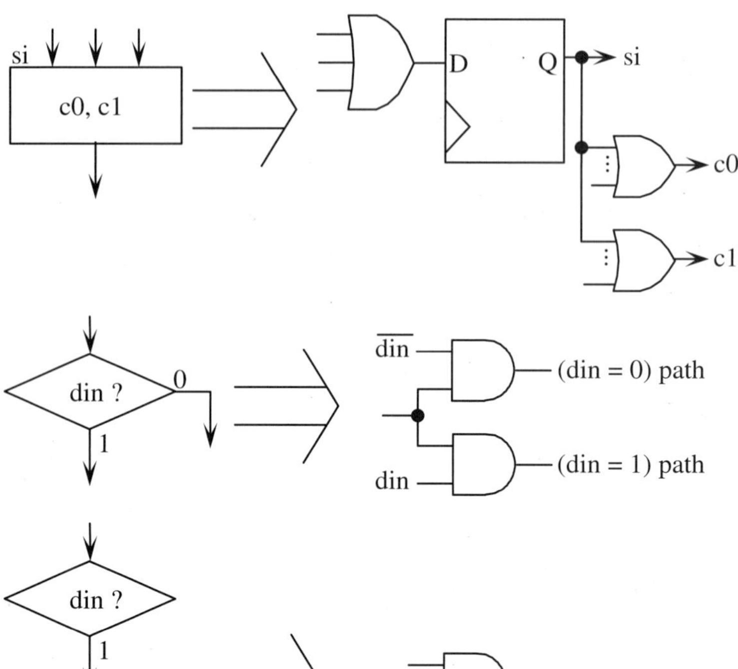



**Figure 5.14**
Conversion from ASM chart elements to digital logic components: (a) a state (with input paths and control signals) converted to a flip-flop and OR gates, (b) a condition check converted to two AND gates, and (c) a conditional output converted to an AND gate.

## 5.2 Automatic Synthesis-Based State Machine Design

For very large designs, manual state machine design becomes unmanageable and error prone. Thus, at some point, it becomes necessary to use automatic state machine design methods. The most commonly used automatic method is one based on a high-level HDL circuit description followed by synthesis using a commercial synthesis tool. In summary, the following simplified 3-step design procedure can be used for automatic synthesis-based state-machine design.

### 5.2.1   Automatic Synthesis-Based Design Procedure

1. Describe the desired operation of the target circuit using *pseudocode*.
2. Convert the pseudocode description into an *HDL program*.
3. Use a *synthesis* tool to synthesize the target circuit from the HDL program.

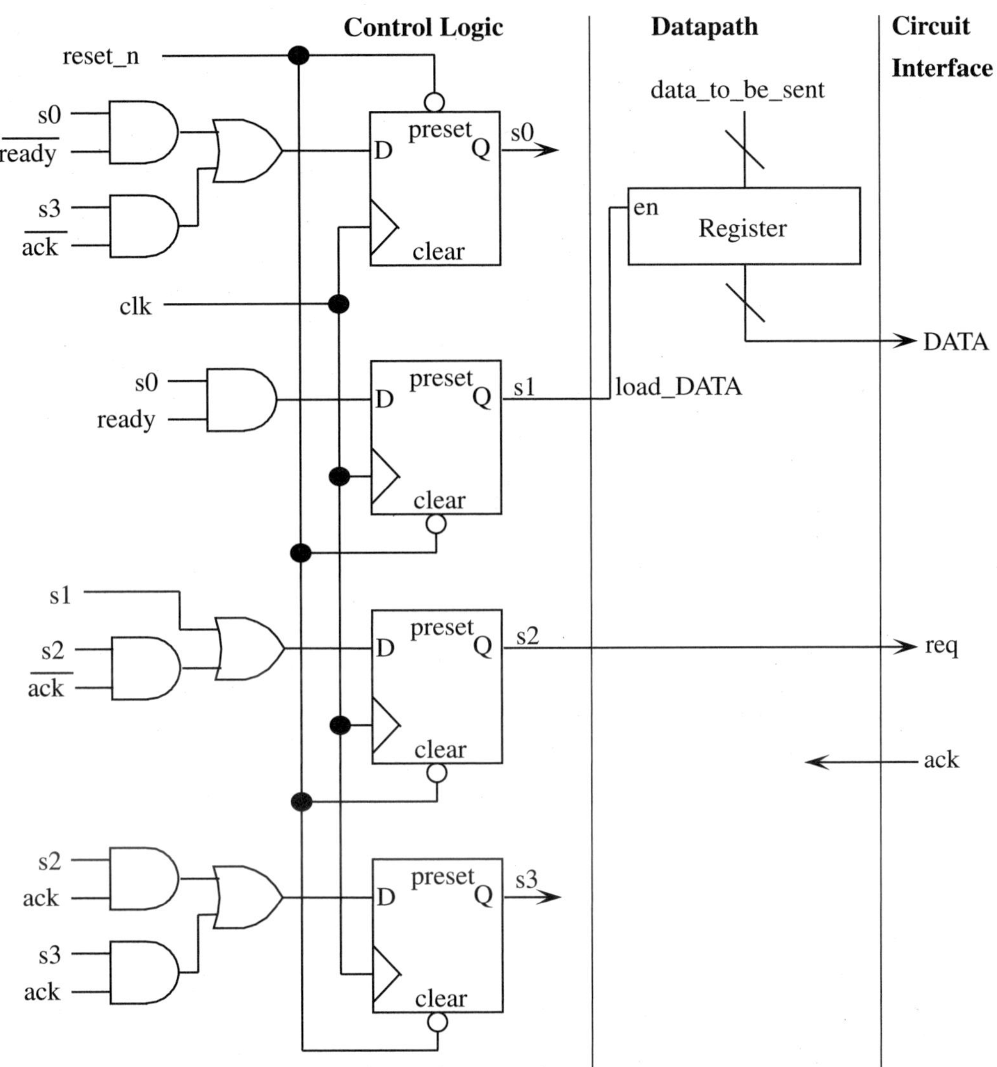



**Figure 5.15**   The control logic, datapath, and circuit interface for the four-phase handshake transmitter circuit.

The circuit that results after synthesis is typically the same as the circuit that results from the manual state machine design method described in the previous section. Thus, by inspecting the synthesized circuit, the user should be able to verify whether the synthesized circuit is similar to the circuit that would be created by manual state machine design. Once the target circuit has been properly synthesized, it can be implemented using an FPGA, ASIC, or other hardware platform, as described in Section 6.1.1.

### 5.2.2  Algorithm to HDL Code Conversion

Let us examine the process of converting a high-level algorithm written in pseudocode into HDL code. Once an algorithm describing the desired behavior of a circuit is converted into HDL code, a synthesis tool can be used to automatically create the necessary datapath and control logic for a state machine that implements the given algorithm (provided that the HDL code has been written in a synthesizable manner, as described in Chapter 4). By using a high-level behavioral description method, a fairly straightforward conversion to HDL syntax can be used for much of the pseudocode. However, the inherently concurrent and continuous nature of hardware must be kept in mind when writing the HDL code. The most important implication of this fact is that a series of sequential operations normally cannot be written in the same manner as the pseudocode. A series of sequential operations typically are implemented as a series of operations executed in consecutive clock cycles of a sequential circuit. Within a single clock cycle, all operations are executed concurrently.

**Recommended Pseudocode Format**

Although pseudocode can, in general, be written in a free format and converted to HDL code, it is recommended that the following pseudocode structure be used in order to facilitate conversion to easily synthesizable HDL code. If necessary, the solution for the application first can be posed using free-format pseudocode and then converted to pseudocode of the form shown here. In this recommended pseudocode structure, there is an initialization step followed by an infinite while loop, which in turn is composed of a series of additional steps.

> *Recommended Pseudocode Structure*:
>
> | | |
> |---|---|
> | Step 0. | Initialize variables; |
> | | While (true) do begin |
> | Step 1. | Step 1 of algorithm; |
> | Step 2. | Step 2 of algorithm; |
> | ... | |
> | Step $k$. | for (i = 0; i < max; i++) ... |
> | ... | |
> | Step $m$. | call to function FUNCTION_1; |
> | ... | |
> | Step $n$. | Step $n$ of algorithm; |
> | | end/ * of while * / |

Various operations can be included in Steps 1 through $n$ of the above pseudocode structure. Pseudocode operations can include simple assignments to variables, arithmetic/logic operations, conditional actions, calls to subroutines/functions, and for/while loops. Assignments to variables, with or without arithmetic/logic operations, can be converted to simple nonblocking HDL assignments. Conditional actions should be converted to **case** blocks whenever appropriate (several values of a variable are being tested) and **if-else** blocks otherwise.

For loop structures such as **for** loops in the pseudocode, two approaches are possible depending on the delay behavior of actions within the loop. In pseudocode, there are loops containing operations that can be performed concurrently and loops containing operations that must be performed as a series of sequential steps with delays between those steps. This is an important difference in HDL. If all operations within a pseudocode loop can be performed concurrently, then the loop can simply be converted to **for** or **loop** constructs within the HDL code. However, if a sequence of delay-separated operations are required within the pseudocode loop, then those actions must be performed as separate steps within the HDL code. Each step of the HDL code is implemented using a state or step case value within a large **case** block (testing the state or step) within an **always** or **process** block activated by the positive edge (or negative edge) of the system clock. If the pseudocode loop is a **for** loop with an initialization step, then that initialization step may need to be performed within a "previous" state or step within the HDL code.

Calls to subroutines and functions can be implemented as branches (changes to the state or step variable). However, there must be a mechanism to return to the main subroutine at the end of the subroutine or function. This can be implemented by saving the state or step number of the step immediately following the subroutine/function call and then restoring this value upon the completion of the subroutine or function.

It may be possible to *convert* a series of sequential pseudocode operations into a set of concurrent HDL operations. This is possible if (1) the pseudocode operations can be executed independently of each other (there are no data or resource dependencies) or (2) some of the pseudocode operations can be converted into other operations so that all converted pseudocode operations can be executed independently of each other.

In some situations, a sequence of blocking assignment statements may be convertible into an equivalent sequence of nonblocking assignment statements. Then, if the original sequence of blocking assignment statements is synthesized, the resulting circuit will (typically) be the same as the manually designed circuit designed from the equivalent sequence of nonblocking assignment statements. An example of this type of situation is the sequence detector code of Example 4.3. Recall that, in that situation, the shift of a variable named prev_data followed by a check of the value of that variable was converted into a shift of prev_data followed by a nonblocking check of the value of the variable *before* the shift.

In other situations, it may not be possible to directly convert a sequence of blocking assignment statements into a corresponding sequence of nonblocking assignment statements. If this type of HDL code is still accepted by the synthesis tool, then the sequence of blocking assignment statements must effectively be partitioned into several sets of nonblocking assignment statements executed over several consecutive clock cycles of a state machine. Some authors refer to this type of translation as conversion of *behavioral Verilog* code into RTL Verilog code.[1]

---

[1]   Refer to the "Resources" section at the end of this chapter.

**Table 5.2:** ▶
Typical pseudocode
operations and
corresponding HDL
code.

| Pseudocode | HDL Code |
| --- | --- |
| Variable assignment | Signal assignment |
| Conditional action | **if-else** or **case** block |
| Sequence of steps | **case**(state) block |
| **for** loop with no delays | **for** or **loop** construct |
| **for** loop with delays | Sequence of states |
| Function call | ret_state ← current state +1; |
|  | state ← location of function |
| Return from function | state ← ret_state |

Typical pseudocode operations and their corresponding HDL constructs are summarized in Table 5.2. Given a complex algorithm to be converted to Verilog, the designer first should identify those operations that can be performed in the same clock cycle of a synchronous sequential circuit. Such operations can be combined into one "step" of the algorithm. This results in an algorithmic description (pseudocode) consisting of a sequence of steps. By converting this pseudocode to the recommended format shown in Table 5.2, subsequent conversion to synthesizable HDL code can be facilitated. The pseudocode, which consists of a sequence of steps, can be converted into a **case**(state) block within an **always** or **process** block, where each **case** instance of the **case** block normally corresponds to one step of the algorithm. The state value (representing the step number) is incremented concurrently with the operations for a single step of the algorithm. Thus, the next time the **always** block is invoked (due to the occurrence of the next positive clock edge), the state value, representing the current algorithm step number, will have been incremented by one.

### Verilog Code Template

Various styles of Verilog coding can be used to implement pseudocode. However, the method recommended in this text is to use the coding method shown below as a "template" for all such algorithms. This template corresponds to the recommended pseudocode format shown in Section 5.2.2.

```
// Template for conversion of complex algorithms to Verilog:
`define FUNCTION_1 z  // z is a "unique" state number > m+x+1
always @(negedge reset_n or posedge clk) begin
  if (~reset_n) begin
    ... // perform initialization operations (Step 0 of pseudocode)
    state <= 0;
  end
  else begin  // on posedge clk
    state <= state + 1;  // auto-increment the "state"
```

```verilog
    case (state)  // steps of algorithm are encoded as "states"
      0: begin
           ... // operations in Step 1 of pseudocode
         end
      1: begin
           ... // operations in Step 2 of pseudocode
         end
         ...
      k-2: begin
           i <= 0;  // for initialization part of "for" loop
         end
      k-1: begin
           ... // 1st set of concurrent operations within "for" loop
         end
      k: begin
           ... // 2nd set of concurrent operations within "for" loop
         end
         ...
      k+x: begin
           ... // last step of "for" loop
           if (i < max)
             state <= k-1;
           i <= i + 1;
         end
         ...
      m+x: begin  // assume that `FUNCTION_1 is called here
           ret_state <= m+x+1;  // state number to return to
           state <= `FUNCTION_1;  // go to `FUNCTION_1
         end
      m+x+1: begin
           ...
      `FUNCTION_1: begin
           ... // operations in first step of `FUNCTION_1
         end
      `FUNCTION_1+1: begin
           ... // operations in second step of `FUNCTION_1
         end
         ...
      `FUNCTION_1+y: begin  // last step of `FUNCTION_1
           state <= ret_state;  // return to calling procedure
         end
         ...
      default: state <= 0;  // Step $n$ (last step) of pseudocode
    endcase
  end  // of else (posedge clk)
end  // of main always block
```

• • • • • • • • • • • • • • • • •

# 5.3  Design Sample: Vending Machine

Several concrete examples of complex algorithm conversion will be presented in the later chapters of this text. All of these examples will use the templates shown above. The Verilog code in Example 4.3 also uses the same type of template as that shown above. However, the sequence-detector Algorithm of Example 4.3 is a special case in that it is a 1-step algorithm (not counting the initialization operations, which can be performed in a reset state, as a separate step).

Suppose that we are given the problem of designing a practical control circuit for an automatic vending machine. Let us assume that this machine only accepts nickels ($n$) dimes ($d$), and quarters ($q$), and keeps a count of the current money *total*. Other inputs to this circuit are a *refund* button and two item choices: a 55-cent *soda* and a 70-cent *water* bottle. It is assumed that only one input signal is present at any given time. The outputs of this circuit are *total*, *beep*, *return_coin*, *eject_soda*, *eject_water*, and *eject_change*. The three "eject" signals are used to control the delivery of a soda, a water bottle, or the remaining money. The *beep* signal is asserted if an item choice is entered before sufficient money has been input, and the *eject_coin* signal is asserted if more money is input than can be stored in the *total* register. The *return_coin* signal is asserted to return the most recent coin entry if the addition of this coin will cause us to overflow the maximum money total that can be maintained. Let us now design this circuit using the automatic and manual state-machine design methods.

### 5.3.1  Automatic State Machine Design for a Vending Machine

Automatic state machine design requires a high-level HDL description for the desired circuit followed by synthesis, simulation, and implementation using an FPGA or ASIC chip. Let us first devise a pseudocode solution for this circuit. In this circuit, there needs to be a *handshaking* interface between two "circuits," the human making the item selection and the machine responding to the human's selection. Whenever there are two independent circuits that need to communicate in this manner, a handshaking protocol is necessary. As discussed in Section 5.1.6, a four-phase handshake circuit can be used for this type of asynchronous communication interface.

The target vending machine circuit will be designed using the four-phase handshaking protocol shown in Figure 5.12(a). The human's item selection, as well as coin input or refund request, will be encoded into a multiple-bit *data* signal. Upon observing $req = 1$, the target circuit will decode the *data* signal to determine the requested action, perform operations based on this request, and then acknowledgement completion of the requested action. As a small optimization, the coin inputs have been encoded into the *data* signal such that the *data* value (when interpreted as an unsigned binary number) is equal to the amount of money input. The following is a pseudocode representation of an algorithm for this circuit.

*Pseudocode Solution for Vending Machine Control Circuit*:

```
Step 0.        total⟨4 : 0⟩ ← 0;  ack ← 0;  beep ← 0;
                   return_coin ← 0;  eject_soda ← 0;  eject_water ← 0;
                   eject_change ← 0;  MAX ← 31;
               While (true) do begin
Step 1.            wait until (req = 1);
Step 2.            case (data) is begin
                       n, d, or q:
                           if (total + data > MAX) then
                               return_coin ← 1;
                           else
                               total ← total + data;
                       soda:
                           if (total < 11) then
                               beep ← 1;
                           else begin
                               total ← total − 11;
                               eject_soda ← 1;
                           end/ * of else * /
                       water:
                           if (total < 14) then
                               beep ← 1;
                           else begin
                               total ← total − 14;
                               eject_water ← 1;
                           end/ * of else * /
                       refund: begin
                               eject_change ← 1;
                               total ← 0;
                           end/ * of case (data = refund) * /
                       end/ * of case (data) * /
Step 3.            ack ← 1 until (req = 0);
Step 4.            ack ← 0;  beep ← 0;  return_coin ← 0;
                   eject_soda ← 0;  eject_water ← 0;  eject_change ← 0;
               end/ * of while * /
```

A few optimizations have been used in this pseudocode. First, all money values have been scaled to 5-cent units since all inputs, outputs, and internal calculations can be done at the 5-cent (nickel) granularity level (5, 10, 25, 55, and 70 cents). This permits us to use fewer bits for the *total* register. Second, although $MAX$ is used just like any other variable in Step 1, it is clear from the use of $MAX$ in Step 2 that $MAX$ can be defined as a constant. Third, the current money total is stored as an unsigned number in order to use fewer bits for the *total* register. Finally, all of the user selections (soda, water, or refund) and coin entries (nickel, dime, or quarter) have been encoded into a single *data* signal. The coin entries have been encoded such that their values are equal to the monetary values of those coins ($n = 1$, $d = 2$, and $q = 5$).

Using the algorithm-to-HDL template shown in Section 5.2, the HDL code for this circuit can be derived from its pseudocode description. Step 0 of the vending-machine pseudocode corresponds to the initialization step. All other steps iterate within a large, infinitely executing loop, with each step of the loop typically corresponding to operations performed within a single state of the vending-machine circuit. This type of pseudocode is used because this type of behavior is representative of the actual behavior of typical synchronous sequential hardware circuits. As stated earlier, a digital circuit executes concurrently and continuously; hardware is active all of the time, provided that power is provided continuously to the circuit.

### Verilog Coding

In the vending machine pseudocode, Step 0 is the initialization step and all of the other steps are executed sequentially within an infinite loop. Thus, in the corresponding Verilog code, the operations in Step 0 can be executed when (reset_n == 0). Also, in this part of the code, the signal (named, for example, state) representing the *current state* of the state machine for this circuit is initialized to 0. Then, on each positive clock edge, a large *case* block testing the value of state is used. The operations in Step 1 of the pseudocode are executed when *state* = 0. The operations in Step 2 of the pseudocode are executed when *state* = 1. Continuing in this manner, it can be seen that Step $k - 1$ of the pseudocode corresponds to state $k$ in the Verilog code. This type of correspondence can be seen in the vending machine Verilog code shown next. Sometimes, however, one step of the pseudocode may need to be separated into several states in the Verilog code if the operations within one step of the pseudocode cannot be executed at the same time (clock cycle). Alternatively, it also may be possible to combine several steps in the pseudocode into the same state in the Verilog code if the operations within several steps of the pseudocode all can be executed concurrently.

```verilog
////////////////////////////////////////////////////////////////
`define TOTAL_BITS 5  // number of bits used in total register
`define MAX 31  // Maximum number representable in total register
`define STATE_BITS 2  // number of bits used in the state machine
`define DATA_BITS 3  // number of bits used to encode user input
`define DATA_n 1      // code for "n" (nickel) user input
`define DATA_d 2      // code for "d" (dime) user input
`define DATA_q 5      // code for "q" (quarter) user input
`define DATA_soda 0   // code for "soda" user input
`define DATA_water 3  // code for "water" user input
`define DATA_refund 4 // code for "refund" user input
```

```verilog
module vending (total, ack, beep, return_coin, eject_soda,
    eject_water, eject_change, req, data, reset_n, clk);
  output [`TOTAL_BITS-1:0] total;  // current money total
  output ack;      // ack part of asynch. communication handshake
  output beep;    // indicates incorrect user action
  output return_coin;  // signal to return most recent coin input
  output eject_soda;    // signal to eject a soda
  output eject_water;   // signal to eject a water bottle
  output eject_change; // signal to eject all money in "total"
  input req;       // req part of asynch. communication handshake
  input [`DATA_BITS-1:0] data;      // encodes input from user
  input reset_n; // asynchronous active-low reset
  input clk;       // clock signal

  // Register definitions for output signals
  reg [`TOTAL_BITS-1:0] total;
  reg ack;
  reg beep;
  reg return_coin;
  reg eject_soda;
  reg eject_water;
  reg eject_change;

  // Register definitions for internal signals
  reg [`STATE_BITS-1:0] state;

  // main always block implementing vending machine algorithm
  always @(negedge reset_n or posedge clk) begin
    if (~reset_n) begin
      total <= 0;  // initialize all register values
      ack <= 0;
      beep <= 0;
      return_coin <= 0;
      eject_soda <= 0;
      eject_water <= 0;
      eject_change <= 0;
      state <= 0;  // initial state to start in
    end
    else begin  // on posedge clk
      state <= state + 1;  // auto-increment state
      case (state)
        0: if (~req)  // wait until (req = 1)
             state <= 0;
```

```verilog
          1: case (data)
               `DATA_n, `DATA_d, `DATA_q:
                 if ((total + data) > `MAX)
                   return_coin <= 1;
                 else
                   total <= total + data;
               `DATA_soda:
                 if (total < 11)
                   beep <= 1;
                 else begin
                   total <= total - 11;
                   eject_soda <= 1;
                 end
               `DATA_water:
                 if (total < 14)
                   beep <= 1;
                 else begin
                   total <= total - 14;
                   eject_water <= 1;
                 end
               `DATA_refund: begin
                 total <= 0;
                 eject_change <= 1;
               end
               default: $display("Data is %0x", data);  // to cover all cases
             endcase
          2: begin
               ack <= 1;
               if (req)  // wait until (req = 0)
                 state <= 2;
             end
          default: begin  // final step: de-assert control signals
             ack <= 0;
             beep <= 0;
             return_coin <= 0;
             eject_soda <= 0;
             eject_water <= 0;
             eject_change <= 0;  // then loop back to state 0
           end  // of case default
         endcase  // of case (state)
       end  // of else (posedge clk)
     end  // of main always block
   endmodule
   ////////////////////////////////////////////////////////////////////
```

### 5.3.2   Manual State Machine Design for a Vending Machine

Manual state machine design of this circuit requires us to follow the five-step design procedure introduced in Section 5.1: pseudocode, RTL program, datapath design, state diagram, and control logic design. In this case, the pseudocode already has been created in Section 5.3.1. Since the pseudocode consists entirely of RTL operations, the RTL program can be considered as equivalent to this pseudocode.

Based on this RTL program, we can derive the datapath shown in Figure 5.16. Note that control signals have been defined for the registers and MUXes based on the RTL operations used in the RTL program. The update signal is used as a simple optimization for the assertion and de-assertion of the three eject signals and the error signal. Using the normal state machine method, there would have to be 2:1 MUXes at the inputs to each of the three eject registers and the error register; the inputs to these 2:1 MUXes would be 0 and 1, selectable with select signals generated from

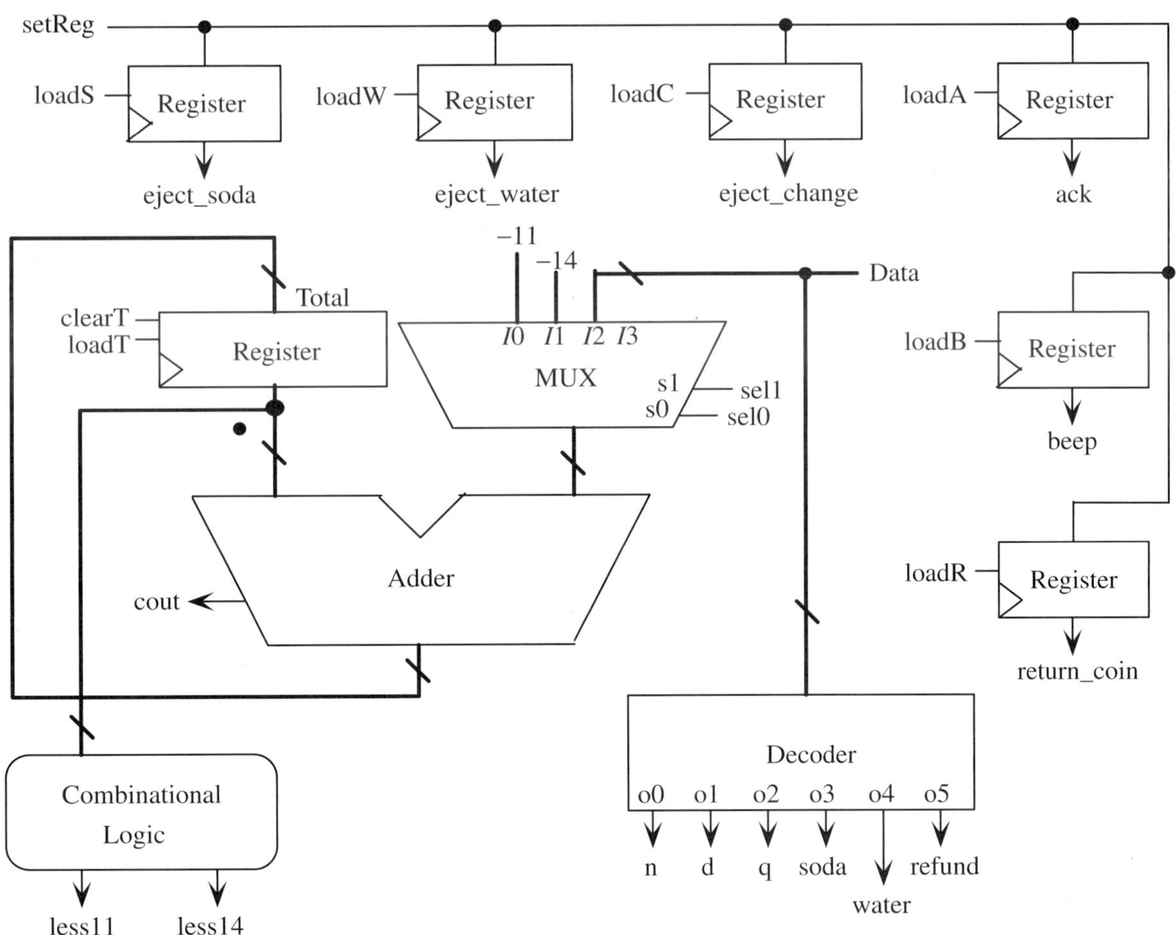


**Figure 5.16**   The datapath corresponding to the Verilog vending module.

the control logic circuit. Status signals also may need to be defined for some of the condition checks used in the RTL program. In this case, the only status signal found to be necessary was cout, the carry-out from the adder.

All other condition checks can be formed from the primary input signals. Based on the datapath and the RTL program, the state diagram for this circuit can be derived as shown in Figure 5.17. The state diagram consists of a state s0 in which all register values are cleared, a state s1 in which the total register, error register, and eject registers are loaded conditionally, and a state s2 in which the eject_soda, eject_water, eject_change, and error registers are cleared (by using the assumption of update = 0 and asserting the proper load signals). In state s2, note that load signals are used instead of clear signals (which typically are asynchronous) in order to ensure that the appropriate eject and error signals are active for exactly one clock period.

It is noted that the error signal only needs to be set when there is an overflow or an underflow out of the adder. Both overflow and underflow conditions can be detected simply by observing the carry out of the adder. If 2's complement arithmetic is used, addition results greater than the bit width of the adder and addition results

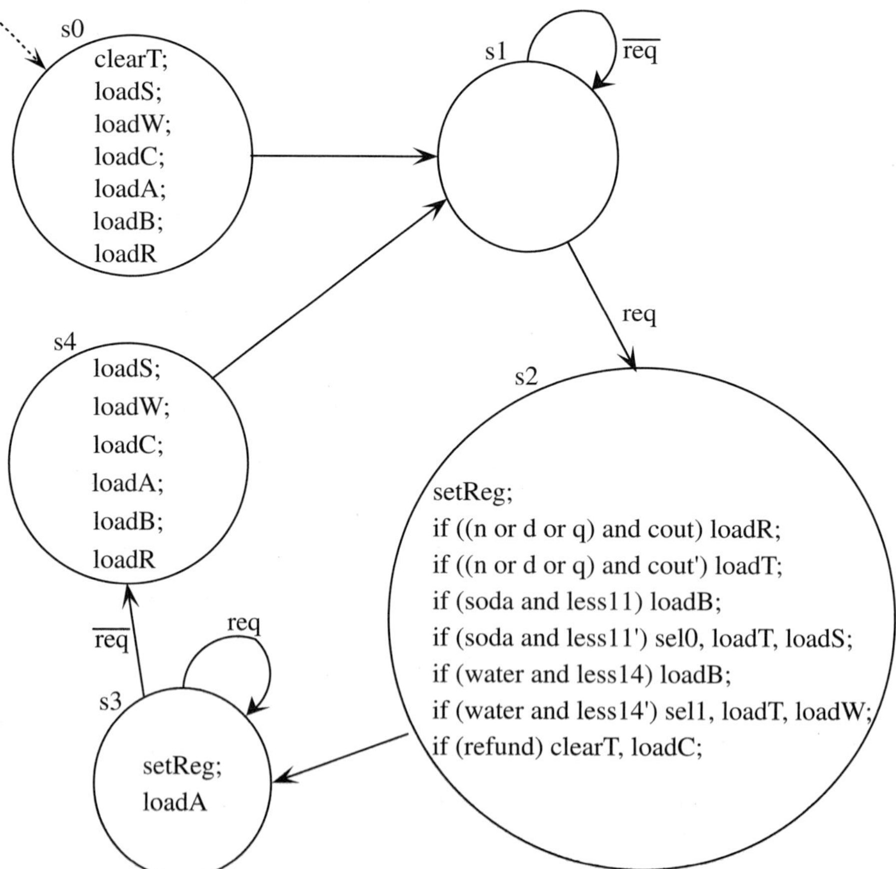


**Figure 5.17**
The state diagram for the vending machine control circuit.

less than zero both will result in a carry-out of the adder. Thus, in state s1, the error register can be set whenever cout $= 1$.

State s1 consists of a set of conditional assertions of control signals. For the error register, we must select a suitable set of MUX select signals, wait for the adder carry-out signal (cout) to become asserted, and then set the new error value to the value of cout. For the total register, we must select a suitable set of MUX select signals, wait for the cout signal, and then load the total register with the adder result if cout $= 0$. Since it turns out that the MUX select signals to be used are the same for the error and total registers, we can use a single set of MUX and 5-bit adder devices. Depending on the combination of input signals present, we must select a different input for the second input to the 5-bit adder used in the datapath. Since the sequence of condition checks used in state $= 0$ of the Verilog code imply a priority on the input signals being checked, this priority must be followed in the state diagram. However, if it is known that a certain set of input signals will never be active simultaneously (such as the n, d, and q inputs), then it also is possible to use an unprioritized check of those input signals—this results in simpler logic in most cases.

Using the datapath of Figure 5.16 and the state diagram of Figure 5.17, we can derive the required control logic as shown in Figure 5.18. There are several different ways to design the control logic. In this case, the one-hot method was used. An encoding method could also be used for the state. Also, since the only signals asserted in state s0 of Figure 5.17 are clear signals, this state could have been implemented using the initial reset_n signal used to clear or set the various registers.

### 5.3.3  Timing Diagram

In order to understand the operation of a synchronous sequential circuit, it is helpful to construct a timing diagram that shows how the various signals in the circuit change with time. Such a timing diagram should correspond with the simulation results for the automatically synthesized circuit (given the same set of input patterns). Comparing simulation results with a manually derived timing diagram is akin to comparing a synthesized circuit with a manually designed circuit.

As an example, let us construct a timing diagram for two operation modes of the vending machine. Suppose that with an initial money total of 130 cents, the user first inserts a dime followed by a quarter. The first coin insertion should result in an update of the money total to 140 cents, and the second coin insertion should result in a "beep" followed by the return of the quarter (since it is assumed that our vending machine cannot maintain a money total larger than 155 cents). The timing diagram resulting from this sequence of user actions is shown in Figure 5.19.

All signals necessary for the proper interpretation of the process to be demonstrated need to be included in the timing diagram. Assuming an initial total value of $0x1$ ita ($= 26$ in decimal $= 130$ cents since each unit is assumed to be equal to five cents), insertion of a dime results in $data = 0x2$ and a positive transition on the $req$ signal. This causes our state machine to transition to state s2, at which point $loadT$ is asserted. The assertion of $loadT$ in state s2 causes the $total$ register to be updated on the positive edge of the $next$ clock cycle. Then, in state s3, $loadA$ is asserted, resulting in an $ack = 1$ signal being sent at the beginning of the next state, which is

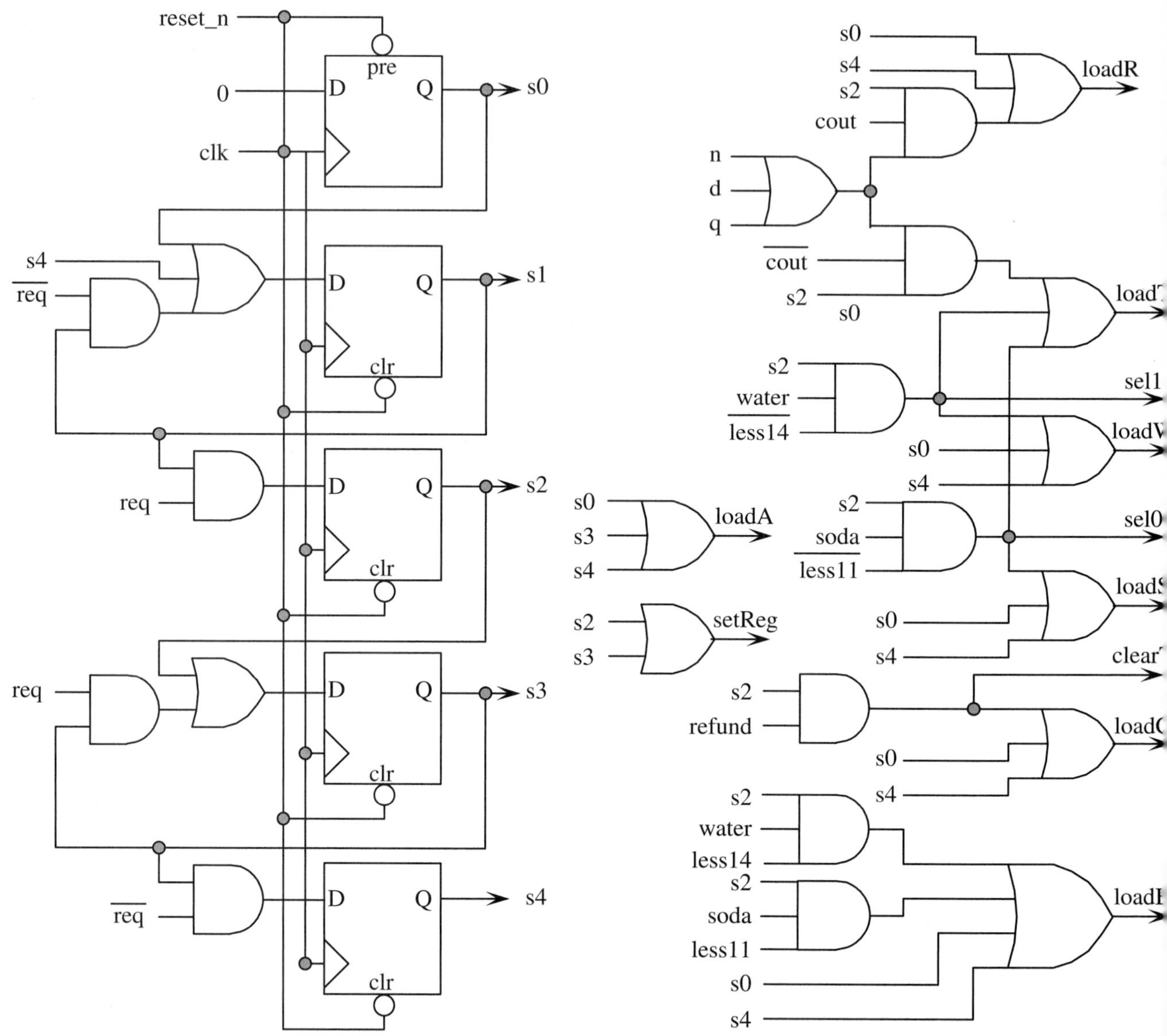

**Figure 5.18**  The control logic for the vending machine control circuit.

still s3 since $req = 1$. Upon seeing $ack = 1$, the input circuit can set $req = 0$, which causes our circuit to transition to state s4. Next, when the quarter is inserted, $req$ gets asserted with $data = 0x5$. The addition of five to $total = 0x1c$ would result in an overflow of the 5-bit $total$ register. Thus, the vending machine circuit observes $cout = 1$ and asserts $load R$ during the next s2 state.

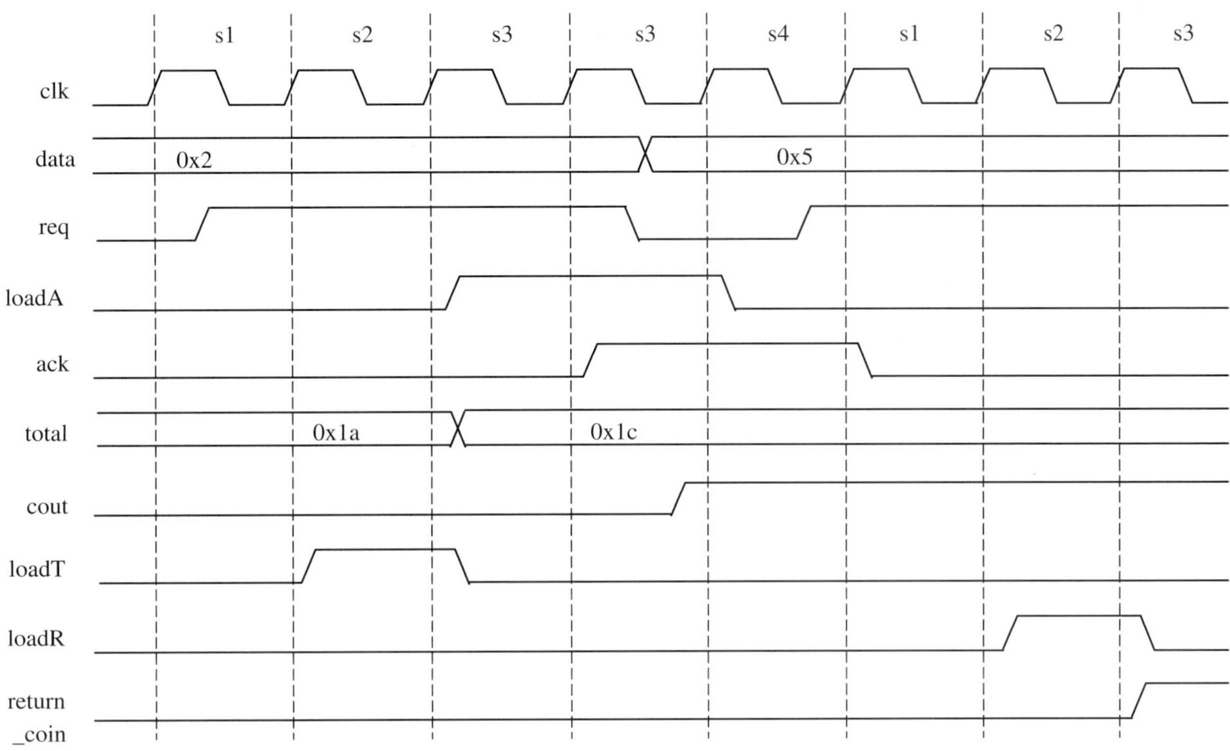



**Figure 5.19**   A timing diagram for the vending machine circuit.

### 5.3.4   Correspondence between Automatic and Manual Designs

If the circuit created by the manual design process is compared to the circuit created by the synthesis tool based on the Verilog code, the two circuits should be found to be logically equivalent. Such a comparison can be made if facilities are provided to view (or otherwise determine) the synthesized circuit—such a facility is provided with some synthesis tools. Many useful facts can be learned from studying the synthesized circuits and comparing them to the expected circuits. For instance, if a "low area" synthesis option is used, ripple-carry adders typically will be created; if a "high speed" synthesis option is used, carry lookahead adders typically will be created. Also, if the target implementation architecture requires the use of certain MSI devices (such as the lookup tables in the Xilinx configurable logic blocks), then the required logic will be mapped to those MSI devices.

Actual comparison of synthesized and manually designed circuits can be an extremely tedious process because of the various optimization techniques used by many synthesis tools. For example, the Synopsys FPGA Express synthesis tool will tend to combine the combinational logic for several MSI devices, such as the MUX and adder in Figure 5.16, in order to shorten the delays through the combinational logic. Also, combinational logic (besides simple state transition logic) will be created for the $D$ inputs as well as the "load enable" inputs to the flip-flops in order to implement the conditional data transfers required (in the state machine design method described above, only the "load enable" inputs are used to implement conditional data transfers).

Due to these practical difficulties in comparing synthesized and manually designed circuits, the designer typically has to rely on simulation results to verify proper circuit design, especially with large circuits. This normally is sufficient, as most commercial synthesis and simulation tools can be trusted to produce reliable results, provided that properly synthesizable Verilog code has been written. However, it still is useful to view the synthesized schematic (if possible) and verify that the synthesized circuit has the general form of the expected circuits. The general layout of the registers produced can be checked. The combinational logic in critical paths can be checked. The number of flip-flops, primary inputs, primary outputs, adders, MUXes, comparators, and other devices created can be compared with the expected numbers of such devices in a manually designed circuit. If the synthesized circuit is not of the desired form, then the circuit either has to be resynthesized using different synthesis options or the Verilog code has to be revised, as described in Chapter 4.

## 5.4 Design Sample: LCD Controller

Let us now investigate the design of a slightly more sophisticated control circuit. This will be a circuit to interface to and control a graphic LCD (liquid crystal display) module. LCDs currently are used in a wide range of appliances, including displays for FAX machines, hand-held electronic toys, and cellular phones. LCD modules range from large high-resolution color graphic-display modules to small low-resolution black-and-white character displays.

Interfacing to and controlling an LCD module can be a nontrivial problem. A simple low-resolution black-and-white character LCD can be controlled by simply sending it a series of bytes in ASCII format. However, graphic displays typically require a sequence of commands followed by a sequence of data bytes. The commands are used to control various modes of operation, position the cursor, and perform initialization operations. The command sequences used to control these LCD displays operate according to an algorithm implemented in software or hardware.

### 5.4.1  Target LCD Module

The target LCD module used in this example will be the HG12601 LCD module, a typical low-cost graphic LCD module suitable for class lab usage and produced by Hynix Semiconductors. The Hynix HG12601 consists of a black-and-white graphic LCD display and two NJR NJU6452A LCD driver chips embedded into the underside of the LCD module. The two NJR (New Japan Radio Co.) NJU6452A chips are used to store the pixel data to be sent to the LCD, set various operation modes of the LCD, and sequence through the pixels of the LCD, thus turning each pixel on or off. There are two LCD driver chips because one chip can store the data for only (and send data to) one-half of the HG12601 LCD. The two NJR LCD driver chips simplify the interface to the actual LCD by automatically performing necessary initialization operations, buffering and then sending the pixel data to the LCD with the correct timing, and permitting an 8-bit microprocessor interface for sending commands and data to the LCD.

The NJR LCD driver chip can interface with two types of microprocessors: Motorola 68xx and Intel 80xx. Thus, for example, an Intel 8080 microprocessor could be used to control the Hynix HG12601 LCD module by sending commands and data to the two NJR LCD driver chips. It also is possible to use any other microprocessor as long as it can emulate the output interface of an Intel 80xx or Motorola 68xx type 8-bit microprocessor. Another possibility is to use the parallel port of a PC to send 8-bit data signals and a few (at least three) control signals to control the HG12601. These methods all involve controlling the LCD module through software.

Since the software for controlling an LCD module is fairly simple, it is possible to design a custom hardware device for controlling the Hynix HG12601 LCD module. If a hardware solution is used, it even is possible to bypass the two NJR NJU6452A LCD driver chips and send commands and data directly to the LCD. However, this would require extra memory for storing the data to be displayed, more output pins, and more sophisticated control of the LCD interface (essentially, performing most of the functions of the original NJU6452A chips).

Let us now consider the design of a hardware LCD interface device for the Hynix HG12601 LCD module with the two NJU6452A LCD driver chips. To do this, we first must make the physical connections to the pins of the HG12601, and then design our LCD interface circuit to execute as a state machine that cycles through several states, outputting the necessary commands and data bytes to the two NJU6452A chips. Table 5.3 shows the I/O pin locations and descriptions for the HG12601 LCD module. There are 20 pins, of which five are used for power. There are eight data

Table 5.3: ▶
I/O pin locations
and brief
descriptions for the
HG12601.

| Pin | Name | Value | Brief Description |
|---|---|---|---|
| 1 | Vss | 0 V | Ground |
| 2 | Vdd | +5 V | LCD supply voltage |
| 3 | Vo | 1–4 V | Variable LCD operating voltage |
| 4 | A0 | H for command | Command or data select signal |
| 5 | /CS1 | L for select | Chip select for IC 1 |
| 6 | /CS2 | L for select | Chip select for IC 2 |
| 7 | CL | 2 KHz | Main clock input |
| 8 | /RD | L for read | Read enable for 80xx microprocessor |
| 9 | /WR | L for write | Write enable for 80xx microprocessor |
| 10–17 | DB0–DB7 | | 8-bit bidirectional data |
| 18 | /RES | | Reset signal (L for 80xx microprocessor) |
| 19 | A | | Anode of LED unit |
| 20 | K | | Cathode of LED unit |

pins (DB0 through DB7), an address pin for selecting command or data mode (A0), a 2 KHz clock input (CL), a reset input (/RES, which should remain low after pulsing high in order to emulate an Intel 80xx interface), two chip select signals for the two NJU6452A chips (/CS1 and /CS2), a read enable signal (/RD), and a write enable signal (/WR).

In order to interface properly to any chip, we must satisfy its electrical and timing specifications (specs). For the NJU6452A chip, these specs can be obtained from the NJU6452A data sheet. From the electrical specs, it can be determined that TTL or CMOS-type voltage levels can be used for the inputs to the NJU6452A chip. For the timing specs, let us assume that the "8080 family interface" mode (for an Intel 80xx type microprocessor interface) is being used. The most important parameters in the timing specs for the NJU6452A chip are the cycle time ($t_{CYC8} \geq$ 1000ns, the time interval between consecutive command or data bytes), data setup time ($t_{DS8} \geq$ 80ns, the time that the data to be written has to be stable *before* the rising edge of the /WR signal), and data hold time ($t_{DH8} \geq$ 10ns, the time that the data to be written has to remain stable *after* the rising edge of the /WR signal). Our Xilinx FPGA must output commands and data to be sent to the NJU6452A chip (and thus to the HG12601 module) while satisfying all of these timing specs. This implies that the clock cycle for sending data cannot exceed 1 MHz (which corresponds to a period of 1 ms = 1000 ns), and that the write data must be stable after as well as before the rising edge of the /WR write enable signal.

The NJU6452A data sheet describes everything needed to use this chip. This chip first is initialized by asserting the /RST input (note that all signals with a slash in their name are active-low signals). Upon asserting the /RST input (connected to the /RES input of the HG12601 module), the NJU6452A chips initialize the display of the LCD and then prepare themselves to receive commands and data from the LCD interface. The LCD display area is partitioned into four "pages" of bits, with 61 columns per page. Each pixel is set to a black or white value, depending on the corresponding bit value. Each data byte can be used to set eight pixels of one column of one page. Thus, two NJU6452A chips can be used to address $32 \times 122$ pixels, since there are $8 \times 61$ pixels per page, four pages per chip, and two chips. The data to be written can be selected by the user in order to display the selected graphic image.

A hardware circuit that can be used to interface to the HG12601 module is shown in Figure 5.20. The LCD interface hardware is implemented using a Xilinx FPGA. The tasks that the FPGA must perform can be determined by examining the data sheet (refer to the end-of-chapter references) for the NJR NJU6452A chip. Essentially, the FPGA must send a series of commands followed by data to the two NJU6452A chips. The data byte sent to a particular page and column lights up the eight pixels in that portion of the LCD according to the data sent (the pixel is ON if the corresponding bit is 1, and OFF if the bit is 0). The sequence of commands and data that must be sent are as follows.

*Pseudocode for LCD Interface*:

Step 1.   For *chip* $\leftarrow$ 1 to 2 do begin
              Enable *chip*

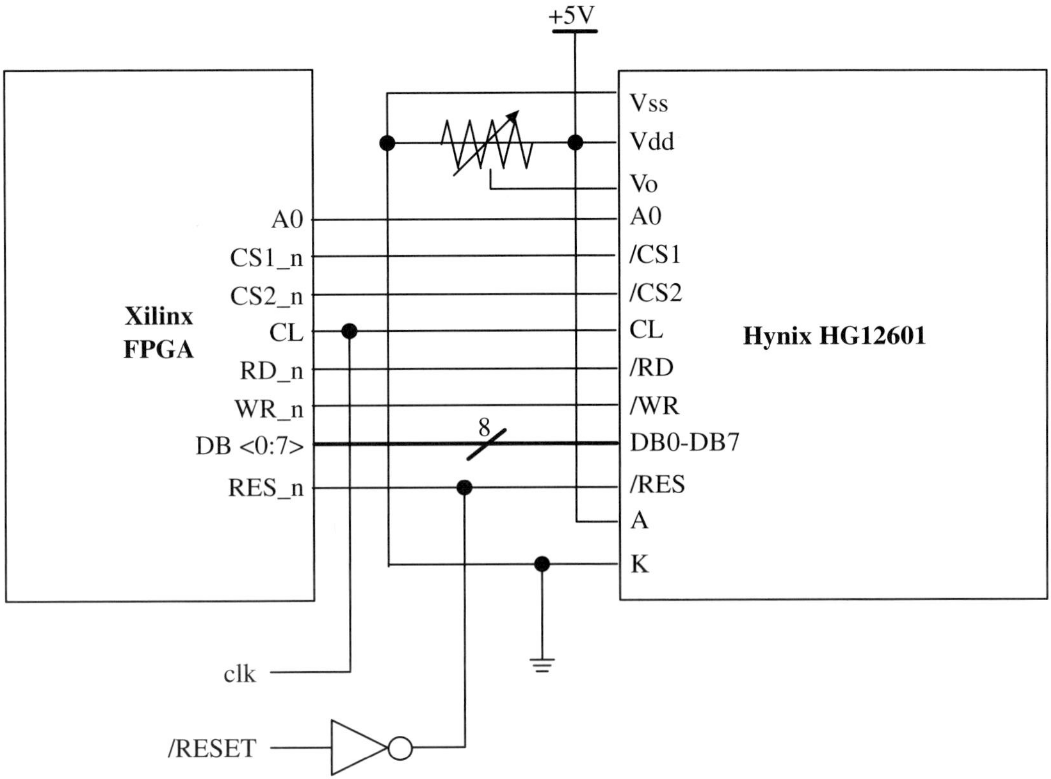


**Figure 5.20**  The hardware interface circuit for the HG12601.

```
                    For page ← 0 to 3 do begin
                        Set column ← 0
                        Set to "read modify write" mode/* auto-increments column */
                        For i ← 0 to 60 do/* 60 = number of columns */
                            Send data byte/* column number is auto-incremented */
                    end/* of for */
                    Disable chip
                end/* of for */
    Step 2.     For chip ← 1 to 2 do begin
                    Enable chip
                    Set display on
                    Disable chip
                end/* of for */
```

### 5.4.2 Verilog Solution

Based on the description of the previous subsection, a Verilog design solution can be devised for the Xilinx FPGA used to implement the LCD interface for the Hynix HG12601 LCD module. The following shows the Verilog code for a possible solution to this problem. Note that the state machine shown in the main **always** block iterates through the sequence of commands and data necessary for displaying pixel data on the HG12601 LCD module, as shown in the pseudocode in the previous subsection.

```verilog
//////////////////////////////////////////////////////////////////////
// MODULE: Example Control Circuit for the Hyundai HG12601 LCD
// Author: Sunggu Lee (with help from Min-Gu Lee and Won-Bae Gil)
// Created: ...
// Last Modified: ...
// Version: 0.3 (lcd3.v)
// Description: Generates the control and data signals required
//   to drive the Hynix HG12601 LCD.  This module is used to
//   draw example letters and diagrams on the target LCD.  This
//   module emulates a Z80-microprocessor interface to the
//   NJU6452A chip, the bit map LCD driver used to display
//   graphics or characters on the HG12601 display.  For now,
//   the pattern displayed is a set of 8 horizontal lines.

// DEFINITIONS
`define DB_BITS      8  // number of bits output on DB
`define PAGE_BITS    2  // log_2 (number of pages in LCD)
`define STATE_BITS   4  // number of bits used in the state mach.
`define DATA_COUNT_BITS 6  // number of bits for data_count
`define NUM_DATA_BYTES  61 // number of data bytes in one chip
`define MAX_PAGE     3  // maximum page number in LCD
`define CS_STAGE     0  // chip select stage of state machine
`define SETUP_STAGE 1   // page and column setup stage
`define DATA_STAGE   7  // data transmit stage
`define RESET_STAGE 9   // reset stage (before next page)
`define FINAL_STAGE 11 // final wrap up stage of state machine
`define CMD_DISPLAY_OFF           8'b10101110 // LCD
`define CMD_PAGE_ADDR             6'b101110   // Instructions
`define CMD_COLUMN_ADDR_0         8'b00000000
`define CMD_READ_MODIFY_WRITE_ON  8'b11100000
`define CMD_READ_MODIFY_WRITE_OFF 8'b11101110
`define CMD_DISPLAY_ON            8'b10101111
`define DATA_LINES 8'b11001100  // data to be sent

// MODULE DECLARATION
module lcd3 (A0, CS1_n, CS2_n, RD_n, WR_n, DB, RES_n, clk);
```

```verilog
output A0;              // H: data, L: instruction code
output CS1_n;           // chip select signal for NJU6452A IC 1
output CS2_n;           // chip select signal for NJU6452A IC 2
output RD_n;            // /RD signal (emulates output of Z80)
output WR_n;            // /WR signal (emulates output of Z80)
output [`DB_BITS-1:0] DB; // data bits
input RES_n;            // reset signal
input clk;             // 2KHz clock input

// SIGNAL DECLARATIONS
// primary inputs and outputs
reg A0;
reg CS1_n;
reg CS2_n;
reg RD_n;
reg WR_n;
reg [`DB_BITS-1:0] DB;
wire RES_n;
wire clk;

// internal signals
reg [`STATE_BITS-1:0] state;    // current state of ASM
reg [`PAGE_BITS-1:0] page;      // current page of display
reg [`DATA_COUNT_BITS-1:0] data_count; // byte position
reg sel_CS1;                    // select CS1 LCD driver chip

// Implements the "pseudocode" for this module
always @(negedge RES_n or posedge clk) begin
  if (~RES_n) begin // upon a RESET low pulse
    A0 <= 0;          // initialize all chip outputs first
    CS1_n <= 1;
    CS2_n <= 1;
    RD_n <= 1;
    WR_n <= 1;
    state <= `CS_STAGE; // set initial state
    page <= 0;          // set initial page setting for LCD
    data_count <= 0;    // set initial byte position for LCD
    sel_CS1 <= 1;       // select CS1 LCD driver chip
  end
  else begin  // upon a positive transition on clk
    state <= state + 1;  // auto-increment state by default
```

```verilog
case (state)
  `CS_STAGE: begin          // Chip Select Stage
     CS1_n <= ~sel_CS1; // select either CS1 or CS2
     CS2_n <= sel_CS1;
     state <= `SETUP_STAGE;
  end
  `SETUP_STAGE: begin  // Setup Stage
     A0 <= 0;       // indicates instruction being sent
     WR_n <= 0;
     DB <= {`CMD_PAGE_ADDR, page};  // set page
  end
  `SETUP_STAGE+1: begin
     WR_n <= 1;   // data should be stable at this time
  end
  `SETUP_STAGE+2: begin
     WR_n <= 0;
     DB <= `CMD_COLUMN_ADDR_0;  // set column to 0
  end
  `SETUP_STAGE+3: begin
     WR_n <= 1;   // data should be stable at this time
  end
  `SETUP_STAGE+4: begin
     WR_n <= 0;
     DB <= `CMD_READ_MODIFY_WRITE_ON;  // write enable
  end
  `SETUP_STAGE+5: begin
     WR_n <= 1;   // data should be stable at this time
     state <= `DATA_STAGE;
  end

  `DATA_STAGE: begin  // Data Stage: send display data
     A0 <= 1;       // used to indicate data being sent
     WR_n <= 0;
     DB <= `DATA_LINES;  // send some data
     data_count <= data_count + 1;
  end
  `DATA_STAGE+1: begin
     WR_n <= 1;   // data should be stable at this time
     if (data_count < `NUM_DATA_BYTES)
       state <= `DATA_STAGE;  // continue sending data
     else begin // all data sent for current chip and page
       state <= `RESET_STAGE;
     end // of else if (data_count == `NUM_DATA_BYTES)
  end // of case `DATA_STAGE+1
```

```verilog
`RESET_STAGE: begin  // Reset before next page
    A0 <= 0;
    WR_n <= 0;
    DB <= `CMD_READ_MODIFY_WRITE_OFF; // reset operation
end
`RESET_STAGE+1: begin
    WR_n <= 1;   // data should be stable at this time
    if (page < `MAX_PAGE)  // have to work on next page
      state <= `SETUP_STAGE;
    else if (sel_CS1) begin // have to work on CS2 chip
      sel_CS1 <= 0;
      state <= `CS_STAGE;
    end
    else begin  // done with all data, pages, and chips
      sel_CS1 <= 1;
      state <= `FINAL_STAGE;  // go to Final Stage
    end
    page <= page + 1;  // done with all data, so page++
    data_count <= 0;   // start at column 0
end // of case `RESET1

`FINAL_STAGE: begin     // Final Stage: set display on
    CS1_n <= ~sel_CS1;
    CS2_n <= sel_CS1;
    A0 <= 0;     // indicates instruction being sent
end
`FINAL_STAGE+1: begin
    WR_n <= 0;
    DB <= `CMD_DISPLAY_ON;
end
`FINAL_STAGE+2: begin
    WR_n <= 1;   // data should be stable at this time
    if (sel_CS1) begin // still have to "finalize" IC2
      state <= `FINAL_STAGE;
      sel_CS1 <= 0;
    end
    else // otherwise, done --> stay in this state
      state <= `FINAL_STAGE+2;
end
```

```
        default: begin    // since all cases should be covered
            WR_n <= 1;
        end // of default case
      endcase // of case (step)
    end // of outermost else
  end // of always loop to compute the next state
endmodule
///////////////////////////////////////////////////////////////
```

In order to verify (at the simulation level) the proper operation of our circuit, it is helpful to use a test bench, also written in Verilog. The test bench for this circuit can consist simply of a low RES_n pulse and a periodic 2-KHz clock signal. The LCD interface circuit then should go through all of the command and data phases necessary to completely download and display its graphic data on the HG12601 LCD. The design of this Verilog test bench code is left to the reader.

# 5.5　Chapter Review

- A sequential circuit is one in which the current outputs depend on the *past* as well as the current inputs. The effect of the past inputs are stored in a register whose output is referred to as the *current state*. A synchronous sequential circuit is one in which the current state only changes at specific times defined by a periodic clock signal.

- The *state machine* method is a systematic procedure for designing general synchronous sequential circuits. This is a 5-step procedure consisting of the design of the (1) pseudocode, (2) RTL (register-transfer level) program, (3) datapath, (4) state machine, and (5) control logic. Each step can be accomplished by using information from the previous steps. The *one-hot* method is a particularly simple and straightforward method for designing the control logic based on the datapath and the state machine. However, since the one-hot method requires one flip-flop for each state, it may be necessary to use an encoding method (such as binary or Gray code encoding) for circuits with large numbers of states.

- For large and/or complex circuits, it is preferable to use an automatic design method based on the synthesis of the target circuit from a high-level behavioral HDL description. It is possible to write a high-level behavioral HDL description in a manner analogous to a computer program, complete with function calls and sequences of operations in each function. Provided that this high-level HDL description is written in a synthesizable manner, the synthesized circuit will be logically equivalent to the same circuit designed in a manual manner using the state machine method. In order to write high-level HDL code that synthesizes into fast and efficient circuits, it is helpful to understand the manual design process and to know the types of circuits that are synthesized from different types of HDL codes.

- The design of modern computer-interface circuits, even a simple interface to an LCD module, typically requires a protocol consisting of a sequence of command and data packets sent to and received from the target device. This command/data sequence must follow the timing requirements of the target device. The methods described in this chapter are useful in writing the high-level synthesizable behavioral HDL code or manually designing the interface circuit for this type of computer-interface protocol circuit,

# 5.6 Resources

[Hayes 1993] and [Wakerly 2002] are two representative textbooks that cover introductory digital logic design. The ASM (algorithmic state machine) method, which has similarities with the state machine method introduced in this chapter, originally was introduced in [Clare 1973]. The application of the ASM method to Verilog code is discussed in [Arnold 1999]. IEEE standards such as [IEEE 2001], mentioned in this and other chapters, can be ordered online through the IEEE web page. The web page http://www.njr.com contains the data sheet for the NJR NJU6452A LCD driver chip.

### 5.6.1   Bibliography


[5.1]  ARNOLD, M. G., *Verilog Digital Computer Design: Algorithms into Hardware*, Prentice Hall, Upper Saddle River, NJ, 1999.

[5.2]  CLARE, C. R., *Designing Logic Systems Using State Machines*, McGraw-Hill, New York, 1973.

[5.3]  HAYES, J. P., *Introduction to Digital Logic Design*, Addison-Wesley, Reading, MA, 1993.

[5.4]   IEEE, *IEEE Standard for Verilog Hardware Description Language 2001*, IEEE Press, New York, 2001. Also available through the IEEE standards web page at http://standards.ieee.org.

[5.5]  "NJU6452A" Data Sheet, New Japan Radio Co., Ltd., http://www.njr.com.

[5.6]  WAKERLY, J. F., *Digital Design: Principles and Practices, 3rd Ed.*, Prentice Hall, Englewood Cliffs, NJ, 2002.

## 5.7 Problems

**P5.1.** Describe the sequence of steps required to successfully obtain a bachelor's degree in electrical engineering. Describe this sequence of steps as an algorithm written in pseudocode. Make any necessary assumptions; define inputs, outputs, and intermediate variables; and write your algorithm much like a typical computer algorithm.

**P5.2.** Write an algorithm, in pseudocode, to describe the operation of an airplane making a successful landing on an airport runway. Make any necessary assumptions; define inputs, outputs, and intermediate variables; and write your algorithm much like a typical computer algorithm.

**P5.3.** Suppose that we wish to design a digital thermostat: a device that turns on or off the heater or air conditioner in order to maintain the room temperature between a lower bound and an upper bound.
  (a) Make any necessary assumptions and write an algorithm (in pseudocode) to describe the desired operation of this target circuit.
  (b) Convert the pseudocode for this circuit into an RTL program.

**P5.4.** Devise a high-level algorithm (written in pseudocode) for a digital hardware circuit used to control an automatic-teller machine that performs three tasks: tells the user the balance of his bank account, permits the user to withdraw an amount of money not greater than the balance on his account, and permits the user to deposit money into his account.

**P5.5.** Summarize the differences between a finite state machine and a state machine as described in this chapter. Construct a simple example to illustrate these differences.

**P5.6.** List and describe the differences between an algorithmic state machine (ASM) and a state machine as described in this chapter. Construct a simple example to illustrate these differences.

**P5.7.** Starting from the state transition table of Table 5.1, derive the logic equations for the binary-encoded factorial circuit control logic.

**P5.8.** Use the 5-step state machine method to design a circuit to control the operation of a micromouse that can find its way through a maze to reach a target point marked with a large X. Use the one-hot method for the control logic design. Make any necessary assumptions and design only the main "processor" for this circuit. Thus, for example, it can be assumed that there is a separate module that can perform the sequence of motor movements necessary to make the micromouse make a 90-degree turn to the right when a maze wall is detected in the forward direction.

**P5.9.** Use the 5-step state machine method to design a circuit that can perform the multiplication of two 8-bit unsigned numbers using shift registers, parallel load registers, counters, adders, and multiplexers. Use the binary encoding method for the control logic design.

**P5.10.** Use the 5-step state machine method to design a circuit that can perform the division of two 8-bit unsigned numbers using shift registers, parallel load registers, counters, adders, and multiplexers. Use the Gray-code encoding method for the control logic design.

**P5.11.** Use the 5-step state machine method to design a digital thermostat device, a circuit that maintains the room temperature between a lower and an upper bound. Use the one-hot method for the control logic design.

**P5.12.** Use the 5-step state machine method to design a circuit to control a pedestrian crosswalk traffic light such that the green "walk" signal blinks at a successfully faster rate as the time for the red "don't walk" signal approaches. Use the control logic design method that is most appropriate for this state machine. State the reasons for your choice.

**P5.13.** Solve Problem P5.12 using the ASM chart method.

**P5.14.** Figure 5.12(b) shows a *two-phase handshake* protocol for asynchronous communication between two independent circuits. For this two-phase handshake protocol, design both the transmitter and receiver circuits using either the state machine design method or the ASM-chart design method.

**P5.15.** Write a synthesizable Verilog program for the pedestrian crosswalk traffic-light control circuit described in Problem P5.12.
   (a) Write a testbench for this circuit and then verify the proper operation of the circuit using a commercial simulation tool.
   (b) Next, synthesize this circuit and verify (via logic equations) that the synthesized circuit is logically equivalent to the manually designed circuit for Problem P5.12.

**P5.16.** Write a synthesizable Verilog program for the multiplier circuit of Problem P5.9 using only addition and shift operations. Next, write another Verilog program that uses the Verilog multiply operator (*) to simplify the program. Synthesize both Verilog programs and compare the two synthesized circuits.

**P5.17.** Use the automatic state machine design method to design the control circuit for a single floor of a two-elevator system with only one set of up and down buttons. Design the pseudocode, its corresponding Verilog program, and the test bench code. Test the Verilog program using simulation, and then synthesize the circuit and test the synthesized circuit using simulation.

**P5.18.** Write and verify (using test bench-based simulation) a synthesizable Verilog program for a money changer that can change quarters into dollar bills and dollar bills into quarters.

**P5.19.** What extra components need to be added to the vending machine circuit designed in Section 5.3 in order to make it a complete circuit for a practical vending machine? Of these extra components, write and verify (using test bench-based simulation) all of the components that can be implemented as digital logic circuits.

**P5.20.** Write and verify (using testbench-based simulation) a synthesizable Verilog program for a coffee vending machine that pours three different types of coffee (i.e., black, with cream, and with both cream and sugar) into a paper cup.

**P5.21.** Modify the LCD interface circuit of Section 5.4 so that the LCD displays a line that ascends at a 45-degree angle in a "wraparound manner"(i.e., when the line reaches the topmost point it starts over at the lowermost point.) Note that the line must continue across the boundaries of the four "pages" of the NJU6452A chip.

**P5.22.** How could the LCD interface circuit of Section 5.4 be modified to display an arbitrary graphical diagram or sequence of characters? Consider the additional devices that are necessary and the additional code required to store and read from those devices.

**P5.23.** Suppose we wish to design an LCD interface circuit for a character LCD display module. Let us assume that this LCD module has an LCD control chip that stores a set of fonts for all possible characters and displays the appropriate character (at the current cursor position) when given the corresponding ASCII code for that character (on the 8-bit DB lines) with $A0 = 0$. Make any other necessary assumptions and write the Verilog code for this character LCD module interface circuit.

**P5.24.** What type of encoding method is most appropriate for the design of the control logic for the LCD interface circuit of Section 5.4? Why? Disregarding the encoding method that you selected, use the binary-encoding method to design the control logic for this circuit.

**P5.25.** Write a test bench for the LCD interface circuit of Section 5.4. Draw a timing diagram (just show the important parts) of the anticipated behavior of the LCD interface circuit with your test bench.

# FPGA and Other Programmable Logic Devices

## Important Concepts

- The concept of a programmable logic device (PLD), the various types of PLDs, and how they can be programmed to form specific digital logic circuits.

- The concept of a field-programmable gate array (FPGA) and how it can be programmed to form specific digital logic circuits.

- How to effectively use FPGAs in conjunction with a hardware description language, such as Verilog, to design and implement large digital logic circuits.

The trend nowadays, and in the near future, is to try to place more and more functionality into a single chip. Digital logic hardware circuits typically can be made more economically by implementing an entire application using a few large chips than many small chips. For large volume circuits, this objective can be met by using a custom-designed *VLSI (very large scale integration)* chip or an *ASIC (application-specific integrated circuit)* chip, in which the target circuit is designed using pre-designed standard-cell libraries of basic components such as NAND and NOR gates. However, since VLSI and ASIC chips require an expensive fabrication process, there is a fabrication delay of a few weeks to a few months, and the initial fabrication cost is fairly expensive. Moreover, the fabrication process must be repeated if an error is discovered in the original design. For these reasons, many small-to medium-volume (and even a few large-volume) hardware designs are implemented using *field-programmable gate array (FPGA)* chips, which are mass produced at the factory and *programmed* by the end-user to form customized circuits. Smaller designs also can be implemented in end-user programmable chips referred to as *PLD (programmable logic device)* chips. This chapter will examine the use of such FPGA and PLD chips.

● ● ● ● ● ● ● ● ● ● ● ● ● ● ● ● ● ●

# 6.1 **Programmable Logic Devices**

A programmable logic device (PLD) is a chip that is mass produced at the factory and then customized by the end-user to form different logic circuits. The circuits that can be implemented using a single PLD chip range from simple combinational-logic circuits to fairly complex sequential state machines. A large variety of PLD (programmable logic device) chips have been proposed over the past few years. Of these, the most well known ones are PLA (programmable logic array), PAL (programmable AND-array logic), GAL (generic-array logic) and PROM (programmable read-only memory) chips. Some designers also view an FPGA as a form of PLD, since an FPGA is also a device programmed by the end-user.

### 6.1.1   Circuit Customization

There are various ways in which circuit customization for PLDs, also referred to as *PLD programming*, can be accomplished. The circuit customization techniques used are applicable to memory devices and FPGAs in addition to PLDs.

The customization of a PLD involves temporarily or permanently creating shorts or opens at specific crosspoints in a matrix of horizontal and vertical wires. The chip can be manufactured to have shorts at all configurable horizontal-to-vertical crosspoints, and certain crosspoints can be changed to the open state during PLD programming by the user. Alternatively, the chip can be manufactured to have opens at all configurable horizontal-to-vertical crosspoints, and certain crosspoints can be shorted together during PLD programming by the user. Also, PLD programming can be performed in a manner that permanently changes all configurable horizontal to vertical crosspoints. Alternatively, the programming can be done in a manner such that the circuit change is only temporary (i.e., the original unconfigured circuit can be retrieved by turning off power to the PLD or actively erasing the configuration information in some manner). This also leads to the concept of *volatile* and *non-volatile* PLD programming methods, depending on whether or not the configured circuit disappears when power is removed from the PLD.

Let us first examine permanent methods for PLD programming. The concept is easiest to understand by analogy to the fuse in an expensive electrical appliance. When there is an accidental surge in current in such an appliance, the fuse, which is essentially a thin filament wire in the direct power path from the electrical power supply (which is encased in a removable module), will burn out and prevent electricity from reaching the appliance. Using *fusible link* technology, a PLD can be constructed to have this type of fuse connecting the horizontal to the vertical wire in a set of programmable crosspoints. Then, by selecting a specific crosspoint and sending a surge of current (or a high voltage) to the fuse at that crosspoint, an open circuit can be created at that crosspoint. In an analogous manner, it is also possible to use *antifuse* technology, which can be thought of as the inverse of the fuse method (i.e., all crosspoints initially have open circuits, and a specific crosspoint can be selected and changed to the "connected" state by sending a surge of current or voltage to that crosspoint).

Programmable devices that undergo permanent changes in circuit structure during programming typically are referred to as *one-time programmable (OTP)* devices. Since antifuse technology is cheaper and more reliable than fuse technology, the earliest PLD devices were created using antifuse technology. However, with advances in transistor technology over the past few decades, modern PLD devices typically use even more reliable methods based on using specially designed transistors at the programmable horizontal-to-vertical crosspoints. Open or short circuits effectively can be created by changing the properties of such transistors to prevent or permit current flow. The physics behind such techniques is quite involved, and the interested reader is referred to textbooks devoted to semiconductor devices. Figure 6.1(a) shows a diagram of a fusible link or antifuse connection.

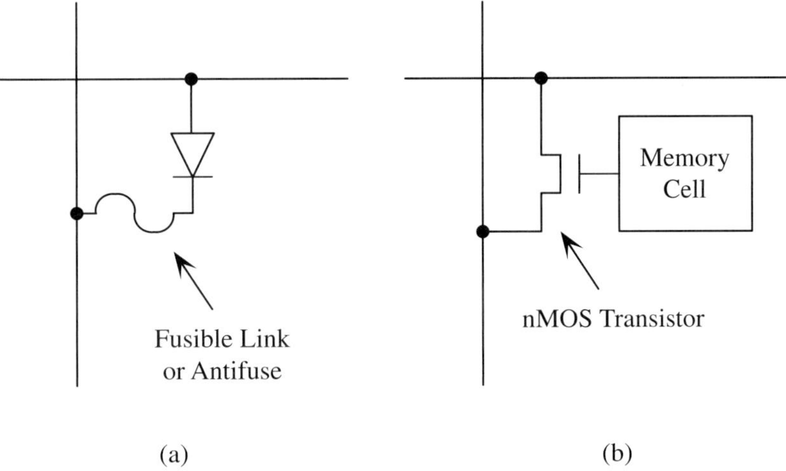


**Figure 6.1**
PLD programming using (a) fusible link or antifuse technology and (b) nMOS transistor switchescontrolled by data stored in memory cells.

Unlike fusible link and antifuse technologies, which create permanent changes in circuit structure, it is possible to use a technology in which the circuit change is only temporary. This results in the capability of reprogramming the same chip many times over, forming different circuits after each reprogramming phase. Figure 6.1(b) shows the most common method for creating this type of reprogrammable circuit connection. At each programmable horizontal-to-vertical crosspoint, there is an nMOS transistor whose gate input is connected to a memory cell. By selecting that memory cell and storing a logic '1' (high voltage) or logic '0' (low voltage) value, the nMOS transistor behaves as a closed or open switch, respectively. If the memory cell uses a volatile memory technology, such as static random-access memory (SRAM), then the PLD is a volatile device which loses its circuit information when power is removed. Alternatively, if the memory cell uses a nonvolatile memory technology, such as flash memory, then the PLD becomes a nonvolatile device in which the programmed circuit remains intact even after power is removed.

From the end-user's viewpoint, a special programming device referred to as a *ROM writer* (also referred to as a *universal programmer* or *PLD programmer*) typically is used to perform the circuit customization for a PLD. The circuit to be programmed into a PLD is described using a hardware description language (or a special memory data-description file format such as the Intel Hexadecimal Format), converted into a downloadable-file format, and then downloaded to the PLD using a ROM writer or a specialized circuit interface. A ROM writer is a device that typically is connected to the parallel port of an x86 personal computer. Although it normally is used for storing memory data into a PROM, EPROM (erasable PROM), or EEPROM (electrically erasable PROM) chip, a ROM writer also can be used to download circuit data into a PLD, since the same technology is used to store horizontal-to-vertical crosspoint connection information.

### 6.1.2 Programmable Logic Arrays

The first device developed specifically as a PLD, and still widely used today as a library part in ASIC designs, is the *programmable logic array* (PLA). A PLA has the general structure shown in Figure 6.2. Note that, as shown in the figure, for ease of notation, a multiple-input AND (or OR) gate can be drawn using a single input line and multiple connections to that single input line (however, all of the multiple connections are actually separate lines). As can be seen from the figure, there is an array of AND gates followed by an array of OR gates. The outputs of the AND gates are product terms, and the outputs of the OR gates are sum-of-product terms. Any sum-of-product expression can be realized using such a structure, provided that the total number of variables is not greater than the number of primary inputs to the PLA, the number of product terms is not greater than the number of product lines available, and the number of sum terms is not greater than the number of OR gates available. An $n \times p \times m$ PLA is used to refer to a PLA with $n$ primary inputs, $p$ product terms, and $m$ output lines (such a PLA has a $2n \times p$ AND plane and a $p \times m$ OR plane).

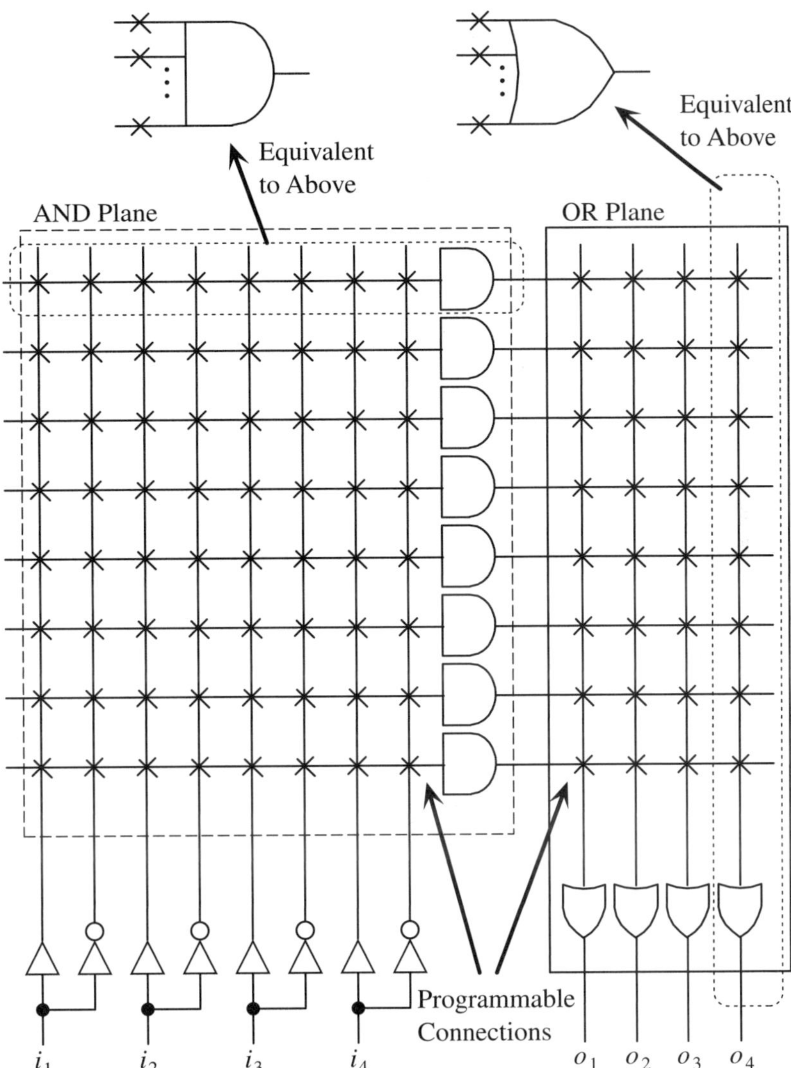



**Figure 6.2**
The general logical
structure of a PLA
(programmable
logic array).

---

**Example 6.1**   **PLA Usage**

Suppose that we wish to use a $6 \times 8 \times 4$ PLA to implement a full-subtracter module, such as that introduced in Example 3.1. This module performs the operation $a - b - borrow\_in$, with the subtraction result assigned to *res* and the borrow out result assigned to *borrow_out*. Substituting $w$ for *borrow_in*, The relevant logic

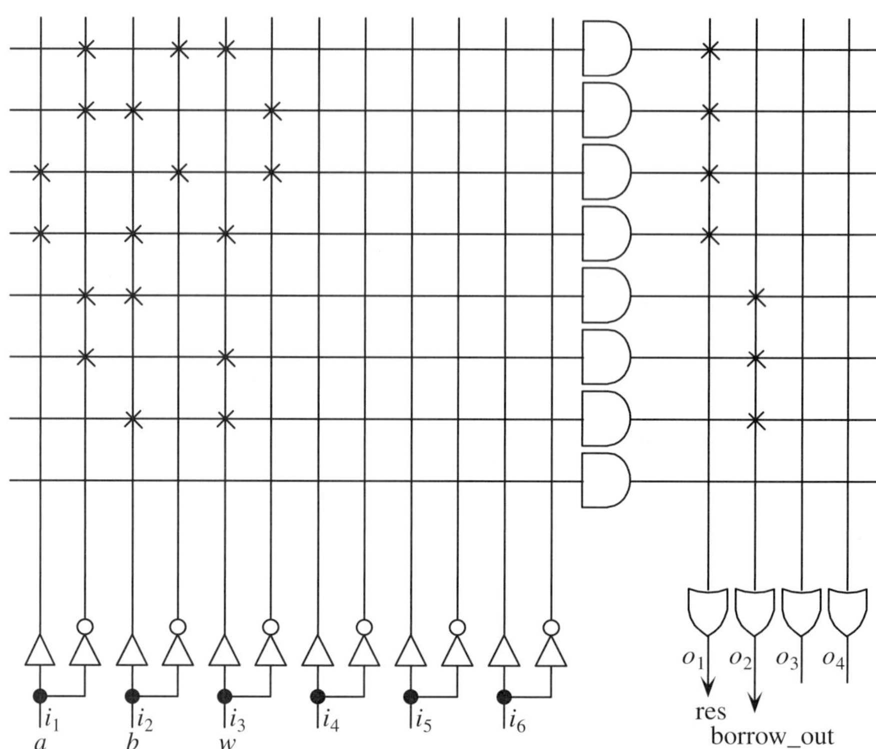



**Figure 6.3**
A $6 \times 8 \times 4$ PLA programmed to implement a full-subtracter circuit.

equations for this module can be summarized as follows.

$$\mathrm{res} = a \oplus b \oplus w$$

$$= a'b'w + a'bw' + ab'w' + abw$$

$$\mathrm{borrow\_out} = a'b + a'w + bw$$

Then, the horizontal-to-vertical and vertical-to-horizontal crosspoints must be programmed to implement these two logic equations. Figure 6.3 shows the $6 \times 8 \times 4$ PLA programmed to implement the logic equations for $res$ and $borrow\_out$. Shorted connections are marked with $\times$ and open connections are left unmarked. There are three primary inputs, seven different product terms, and two primary outputs. Any inputs, product terms, or outputs that do not need to be used simply can be left unprogrammed.

For efficiency of hardware usage and high performance, a PLA should be constructed in a compact rectangular portion of the silicon die. The PLA should be designed so that AND gates with possibly large numbers of inputs (in the AND plane) and OR gates with possibly large numbers of inputs (in the OR plane) can be

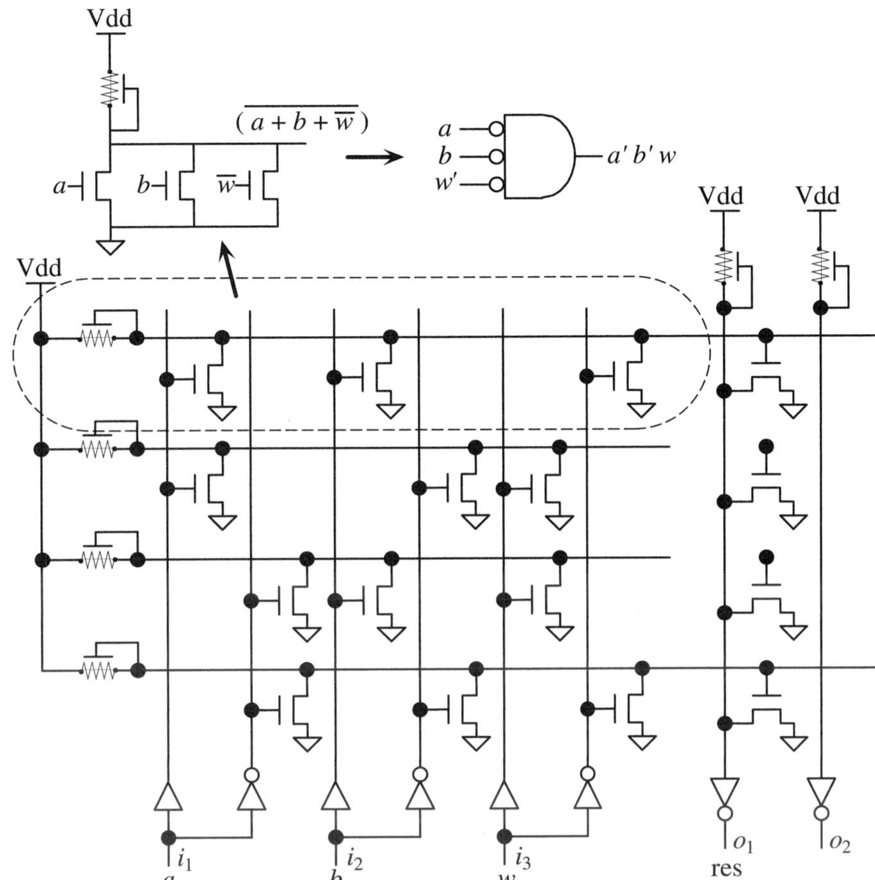

**Figure 6.4**
An nMOS transistor realization of a PLA circuit.

constructed. To understand how all of these requirements can be met, it is helpful to view the transistor design for a PLA circuit. Several different types of designs have been proposed. However, most of them are based on a design similar to the NOR-NOR-invert nMOS transistor implementation shown in Figure 6.4. Note that this figure shows a $3 \times 4 \times 2$ PLA configured to implement part of the full-subtracter circuit (the *res* logic equation part).

Figure 6.4 shows how a NOR gate with a large number of possible inputs can be laid out in a linear one-dimensional structure. Interestingly, NOR gates are used in both the AND plane and the OR plane. This can be done because a NOR gate is a universal gate. Thus, sets of NOR gates can be used to implement an AND gate or an OR gate. In this case, a NOR gate in the AND plane can be viewed as an AND gate with inverted inputs. Likewise, a NOR gate in the OR plane can be viewed as an OR gate with inverted outputs. Thus, a NOR-NOR-invert topology becomes equivalent to an invert-AND-OR topology (NOR = invert-AND and NOR-invert = OR).

### 6.1.3  Programmable Read-Only Memories

As implied by its name, a ROM (read-only memory) is basically a memory device. The manufacturer fabricates the ROM as a silicon chip using a sequence of chemical processes. By customizing a "mask" (similar to a photomask used in photo processing) based on memory data provided by a customer, the ROM can be manufactured to store a specific set of memory data. There is a set of address lines used to point to a specific data location. Then, by asserting an output enable signal (and perhaps a chip select signal), the data at that location can be observed through the data lines. A PROM (programmable read-only memory) is a ROM in which the customization of the chip (to store memory data) can be done by the end-user instead of the chip manufacturer.

However, a memory device as described here also can be considered as a programmable combinational-logic device. Primary inputs can be connected to the address lines, primary outputs can be connected to the output data lines, and specific combinational-logic functions can be implemented by storing the appropriate data into the PROM. In order to determine what data needs to be stored in the PROM, it is helpful to note that this PROM structure can be viewed as a special type of PLA structure: one in which only the OR plane can be programmed, the AND plane is fixed, and the product terms corresponding to all possible minterms must be included in the PLA. Figure 6.5 shows a PROM drawn as a PLD.

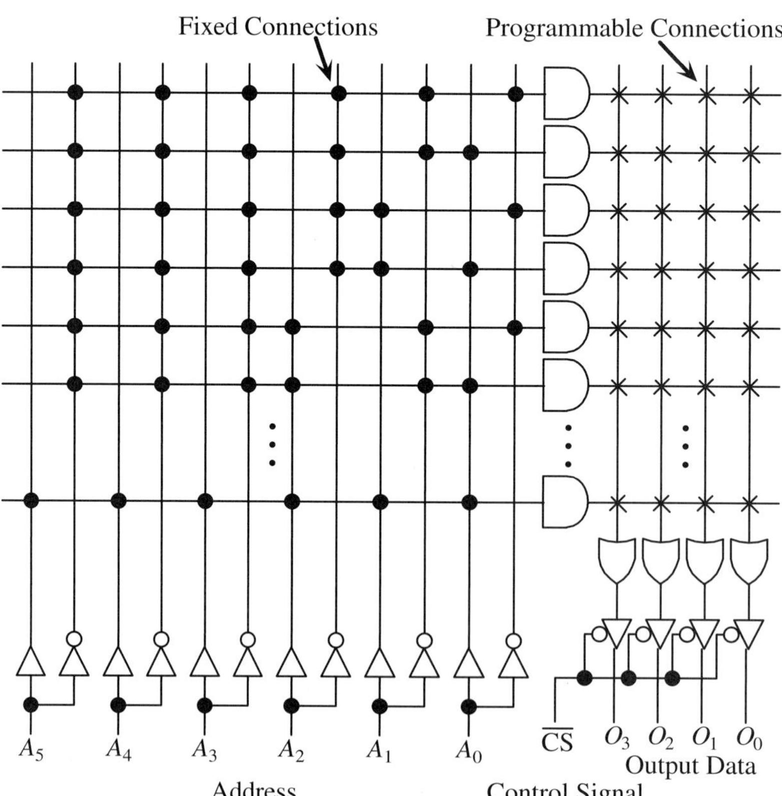


**Figure 6.5**
A PROM drawn as a programmable combinational-logic device.

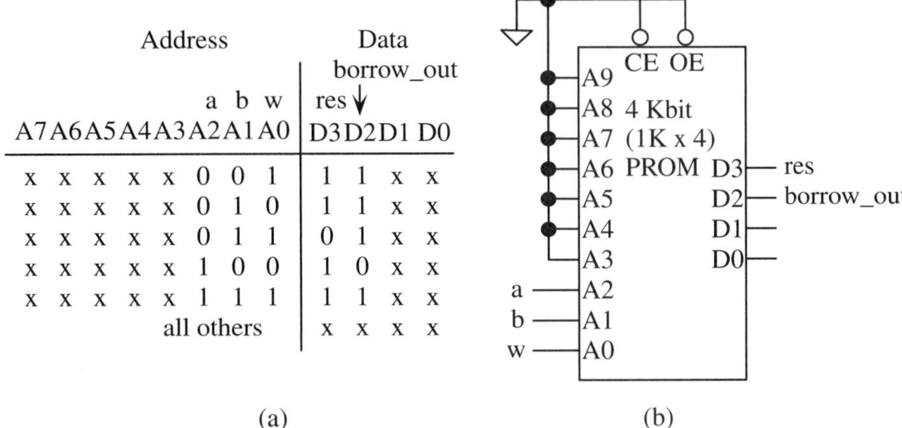



**Figure 6.6**
A 4 K-bit (1-K × 4) PROM used to implement a full subtracter: (a) the contents of the memory and (b) the physical connections.

---

**Example 6.2**   **Using PROM as a PLD**

In order to use a PROM as a PLD, we must program the PROM to store the "truth table" corresponding to the combinational logic function that we wish to implement. Since all possible minterms will be present in the PROM, there is no need to produce simplified logic equations. Next, we must assign primary input variables to the individual address lines of the PROM and primary output variables to the individual data lines of the PROM. Any unused data lines can be left unconnected. However, unused address lines should be connected to ground or Vdd (the chip select and output enable signals should also be asserted). The reason for this is that if chip inputs are left unconnected, then it is possible for the chip to malfunction due to noise entering the chip through the unconnected chip inputs.

Let us use a 4 K-bit (1-K × 4 format) PROM chip to implement the full subtracter of Example 6.1. In order to show the complete solution, we must present the data to be stored in the PROM and the physical connections to be made to the PROM chip. The data to be stored simply can be based on the truth table for the logic functions to be implemented. Ideally, it should be listed in an address-data table format. Figure 6.6 shows the complete solution for this PROM chip. Note that there are ten address lines and four data lines, since it is a 4 K-bit PROM in 1-K ($= 2^{10}$) × 4 format. Note also that all unused address lines have been connected to ground, and the $\overline{CS}$ and $\overline{OE}$ lines have also been connected to ground to permanently enable the PROM. chip.

---

### 6.1.4   Programmable And-Array Logic

Many different types of PLDs have been proposed by various manufacturers over the years. Although a PLA is an extremely useful device, it is costly to produce a PLA as a separate chip because of the large number of connection points that have to be made programmable. A PROM only requires that the connections in its OR plane be made programmable. However, a PROM also requires $2^n$ product lines, where $n$

is the number of address lines (or primary inputs to the PLD), which can become an extremely large number given a large $n$. The various PLDs proposed over the years have attempted to address these shortcomings in addition to adding other useful capabilities. The extra capabilities that have been added include programmable flip-flops, tri-state buffers, feedback paths, bidirectional ports, and the capability of operating in several "modes" (combinational, sequential, tri-state, etc.), depending on the requirements of the user. Most of these advanced PLDs are variations of a type of PLD referred to as a *programmable AND-array logic (PAL)*.[1]

As implied by its name, a PAL is a PLD with a programmable AND plane and a fixed (nonprogrammable) OR plane. By making the OR plane fixed, the number of connection points that have to be made programmable is reduced drastically. This results in a device that is much cheaper to manufacture. PAL devices are manufactured in many different configurations. For example, a PAL16L4 chip is a PAL with 16 primary inputs and four active-low primary outputs. The primary outputs are produced from a set of OR gates (or NOR gates), each with a fixed number of inputs. For this device, since there are only active-low primary outputs, in order to implement a logic function with the output $F$ directly, we should either formulate the logic equation for $F'$ and implement that or use an extra inverter in order to produce a positive logic $F$ output. This is illustrated in the following example.

**Example 6.3**    **PAL Programming**

Let us use a PAL18L4 chip to implement the full-subtracter module of Example 6.1. By referring to the data sheet for the PAL18L4, it can be determined that this is a chip with 18 primary inputs and four active-low primary outputs. Since this PAL chip has only active-low outputs, in order to generate the *res* and *borrow_out* outputs directly, it is necessary to implement the inverse functions *res'* and *borrow_out'*. The equations for *res'* and *borrow_out'* can be derived by using DeMorgan's Theorem or by "circling the 0's" in the K-maps for the logic functions *res* and *borrow_out*. Using either method, we obtain:

$$\text{res}' = a'b'w' + a'bw + ab'w + abw'$$

$$\text{borrow\_out}' = ab' + aw' + b'w'$$

Based on these equations, the programmable connections in the PAL18L4 chip can be programmed as shown in Figure 6.7. Note how the product terms produced are those for the inverse functions *res'* and *borrow_out'*, even though the outputs produced are *res* and *borrow_out*.

---

[1]    Information on the other types of PLDs available can be obtained by referring to introductory digital-logic textbooks and the web pages of modern PLD manufacturers.

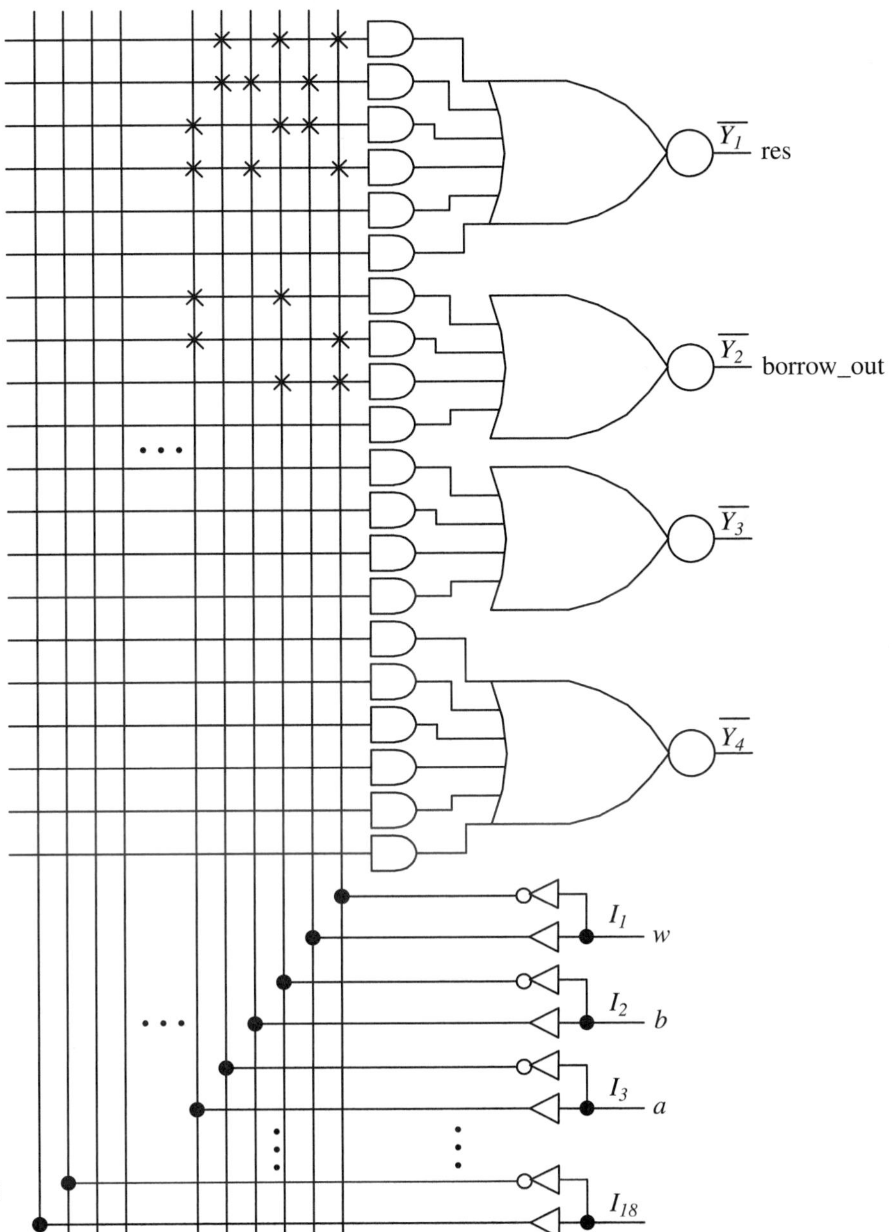



**Figure 6.7**
A PAL18L4 programmed to implement a full subtracter.

## 6.2  Field-Programmable Gate Arrays

FPGAs are the most complex types of chips that can be programmed by the end-user to form customized circuits. There are even FPGA chips with several million transistors, rivalling the size, circuit density, and switching speed of VLSI and ASIC chips. As its name implies, an FPGA is a field-programmable version of a *gate array*. Thus, in order to understand FPGAs, it is helpful to study gate arrays first.

### 6.2.1  Gate Arrays

A custom-designed VLSI chip is designed from scratch by specifying the shapes for each of the structures used to form transistors, wires, and connections in the chip. This permits the designer to customize each transistor so that it has the exact analog and digital characteristics (delay, signal strength, voltage amplification, etc.) required for his design. However, this also makes the design process extremely tedious and time-consuming. An ASIC chip is designed by connecting together predesigned structures (logic gates, common combinational and sequential devices, etc.) from a standard library in order to create the target circuit. Although this limits the ability of the designer to create chips with extremely high performance or finely tuned electrical characteristics, it does result in a significantly faster design process.

A gate array is a type of ASIC (application-specific integrated circuit). In a gate array, there is a rectangular matrix of predesigned structures (cells) laid out a in a two-dimensional array topology, with sets of vertical and horizontal wires between adjacent cells. The designer creates his circuit by customizing each cell to form one of a set of possible devices and specifying connections to the vertical and horizontal wires in the "routing channels" between cells. The general structure of a gate array is shown in Figure 6.8.

Each cell in the gate array can be customized to become one of several possible SSI or MSI-level logic gates or devices. This customization is achieved by connecting or disconnecting specific wires in the programmable cell, such as is done in the PLD programming process discussed in Section 6.1.1. In a like manner, connections can be made from the cells to the wires in the routing channels to form larger circuits. This type of design process results in the extremely fast and reliable design of ASIC chips (at the cost of resource, power, and performance losses due to the use of predesigned cells and routing wires, many of which will not even be used).

---

**Example 6.4**   **Customizing a Gate Array**

Suppose we wish to use a gate array to implement the logic function

$$f = (ab' + a'be)c' + bcd' \tag{6.1}$$

Let us also assume that the gate array has one 4:1 multiplexer (MUX), one inverter at each cell location, and a bunch of horizontal or vertical wires between each adjacent pair of cells. This type of gate array, customized to realize the above equation, is shown in Figure 6.9. The customization used in this figure is explained next.

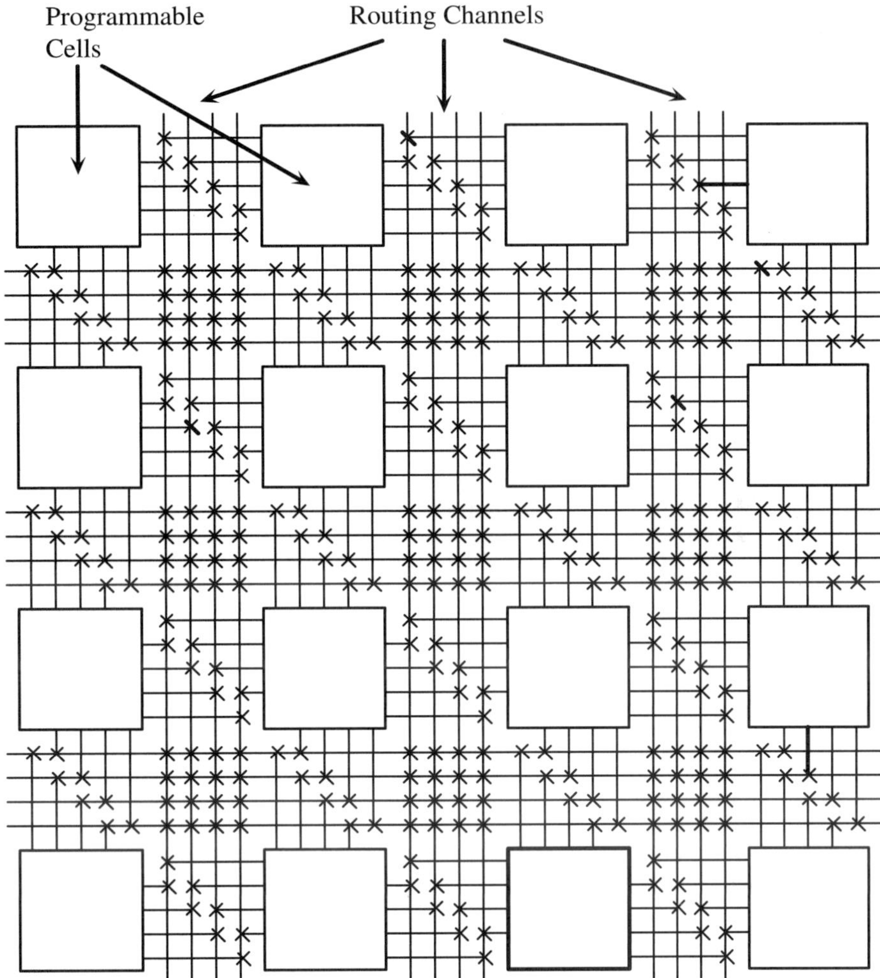


**Figure 6.8**
The basic structure
of a gate array.

First, we must make use of the fact that a multiplexer can be used to implement an arbitrary combinational-logic function, as long as the function does not have too many inputs. More specifically, in introductory digital logic, it is taught that a $2^k : 1$ MUX can implement any combinational logic function of up to $k+1$ variables. Thus, a 4:1 MUX can be used to implement functions of up to three input variables since $4 = 2^2$ implies that $k = 2$. The function $ab'$ can be implemented by connecting $a$ and $b$ to the select inputs and connecting $(0, 0, 1, 0)$ to the $(I0, I1, I2, I3)$ inputs of a 4:1 MUX. Furthermore, the function $(ab' + a'be)$ can be implemented by connecting $a$ and $b$ to the select inputs and $(0, e, 1, 0)$ to the $(I0, I1, I2, I3)$ inputs of a 4:1 MUX.

Next, to implement the function $(ab' + a'be)c'$, we must use a second 4:1 MUX to implement $K_1c'$, where $K_1 = (ab' + a'be)$, the output of the first 4:1 MUX. Thus, the output of one MUX must be connected to an input of another MUX residing in a separate cell. This can be accomplished by connecting the output of the first MUX

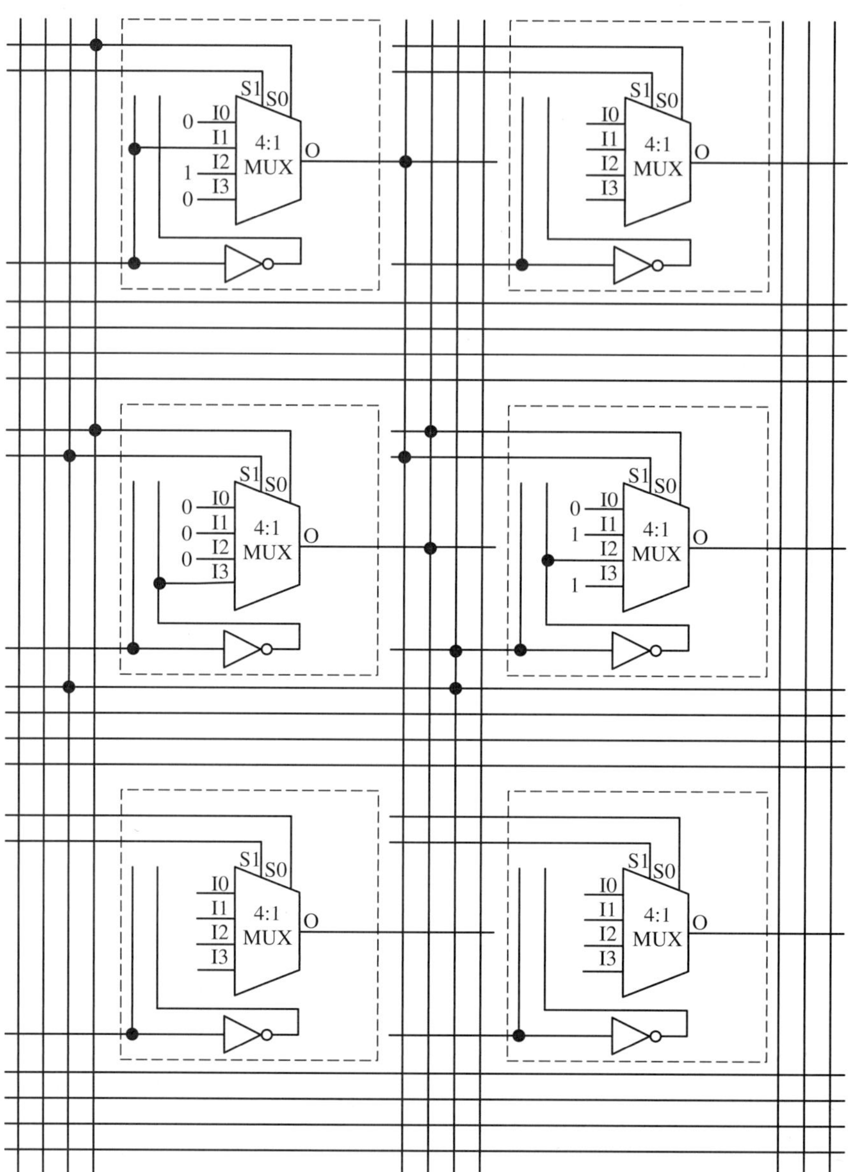

**Figure 6.9**
Customization of a gate array to form $f = (ab' + a'be)c' + bcd'$.

to one of the horizontal or vertical wires running between adjacent pairs of cells and then connecting that same wire to the input of the second MUX. Continuing in this manner, it can be seen that the entire function $f$ can be implemented using three cells and connections between those three cells. This type of circuit customization, shown in Figure 6.9, involves storing 1's and 0's at specific locations and breaking or creating connections between specific horizontal and vertical wires.

### 6.2.2   FPGA Overview

A field-programmable gate array (FPGA) is a chip with predesigned and fabricated cells and routing wires that is customized by creating (or destroying) connections within the cells and between the routing wires and cells. The technology used for FPGA circuit customization is the same as that used for PLD programming, as discussed in Section 6.1.1.

### 6.2.3   Xilinx FPGA

As an example of an actual commercial FPGA chip, let us examine the XC2S200 Spartan-II family FPGA chip produced by Xilinx Corp. This chip is a 2.5 V CMOS device with the equivalent of 200,000 logic gates, a separate 56 K bit RAM memory, and 140-284 user I/O pins (depending on the package type). As with most Xilinx FPGAs, it uses SRAM cells to store the configuration information and is thus a volatile programmable chip that can undergo an almost unlimited number of reprogramming cycles.

Figure 6.10 shows a block diagram of the internal structure of a Spartan-II family FPGA chip. In the middle is a matrix of configurable logic blocks (CLBs), which

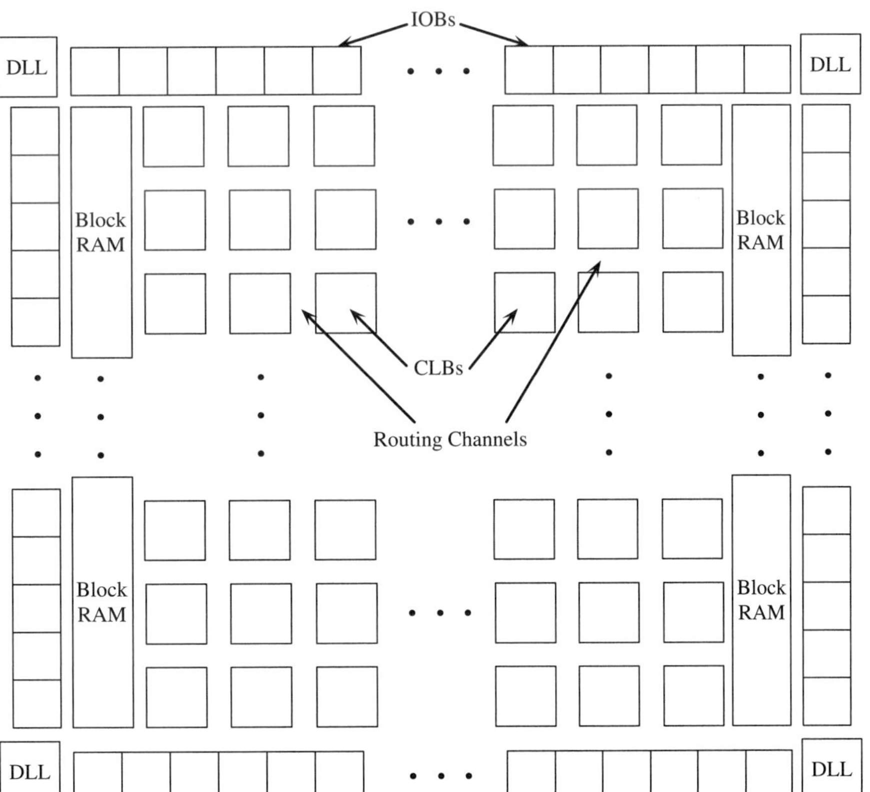


**Figure 6.10**
Block diagram of a Spartan-II family FPGA [Xilinx 2000].

correspond to the array of configurable cells in a gate array ASIC. In the channels between adjacent CLB columns and CLB rows are sets of routing wires that optionally can be connected during the FPGA configuration process.

The Xilinx Spartan-II family FPGA also has other components that have been added to enhance its capabilities. The blocks labeled "Block RAM" correspond to pretested SRAM structures which can be utilized as part of the user's FPGA design. If the user requires more memory than that available from these Block RAM cells, then some of the CLB cells can be configured and used as extra RAM cells. The four blocks labeled "DLL" are delay-locked loop circuits that can be used to synchronize global clock signals by delaying certain clock signal edges so that all clock signal edges arrive at D flip-flops at the same time instant. Finally, the I/O logic cells include I/O block (IOB) cells that can be configured to be compatible with various I/O interface methods; serve as input, output, or bidirectional I/O pins with or without tri-state capability; and support *boundary scan*[2] for chip configuration and debugging.

The main configurable logic in the Xilinx Spartan-II family FPGA is contained in the matrix of CLB (configurable-logic block) cells. Each Xilinx CLB consists of several look-up tables (LUTs,—i.e., small PROMs), multiplexers (MUXes), D flip-flops, and other random logic. The Xilinx Spartan-II CLB, in particular, consists of four logic cells arranged in two slices. One slice of the Spartan-II CLB is shown in Figure 6.11. Each 4-input LUT can be configured to implement an arbitrary 4-input combinational logic function or a $16 \times 1$ synchronous RAM cell. The LUTs in a CLB can be combined to form larger functions or larger synchronous RAM cells. There is an array of $28 \times 42 = 1,176$ such CLBs in the XC2S200 chip.

The other major configurable portion of the Xilinx Spartan-II family FPGA is the Block RAM cell. As shown in Figure 6.12, each Block RAM cell is a fully synchronous dual-ported 4096-bit RAM with independent read/write control for both ports. Thus, a Block RAM cell can be used to implement a normal read/write RAM or a FIFO (first-in-first-out) queue.

### 6.2.4  FPGA Configuration

Let us now study the details of how a user design can be implemented in an FPGA chip. FPGA companies typically provide extensive software support to enable users to easily design and implement hardware designs using the targeted FPGA chips. The software typically available includes tools for schematic design entry, language-based design entry, logic simulation, synthesis using the available programmable logic gates and connections (a synthesis tool *generates* a gate-level logic design from a high-level structural or behavioral description of a circuit), and the creation of a download file that can be used to program the targeted FPGA chip. Such tools free the user from having to derive and map sophisticated logic diagrams to the logic gates and routing channels of the target FPGA.

However, even with the FPGA software described here, additional hardware support is necessary before an FPGA chip actually can be programmed. An FPGA

---

[2]  Boundary scan is an interesting standard method for probing the I/O pin values of a chip. It will be described in detail in Section .

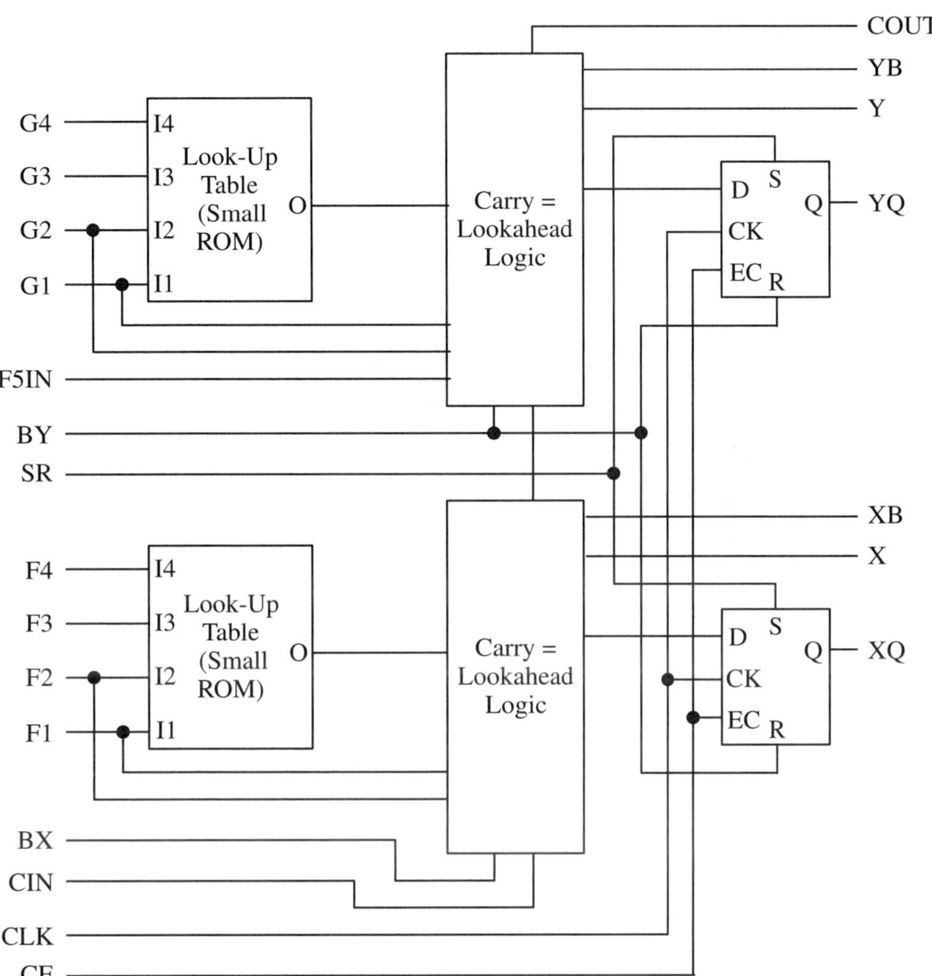


**Figure 6.11**
Block diagram of one slice of a Spartan-II CLB (two slices in one CLB) [Xilinx 2000].

chip typically has *mode* pins (which can be set to programming mode), and several other pins that can be used to actually enter a design into the FPGA chip. The mode pins must be set to specific values and a protocol must be followed in order to download a design into the FPGA chip. If the FPGA chip uses a nonvolatile technology for programming (setting the wire connections inside the chip), then a separate FPGA programming device (similar in function to a PROM programmer) can be used to set the mode pin and follow the protocol required to download a design into the FPGA chip. However, if the FPGA chip uses a volatile technology, then the design must somehow be entered into the chip (while it is inserted in the application board) each time the chip is powered on.

The programming of an FPGA chip (more typically referred to as *FPGA config-uration*), can be accomplished using one of several possible modes. First, data can be entered serially or in parallel. Second, data can be entered synchronously, timed to

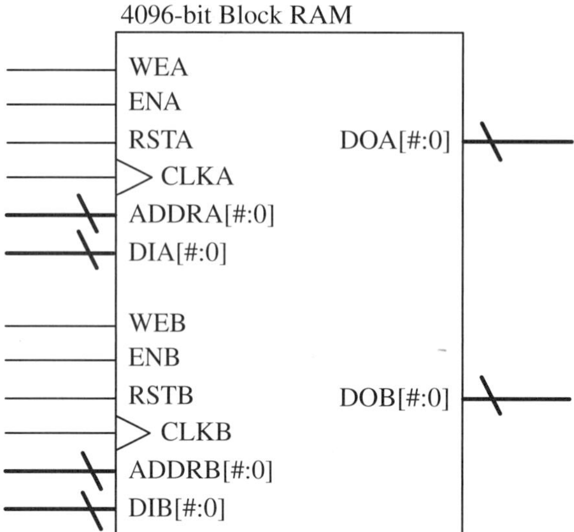


**Figure 6.12**
Pin-outs for the
Xilinx Spartan-II
family Block RAM
cell [Xilinx 2000].

a clock signal, or asynchronously with appropriate handshaking signals. Third, and most importantly, the FPGA can be configured in *master mode*, in which the FPGA is the initiator and controller of the configuration process, or in *slave mode*, in which an external device controls the configuration process. Not all combinations of the possible configuration modes are supported by all types of FPGA chips. Typically, the fastest configuration method uses a parallel, synchronous, slave configuration mode. However, since the size of a configuration file is on the order of about 2 M-bits, assuming a configuration clock of 2 MHz (which is very slow), the total configuration time is on the order of about 1 second, even if serial configuration is being used. Thus, for most applications, serial configuration modes offer sufficiently fast configuration. A configuration mode supported in some of the newer FPGAs is *boundary-scan* mode. In this mode, which can be considered to be a variant of slave serial synchronous mode, the FPGA is configured using the I/O pins normally reserved for boundary-scan testing of the chip (discussed in Section 6.2.6). An example of boundary-scan mode configuration will be discussed in Section 6.2.5.

Typically, however, the simplest method of FPGA configuration is to use a serial, synchronous, master mode in which the FPGA receives its data from a small serial PROM. With this method, the user simply has to program the PROM using a PROM programmer, connect the PROM to the FPGA chip using a simple and standard hardware circuit, and provide power to the FPGA and PROM. The FPGA then enters the master configuration mode, generates a clock signal, and requests and receives all of the data that it needs from the PROM. The main drawback of this method is that these PROM chips only can be programmed once. Alternatively, a serial EEPROM (electrically erasable PROMs, which can be reprogrammed multiple times using a PC) can be used. However, such serial EEPROM chips can wind up costing as much or more than the FPGAs.

An alternative FPGA configuration method is to use a serial, synchronous, slave mode. In this mode, the FPGA is configured by an external microprocessor or another FPGA in a daisy-chain configuration. When an external microprocessor is used, the microprocessor must generate the serial input data signal, the configuration clock signal, and other configuration-related signals by outputting data on an appropriate output port. Thus, the microprocessor must execute a program to output the appropriate sets of bits on the appropriate output port in order to generate the configuration-related signals that are required. The FPGA data sheet typically contains detailed information on the sequence of configuration-related signals and the configuration data format that is required.

Two PC-based configuration methods are discussed in Section 6.2.5. In the first method, the Xilinx XC2S200 chip is configured directly from a PC using a special parallel cable and slave-serial configuration mode. In the second method, a serial EEPROM (electrically erasable PROM) is first loaded with configuration data using boundary-scan configuration mode. Then the XC2S200 configures itself from the data in the serial EEPROM using master-serial configuration mode. Both of these methods require support from Xilinx software and hardware (the parallel cable). However, it also is possible to design and implement a custom configuration interface based on the information provided in the data sheets regarding the configuration data format and protocol. Such a custom configuration method, which can be useful for learning about the configuration process, is discussed in the reference [Lawman 1995].

### 6.2.5   Xilinx Spartan-II FPGA Configuration Example

Let us now consider, in detail, how an actual FPGA chip (the XC2S200 chip from the Xilinx Spartan-II family) can be configured. The Xilinx CAD software generates a downloadable binary file from a user-entered design. This binary file has a specific format consisting of a preamble, the configuration data, and a postamble. Since this binary file is generated for us, however, we do not need to be concerned with the specifics of this format (the format is described in the relevant data sheet). The downloadable binary file, with an extension of ".bit," must then be used to program the connections in the FPGA in order to create the desired circuit (i.e, the FPGA must be *configured*).

#### Configuration Modes

Of the possible configuration modes, Xilinx Spartan-II devices support the slave serial, master serial, slave parallel, and boundary scan modes. All configuration modes supported are synchronous modes. The slave serial, master serial, and slave parallel modes operate in the general manner discussed earlier. The boundary-scan configuration mode can be considered as a special variant of slave-serial mode configuration in which dedicated boundary scan I/O pins (TMS, TDI, TDO, and TCK) are used instead of the regular configuration I/O pins. Three mode pins ($M0$, $M1$, $M2$) are used to select among the four possible configuration modes. Although not necessary in general, it is possible to specify that all I/O pins should be

| Configuration Mode | Pre-Configuration Pull-Ups | $M0$ | $M1$ | $M2$ |
|---|---|---|---|---|
| Master serial | No | 0 | 0 | 0 |
|  | Yes | 0 | 0 | 1 |
| Slave parallel | Yes | 0 | 1 | 0 |
|  | No | 0 | 1 | 1 |
| Boundary scan | Yes | 1 | 0 | 0 |
|  | No | 1 | 0 | 1 |
| Slave serial | Yes | 1 | 1 | 0 |
|  | No | 1 | 1 | 1 |

pulled up or left floating prior to configuration. The mode pin values required for these configuration modes are shown in Table 6.1.

Figure 6.13 shows the circuit diagram for the connections that can be used to configure a set of two Spartan-II FPGAs connected in a daisy chain manner. More Xilinx chips could be appended to this chain by using connections similar to the "slave" device shown. This diagram illustrates two configuration modes: the master serial and slave serial modes. The "master" Spartan-II device generates the configuration clock signal (CCLK), which enters the PROM device (holding the configuration data) and causes the PROM to output data from consecutive address locations starting from address 0. The data output by the PROM (on the DATA pin) is entered into the master Spartan-II device through its DIN pin. Once the master Spartan-II device receives all of its configuration data, consecutive data bits are sent out on its DOUT pin. These later data bits are used to configure the slave Spartan-II device. Note that the master Spartan-II device continues generating the CCLK signal so that the slave Spartan-II device can receive its configuration data synchronized to the CCLK signal.

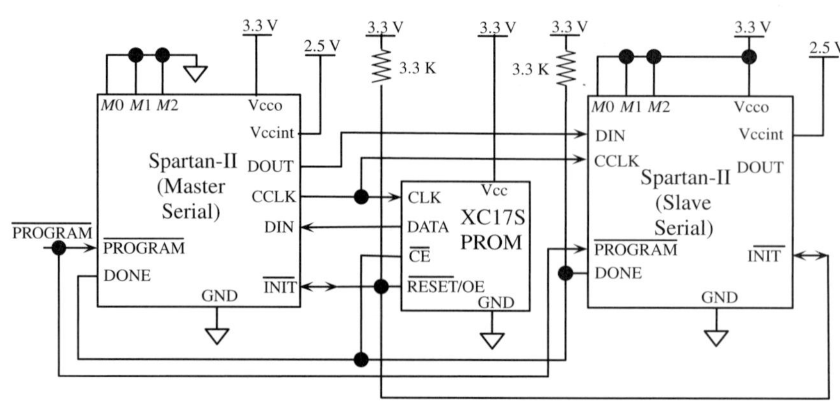


Figure 6.13
Block diagram of a circuit used to configure two Xilinx Spartan-II chips connected in a daisy chain manner. [Xilinx 2000]

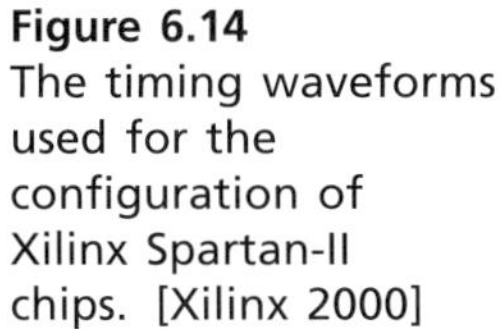

**Figure 6.14**
The timing waveforms used for the configuration of Xilinx Spartan-II chips. [Xilinx 2000]

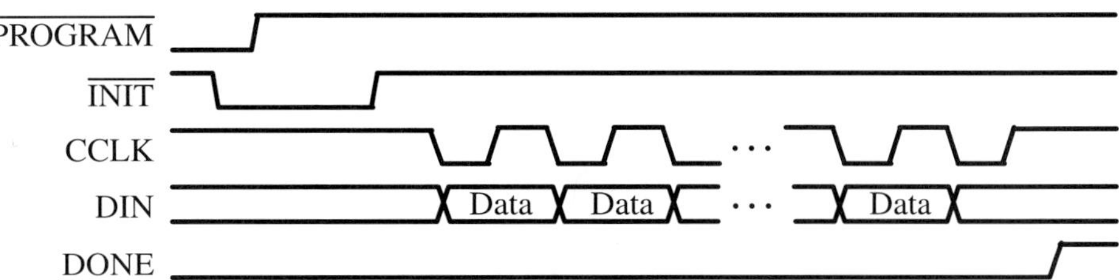


Besides the CCLK, DIN, DOUT, and mode pins, the other configuration-related pins used are $\overline{\text{PROGRAM}}$, $\overline{\text{INIT}}$ and DONE. When the $\overline{\text{PROGRAM}}$ pin is asserted (made equal to 0), the configuration memory (circuit connection data) in the FPGA is erased. The $\overline{\text{PROGRAM}}$ pin must be asserted for a minimum of 300 ns. Then, when the $\overline{\text{PROGRAM}}$ pin is pulled high, the $\overline{\text{INIT}}$ pin is asserted (made equal to 0 for a maximum of 100 $\mu$s) to indicate that the configuration memory is being cleared. After the configuration memory has been cleared completely, the $\overline{\text{INIT}}$ pin is deasserted (pulled high). At this time, the mode pins are sampled to determine the configuration mode, the CCLK signal starts oscillating in a periodic manner, and the configuration data is entered through the DIN pin in synch with the rising edge of the CCLK signal. The DONE signal is de-asserted (made equal to 0) at the start of configuration, and is asserted (made equal to 1) when all configuration data bits have been entered into the FPGA.

The configuration process can be initiated by asserting the $\overline{\text{PROGRAM}}$ pin or by initially applying power to the FPGA. Figure 6.14 shows the timing waveforms for the configuration process, starting from when power initially is applied to the FPGA device, resulting in a power-on reset. As can be seen, first the FPGA's connection information (configuration memory) is cleared (by asserting $\overline{\text{PROGRAM}}$). The FPGA indicates that its configuration memory is in the process of being cleared by asserting $\overline{\text{INIT}}$. When the FPGA's configuration memory has been cleared completely, $\overline{\text{INIT}}$ returns to the HIGH state. Then, after a short delay, the CCLK pin starts to oscillate in a periodic manner and configuration data is entered on DIN. After the last configuration data bit has been entered on DIN, the FPGA performs a CRC (cyclic redundancy check)[3] to verify that there are no errors in the configuration data. If the CRC check is passed, then the DONE pin becomes asserted, thereby indicating a successful configuration.

### Configuration using a PC Interface

There are several ways to interface a PC to an FPGA chip such that the FPGA can be configured from the host PC. The main alternatives would be to use an I/O

---

[3]   A CRC (cyclic redundancy check) is a standard commonly used method for detecting if data has been corrupted. When the data initially is created, several check bits are inserted. Then, during the CRC check, the check bits observed are compared with the proper check bits for the observed data. If the observed and expected check bits are identical, then it is highly unlikely that the data has been corrupted.

bus such as the PCI bus, to use an I/O interface chip such as the 8255, or to use one of the PC's external ports. The first alternative, using an I/O bus such as the PCI bus, requires detailed knowledge of the I/O bus specifications and the creation of a custom-designed board for the I/O interface. The second alternative, using the 8255 I/O interface chip, is a much simpler and more flexible interface method. However, this method also requires the creation of a custom-designed board that must be inserted into the PC's motherboard (the main board with the CPU and memory chips). With both methods, inserting a custom-designed board into a PC's motherboard can be extremely risky, as a short in the custom-designed board can permanently damage the host PC (it is also quite a cumbersome process). Thus, the third alternative, involving the use of one of the PC's external ports, will be investigated in this section.

Xilinx provides special cables for use with PC-based configuration. These cables (MultiLINX, Parallel Cable III, and Parallel Cable IV) connect to the USB, serial, or parallel port on the PC side and pin headers on the application board side (which connect to the FPGA chip). The Xilinx FPGA software provides tools to download configuration data through these parallel port cables to Xilinx FPGA chips or serial EEPROM chips connected to Xilinx FPGAs.

| *Example 6.5* | **PC-Based Slave-Serial Configuration Mode** |

Let us consider PC-based configuration for a Xilinx XC2S200 Spartan-II FPGA. For simplicity, we will use the slave-serial mode of configuration (which requires mode pin settings of $(M0, M1, M2) = (1, 1, 1)$) and a custom-built parallel-port cable. Connections need to be made from the parallel-port cable to the CCLK, DONE, DIN, and $\overline{\text{PROGRAM}}$ pins of the XC2S200 chip. In addition, connections to ground and the reference voltage (typically 5 V) are required. Figure 6.15 shows all of

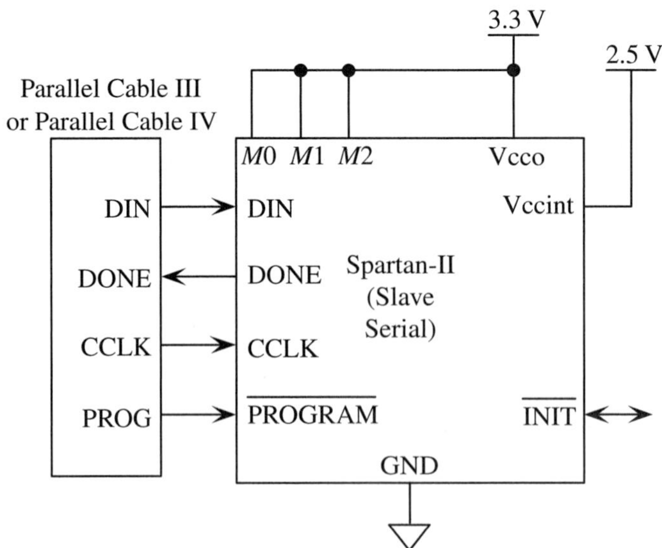


**Figure 6.15**
Connections required for PC-based slave-serial configuration mode of the Xilinx XC2S200 chip [Xilinx 2000].

these connections from the parallel-port cable to configuration-related pins on the XC2S200 chip.

Then the download file (bit file) must somehow be entered into the XC2S200 chip. This can be done using the iMPACT Programmer tool in the Xilinx FPGA software.

---

| **Example 6.6** | **PC-Based Boundary-Scan Configuration Mode** |
|---|---|

If the PC-based slave-serial configuration mode is used to configure the XC2S200 FPGA directly, then configuration memory data will be completely lost (and the configured circuit will cease to exist) once power is removed from the application board. Thus, an alternative method is to configure a serial EEPROM chip connected to the XC2S200. Then, since the EEPROM chip is nonvolatile, configuration memory data will be retained even if power is removed from the application board—once power is restored, the FPGA must reconfigure itself from the data stored in the serial EEPROM. In order to do this, the FPGA's configuration mode must be set to *master* serial mode $((M0, M1, M2) = (0, 0, 0))$.

Data can be downloaded into the serial EEPROM chip suggested for use with the XC2S200 chip (the Xilinx XC18v02 serial EEPROM) using the boundary-scan mode. In order to do this, it is possible to use the same parallel-port cable as the cable used for PC-based slave-serial mode configuration. All that is required is to change the connections to configuration-related pins to those used with boundary scan (TCK, TDO, TDI, TMS). Figure 6.16 shows the connections required for this type of configuration scheme.

As before, the iMPACT Programmer tool can be used to download the configuration file (bit file) for the target circuit. However, during the "Implementation" step used to create the download file, it is important to state that JTAG Clock should be used instead of CCLK for configuration.

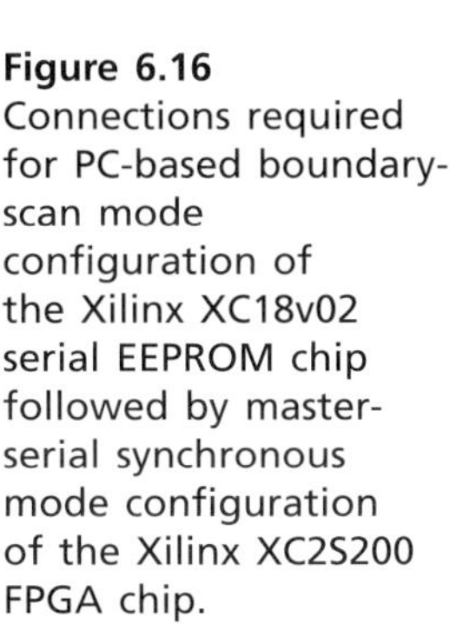

**Figure 6.16**
Connections required for PC-based boundary-scan mode configuration of the Xilinx XC18v02 serial EEPROM chip followed by master-serial synchronous mode configuration of the Xilinx XC2S200 FPGA chip.

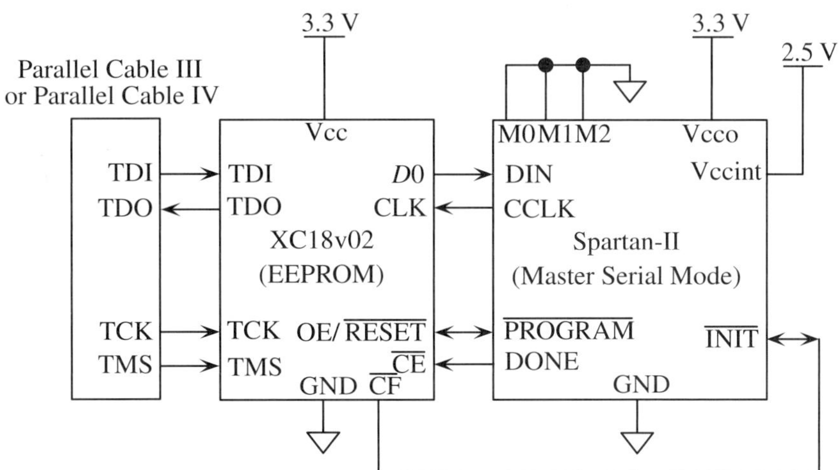

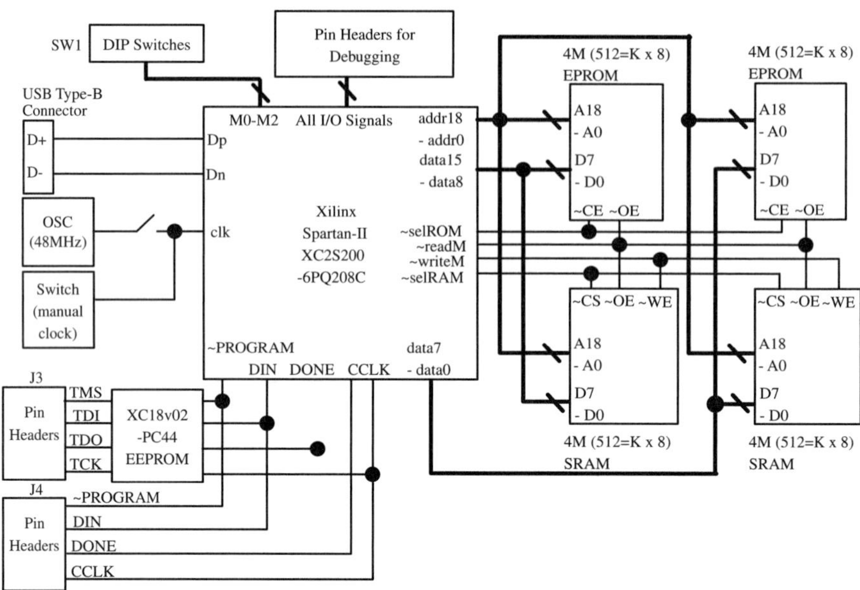


**Figure 6.17**
Block diagram for
an FPGA prototyping
board design.

In order to experiment with various configuration options and to serve as a test platform for the designs to be presented in later chapters, the author designed and implemented a printed circuit board (PCB). The board that the author implemented includes a USB receptacle (the USB I/O interface is discussed in Chapter 7) for experimenting with USB I/O interface designs, SRAM and EPROM chips for experimenting with the RISC CPU designs, and pin headers for all user I/O pins of the XC2S200 FPGA in order to facilitate testing and debugging. The XC2S200 chip can be programmed using slave serial synchronous mode directly, or it can be programmed (in master serial synchronous mode) using data stored in the XC18v02 EEPROM, which can in turn be programmed from a PC using boundary scan mode. Figure 6.17 shows a block diagram of this test platform board design.

### 6.2.6 Boundary Scan

*Boundary scan* refers to a method for probing the I/O pin values of chips in a circuit board. Before the advent of boundary scan, the only way to probe the I/O pin values of chips in a circuit board was to make actual connections to all pins of interest using alligator clips or some other wire connection method. For production testing of large quantities of boards, a "bed of nails" was used to connect to all I/O pins of interest. Obviously, this type of method is very expensive and unreliable. Thus, boundary scan was proposed as an alternative method. In boundary scan, the circuitry in the I/O buffers attached to all I/O pins includes multiplexers and flip-flops that can be configured to form one long shift register. Then, if all of the chips are connected in a daisy-chained manner, as shown in Figure 6.18, probing the I/O pin values of all chips simply requires the I/O values to be stored in their respective I/O buffer flip-flops and then shifted out in a serial manner.

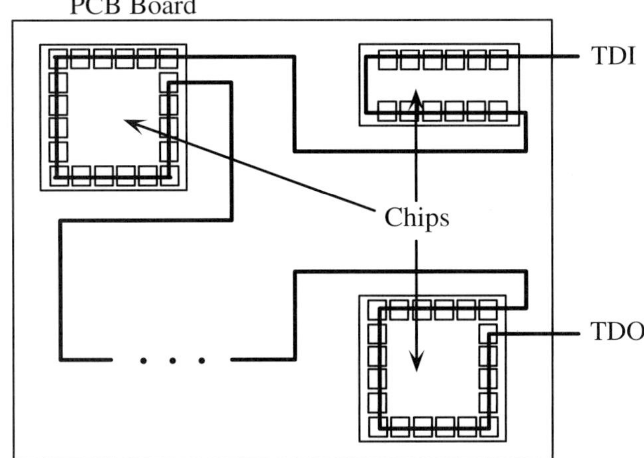



**Figure 6.18**
Block diagram showing chips connected in a boundary scan chain.

Additionally, test data can be entered into all chips by serially shifting in all of the test data. Then, when all of the test-data bits have reached their respective I/O pins, the mode can be changed from test mode to normal mode (in normal mode, the shift register used for testing is no longer active, and the I/O buffers are expected to provide data to the insides of their chips).

Figure 6.19 shows the design for the I/O buffers of the Xilinx XC2S200 chip, which supports boundary scan. As can be seen, when the SHIFT/CAPTURE control sig-

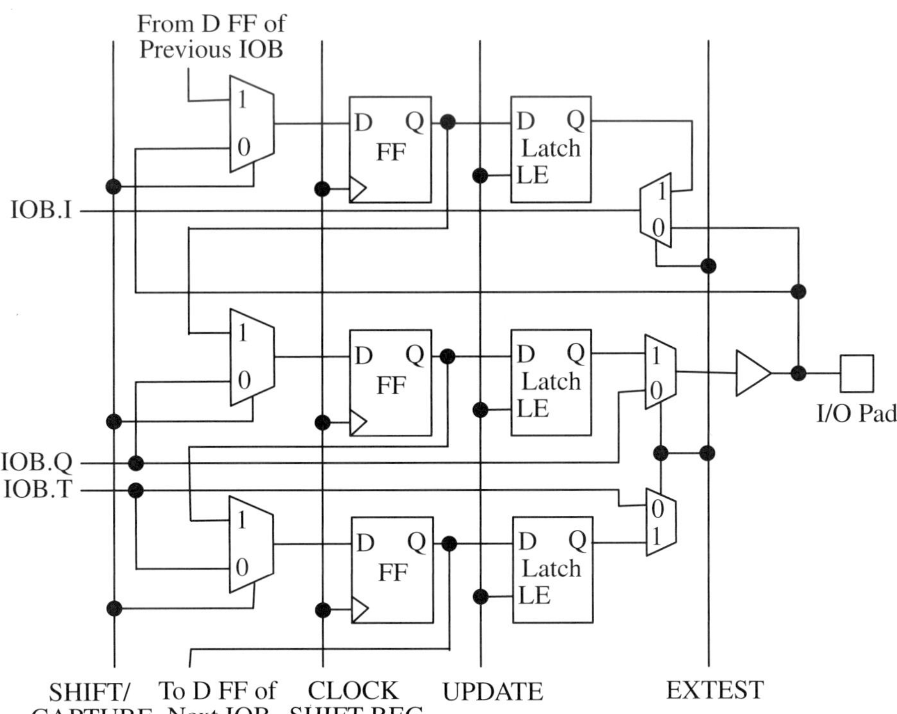



**Figure 6.19**
Xilinx XC2S200 I/O buffer design supporting boundary scan. [Xilinx 2000]

nal is set to 0 (normal mode), the input signal from the I/O pad is clocked into a D flip-flop (and entered into the inside of the chip if EXTEST = 0) or an output signal from the inside of the chip is clocked into a D flip-flop (and routed to the I/O pad if EXTEST = 0). Then, when the SHIFT/CAPTURE control signal is set to 1 (test mode), all of the D flip-flops in all of the I/O buffer blocks are chained into one long shift register. By clocking this shift register, it is possible to observe all I/O pad values through a single shift-out port (the TDO pin).

The version of boundary scan most commonly used today is based on the IEEE 1149.1 standard. This standard was proposed by the Joint Test Action Group (JTAG), a consortium of companies and individuals, in the late 1980s. The standard primarily defines a set of four special I/O pins (the TAP (test access port)), an instruction register, test-data registers, and instructions used to support various boundary-scan operations. The TAP (test access port) pins consist of TCK (testing clock), TMS (for selecting test mode), TDI (for entering serial test input data), and TDO (for observing serial test output data).

Although intended for use in production testing of circuit boards, boundary scan can also be used for other purposes. Xilinx uses the boundary scan architecture to enter configuration data serially into a set of FPGA and PROM chips that can be connected in a daisy-chain manner (the TDO pin of one chip is connected to the TDI pin of the next chip).

## 6.3 Chapter Review

- A *programmable logic device (PLD)* is a mass-produced chip in which there are numerous horizontal and vertical connection wires, and the end-user can create a customized circuit in the PLD by programming the horizontal-to-vertical crosspoints as open or short circuits. *one-time programmable (OTP)* PLD chips use *fusible link* or *antifuse* technology to create opens or shorts at selected crosspoints in the PLD chip—such changes are permanent. On the other hand, in a multiple-time programmable PLD chip, binary data is stored at all programmable crosspoints. Depending on the data stored at a crosspoint, there will be an open or a short circuit at that crosspoint. Such a PLD chip can be volatile or non-volatile, depending on whether volatile or nonvolatile memory is used to store circuit connection data.

- The various types of PLDs available include PLAs (programmable logic arrays), PROMs (programmable read-only memories), and PALs (programmable AND-array logic). A PLA consists of a matrix of AND gates (referred to as the AND plane) followed by a matrix of (OR gates referred to as the OR plane). In an $n \times p \times m$ PLA structure, there are $n$ primary inputs (which enter the AND plane), $p$ product terms (which are the outputs of the AND gates in the AND plane), and $m$ primary outputs (which are the outputs of the OR gates in the OR plane). The crosspoints in both the AND plane and the OR plane can be programmed to form any sum-of-products expression with not more than $n$ variables and not more than $p$ product terms.  A PROM can be considered as a PLA with a fixed AND plane and a programmable OR plane in which the AND plane produces all possible minterms for the given set of primary input variables.  A PAL is a variant of a PLA with a fixed OR plane and a programmable AND plane.

■ An *field-programmable gate array (FPGA)* is a user-programmable version of a *gate array*.  A gate array is a type of ASIC in which there is a two-dimensional matrix of programmable "cells" with routing channels in between adjacent cells. Each cell can be customized to form a simple logic function (such as a logic gate function, 1-bit full adder, flip-flop, etc.)  and inputs and outputs of individual cells can be connected together by making connections to specific wires in the routing channels.  This type of circuit customization can be achieved using the same methods used for the programming of PLD chips.

■ FPGA companies typically provide extensive software and hardware support for the programming of FPGA chips.  Software support typically includes schematic and language-based design entry tools, simulation tools, synthesis tools, and tools to generate the FPGA configuration download file.  Once circuit information has been prepared in the form of a download file, it can be used to program an FPGA by using a special FPGA programmer device or a custom-designed FPGA programming interface circuit.  An FPGA can be programmed using combinations of slave or master mode, serial or parallel mode, and synchronous or asynchronous mode.  In addition, many of the newer FPGAs support configuration using *boundary scan* mode, which is a variant of slave, serial, synchronous mode originally devised as a standard method for observing and setting the I/O pin values of a chip.

# 6.4 Resources

[Wakerly 2002] and [Hayes 1993] are textbooks on digital-logic design that include discussions on PLDs. [Muller 1988] is a book on semiconductors that explains the physics involved in the operation of various types of transistors used in integrated circuits. [Lattice 1996] and [Xilinx 2000] are data books for FPGA devices produced by Lattice Semiconductor and Xilinx, respectively.  Up-to-date technical information about PLDs and FPGAs manufactured by various companies can be obtained by referring to the web pages for those companies. [Lawman 1995] presents a useful method for configuring Xilinx FPGAs using an x86 PC's parallel-port interface. [Peacock 1999] provides useful information about the x86 PC parallel-port interface.

### 6.4.1  Bibliography

[6.1]  HAYES, J. P., *Introduction to Digital Logic Design*, Addison-Wesley, Reading, MA, 1993.

[6.2]  LATTICE SEMICONDUCTOR, *Lattice Semiconductor Data Book*, Lattice Semiconductor Corp., Hillsboro, 1996.

[6.3] LAWMAN, G., "Configuring FPGAs over a Processor Bus," an Xilinx application note found in http://www.xilinx.com/apps/4000.htm#appnotes, September 1995.

[6.4] MULLER, R. S., *Device Electronics for Integrated Circuits, 2nd Ed.*, John Wiley and Sons, New York, 1988.

[6.5] PEACOCK, C., "Interfacing the Standard parallel Port," an application note found in http://www.senet.com.au/~cpeacock, April 1999.

[6.6] WAKERLY, J. F., *Digital Design: Principles and Practices, 3rd Ed.*, Prentice Hall, Englewood Cliffs, NJ, 2002.

[6.7] XILINX CORP., *The Programmable Logic Data Book 2000*.

[6.8] http://www.altera.com, home page for Altera Corp. (supplier of FPGA chips and software).

[6.9] http://www.xilinx.com, home page for Xilinx Corp. (supplier of FPGA chips and software).

●  ●  ●  ●  ●  ●  ●  ●  ●  ●  ●  ●  ●  ●  ●

# 6.5 PROBLEMS

**P6.1.** As stated in Section 6.1.3, a ROM can be used to implement arbitrary combinational logic circuits. Can a *RAM* (random access memory, a read/write memory device) also be used to implement arbitrary combinational logic circuits? Can a RAM be used as a PLD? What are the main difficulties with using a RAM to implement arbitrary combinational-logic circuits?

**P6.2.** Using a diagram similar to that shown in Figure 6.3, show how an $8 \times 10 \times 4$ PLA (and a few D flip-flops) can be used to implement a 4-bit Gray code counter.

**P6.3.** Using a diagram similar to that shown in Figure 6.7, show how a PAL18L4, with the addition of a few D flip-flops, can be used to implement a 4-bit binary counter. Be sure to show the derivation of all logic equations used. Note that inverse logic functions may have to be used because the PAL18L4 produces active-low output signals.

**P6.4.** Using a diagram similar to that shown in Figure 6.6, show how a 128 K-bit (16-K $\times$ 8) PROM can be used to implement a 12-bit even-parity checker. This circuit outputs a '1' if there are an even number of '1' bits in the 12-bit input data.

**P6.5.** Show how a PAL18L4 and D flip-flops can be used to implement a sequence detector circuit for the sequence 101101. This circuit should output a '1' whenever the target sequence (101101) is observed in the serial data input stream. During all other times, the output should be '0.' Note that overlapped sequences are possible; thus, the input stream ... 0101101101 ... should produce the output stream ... 0000001001 ....

**P6.6.** Show how one $4 \times 6 \times 4$ PLA and several D flip-flops can be used to implement a sequence generator for the sequence 01111110.

**P6.7.** Show how a large EPROM chip can be used to implement a combinational multiplier circuit that multiplies two 8-bit numbers. What is the minimum size (and data configuration: $n \times m$) required for the EPROM if we wish to use only one EPROM chip? Be sure to show all connections to the EPROM and the contents of the EPROM. For brevity, just a few of the EPROM contents can be shown, as long as the general trend and the other data contents can be inferred from the data (and the comments) shown.

**P6.8.** Show how a 256 K-bit (32K $\times$ 8) PROM can be used to implement a combinational divider circuit that divides a 8-bit 2's complement binary number (the dividend) by an 4-bit 2's complement number (the divisor) to produce an 4-bit quotient and an 4-bit remainder. Be sure to show all connections to the PROM and the contents of the PROM.

**P6.9.** For a circuit such as a combinational multiplier, is it simpler to use a PROM or a PLA? What are the difficulties with using a PLA to implement this type of circuit? What are the advantages of using a PLA to implement this type of circuit? What types of circuits can be implemented more easily with PROMs, and what types of circuits can be implemented more easily with PLAs?

**P6.10.** Let us assume that we are given a gate array of the type described in Example 6.4. Show how this gate array can be customized to implement a 1-bit full adder circuit. Start with the truth table for the 1-bit full adder, derive the simplified logic equations for the *sum* and *cout* (carry out) outputs, and then show how these logic equations can be mapped onto the cells and routing wires in the target gate array.

**P6.11.** What is a look-up table (LUT)? How can an LUT be used to implement an arbitrary combinational logic circuit? What are the limits on the types of combinational logic circuits that an LUT (of a specific size) can implement?

**P6.12.** What types of SSI and MSI logic devices can be found in a configurable logic block (cell) of a Xilinx FPGA? Why are these devices (and why not other devices) used in the configurable logic block?

**P6.13.** Suppose we wish to use a Xilinx Spartan-II FPGA to implement a CPU circuit. What common computer component can best be implemented using the Block RAM cell in the Xilinx Spartan-II? (*Hint*: It is in the path between the CPU and the main memory.) How can a Block RAM cell be used to implement this computer component? (Note: Be specific. Show the signals tied to the input and output pins of the Block RAM, etc.)

**P6.14.** What is the main difference between the slave and master modes of FPGA configuration? From the viewpoint of the end-user, which mode is easier to use and why?

**P6.15.** What is the main difference between the synchronous and asynchronous modes of FPGA configuration? What types of signals and what type of protocol is required in order to use an asynchronous mode of FPGA configuration?

**P6.16.** What are the advantages and disadvantages of using boundary scan for all of the chips comprising a hardware circuit board?  Xilinx has adopted boundary scan as a method for configuring its newer FPGA chips.  Describe the modifications that must be made to the basic architecture of an FPGA chip in order to enable configuration using boundary scan.

**P6.17.** Study the "configuration" section of the data sheet for a Xilinx Spartan-II FPGA chip.  List the exact sequence of bytes that must be sent to the FPGA chip in order to properly configure it, assuming an XC2S200 FPGA. Next, assuming that the XC2S200 FPGA is configured using a custom-designed hardware interface circuit, describe the exact sequence of electrical signals that must be sent to and received from the FPGA chip in order to configure it properly in a parallel, synchronous, slave configuration mode.  Draw a timing diagram in order to aid in your description.

**P6.18.** Write a Verilog program for a circuit used to generate the configuration-related signals shown in Figure 6.14.  For simplicity, it can be assumed that the data portion of the configuration download file consists of the repeated byte sequence 01100110.

**P6.19.** Write a test bench, in Verilog, for the configuration interface circuit of Problem P6.18. This test bench should include all of the expected responses from the FPGA chip and verification of all signal values generated by the configuration interface circuit. If possible, simulate the test bench and the configuration interface circuit using a commercial simulation tool.

**P6.20.** It is possible to generate all configuration- related signals using a PC's parallel port interface and an appropriate program running on the PC. In fact, this type of program is what is used in the iMPACT Programmer tool, provided as part of the Xilinx FPGA sofware, as described in Example 6.5.  Referring to this example and the material provided on the internet, such as [Lawman 1995] and [Peacock 1999], write a C or C++ configuration download program for the XC2S200 FPGA that works with the PC's parallel port interface.

# CHAPTER 7

# Design of a USB Protocol Analyzer

## Important Concepts

- The basic features and techniques used in the Universal Serial Bus I/O protocol.
- The concepts of *differential signaling*, *NRZI encoding*, *CRC codes* used for error checking, *bit stuffing*, and *destuffing*.
- How to design a Verilog circuit that "understands" and works with a modern PC interface such as the Universal Serial Bus.

Universal Serial Bus (USB) is an example of a modern commonly available peripheral interface bus. USB, IEEE 1394 (also known as FireWire), and PCI-X are expected to become the dominant methods for communicating with peripheral and network devices, replacing the serial, parallel, keyboard, mouse, audio, game ports, and other interfaces currently cluttering up the backs of PC cases. Of these new modern interfaces, PCI-X is expected to be used mostly for communication within the PC chassis, IEEE 1394 is expected to be used mostly for peripherals requiring high-bandwidth connections to the PC (such as video cameras and web cams), and USB is expected to be a low-cost solution for peripherals requiring low- to medium-bandwidth connections to the PC (such as the keyboard, mouse, scanner, printer, digital camera, etc.). USB connections now come standard with all IBM-compatible PCs.

There are two common versions of the USB specification in use today. The USB1.1 specification supports low-speed (1.5 Mbps) and full-speed (12 Mbps) devices. The newer USB2.0 specification, which is backwards-compatible with the USB1.1 specification, also supports high-speed devices, which can support data communication at 480 Mbps. With the use of USB2.0 high-speed mode, USB can compete with IEEE 1394 for some types of high-bandwidth devices (such as external hard disks).

Like most modern communication interface mechanisms developed for the PC, USB has the following features:

- It is *hot-pluggable*, which means that I/O devices can be attached and automatically configured while the PC is running.

- There is a single connector type, which can be used to connect to any I/O device.

- It is a high-performance serial bus, with USB1.1 devices capable of 12 Mbps and USB2.0 devices capable of 480 Mbps.

- It supports a large number of devices (up to 127 devices for each root hub port).

- It provides automatic error detection and recovery (retrying a transaction, until a timeout period expires, whenever incorrect data delivery is detected).

- Cable power and power management features are available. 5.0 V DC power is available from the USB cable (with a current of 100 to 500 mA), and USB devices automatically enter a *suspend* state after 3 ms of no bus activity.

- It is completely external to the PC, which means that there is no need to open up the PC case or design special I/O cards in order to interface with the PC.

USB high-speed mode (capable of 480 Mbps) and most other modern communication interfaces typically use a method known as *low-voltage differential signalling* in order to transfer data across the wire. In this method, a single digital signal (0 or 1) is sent using a pair of positive-voltage and negative-voltage wires. The actual signal sent is determined based on the voltage difference between the two wires. For high-speed transmission, the voltage differences used can be quite small. This type of method permits extremely high-speed transmission and is highly resistant to noise that can distort the voltage values on the wires as they are transmitted to their destinations. However, the low-voltage differential signalling method requires the use of an operational amplifier and other analog circuitry to be able to detect the digital signals being transmitted on the positive-and negative-voltage wire pair.

Unlike the above type of method, USB low-speed (1.5 Mbps) and USB full-speed (12 Mbps) modes use a signalling method that permits the use of a fully digital circuit to determine the actual digital signals being transmitted. Although USB low-speed and full-speed modes also use differential voltage signalling, the USB1.1 specification states that both the positive-and negative-voltage wires will maintain a voltage of at least 2.0 V for a *HIGH* value and a voltage of at most 0.8 V for a *LOW* value. Thus, with these voltage levels, a fully digital circuit can be used to determine the digital signal being transmitted by the positive-and negative-voltage wire pair. USB (in particular, the full-speed mode of the USB1.1 specification) was chosen as our design target because it is a modern communication interface for which a completely digital solution is possible.

●  ●  ●  ●  ●  ●  ●  ●  ●  ●  ●  ●  ● ● ●

# 7.1  Overview Of USB Full-Speed Mode

This chapter will present a design for a USB1.1 full-speed interface circuit. In order to understand this design, it is first necessary to understand the USB1.1 full-speed interface protocol. However, since the main objective of this textbook is to teach advanced digital logic design, and not USB interface design, only the parts of the USB protocol required to understand the design problem and solution adopted will be presented in this chapter. The interested reader is referred to the USB references listed at the end of this chapter for a more complete treatment of the USB protocol.

USB, as its name implies, is a "bus". All requests and responses are broadcast as in a traditional bus. This facilitates central control and arbitration. However, the physical structure of a USB "bus" is different from a traditional bus, in which there is a common set of signal lines (the bus) that is used to link all devices on the bus. At any given time instant, the bus has to be time-shared among all devices that wish to send data on the bus. This type of structure has the disadvantage of not being scalable—as more and more devices are attached to the bus, the performance of the bus (for each device) slows down due to contention for the bus.

USB devices connect to a USB host (typically a PC) in a tree-like manner, as shown in Figure 7.1. At the root of that tree is the *root hub*, installed in the motherboard of most modern PCs. Upstream connections are made through *Series A* USB connections, while downstream connections are made through *Series B* USB connections. Only *hubs* and devices with *hub* functionality can have downstream connections. The entire tree of devices connected to a single root hub must consist of 127 or fewer devices. The hardware associated with a USB system consists of the *host controller, root hub, hub*, and *devices*. The USB software consists of the *host controller driver, USB driver*, and *USB device drivers (client drivers)* for each different device type.

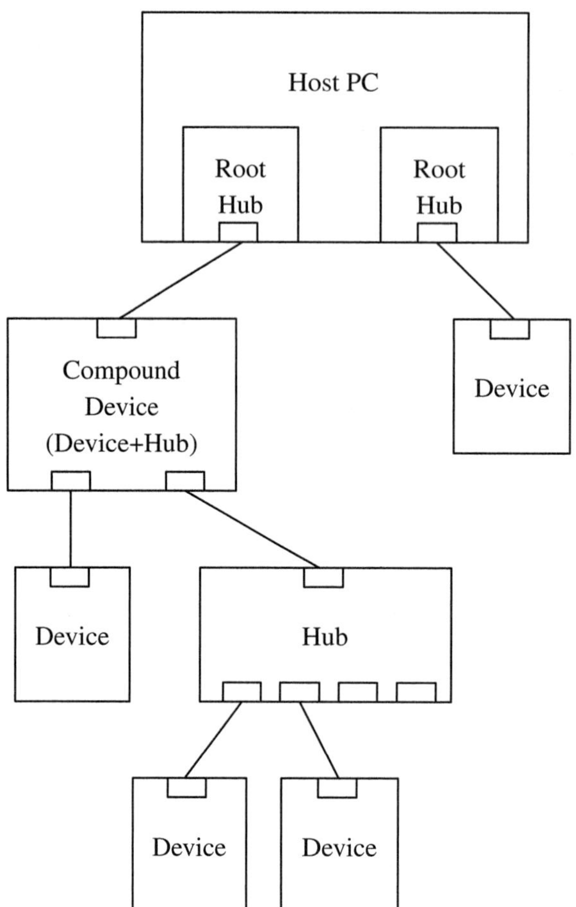



**Figure 7.1**
Example connection
of USB devices to a
PC.

### 7.1.1  Packet Transfer Protocol

During a data transfer, the USB system functions in the following overall manner. Suppose that a user wishes to send some data (e.g., a JPEG-encoded photograph) from a device (e.g., a digital camera) to the PC. The user would first start up the interface software (for that device) on the PC. Then, when the user clicks on the "download photo" command, the following sequence of events occur. The interface software (through the USB client driver for that device) makes a call to the operating system (via an I/O request packet), at the same time supplying a memory buffer used to store data when transferring data to and from the USB device. When the I/O request packet is received from the USB client driver, the *USB driver* organizes the request into individual transactions that can be executed during a series of 1 ms frames. Then, the *USB host controller driver* performs the task of scheduling the transactions to be performed over the USB.

The *host controller* receives each transaction request and prepares a transaction (which is composed of a series of USB packets) based on the contents of the transfer descriptor built by the USB host controller driver. Then, an *IN* packet is sent to the device through the root hub and any other hubs in the path from the PC to the device. The *IN* packet instructs the device to send the PC some data. Then, in response to this packet, the device sends a *DATA*[1] packet through the hubs and root hub to the host controller, which then deposits the data in the memory buffer which has been reserved for that data. The USB device driver then detects the arrival of the data and informs the interface software, which can then display that data to the user. At the same time, the host controller sends a *HANDSHAKE* packet, through the root hub and other hubs, to the device. This completes the transaction. Note that several transactions may be necessary if the data to be transferred is large.

Although the above process may sound fairly complicated, the reader does not have to understand this process in intimate detail in order to understand the USB protocol analyzer design. Viewed in a simple context, all that is occurring is that there is a sequence of packets being transmitted back and forth between the host (typically a PC) and the device (e.g., a digital camera). This sequence of packet transfers occurs according to the USB protocol, which logically groups the packets into several different types of transactions. In fact, if the USB protocol analyzer design is implemented as a hardware prototype, then it could be used to observe the different types of USB packets actually being sent back and forth between a PC and USB devices, thereby leading to a better understanding of the USB protocol.

### 7.1.2  Initialization Sequence

When a USB device initially is connected to a hub or root hub, the device is recognized and configured in the following manner. A simple signalling mechanism (described in the subsection below) is used to inform the hub (if the device is connected to a hub) or root hub (if the device is connected directly to the root hub) of the attachment of the new USB device. The new USB device then is detected by the USB host controller driver in the host PC during the next periodically generated "inquiry" transaction. The port (to which the device is attached) is then enabled so that the hub (or root hub) will pass bus traffic to the device. Configuration software then issues a *ResetPort* request, forcing the device into its default initial state. A series of *SETUP* packets, followed by *IN*, *OUT*, and *DATA* packets then are used to configure the device. This configuration process assigns a unique address to the USB device, reads descriptors from the device to determine the characteristics and capabilities of the device, and reserves any USB resources requested by the device.

---

[1]  Actually, there are two types of *DATA* packets: *DATA0* and *DATA1*. On every data transfer, the sending side alternates the transmission of DATA0 and DATA1 packets. This method of alternating DATA0 and DATA1 type packets is used so that errors in data transmission easily can be detected and compensated for, as described in [Anderson 1997].

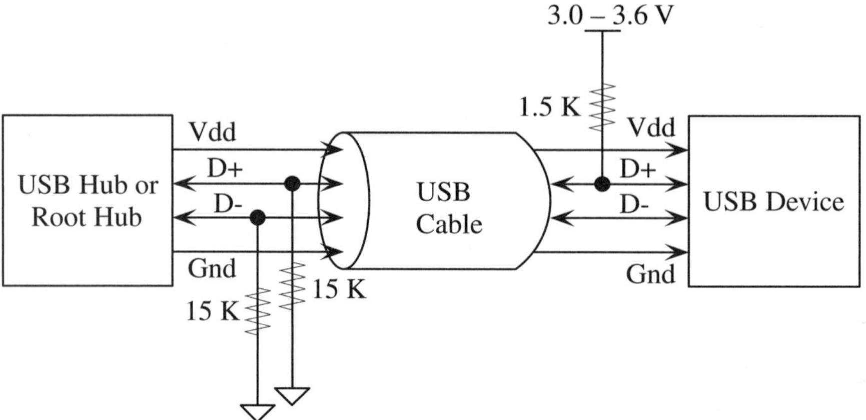


**Figure 7.2**
USB full-speed mode connection between a hub and a USB device.

### 7.1.3   Physical Layer Interface

A USB cable consists of four wires: Vdd (red wire), D− (white wire), D+ (green wire), and ground (black wire). For full-speed mode (with a peak data rate of 12 Mbps), this cable must be connected at the hub (refers to both the hub or root hub) port and the device port as shown in Figure 7.2. For low-speed mode (with a peak data rate of 1.5 Mbps), the only difference is that the D− wire is connected to the pullup resistor (instead of the D+ wire) in Figure 7.2. Because of the 15-KOhm pulldown resistors in the hub port, the D+ and D− wires will both be at 0 V when no device is attached to the other end of the USB cable. However, when a full-speed mode USB device is attached, the D+ wire will be pulled up (to a voltage greater than 2.0 V) because of the smaller 1.5-KOhm pullup resistor tied to the D+ wire in the device port. When this (D+ = HIGH, D− = LOW) state is maintained for longer than 2.5 $\mu$s, the hub detects the attachment of a full-speed USB device. After the hub has detected the device attachment and informed the host, the hub will *reset* the device by setting (D+ = LOW, D− = LOW) for greater than 10 ms. After this time, the USB connection enters into its *idle* state for full-speed mode (D+ = HIGH, D− = LOW). Then, USB transactions can be sent and received by sending and receiving packets in the manner described in the following subsections.

After the initial device-attachment sequence described above, a USB connection operates as follows. USB signals are sent in both *differential mode* and *single-ended mode*. In differential mode, the USB connection is either in the *J state* or the *K state*. For a full-speed USB connection, the *J* state is indicated by D+ = HIGH and D− = LOW, and the *K* state is indicated by D+ = LOW and D− = HIGH (the inverse values for low-speed mode). The single-ended mode is used only for the EOP (end-of-packet) indication. An EOP is generated by setting (D+ = LOW, D− = LOW) for two bit times, followed by the *J* state. An EOP is recognized if D+ = LOW, D− = LOW) is detected for one or more bit times.

When two independent systems (such as a USB device and a PC) communicate, there must be a method for detecting when each data bit sent by the transmitting side can be sampled by the receiving side. In an asynchronous system, this is accomplished with a *handshaking* mechanism, in which a *request (REQ)* signal is sent out at the same time that the data signal is sent out, and an *acknowledge (ACK)* signal is returned by the receiver to indicate correct receipt of the data. In a synchronous system, the data must be sent out *synchronized* to a *clock* signal. The clock signal can be sent out on a separate wire, or it can be combined with the data signal on the same wire. To send out a clock signal and a data signal using a single wire, the clock frequency must be agreed upon in advance. Even with a known clock frequency; however, two independent systems will operate using different clocks with different phases and slightly varying frequencies. Thus, a mechanism must be present to synchronize the phases and frequencies of the clocks in the two independent systems. This can be accomplished by ensuring that sufficient numbers of signal transitions occur on the transmitting wire and then dynamically adjusting the phase and frequency of the clock at the receiving side in response to the observed signal transitions.

USB uses a synchronous communication mechanism. In order to synchronize the transmitting and receiving side without the use of a separate clock signal, a SYNC pattern is sent out at the beginning of each data packet and *Non-Return-to-Zero-Inverted (NRZI) encoding* is used. The SYNC pattern consists of a sequence of $J, K, J, K, \ldots$ states, which can be used by the receiving side to synchronize the phase and frequency of the receiving clock.

NRZI encoding is a method used to send a clock signal and a data signal on a single wire. In this type of method, the clock frequency must be agreed upon in advance, which in the case of USB full-speed mode is 12 MHz. In NRZI encoding, a '0' bit is indicated by a transition on the transmitting wire, while a '1' bit is indicated by no transition on the transmitting wire. Thus, the SYNC signal corresponds to a series of '0' bits. To be more precise, the SYNC signal corresponds to seven '0' bits followed by one '1' bit. Figure 7.3 shows an example of the NRZI-encoded D+ and D− signal values and the corresponding binary values for an example USB packet.

Since all real clocks have slight inaccuracies, the receiver's clock must be resynchronized with the incoming data every now and then. However, if a long data packet is sent and the data packet contains a long sequence of '1' bits, there will be no signal transition in the NRZI data stream. Thus, the receiver's clock cannot be resynchronized. In order to avoid this scenario, *bit stuffing* is used. In

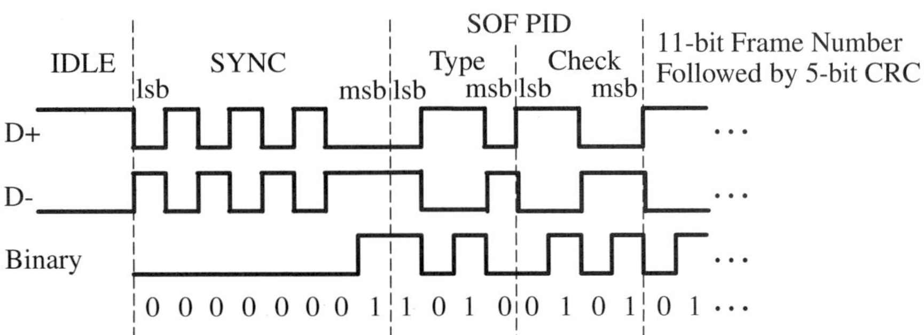



**Figure 7.3**
NRZI-encoded USB signals and the corresponding binary values for an SOF packet.

this method, whenever there is a sequence of six consecutive 1-bits, a '0' bit is inserted into the data stream at the transmitting side. Then, at the receiving side, whenever a sequence of six consecutive 1-bits followed by a '0' bit is detected, the '0' bit is construed as a "stuffed" bit and removed before the data is processed further. Since the '0' bit will be stuffed into the data stream regardless of whether there are exactly six consecutive 1-bits or more than six consecutive 1-bits, this bit stuffing technique works for all cases.

### 7.1.4  USB Packets

All USB control messages and data transfers occur as a series of USB packets transmitted on the USB cable. On a higher level, all USB communication is based on transferring data at regular 1 ms intervals called *frames*. Within each frame, there are *transactions* generated by the USB driver for performing various tasks (such as sending isochronous audio data to a USB speaker, receiving data from an external USB hard disk, etc.). Each transaction, in turn, is composed of a sequence of USB packets. Isochronous transfers (in which data must be sent at a constant data rate), bulktransfers (in which large blocks of data must be sent), interrupt transfers (in which an interrupt request is sent whenever some data needs to be transferred), and control transfers (in which control messages are sent) can be accomplished using various types of transactions. Regardless of the communication type, however, all USB communication uses a small set of basic USB packet types.

There are ten basic packet types, classified into *TOKEN*, *DATA*, *HANDSHAKE*, and *SPECIAL* packets depending on their usage. Packets are broadcast to all devices and hubs connected to a USB "tree." The appropriate device or hub responds to the packet if it is addressed to that device or hub. Token packets, which identify the transaction being initiated and, thus, also indicate the data that follows, are sent out at the beginning of a transaction and consist of START-OF-FRAME (*SOF*), *SETUP*, *OUT*, and *IN* packets. Data packets, which are used to send the actual data, consist of *DATA0* and *DATA1* packets. The first data packet to be sent by a device (or the host) is a DATA0 packet. The next data packet to be sent out is then chosen to be a DATA1 packet. After that, a DATA0 packet is sent out, and so on. Two types of data packets are used in this alternating manner so that the receiving side can determine if a data packet has been lost. Handshake packets, which are sent by a device (or the host) after a data packet is received, consist of *ACK* (acknowledge correct data receipt), *NAK* (no acknowledge, used to inform the host that the device is temporarily unable to accept or return data), and *STALL* (used to inform the host that the device is unable to complete the transfer) packets. Finally, the only special packet defined is the *preamble* packet, used to notify all hubs that a low-speed transaction will follow. Hubs respond to a preamble packet by enabling their low-speed ports, while all other devices ignore the preamble packet.

Figure 7.3 shows the basic format of a USB packet. All packet fields are sent out with the *lsb* (least significant bit) first. First, a SYNC byte (10000000) is sent out as "00000001" (since the *lsb* is sent out first). Next is the "Packet ID" field, which identifies the type of packet being sent. Although only four bits are required for the ten different packet types, eight bits are used for the Packet ID field in order

| Packet Category | Packet Type | PID Type Subfield | PID Check Subfield |
|---|---|---|---|
| Token | Start of Frame (SOF) | 0101 | 1010 |
| | Setup Packet (SETUP) | 1101 | 0010 |
| | Input Packet (IN) | 0001 | 1110 |
| | Output Packet (OUT) | 1001 | 0110 |
| Data | Data0 (DATA0) | 0011 | 1100 |
| | Data1 (DATA1) | 1011 | 0100 |
| Handshake | Acknowledge (ACK) | 0010 | 1101 |
| | Negative Acknowledge (NAK) | 1010 | 0101 |
| | Endpoint Stall (STALL) | 1110 | 0001 |
| | No Response Yet (NYET) | 0110 | 1001 |
| Special | Preamble (PRE) | 1100 | 0011 |

to provide error checking. As shown in Figure 7.3, the "Packet ID" consists of a "Type" subfield followed by a "Check" subfield (with the Check subfield corresponding to the 1's - complement form of the Type subfield). Table 7.1 shows the Packet ID values for all packet types. After the Packet ID field is a variable-length "Packet-Specific Information" field. The Packet-Specific Information field is eleven bits for token packets, 0 to 1023 bytes for data packets, and empty for handshake and special packets. After the Packet-Specific Inforamtion field is a 5-bit or 16-bit *Cyclic Redundancy Check (CRC)* field. The CRC is used as a method of checking for errors in the Packet-Specific Inforamtion field. The generation and checking of CRC bits is covered in Section 7.1.5. Token packets use a 5-bit CRC, data packets use a 16-bit CRC, and handshake and special packets do not have a CRC field. Finally, all packets (except for preamble packets) are terminated with an end-of-packet (EOP) indicator, which consists of LOW values for two bit cycles in both the D+ and D− wires (this is referred to as a *single-ended mode* signal, since the D+ and D− are *not* complements of each other, as in a normal differential signal pair), followed by at least one *Idle* state (which is indicated by a *J* state, i.e., high-D+ and low-D− values).

### 7.1.5   Cyclic Redundancy Checks

The *cyclic redundancy check (CRC)* is a commonly used method for checking binary data for the presence of errors. Errors can occur in binary data (especially when it is transmitted from one physical location to another) due to background radiation from the environment, electromagnetic noise, weather effects, and a variety of other reasons. Such errors are manifested as bit flips, missing bit sequences, spurious inserted bit sequences, stuck-at-1 or stuck-at-0 bit streams, etc.

The CRC method works on the concept of division of polynomials. A binary data stream of length $n$ can be considered as an $n$-order polynomial function (i.e., $c_n x^n + c_{n-1} x^{n-1} + \cdots + c_1 x^1 + c_0$), where $x$ is a variable and $c_i$ are constants equal to either 1 or 0 ($c_i$ is set to 1 if the $i$-th bit in the $n$-bit data stream is a 1). Likewise, we can represent a $k$-order polynomial (in which the constant parameters are restricted to 1 and 0 and the same variable $x$ is used) as a $k$-bit binary word referred to as the *generator polynomial*. Then, if the $n$-order "data stream" polynomial is divided by the $k$-order generator polynomial, there will be a $(n-k)$-order result polynomial and a $k$-order remainder polynomial. This $k$-order remainder polynomial (referred to as the *residue*), encapsulates the $n$-bit data stream in a compact $k$-bit format, assuming that $k$ is much smaller than $n$.

Division of polynomials with binary constants can be implemented in hardware using a simple, modified shift-register structure referred to as an *linear feedback shift register (LFSR)*. An LFSR is simply a modified shift register structure in which some of the D flip-flops in the shift-register receive their inputs from the outputs of exclusive-OR gates instead of adjacent D flip-flops. Figure 7.4 shows an example of such a structure and the $k$-bit generator polynomial implied by that structure. As can be seen from the figure, the LFSR is essentially a shift register (left shift in this case) in which exclusive-OR gates are placed in front of all flip-flop locations, except the *lsb* (least significant bit), corresponding to the 1-bits in the binary representation of the generator polynomial. The output of the LFSRs *msb* (most significant bit) is exclusive-ORed with the serial-input data stream (entered with its *lsb*, first). The output of this exclusive-OR gate is fed back to the input of the rightmost (*lsb* position) flip-flop and the other exclusive-OR gates. The $n$-bit data stream is entered sequentially from the left, with the *lsb* of the data entered first. The data that remains in the set of $k$ flip-flops forming the LFSR after all $n$ data bits have been input corresponds to the *residue* (also referred to as the *remainder*) formed from the division of the $n$-bit data polynomial by the $k$-bit generator polynomial. The theory behind why this works is explained in [Siewiorek 1998] and other books that cover coding theory.

By using the above polynomial division procedure and a generator polynomial of a specific type and length (say $k$), the binary data is mapped to a $k$-bit sequence in such a manner that different binary-data sequences are most likely mapped to different $k$-bit sequences (residues). Given a set of $l$ different $n$-bit binary sequences, the probability distribution of the corresponding $k$-bit residues will tend to be a *uniform random* distribution, regardless of the distribution of the original $l$ binary-input sequences, *provided* that a properly chosen generator polynomial is used. If $k$ is less than $n$ (which is expected to be the case most of the time), then it is possible for two or more $n$-bit input data streams to result in the same $k$-bit residue. This is referred to as

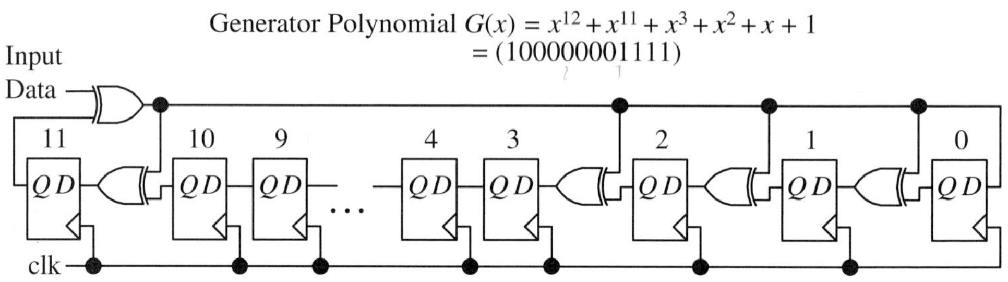


**Figure 7.4**
An example of an LFSR circuit and the CRC generator polynomial used in that circuit.

*aliasing.* Due to the possibility of aliasing, it is possible that an erroneous-input data stream will go undetected, since it may result in the same residue as a correct data stream. However, by using a properly chosen generator polynomial, this property is minimized, since the $k$-bit residues tend to be have a uniform random probability distribution. There are several standard generator polynomials (published in standards publications, textbooks, and other publicly available sources) with this desired property for various bit lengths.

The USB specification uses this CRC error detection method in a slightly modified manner. It is assumed that an LFSR, of a type similar to that shown in Figure 7.4, is used for division of the data polynomial by the generator polynomial. For the 5-bit CRC check of token packets, the 5-bit generator polynomial used is $G_5(x) = x^5 + x^2 + 1$ ($= 00101$ as a 5-bit vector; the $x^5$ position is implicit). For the 16-bit CRC check of data packets, the 16-bit generator polynomial used is $G_{16}(x) = x^{16} + x^{15} + x^2 + 1$ ($= 1000000000000101$ as a 16-bit vector; the $x^{16}$ position is implicit).

For both the 5-bit and 16-bit CRC, the following CRC checking procedure is used. First, the LFSR is initialized to a value consisting of all 1-bits. Then, the *msb* of the LFSR is exclusive-ORed with the serial input data as it is entered from the right side (let us refer to this result as check msb). Next, the LFSR is shifted left (with a 0 shifted into the *lsb* position). Then if check msb = 1, the LFSR is exclusive-ORed with the bit-vector form of the generator polynomial. Then the *msb* of the LFSR is exclusive-ORed with the next input data bit, and this whole process is repeated until all-data bits have been consumed. Finally, the *residue* (the bits remaining in the LSFR) is compared with an "expected residue." If the residue is equal to the expected residue, the data is assumed to be error-free. Otherwise, there is definitely an error in the input data. This method is guaranteed to detect all single-and double-bit errors. All data except for the PID field is checked in the CRC. The expected residue is 01100 for the 5-bit CRC check and 1000000000001101 for the 16-bit CRC check.

For CRC generation, the following procedure is used to protect all data bits (except for the PID field, which does not need to be checked since it includes a PID Check subfield). As in CRC checking, the LFSR is seeded with an all 1-bits pattern. The data bits are sent out with the *lsb* first. For each data bit to be protected, the same shift and exclusive-OR procedure is used as in the CRC checking process. However, after the last protected data bit has been sent and the LFSR has been shifted and (optionally) exclusive-ORed with the CRC generator polynomial, the data remaining in the LFSR (residue) is inverted to form the CRC Check field. Finally, the CRC Check field is sent out (*msb* first), following the data.


### 7.1.6   Observation of Actual USB Signals

The design problem to be addressed in this chapter is to build a USB full-speed mode protocol analyzer. Thus, our circuit must "monitor" the signals traversing an actual USB cable, and then analyze those signals to determine the exact packets that are being sent and the meaning of those packets.

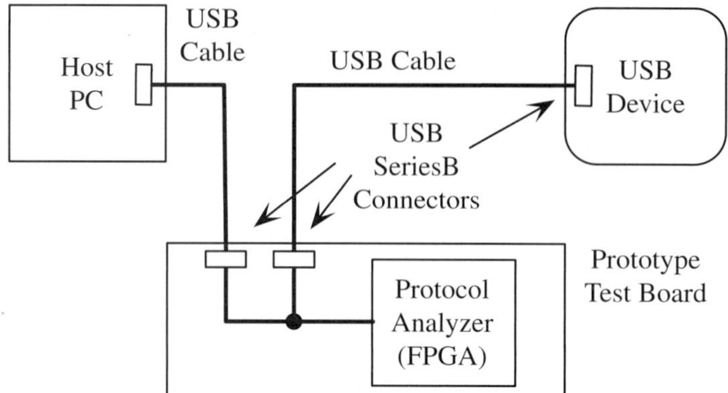



**Figure 7.5**
The circuit used to break a USB connection for observation.

Since we are working with an existing circuit (the PC) that outputs signals used by our target circuit (the USB protocol analyzer), it will be extremely helpful to observe those actual signals. Thus, let us use the configuration shown in Figure 7.5 to "break" the USB connection between a PC and a USB device and then observe the signals on the wires at the break point using an oscilloscope or a logic analyzer. Figure 7.6 shows the D+ and D− signals observed during the SYNC field of a USB packet. Clearly, the two signals are *not* even close to being square waves. This is to be expected, since signals cannot change instantaneously in real systems. Thus, at high frequencies, signals that ideally should be square waves tend to look more like sine waves. While the two signals are close to being inverses of each other, they are not *exact* inverse waveforms.

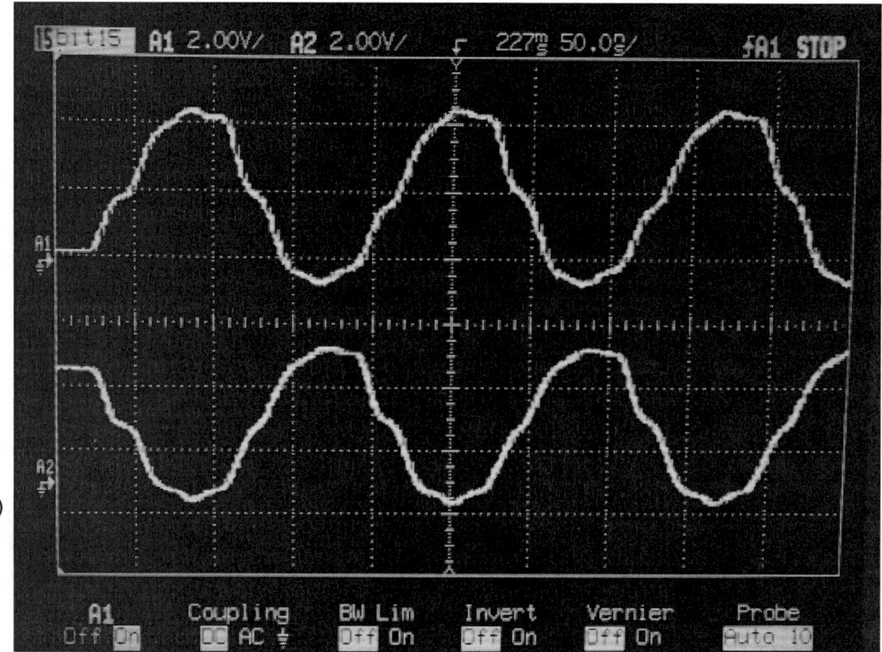



**Figure 7.6**
The actual (oscilloscope) output for the D+ and D− signals observed during the SYNC field of a USB packet.

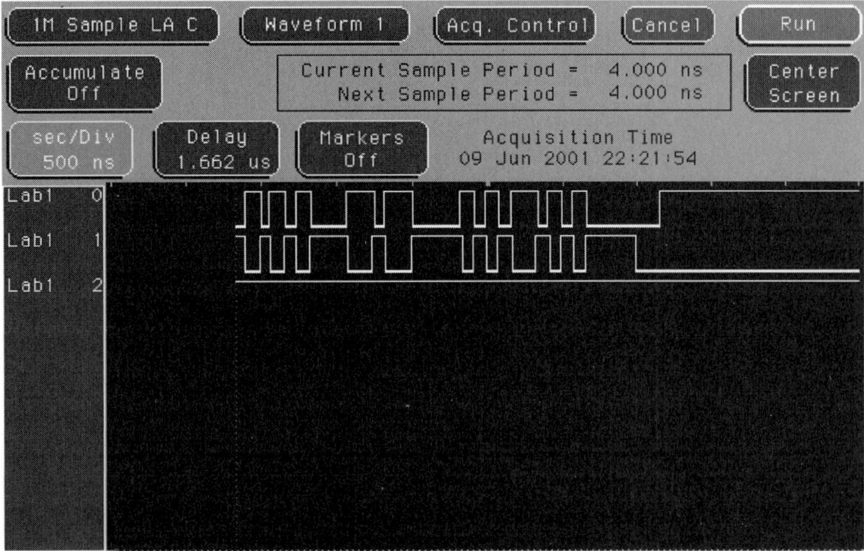


**Figure 7.7**
The logic output for the D+ and D− signals observed during the SYNC field of a USB packet.

A logic analyzer is a device that is used to observe the digital (0 or 1) values observed on a set of wires. Thus, the logic analyzer takes an analog signal, such as those observed with the oscilloscope, and produces a digital signal by digitization (i.e., the signal is interpreted as a '1' or a '0' depending on whether the voltage of the observed signal is greater or less, respectively, than a certain threshold value). Figure 7.7 shows the USB D+ and D− signals observed using a logic analyzer.

The pattern shown is for an SOF packet, including the initial SYNC field. There are several interesting observations to be made. First, the D− signal is *not* an exact digital inverse of the D+ signal. This fact is not too surprising since the exact durations of the '1' and '0' periods in the D+ and D− signals are dependent on the threshold value used for digitization, and the crossover point for the analog D+ and D− signals varies slightly for each data point (as can be seen from the oscilloscope output in Figure 7.6). However, a more surprising fact is that the lengths of the '1' and '0' pulses (during the SYNC field phase, when they should both be the same) are *significantly* different. In fact, when measured using a ruler, it was found that, for the D+ signal, the longest '1' pulse was about 117 ns while the shortest '0' pulse was about 50 ns (a difference of over two times)! This observed difference is extremely important since the receiving side must use the SYNC field to synchronize its data sampling clock. Another important observation is that the "period" of the SYNC field is approximately 6 MHz, as it should be since full-speed USB data should be sent out at a frequency of 12 MHz. (The length of a '0' pulse and the following '1' pulse within a SYNC field is, for example, 50 ns + 117 ns = 167 ns, 1.0/167 ns = 6 MHz.) Ideally, the SYNC field of the D+ signal should be sampled in the middle of each '1' and '0' period. The phase-locking mechanism required for this will be discussed in Section 7.2.2.

## 7.2 Design Overview

The design problem being addressed is to design a *protocol analyzer* for the full-speed mode of USB. A protocol analyzer essentially monitors all activity on a bus and then presents that data to the user in terms of the types of packets transferred and the meaning of those packets. There exist commercial protocol analyzers for USB (refer to [Hyde 1999]) as well other common bus and network interfaces. In designing a protocol analyzer for the USB full-speed mode, we will be designing a fully digital circuit that "understands" the USB specification and undergoes state transitions that encode the knowledge of the current state of the transactions between the various parties involved in a USB data transfer. It is then only a small additional step to design a digital circuit to enable custom-made devices to communicate with the PC via USB (i.e., to build a USB device controller) . Designing a protocol analyzer for USB also will prove invaluable in that it will help the designer learn how to design circuits that interface and work with a modern communication interface on a commercial device.

### 7.2.1 State Machine

In order to be able to decode the incoming USB data, form the proper packet types, check the packet ID and data for transmission errors, and output the packet type, data, and transmission-status information to the outside world, it is necessary to design a state machine. As described in Chapter 5, a state machine essentially consists of a set of states (encoded using flip-flops) and transitions between those states governed by the values of several status (or condition) bits.

In the case of the USB-protocol analyzer design problem, the state should encode the current interpretation of the past USB packets observed thus far, and a state transition should occur whenever the receipt of a new bit changes the current interpretation. Initially, we should be expecting a SYNC field, since that denotes the beginning of all types of packets. Once the SYNC field has been received correctly, we should be expecting a Packet ID field. Then, if the type of the packet, determined from the packet ID, is a token or data packet, we should be expecting a Data field, followed by a CRC field, followed by an EOP indicator. If the packet is a handshake packet, an EOP indicator should immediately follow the packet ID field. Finally, if the packet is a preamble packet, then an Idle (*J* state) field followed by a low-speed packet is to be expected. However, for this design, only the full-speed mode will be supported, and the preamble packet thus will be ignored.

### 7.2.2 Subcircuit Partitioning

Using a hierarchical design method, let us design the circuit as a set of subcircuits which are linked together to form the overall circuit. The following subcircuits will be used.

- usb_pll: a digital phase-locked loop subcircuit that creates a clock signal whose phase is adjusted (over several clock cycles) based on the observed D+ signal.

- nrzi_bin: converts the NRZI-encoded D+ signal (ignoring the D− signal, which is the complement of the D+ signal) into a binary signal.

- crc5: a 5-bit CRC checker used to check the 5-bit CRC in a token packet for errors.

- crc16: a 16-bit CRC checker used to check the 16-bit CRC in a data packet for errors.

- pidr: a PID recognizer subcircuit used to determine the PID type of a token packet by reading the PID Type field (and the corresponding PID Check field).

- usb_interface: implements the state machine for the operation of the protocol analyzer; this subcircuit implements the bulk of the target circuit.

- usb_pa: the top-level circuit combining all of the lower-level subcircuits used to form the complete circuit.

- tb_usb_pa: the test bench for this circuit, which applies test vectors to the usb_pa circuit and checks the results of the application of those test vectors.

The main purpose of the circuit partitioning is to simplify the design, and to identify and implement separate, complete tasks as separate subcircuits. The details of the implementations of the above subcircuits are given below.

### Digital Phase-Locked Loop (Subcircuit usb_pll)

From the discussion of actual observed USB signals in Section 7.1.6, it is clear that a mechanism is required to *lock* onto the phase of the clock signal embedded in the received USB signal in order to properly interpret the NRZI data stream received. Since the clock information is embedded in the up and down transitions of the D+ and D− signals, it is necessary to closely monitor the D+ (and/or D−) signal in order to determine exactly *when* transitions occur. The *USB receive clock*, which is a 12-MHz clock signal (assuming USB full-speed mode) used for sampling the received NRZI signal, should iterate in a periodic manner, with the timing of the transitions (the phase) determined by the transition points of the D+ (and/or D−) signal. Assuming data sampling on rising clock edges, the USB receive clock should transition from 0 to 1 in the *middle* of any potential transitions in D+ and D−. Thus, since the clock used to generate the D+ and D− signals at the transmitting side tends to drift in phase, the USB receive clock must also adjust its phase based on the transitions observed in the D+ (and/or D−) signal.

Essentially, a phase-locked loop (PLL) circuit is required to generate the 12 MHz USB receive clock. The basic function of a PLL is to adjust the phase of an internal clock signal so that it matches the phase of an observed external signal (i.e., when there is a transition in the external signal, there should also be a transition in the internal clock signal at approximately the same time instant). PLL circuits normally are constructed as analog circuits that adjust their clock phases slowly (incrementally) based on a base external signal. However, it is also possible to construct a digital PLL circuit by using a sampling frequency of several times the required clock frequency.

### NRZI-to-Binary Converter (Subcircuit `nrzi_bin`)

Module `nrzi_bin` converts the NRZI-encoded USB signal D+ (denoted as `Dp_rec`) into a normal binary signal (denoted as `Dp_bin`). This conversion permits the rest of the circuit to work with binary signals directly. Note that although only `Dp_rec` is used for the NRZI-to-binary conversion, both `Dp_rec` and `Dn_rec` (representing the D− signal) are needed as inputs in order to observe the EOP (at which time `Dp_bin` is set to 0). If a transition is detected in the `Dp_rec` NRZI signal, the `Dp_bin` binary signal is set to 0; otherwise, it is set to 1.

### CRC Checker Subcircuits (`crc5` and `crc16`)

Since a cyclic redundancy check (CRC) is used in the full-speed USB1.1 bus protocol, the protocol analyzer must recognize the CRC codes being sent across the USB wires. In addition, it would be desirable for the protocol analyzer to be able to determine whether any errors are indicated by the CRC codes observed. There are two types of CRC codes used in USB1.1. A 5-bit CRC is used to protect the non-PID information fields in token packets, and a 16-bit CRC is used to protect the data field in data packets.

### Packet ID Recognizer (Subcircuit `pidr`)

The PID field of a token packet consists of a PID subfield and a PID Check subfield, which is formed by inverting the bits of the PID subfield. The PID should be deciphered immediately following a SYNC field. However, the first few bits of the SYNC field, which is used to resynchronize the receiver's clock to the sending data, may not be decoded properly due to the resynchronization delay. Thus, the USB1.1 specification states that only the last five bits of the SYNC field, consisting of 10000 (since the *lsb* is sent first), need to be recognized by the receiver side. The `pidr` subcircuit implements a PID recognizer based on this specification. Note that PID values (some them incorrect) may be output spuriously by this subcircuit, since it looks for PID fields followed the abbreviated 5-bit SYNC field of 10000, for which an identical pattern can be present in data, token, or CRC fields. Thus, the higher-level circuit which uses this PID recognizer is responsible for determining exactly *when* to use the PID values output by this subcircuit.

### State Machine Subcircuit (Subcircuit `usb_interface`)

The `usb_interface` subcircuit implements the state machine for the USB1.1 protocol-analyzer. The bulk of the protocol analyzer code is contained in this subcircuit. This subcircuit is written using several blocks. One block is used to receive the serial USB data (which has been decoded from NRZI format into binary format by the `nrzi_bin` subcircuit), into a 16-bit register while performing *destuffing* of the data (removing a 0-bit following six consecutive 1-bits, since that 0-bit is a *stuffed bit* sent out by the transmitter side in order to force transitions in the NRZI bit stream). Several blocks are used to control reading from and writing to the FIFO register, which stores the protocol data observed so that they can be read by an external circuit at any time. The `main` block, however, is used to implement a state machine that transitions among several states which represent the USB transactions and USB states observed by the protocol analyzer. This state information is used to store the perceived protocol data into the FIFO buffer upon observing the end of the EOP sequence.

### Top-Level Circuit (usb_pa)

The usb_pa top-level circuit unites all of the previous subcircuits to form the overall USB1.1 protocol analyzer circuit. Besides providing connections between the lower-level subcircuits, it creates a one-clock-cycle delayed form of the USB D+ and D− signals so that they can be observed in the same clock cycle as the Dbinary signal, which is the binary-data value decoded by the nrzi_bin subcircuit (this subcircuit outputs the decoded binary value after a delay of one clock cycle).

### Test-Bench Code (Circuit tb_usb_pa)

The circuit tb_usb_pa is a test-bench. It does not need to be synthesized into a working hardware circuit. It simply is used to verify the proper operation of the usb_pa circuit using simulation. Thus, signals, print statements, and delay statements can be used liberally throughout the code for this circuit. There are three main parts to a test bench. First, the test bench must provide connections to the circuit being tested (typically referred to as the *Unit Under Test (UUT)*). The inputs of the UUT are connected to signals which must be closely controlled to form the test vectors. The outputs of the UUT are connected to other signals which must be closely monitored by this test-bench circuit in order to detect the presence or absence of simulation errors. Second, the test bench applies test vectors to the UUT using a Verilog **always** block. Third, the test bench observes the outputs of the UUT using another **always** block—this block should include checks to determine if the UUT is executing properly. Along with the checks in this block, the simulation output waveforms (or table of simulation data) should be examined closely to determine if the circuit actually is performing as desired.

Test bench circuits and circuits used for describing hardware do not need to be written using the same guidelines as programming languages. Code for a hardware circuit should be written with an emphasis on the quality (speed and hardware savings) of the circuit that can be synthesized from that code. Test bench code should be written in order to maximize the ability to generate test vectors and check the results of those test vectors in as general a manner as is necessary. Writing such code in an elegant, concise manner is secondary to such concerns. Thus, the test bench tb_usb_pa simply is written using a long sequence of delay and signal assignments for the test generation block, and likewise for the test result checking block.

## 7.3  Verilog Solution

The Verilog design for the USB protocol analyzer circuit is split into the modules listed in Section 7.2.2. In order to write the Verilog code in a readable manner, several constant definitions are used, with descriptive names used for the constants.

The following is a file named `defs.vh` that includes definitions used by two or more modules. By using the Verilog **'define** statement and all capital letters for constant names, constants easily can be identified in Verilog module descriptions, as those names written in all capital letters with a ' (backquote) prefix. Although test benches ideally should be written and tested for all submodules created, such submodule test benches are not shown here for the sake of brevity. However, it is recommended that the reader follow the practice of writing and testing test benches for all submodules in a hierarchical Verilog code description.

The following are the meanings of the constants used in `defs.vh`.

1. 'CRC5_OK and 'CRC16_OK are the 5-bit and 16-bit expected cyclic-redundancy check (CRC) *residues* (i.e., if the data remaining in the 5-bit (or 16-bit) CRC register after the shift operations required for the CRC is equal to 'CRC5_OK (or 'CRC16_OK), then the data is determined to be error-free).

2. The 'IDLE through 'HANDSHAKE constants are used to indicate the current state of the state machine used by the protocol analyzer. These constant names are meant to be descriptive of the type of packet being processed in a particular state.

3. 'SOF_PID through 'NYET_PID are the 4-bit packet identifier (PID) codes as defined by the USB specification. The constants 'IDLE_PID and 'ERROR_PID are extra codes defined by the author to facilitate the coding of this protocol analyzer.

4. 'IDLE_TRANSACTION through 'IN_TRANSACTION are used to indicate the current USB transaction being observed on the USB cable.

5. 'NORMAL through 'ERROR_PID_UNKNOWN are used to indicate the normal/error status of the current USB packet being observed.

6. 'FIFO_WIDTH through 'FIFO_EOD are constants used in the implementation of the FIFO buffer, which stores the USB packet information before it is read by the monitoring PC.

```
//////////////////////////////////////////////////////////////
// File defs.vh:
// COMMON DEFINITIONS
'define CRC5_OK   5'b0_1100 // the expected 5-bit CRC residue
'define CRC16_OK  16'b1000_0000_0000_1101 // expected residue
'define SHIFT_DEPTH 16      // shift register for delayed data
'define IDLE      2'b00     // the states of the state machine
'define TOKEN     2'b01
'define DATA      2'b11
'define HANDSHAKE 2'b10
'define SOF_PID     4'b0101 // PID types (= packet types)
'define SETUP_PID   4'b1101
'define OUT_PID     4'b0001
'define IN_PID      4'b1001
'define DATA0_PID   4'b0011
'define DATA1_PID   4'b1011
'define ACK_PID     4'b0010
'define NACK_PID    4'b1010
```

```verilog
`define STALL_PID   4'b1110
`define NYET_PID    4'b0110
`define IDLE_PID    4'b0000  // should be != existing codes
`define ERROR_PID   4'b1111  // should be != existing codes
`define IDLE_TRANSACTION    2'b00   // the transaction types
`define SETUP_TRANSACTION   2'b01
`define OUT_TRANSACTION     2'b11
`define IN_TRANSACTION      2'b10
`define NORMAL              2'b00   // error status codes
`define ERROR_CRC5          2'b01
`define ERROR_CRC16         2'b10
`define ERROR_PID_UNKNOWN   2'b11
// DEFINITIONS for parallel interface to protocol analyzer
`define FIFO_WIDTH 8    // bits output in parallel from FIFO
`define FIFO_DEPTH 64   // num. words stored in the FIFO buffer
`define FIFO_BITS 6     // should be equal to log(FIFO_DEPTH)
`define FIFO_EOD 8'b0111_1111   // end-of-data marker for FIFO
/////////////////////////////////////////////////////////////////
```

### 7.3.1  Digital Phase-Locked Loop

The approach used for the implementation of a digital phase-locked loop circuit is based on the use of an adjustable counter. This code is based on the digital phase-locked loop design presented in [Zeidman]. An alternative approach for the implementation of a digital phase-locked loop, based on the use of a state machine, is discussed in a white paper in [USB 2002]. The intent is to synchronize usb_clk_out (the USB receive clock) with changes in the observed USB D+ signal such that there is an upward transition in usb_clk_out in the *middle* of any potential changes in the D+ signal. This is required in order to ensure that the D+ value is reliably sampled. Note that the initial D+ value will not be in phase with the usb_clk_out value, and thus, needs to be synchronized by using the SYNC (NRZI 8'b1000_0000) pattern. All transitions in the D+ signal (NRZI 0-bits) can be used to readjust the usb_clk_out signal in order to ensure reliable data sampling.

The code for the usb_pll digital phase-locked loop module is written as follows. The D+ signal (we could also use the D− signal) is sampled using a $4\times$ sampling clock (48-MHz clock), A 2-bit counter is operated using the 48-MHz clock. The *msb* of the 2-bit counter is used as the 12 MHz USB receive clock, and the input NRZI signal is obtained by sampling the D+ signal on the positive transition of the USB receive clock. In order to ensure that NRZI sampling occurs in the *middle* part of the input NRZI signal (i.e., to achieve phase locking), the following adjustments are made to the counting sequence of the 2-bit counter. If no transition is detected in the D+ signal, the state (counter output) transitions according to a normal 2-bit binary-counting sequence. If a transition is detected in the D+ signal while in state 3, then the next state is state 0 (as in the normal binary- counting sequence). However, if a D+ transition is detected while in state 0, then the next state (for one more clock cycle) is state 0 (in order to lengthen the USB receive clock). If a D+ transition is detected while in state 2, then the next state is state 0 (in order to shorten the USB receive clock). Finally, if a D+ transition is detected while in state 1, then the next state is state 1 again (this lengths the USB receive clock). In this case, note that the use of state 1 for the next state (instead of state 0) results in a "slow" phase adjustment.

```verilog
/////////////////////////////////////////////////////////////////
// MODULE usb_pll: Digital Phase-Locked Loop (PLL) Circuit
// Author: Sunggu Lee
// Date Created: ...
// Last Modified: ...
// Note: Based on code presented in [Zeidman 1999], but
//    significantly modified in order to customize it for use
//    with the USB1.1 protocol analyzer.
// Description: Produces a phase-adjusted 12MHz USB clock based
//    on a 4x sampling (48MHz) of the observed USB1.1 D+ NRZI
//    signal during the SYNC phase. Attempts to produce positive
//    transitions in usb_clk_out in the middle of changes in D+.

// GLOBAL DEFINITIONS
`include "defs.vh"

// LOCAL DEFINITIONS
`define COUNT_BITS 2  // bits used in adjustable counter

// TOP MODULE
module usb_pll (usb_clk_out, reset_n, sample_clk, Dp);

output usb_clk_out;   // 12MHz USB clock derived from NRZI Dp
input reset_n;        // active-low system-wide reset
input sample_clk;     // Fast system clock (48MHz)
input Dp;             // the D+ input of the USB1.1 bus

// SIGNAL DECLARATIONS for OUTPUTS
wire usb_clk_out;

// SIGNAL DECLARATIONS for INTERNAL SIGNALS
reg [`COUNT_BITS-1:0] count; // Adjustable counter used to
                             //   lock on to transitions in D+
reg prev_Dp;                 // Dp in previous clock cycle

// ASSIGN STATEMENTS
assign usb_clk_out = count[`COUNT_BITS-1]; // {\it msb} of count
```

```verilog
// Main always block
always @(negedge reset_n or posedge sample_clk) begin
  if (~reset_n) begin
    count <= 0;     // initialize count used for usb_clk_out
    prev_Dp <= 1;   // IDLE mode of USB full speed has Dp = 1
  end
  else begin
    // Check for a transition (both pos. and neg.) in Dp
    if ((~prev_Dp & Dp) || (prev_Dp & ~Dp)) begin
      case (count)
        2'b00: count <= 2'b00;
        2'b01: count <= 2'b01;  // used to slowly adjust phase
        2'b10: count <= 2'b00;
        default: count <= 2'b00;
      endcase  // of case
    end
    else // if no transition in Dp
      count <= count + 1;  // assumes wraparound from max to 0

    // Store previous Dp input to find rising and falling edges
    prev_Dp <= Dp;
  end // of outermost else
end // of always


endmodule // of PLL for full-speed USB 1.1 interface
/////////////////////////////////////////////////////////////
```

The following is a test bench that can be used to test the usb_pll digital phase-locked loop module. Since this module simply outputs a clock signal that slowly locks onto the phase of another input signal, it is tested in a manner that is somewhat different from previous test benches. The Dp signal input is generated based on the actual observed (measured) data from a USB1.1 cable. The proper operation of this module can be verified by observing the output signal waveform. It is working correctly if the output clock usb_clk_out changes value at approximately the same time as the Dp signal after the circuit has had a chance to observe changes in the Dp signal for a few sampling clock cycles (i.e., it is working correctly if the *phase* of the usb_clk_out signal slowly *locks* onto the *phase* of the Dp signal).

```verilog
/////////////////////////////////////////////////////////////
// MODULE tb_usb_pll: Test Bench for "usb_pll" // Author: Sunggu
Lee // Date Created: ...  // Last Modified: ...  // Description:
Tests the "usb_pll" digital phase-locked loop.

// COMMON DEFINITIONS
`include "defs.vh"
```

```verilog
// LOCAL DEFINITIONS
`timescale 1ps/1ps      // 1ps accuracy &  1ps precision
`define PERIOD0 20833  // sampling clock period (1/48MHz)
`define PERIOD1 83333  // USB1.1 full-speed clock (1/12MHz)
`define TEST_CYCLES 4  // number of separate tests used

// MODULE DECLARATION
module tb_usb_pll ();  // empty parameter list for test bench

  // SIGNAL DECLARATIONS FOR CONNECTIONS TO UUT
  wire usb_clk_out; // desired phase-locked 12MHz USB clock
  reg reset_n;       // active-low system-wide RESET
  reg sys_clk;       // 48MHz clock used for sampling
  reg Dp;            // the D+ input of the USB1.1 bus

  // SIGNALS USED TO CONTROL THE TEST BENCH OPERATION
  integer error_count = 0; // number of errors encountered
  integer i;               // index variable

  // Instantiate the UUT (unit under test)
  usb_pll uut (usb_clk_out, reset_n, sys_clk, Dp);

  // generate the reset_n and sys_clk signals
  initial begin
    sys_clk = 0;       // 48MHz (approximately) sampling clock
    reset_n = 1;       // create a reset pulse
    reset_n = #(`PERIOD1/4) 0;
    reset_n = #`PERIOD1 1;     // `PERIOD1 length '0' pulse
  end

  // generate the clock
  always #(`PERIOD0/2) sys_clk = ~sys_clk;

  // test vector generation process -- check results manually
  initial begin  // to generate test vectors
    $display("Start simulation.");
    Dp = 1;  // IDLE mode
    #(`PERIOD1 * 2);  // wait longer than reset_n pulse
    #(`PERIOD1 * 8);  // test it with no Dp value changes
    $display("Check usb_clk_out pattern 0 at %0t.", $time);
```

```verilog
      // generate five pulses in phase with initial usb_clk_out
      #(`PERIOD0 / 4);  // wait to get in phase
      for (i = 0;  i < 5;  i = i + 1) begin
        Dp = ~Dp;   // note: maximum Dp change rate is half of
         #`PERIOD1;  //    the USB1.1 clock rate = 6MHz
      end // of for
      $display("Check usb_clk_out pattern 1 at %0t.", $time);

      // generate several more pulses starting out of phase
      #(`PERIOD1 / 2);  // wait to get out of phase
      for (i = 0;  i < 5;  i = i + 1) begin
        Dp = ~Dp;   // note: maximum Dp change rate is half of
         #`PERIOD1;  //    the USB1.1 clock rate = 6MHz
      end // of for
      $display("Check usb_clk_out phase change 1 at %0t.", $time);
      Dp = ~Dp;
   #(`PERIOD1 * 37 / 84);  // wait to get out of phase
      for (i = 0;  i < 5;  i = i + 1) begin
        Dp = ~Dp;   // note: maximum Dp change rate is half of
         #`PERIOD1;  //    the USB1.1 clock rate = 6MHz
      end // of for
      $display("Check usb_clk_out phase change 2 at %0t.", $time);
   #(`PERIOD1 * 105 / 84);  // wait to get out of phase
      for (i = 0;  i < 5;  i = i + 1) begin
        Dp = ~Dp;   // note: maximum Dp change rate is half of
         #`PERIOD1;  //    the USB1.1 clock rate = 6MHz
      end // of for
      $display("Check usb_clk_out phase change 3 at %0t.", $time);

      #`PERIOD1;  Dp = 1;   // IDLE mode
      $display("Done with simulations.");
      $finish;
   end  // of main test vector generation process

endmodule
//////////////////////////////////////////////////////////////////
```

### 7.3.2  NRZI-to-Binary Converter

The `nrzi_bin` module converts a USB signal, which uses NRZI encoding, to a normal binary signal so that it can be processed more easily in the rest of the USB protocol analyzer circuit.  The single output `Dp_bin` is the translated binary signal while the `Dp_rec` and `Dn_rec` input signals are the D+ and D− USB signals observed on the USB cable.  The current version of this module simply observes the `Dp_rec` signal and assigns `Dp_bin` a value of 1 if there is a transition in `Dp_rec` and 0 otherwise.

The 12-MHz-phase adjusted clock signal `usb_clk_out` is connected to this module's `clk` input and used to sample the `Dp_rec` signal.

```verilog
///////////////////////////////////////////////////////////
// MODULE nrzi_bin: NRZI-to-binary converter
// Author: Sunggu Lee
// Date Created: ...
// Last Modified: ...
// Description: Changes NRZI USB signal (Dp_rec, Dn_rec)
//   to a normal binary signal (Dp_bin).  Note that we assign
//   Dp_bin = 0 if an EOP is received on (Dp_rec, Dn_rec).

// GLOBAL DEFINITIONS
`include "defs.vh"

module nrzi_bin (Dp_bin, reset_n, clk, Dp_rec, Dn_rec);

output Dp_bin;   // the binary-encoded Dp signal to be output
input reset_n;   // active-low RESET signal
input clk;       // clock used to sample (Dp_rec, Dn_rec)
input Dp_rec;    // NRZI-encoded USB (Dp, Dn) signals received
input Dn_rec;

// SIGNAL DECLARATIONS
reg Dp_bin;       // output described above
reg Dp_rec_del;   // one-clock cycle delayed form of Dp_rec

// Main subroutine
always @(negedge reset_n or posedge clk) begin
  if (~reset_n) begin
    Dp_bin <= 0;
  end
  else begin
    if (~Dp_rec && ~Dn_rec) // if EOP, send out 0 as default
      Dp_bin <= 0;
    else if (Dp_rec_del != Dp_rec)  // if (Dp_rec has changed)
      Dp_bin <= 0;
    else            // if (Dp_rec has NOT changed), output is 1
      Dp_bin <= 1; // NOTE: includes high-Z (during simulation)!
  end
end
```

```verilog
// subroutine to generate 1-clock cycle delayed Dp_rec signal
always @(negedge reset_n or posedge clk) begin
  if (~reset_n)
    Dp_rec_del <= 1;
  else
    Dp_rec_del <= Dp_rec;
end


endmodule // of module nrzi_bin
////////////////////////////////////////////////////////////////
```

The following is a test bench that can be used to verify the proper operation of the nrzi_bin module. This test bench simply creates a series of different NRZI patterns and then verifies that the nrzi_bin module has converted them properly into binary format.

```verilog
////////////////////////////////////////////////////////////////
// MODULE tb_nrzi_bin: Test Bench for "nrzi_bin"
// Author: Sunggu Lee
// Date Created: ...
// Last Modified: ...
// Description: Tests the "nrzi_bin" NRZI-to-binary converter

// LOCAL DEFINITIONS
`timescale 1ns/1ns    // 1ns simulation accuracy & 1ns precision
`define DELAY    100 // delay value used in testing (100 ns)

// Test bench module definition
module tb_nrzi_bin ();
  // SIGNAL DEFINITIONS
  // Signals used to connect to the UUT (unit under test)
  wire Dp_bin;  // the binary-encoded Dp signal to be output
  reg reset_n;  // active-low RESET signal
  reg clk;      // clock used to sample (Dp_rec, Dn_rec)
  reg Dp_rec;   // NRZI-encoded USB (Dp, Dn) signals received
  reg Dn_rec;
  // Internal signals
  integer error_count = 0;   // total number of errors found
  integer i;                 // test vector generation index

  // Connect to the UUT (unit under test)
  nrzi_bin uut (Dp_bin, reset_n, clk, Dp_rec, Dn_rec);
```

```verilog
// generate the reset_n and clk signals
initial begin
  clk = 0;                    // initialize clock signal
  reset_n = 1;                // create a reset pulse
  reset_n = #(`DELAY/4) 0; // `DELAY/4 time units at 1 value
  reset_n = #`DELAY 1;      // 0 pulse of length `DELAY
end
always #(`DELAY/2) clk = ~clk;  // periodic clock signal

// Generate the test vectors and check the results
initial begin
  $display("Start simulation.");
  Dp_rec = 1;  Dn_rec = 0;  // note: differential signal
  #(`DELAY * 2);  // initial delay longer than reset pulse

  // generate a SYNC packet (8'b1000_000), {\it lsb} first
  for (i = 0;  i < 7;  i = i + 1) begin
    Dp_rec = ~Dp_rec;  Dn_rec = ~Dn_rec;
    #`DELAY;    // change in (Dp_rec, Dn_rec) --> 0 bit
    error_check(1'b0, Dp_bin);
  end // of for loop
  #`DELAY;      // no change in (Dp_rec, Dn_rec) --> 1 bit
  error_check(1'b1, Dp_bin);

  // generate and check a sequence of 1 bits
  for (i = 0;  i < 3;  i = i + 1) begin  // four 1 bits
    #`DELAY;    // no change in (Dp_rec, Dn_rec) --> 1 bit
    error_check(1'b1, Dp_bin);
  end // of for

  // generate and check 8'b1100_1100 sequence, {\it lsb} first
  for (i = 0;  i < 2;  i = i + 1) begin
    Dp_rec = ~Dp_rec;  Dn_rec = ~Dn_rec;
    #`DELAY;    // change in (Dp_rec, Dn_rec) --> 0 bit
    error_check(1'b0, Dp_bin);
    Dp_rec = ~Dp_rec;  Dn_rec = ~Dn_rec;
    #`DELAY;    // change in (Dp_rec, Dn_rec) --> 0 bit
    error_check(1'b0, Dp_bin);
    #`DELAY;    // no change in (Dp_rec, Dn_rec) --> 1 bit
    error_check(1'b1, Dp_bin);
    #`DELAY;    // no change in (Dp_rec, Dn_rec) --> 1 bit
    error_check(1'b1, Dp_bin);
  end // of for
```

```
    // check proper operation with EOP
    Dp_rec = 0;  Dn_rec = 0;    // single-ended 0
    #(`DELAY * 2);
    error_check(1'b0, Dp_bin); // output binary 0 with EOP


    #`DELAY;  // done with tests
    if (error_count > 0)
      $display("ERROR: There were %0d errors!", error_count);
    else
      $display("Simulation complete with NO errors.");
    $finish;            // terminate the simulation
  end  // of always block for test vector generation and checking


  // Error checking task
  task error_check;
    input expected_bin;  // the expected binary result
    input actual_bin;    // the actual binary result
  begin
    if (expected_bin !== actual_bin) begin
      error_count = error_count + 1;
      $display("ERROR: expected = %0b, actual = %0b at time %0t.",
             expected_bin, actual_bin, $time);
    end  // of if
  end  // of outermost begin for error checking task
  endtask


endmodule
////////////////////////////////////////////////////////////////
```

### 7.3.3  CRC Checker Submodules

The following two programs show the Verilog code for the 5-bit cyclic redundancy
check (CRC) and 16-bit CRC checker modules.  The comments at the beginning
of the codes for these modules include descriptions of the algorithms used for
error checking.

```
////////////////////////////////////////////////////////////////
// MODULE crc5: 5-bit CRC Checker
// Author: Sunggu Lee
// Date Created: ...
// Last Modified: ...
// Description: Implements the 5-bit CRC module for USB 1.1.
//    Based on the CRC code in pp. 273-275 of [Zeidman 1999].
//    The algorithm used for this CRC check follows the method
//    outlined in pp. 158-159 of the USB Specification
```

```verilog
//     (Revision 1.1, Sep. 23, 1998):
//        (1) Initialize 5-bit LFSR to all 1's;
//        (2) Repeat for all data bits to be protected:
//           (2a) Take EX-OR of LFSR's msb and input data bit;
//           (2b) Shift LFSR left one bit (shift in 0 into {\it lsb});
//           (2c) If EX-OR result in (2a) was 1,
//                 then LFSR <-- LFSR EX-OR 5'b0_0101;
//        (3) If residue in LFSR = 5'b0_1100 (expected residue),
//                 then there is NO error.

// GLOBAL DEFINITIONS
`include "defs.vh"

// LOCAL DEFINITIONS
`define TAP5   5'b00101  // G(X) = X^5 + X^2 + 1

// begin module definition
module crc5 (data, reset_n, clk, en, din, eop_done);

output [4:0] data;      // parallel output of the LFSR
input reset_n;                  // active-low RESET
input clk;                      // clock used for shifting
input en;                       // enables shifting
input din;                      // the serial input to the LFSR
input eop_done;                 // done with CRC check

// SIGNAL DEFINITIONS
reg [4:0] data;         // output defined above

wire check_msb; // stores EX-OR of LFSR's msb and data bit

// main subroutine to perform the USB 5-bit CRC check
assign check_msb = data[4] ^ din;   // msb EX-OR din
always @(negedge reset_n or posedge clk) begin
  if (~reset_n) begin
    data <= 5'b1_1111;
  end
  else begin
    if (en) begin
      if (check_msb) // check if ((msb of LFSR) EX-OR din) = 1
        data <= (data << 1) ^ `TAP5;
      else
        data <= data << 1;
    end
```

```verilog
      else if (eop_done)  // re-initialize LFSR for next check
        data <= 5'b1_1111;
    end // of outermost else
end // of always

endmodule
///////////////////////////////////////////////////////////////

///////////////////////////////////////////////////////////////
// MODULE crc16: 16-bit CRC Checker
// Author: Sunggu Lee
// Date Created: ...
// Last Modified: ...
// Description: Based on the same algorithm as that used in
//   module crc5, with the only exceptions being the use of
//   the 16-bit generator polynomial 16'b1000_0000_0000_0101
//   and the 16-bit expected residue 16'b1000_0000_0000_1101.

// GLOBAL DEFINITIONS
`include "defs.vh"

// LOCAL DEFINITIONS
`define TAP16   16'b1000_0000_0000_0101 // = G(X)

module crc16 (data, reset_n, clk, en, din, eop_done);

output [15:0] data; // parallel output of LFSR
input reset_n;              // active-low RESET
input clk;                  // clock used for shifting
input en;                   // enables shifting
input din;                  // the serial input to the LFSR
input eop_done;             // done with check

// SIGNAL DEFINITIONS
reg [15:0] data;     // output defined above

wire check_msb; // holds EX-OR of LFSR's msb and input data bit

// main subroutine for this module
assign check_msb = data[15] ^ din;  // msb EX-OR din
always @(negedge reset_n or posedge clk) begin
  if (~reset_n) begin
    data <= 16'b1111_1111_1111_1111;
  end
```

```verilog
    else begin
      if (en) begin
        if (check_msb)        // check if (msb EX-OR din) == 1
          data <= (data << 1) ^ `TAP16;
        else
          data <= data << 1;
      end
      else if (eop_done)    // re-initialize LFSR for next check
        data <= 16'b1111_1111_1111_1111;
    end // of outermost else
end // of always block


endmodule
/////////////////////////////////////////////////////////////////
```

The following test bench is used to test both the crc5 and crc16 modules. A single test bench is used for these two modules, because they are short and coded very similarly.

```verilog
/////////////////////////////////////////////////////////////////
// MODULE tb_crc: Test Bench for "crc5" and "crc16"
// Author: Sunggu Lee
// Date Created: ...
// Last Modified: ...
// Description: Tests the "crc5" and "crc16" Cyclic Redundancy
//   Code (CRC) checker modules.

// GLOBAL DEFINITIONS
`include "defs.vh"

// LOCAL DEFINITIONS
`timescale 1ns/1ns    // 1ns simulation accuracy & 1ns precision
`define DELAY    100 // delay value used in testing (100 ns)

// Test bench module definition
module tb_crc ();
  // SIGNAL DEFINITIONS
  // Signals used to connect to the UUTs (units under test)
  wire [4:0] data5;    // data output from crc5
  wire [15:0] data16; // data output from crc16
  reg reset_n;   // active-low RESET signal
  reg en5;        // the enable input to the crc5 module
  reg en16;       // the enable input to the crc16 module
  reg clk;        // clock signal used for shifting
  reg din;        // the binary form of the observed serial data
  reg eop_done; // indicates the end of the EOP indicator
```

```verilog
// Internal signals
integer error_count = 0; // total number of errors found
integer i;               // index value
reg [31:0] temp;         // used for test vector generation

// Connect to the two UUTs (units under test)
crc5 uut1 (data5, reset_n, clk, en5, din, eop_done);
crc16 uut2 (data16, reset_n, clk, en16, din, eop_done);

// generate the reset_n and clk signals
initial begin
  clk = 0;                   // initialize clock signal
  reset_n = 1;               // create a reset pulse
  reset_n = #(`DELAY/4) 0;   // `DELAY/4 time units at 1 value
  reset_n = #`DELAY 1;       // 0 pulse of length `DELAY
end
always #(`DELAY/2) clk = ~clk;  // periodic clock signal

// Generate the test vectors and check the results
initial begin
  $display("Start simulation.");
  en5 = 0;  en16 = 0;  din = 0; eop_done = 0;  // init.
  #(`DELAY * 2);   // initial delay longer than reset pulse

  // border case: data input = all 0's
  din = 0;  en5 = 1;  en16 = 1;
  for (i = 0;  i < 16;  i = i + 1)
    #`DELAY;
  error_check5(5'b0_0001, data5);   // pre-compute all
  error_check16(16'h800d, data16); //   expected values
  en5 = 0;  en16 = 0;  eop_done = 1;
  #(`DELAY * 2);   // reset in preparation for next test

  // border case: data input = all 1's
  eop_done = 0;  din = 1;  en5 = 1;  en16 = 1;
  for (i = 0;  i < 16;  i = i + 1)
    #`DELAY;
  error_check5(5'b0_1010, data5);   // pre-compute all
  error_check16(16'h0000, data16); //   expected values
  en5 = 0;  en16 = 0;  eop_done = 1;
  #(`DELAY * 2);   // reset in preparation for next test
```

```verilog
// crc5 check: normal case with correct CRC code
temp = 'he083;  // input data = {crc_code , 11'h083}
eop_done = 0;  en5 = 1;  en16 = 0;
for (i = 0;  i < 16;  i = i + 1) begin
  din = temp[0];
  temp = temp >> 1;
   #`DELAY;
end // of for
error_check5(`CRC5_OK, data5);  // result should be OK
en5 = 0;  en16 = 0;  eop_done = 1;
#(`DELAY * 2);   // reset in preparation for next test

// crc16 check: normal case with correct CRC code
temp = 'hffabbf30;  // input data = {crc code, 16'hbf30}
eop_done = 0;  en5 = 0;  en16 = 1;
for (i = 0;  i < 32;  i = i + 1) begin
  din = temp[0];
  temp = temp >> 1;
   #`DELAY;
end // of for
error_check16(`CRC16_OK, data16); // result should be OK
en5 = 0;  en16 = 0;  eop_done = 1;
#(`DELAY * 2);   // reset in preparation for next test

// normal crc5 and crc16 checks with incorrect CRC codes
temp = 'habcd1234;
eop_done = 0;  en5 = 1;  en16 = 1;
for (i = 0;  i < 32;  i = i + 1) begin
  din = temp[0];
  temp = temp >> 1;
   #`DELAY;
end // of for
error_check5(5'b1_1010, data5);   // pre-computed value
error_check16(16'hdd7c, data16); // pre-computed value

#`DELAY;  // done with tests
if (error_count > 0)
  $display("ERROR: There were %0d errors!", error_count);
else
  $display("Simulation complete with NO errors.");
 $finish;            // terminate the simulation
end  // of always block for test vector generation and checking
```

```verilog
    // Error checking tasks
    task error_check5;
      input [4:0] expected;  // expected value
      input [4:0] actual;    // value produced by circuit
    begin
      if (expected !== actual) begin
        error_count = error_count + 1;
        $display("ERROR: expected = %0b, actual = %0b at time %0t.",
                expected, actual, $time);
      end  // of if
    end  // of outermost begin for error checking task
    endtask
    task error_check16;
      input [15:0] expected;  // expected value
      input [15:0] actual;    // value produced by circuit
    begin
      if (expected !== actual) begin
        error_count = error_count + 1;
        $display("ERROR: expected = %0b, actual = %0b at time %0t.",
                expected, actual, $time);
      end  // of if
    end  // of outermost begin for error checking task
    endtask

endmodule
/////////////////////////////////////////////////////////////////////
```

### 7.3.4  Packet ID Recognizer

The `pidr` module is a packet identifier (PID) recognizer that outputs the PID code, using the `pid` 4-bit output signal, when a valid PID code is observed followed a SYNC pattern in the serial USB input data stream.  The serial USB input data stream is stored in binary form in the `data` vector. When a SYNC pattern is observed in the least significant eight bits, the next set of eight bits are checked for a valid PID pattern, which consists of a 4-bit PID code followed by its inverse pattern.

When a valid PID pattern is observed, the corresponding PID code is output on the `pid` output. If an invalid PID pattern is observed, `ERROR_PID is assigned to `pid`. This PID code output is maintained for two clock cycles before output pid = `IDLE_PID.

```verilog
/////////////////////////////////////////////////////////////////////
// MODULE pidr: PID Recognizer for USB1.1
// Author: Sunggu Lee
// Date Created: ...
// Last Modified: ...
```

```verilog
// Description: Outputs the PID code when SYNC is detected.
//    Based in part on the PID recognizer design written by
//    Young-Uk Park.

`include "defs.vh"          // common definitions

`define SYNC 5'b1_0000 // local definition for SYNC character;
          // this definition allows for spurious 1 pulses due to
          // the PLL not locking to the clock at the beginning

module pidr ( pid, reset_n, clk, data );

output [3:0] pid;   // the recognized PID value
input reset_n;        // active-low RESET
input clk;            // used to clock in the data
input [`SHIFT_DEPTH-1:0] data; // most recent serial data bits

// SIGNAL DECLARATIONS
reg [3:0] pid;        // output defined above

reg hold_pid;         // hold current PID for 1 more clock cycle

// main code
always @(negedge reset_n or posedge clk) begin
  if (~reset_n) begin
    pid <= `IDLE_PID;
    hold_pid <= 0;
  end
  else begin        // reset_n is high and (posedge clk)
    if (data[7:3] == `SYNC) begin
          // SYNC has been received into lower 8 bits of data.
          // Note that only 5 bits of the SYNC are checked.
        hold_pid <= 1; // hold PID value for 1 more clock period
        case (data[15:8])   // also looks at the PID check value
          8'b10100101 :  pid <= data[11:8];  // SOF
          8'b00101101 :  pid <= data[11:8];  // SETUP
          8'b11100001 :  pid <= data[11:8];  // OUT
          8'b01101001 :  pid <= data[11:8];  // IN
          8'b11000011 :  pid <= data[11:8];  // DATA0
          8'b01001011 :  pid <= data[11:8];  // DATA1
          8'b11010010 :  pid <= data[11:8];  // ACK
          8'b01011010 :  pid <= data[11:8];  // NACK
```

```verilog
        8'b00011110 :  pid <= data[11:8];  // STALL
        8'b10010110 :  pid <= data[11:8];  // NYET
        default     :  pid <= `ERROR_PID;
      endcase
    end
    else if (hold_pid)
      hold_pid <= 0; // release PID value on next clock cycle
    else
      pid <= `IDLE_PID; // no SYNC signal detected
  end // of outermost else
end // of always block

endmodule
//////////////////////////////////////////////////////////////////
```

The following is a test bench that can be used to test the pidr PID recognizer module.

```verilog
//////////////////////////////////////////////////////////////////
// MODULE tb_pidr: Test Bench for "pidr"
// Author: Sunggu Lee
// Date Created: ...
// Last Modified: ...
// Description: Tests the "pidr" PID recognizer module.

// GLOBAL DEFINITIONS
`include "defs.vh"

// LOCAL DEFINITIONS
`timescale 1ns/1ns    // 1ns simulation accuracy & 1ns precision
`define DELAY    100 // delay value used in testing (100 ns)

// Test bench module definition
module tb_pidr ();
  // SIGNAL DEFINITIONS
  // Signals used to connect to the UUT (unit under test)
  wire [3:0] pid;                // the PID value output by pidr
  reg reset_n;  // active-low RESET signal
  reg clk;      // used to clock in the data
  reg [`SHIFT_DEPTH-1:0] data; // data to be checked
  // Internal signals
  integer error_count = 0;    // total number of errors found

  // Connect to the two UUTs (units under test)
  pidr uut ( pid, reset_n, clk, data );
```

```verilog
// generate the reset_n and clk signals
initial begin
  clk = 0;                      // initialize clock signal
  reset_n = 1;                  // create a reset pulse
  reset_n = #(`DELAY/4) 0; // `DELAY/4 time units at 1 value
  reset_n = #`DELAY 1;     // 0 pulse of length `DELAY
end
always #(`DELAY/2) clk = ~clk;   // periodic clock signal

// Generate the test vectors and check the results
initial begin
  $display("Start simulation.");
  #(`DELAY * 2);   // initial delay longer than reset pulse
  error_check(`IDLE_PID, pid);

  // border cases: invalid SYNC character
  data = 'b0;  // all 0 case
  #`DELAY;
  error_check(`IDLE_PID, pid);
  data = 'b1111_1111_1111_1111;  // 16 1-bits
  #`DELAY;
  error_check(`IDLE_PID, pid);

  // border cases: valid SYNC, invalid PID
  data = 'b0000_0000_1000_0000;  // PID part = all 0's
  #`DELAY;
  error_check(`ERROR_PID, pid);
  data = 'b1111_1111_1000_0000;  // PID part = all 1's
  #`DELAY;
  error_check(`ERROR_PID, pid);
  data = 'b1100_1100_1000_0000;  // PID check != PID
  #`DELAY;
  error_check(`ERROR_PID, pid);

  // normal cases: valid SYNC, valid PID
  data = 'b1010_0101_1000_0000;
  #`DELAY;
  error_check(`SOF_PID, pid);
  data = 'b0010_1101_1000_0000;
  #`DELAY;
  error_check(`SETUP_PID, pid);
  data = 'b1101_0010_1000_0000;
  #`DELAY;
  error_check(`ACK_PID, pid);
```

```
    #`DELAY;  // done with tests
    if (error_count > 0)
      $display("ERROR: There were %0d errors!", error_count);
    else
      $display("Simulation complete with NO errors.");
    $finish;            // terminate the simulation
  end  // of always block for test vector generation and checking

  // Error checking task
  task error_check;
    input [3:0] expected;  // the expected result
    input [3:0] actual;    // the actual result
  begin
    if (expected !== actual) begin
      error_count = error_count + 1;
      $display("ERROR: expected = %0b, actual = %0b at time %0t.",
           expected, actual, $time);
    end  // of if
  end  // of outermost begin for error checking task
  endtask

endmodule
///////////////////////////////////////////////////////////////////
```

### 7.3.5   State Machine Subcircuit

The `usb_interface` module implements the bulk of the code for this USB protocol analyzer circuit. It consists of a several state machines that are used to control various parts of the protocol analyzer circuit. Since the `usb_interface` module implements the main bulk of the protocol analyzer circuit, this module will be tested only after it is included as a subcircuit of the protocol analyzer circuit. This is an exception to the general rule regarding the testing of all modules with separate test benches before inclusion into higher level circuits.

The beginning part of the `usb_interface` module definition is shown below. `DONT_RECORD_SOF` is a constant parameter that, if set to 1, results in USB *start-of-frame (SOF)* packets not being recorded in the FIFO buffer used to record USB packet information. This feature allows the user to avoid recording long sequences of SOF packets in the FIFO buffer, which has a limited size. The output `fifo_data` is used to send information about a USB packet when the `read_fifo_n` input (active-low) is asserted. The `eop_detected` output is asserted when the end of a USB packet (both D+ and D− low) is detected. The `eop_done` output is a one-clock-cycle delayed version of `eop_detected`.

A "word" of information about the current USB packet consists of the `transaction` (the current USB transaction), `status` (the status of the current USB packet: normal or error), and `pid` (the USB packet ID code). There are signals used to compute and store each of these values before they are stored in the FIFO buffer. The state (a 2-bit signal), maintains information about the current state of the state machine used to control the processing of the USB packet being observed.

Other variables required by the state machine are declared as **reg** or **wire** signals. Signals which must be assigned values within an **always** or **initial** block are declared as type **reg**, while others, including those signals produced as outputs of a submodule, are declared as type **wire**. The data signal holds the serial bit data observed on the USB cable and the destuff and reg_destuff signals are used for removing stuffed 0-bits in the USB data stream. There are also several signals used to work with the 5-bit and 16-bit CRC error-checking bits in the USB data stream, signals used to work with the FIFO buffer storing information about the USB packets observed thus far, and a few miscellaneous signals.

```
//////////////////////////////////////////////////////////////////
// MODULE usb_interface: USB Bus Interface
// Author: Sunggu Lee
// Date Created: ...
// Last Modified: ...
// Description: Implements the USB interface for the protocol
//    analyzer.  All transaction, and status (normal or error)
//    information is recorded.  The packet type, transaction,
//    status, and data are sent out in consecutive 8-bit
//    bytes so that they can be read by any upper-level module.

// COMMON DEFINITIONS
`include "defs.vh"

// LOCAL DEFINITIONS
`define DONT_RECORD_SOF 0 // Change to 1 to skip SOF packets

// MODULE DECLARATION
module usb_interface (fifo_data, eop_detected, eop_done,
            reset_n, clk, data_in, Dp, Dn, read_fifo_n);

output [`FIFO_WIDTH-1:0] fifo_data;  // FIFO data to be output
output eop_detected; // EOP indicator
output eop_done;       // indicates the end of the EOP sequence
input reset_n;         // active-low RESET signal
input clk;             // phase-locked full-speed USB clock
input data_in;         // binary form of the NRZI (D+, D$-$) pair
input Dp, Dn;          // the D+ and D$-$ inputs from USB interface
input read_fifo_n;     // FIFO data read request (active LOW)

// SIGNAL DECLARATIONS
// module outputs
reg [`FIFO_WIDTH-1:0] fifo_data;
```

```verilog
reg eop_detected;
reg eop_done;

// state variables
reg [1:0] state;
reg [1:0] transaction;
reg [1:0] status;
wire [3:0] pid;                  // used to hold the PID code

// other variables
reg [`SHIFT_DEPTH-1:0] data; // holds delayed data
reg destuff;                     // for removing "stuffed" 0-bits
reg [4:0] reg_destuff;           // used to check for 6'b11_1111
reg check_crc5;                  // used for the 5-bit CRC check
wire en_check_crc5;
wire [4:0] crc5_data;
reg check_crc16;                 // used for the 16-bit CRC check
wire en_check_crc16;
wire [15:0]  crc16_data;
reg [3:0] count;                 // number of data bits read
reg store_pid;
reg [`FIFO_WIDTH-1:0] fifo[0:`FIFO_DEPTH-1]; // FIFO of data
                       // to be output to parallel interface
reg write_fifo_n;
reg fifo_write_update;
reg fifo_read_update;
reg write_eod;
reg [`FIFO_BITS-1:0] write_ptr;
reg [`FIFO_BITS-1:0] read_ptr;
wire ignore_transaction;
//////////////////////////////////////////////////////////////////
```

The following **always** block is used to store the USB bits observed on the USB wire. These bits are stored in binary form (translated from their original NRZI format) after stuffed 0-bits have been removed.[2] In the following code, the data signal is shifted right continually, with the newly observed USB data bit shifted into the *msb* position of data. At the same, reg_stuff is shifted right in a similar manner. Then, when six consecutive 1-bits are observed in data, the destuff signal is set to 1 and the next new USB data bit is skipped (not stored into data). This new USB data bit should be a 0-bit; otherwise, an error is indicated.

---

[2]  Recall that *stuffed* 0-bits are bits inserted into the bit stream by the transmitting side in order to prevent excessively long sequences of 1-bits, since such sequences can result in a loss of data synchronization. Section 7.1.3 describes NRZI encoding in detail.

```verilog
////////////////////////////////////////////////////////////////
// Always block to store data while skipping stuffed bits
always @(negedge reset_n or posedge clk) begin
  if (~reset_n) begin
    data <= 'b0;
    reg_destuff <= 'b0;
    destuff <= 0;
  end
  else if (destuff) begin // skip over stuffed bit
    reg_destuff <= {data_in, reg_destuff[4:1]};
    destuff <= 0;
    $display("Skipping over stuffed 0-bit, time %0t.", $time);
    if (data_in != 0)
      $display("ERROR!  Stuffed bit is 1, time %0t.", $time);
  end
  else begin // normal data bit
    if ( (state != `IDLE) &&   // Don't destuff in IDLE state
        ({data_in, reg_destuff} == 6'b11_1111) )
      destuff <= 1;  // skip over next data_in bit
    reg_destuff <= {data_in, reg_destuff[4:1]};
    data <= {data_in, data[`SHIFT_DEPTH-1:1]};
      // shift right with msb = most recent data
  end
end // of always process for data storage with destuffing
////////////////////////////////////////////////////////////////
```

The following set of three **always** blocks implements procedures for reading from and writing to the FIFO buffer. Upon a falling transition of the read_fifo_n signal, the **always** block implementing the FIFO read process is activated. In this process, if the read_ptr is less than the write_ptr, which indicates that there is valid data in the FIFO, then the FIFO data pointed to by read_ptr is returned in signal fifo_data, and fifo_read_update is set to 1. Otherwise, fifo_data is set to a sequence of eight 1-bits, and fifo_read_update is set to 0. Note that a sequence of eight 1-bits is not a valid FIFO word, since a valid PID is never 4'b1111. If fifo_read_update is asserted, the read_ptr for the FIFO buffer is incremented upon every rising transition of the read_fifo_n signal. Likewise, if fifo_write_update is asserted, the write_ptr for the FIFO buffer is incremented upon every rising transition of the write_fifo_n signal. It is not necessary to write the actual FIFO data during this process, as this will be done by another process. The fifo_write_update signal is set to 0 when a full FIFO buffer is detected. The current Verilog code version does not use a circular FIFO buffer.

```verilog
////////////////////////////////////////////////////////////
// Implements the read process for the FIFO buffer
always @(negedge reset_n or negedge read_fifo_n) begin
  if (~reset_n) begin
    fifo_data <= 8'b1111_1111;    // 8 should = `FIFO_WIDTH
    fifo_read_update <= 0;
  end
  else begin  // on negative edge of read_fifo_n
    if (read_ptr != write_ptr) begin // read_ptr < write_ptr
      fifo_data <= fifo[read_ptr];   // output FIFO data
      fifo_read_update <= 1;
    end
    else begin                       // valid data not written yet
      fifo_data <= 8'b1111_1111;  // 8 should = `FIFO_WIDTH
      fifo_read_update <= 0;
    end
  end // of outermost else
end // of always block for reading FIFO

// Updates the read_ptr for the FIFO buffer
always @(negedge reset_n or posedge read_fifo_n) begin
  if (~reset_n)
    read_ptr <= 0;
  else begin  // on positive edge of read_fifo_n
    if (fifo_read_update)
      read_ptr <= read_ptr + 1;
  end
end // of always block for updating read_ptr

// Updates the write_ptr for the FIFO buffer
always @(negedge reset_n or posedge write_fifo_n) begin
  if (~reset_n) begin
    write_ptr <= 0;
    fifo_write_update <= 1;
  end
  else begin  // on positive edge or write_fifo_n
    if (fifo_write_update)
      write_ptr <= write_ptr + 1;
    if (write_ptr == `FIFO_DEPTH-1)
      fifo_write_update <= 0;
  end
end // of always block for updating write_ptr
////////////////////////////////////////////////////////////
```

The following code shows the alias definition for the `ignore_transaction` signal and the beginning part of the **always** block for the main state machine of this protocol analyzer circuit. After the initialization phase, this **always** block operates in phases as indicated by the `state` variable. These phases are active—provided that a stuffed 0-bit is not being processed in the USB data stream. In the `IDLE` phase (with `state` = `IDLE`), a `count` variable is incremented with each clock cycle. When `count` reaches four, the `write_eod` signal is checked, and an end-of-data marker is written into the FIFO buffer if `write_eod` is asserted. Then, when `count` reaches 15, the `pid` signal is checked for a valid PID code.

If `pid` is not `IDLE_PID` (used with error conditions), then the PID code is interpreted. If the PID code is SOF, SETUP, OUT, or IN, then the next phase is set to the `TOKEN` phase, `check_crc5` is asserted, and `transaction` is set to the appropriate transaction type. If the PID code is DATA0 or DATA1, then the next phase is set to the `DATA` phase and `check_crc16` is asserted. For all other PID codes, the next phase is set to the `HANDSHAKE` phase.

```verilog
///////////////////////////////////////////////////////////////
// Condition for ignoring the recording of transaction data
assign ignore_transaction =
    (transaction == `IDLE_TRANSACTION) & (`DONT_RECORD_SOF);

// State machine used to control overall operation of circuit
always @(negedge reset_n or posedge clk) begin
  if (~reset_n) begin
    state <= `IDLE;
    transaction <= `IDLE_TRANSACTION;
    status <= `NORMAL;
    check_crc5 <= 0;
    check_crc16 <= 0;
    store_pid <= 1;
    write_fifo_n <= 1;
    write_eod <= 0;
    count <= 0;
  end
  else if (!destuff) begin  // only in non-destuff mode
    case (state)
      `IDLE:  begin
        if (count == 4'b0100) begin
          if (write_eod) begin  // To output end-of-data marker
            fifo[write_ptr] <= `FIFO_EOD; // end-of-data marker
            write_fifo_n <= 0;
            write_eod <= 0;
          end
        end
        else if (count == 4'b0101)
          write_fifo_n <= 1;
```

```verilog
          if (count == 4'b1111) begin   // start PID check
            if (pid != `IDLE_PID) begin // valid SYNC and PID
              count <= 0;       // reset count of bits observed
              store_pid <= 1;   // store PID value in FIFO next
              case (pid)
                `SOF_PID:  begin
                  state <= `TOKEN;
                  check_crc5 <= 1;
                  transaction <= `IDLE_TRANSACTION;
                end
                `SETUP_PID: begin
                  state <= `TOKEN;
                  check_crc5 <= 1;
                  transaction <= `SETUP_TRANSACTION;
                end
                `OUT_PID:  begin
                  state <= `TOKEN;
                  check_crc5 <= 1;
                  transaction <= `OUT_TRANSACTION;
                end
                `IN_PID:  begin
                  state <= `TOKEN;
                  check_crc5 <= 1;
                  transaction <= `IN_TRANSACTION;
                end
                `DATA0_PID:  begin
                  state <= `DATA;
                  check_crc16 <= 1;
                end
                `DATA1_PID:  begin
                  state <= `DATA;
                  check_crc16 <= 1;
                end
                `ACK_PID: begin
                  state <= `HANDSHAKE;
                end
                `NACK_PID: begin
                  state <= `HANDSHAKE;
                end
                `STALL_PID: begin
                  state <= `HANDSHAKE;
                end
                `NYET_PID:  begin
                  state <= `HANDSHAKE;
                end
```

```
                    default: begin
                       state <= `HANDSHAKE;
                       status <= `ERROR_PID_UNKNOWN;
                    end
                  endcase
               end  // of if (pid != `IDLE_PID)
            end  // of else if (count == 4'b1111)
            else begin
               count <= count + 1;
            end
         end  // of case (state == `IDLE)
//////////////////////////////////////////////////////////////////
```

The `TOKEN phase is used with token packets, which are SOF packets or preambles
to data packets. Since SOF packets should not be recorded if ignore_transaction is
set, this parameter is checked first. Then, if store_pid is asserted, the current PID,
transaction, and status information are stored in the FIFO; and store_pid is set to 0,
indicating that data should be stored in the FIFO next. If store_pid is 0 and count
(the count of valid USB data bits) is 0 or 8, the data portion is stored in the FIFO.
Note that a slightly delayed part of the data stream is stored in the FIFO due to the
need to synchronize with the portion of the protocol analyzer circuit detecting the
end-of-packet (EOP). The last portion of the code for this phase looks for the (EOP)
indicator and the end of the EOP indicator in order to check the CRC values for
the current USB packet. The CRC error status is updated, and the next phase is
set to `IDLE.

```
         //////////////////////////////////////////////////////////////////
         `TOKEN:  begin  // case (state == `TOKEN)
            if (~ignore_transaction) begin
               if (store_pid) begin // first, store PID info in FIFO
                  fifo[write_ptr] <= {pid, transaction, status};
                  write_fifo_n <= 0;
                  store_pid <= 0;      // next, store data part in FIFO
               end
               else if (count[2:0] == 3'b000) begin
                  fifo[write_ptr] <= data[13:6]; // 2-cycle delayed
                  write_fifo_n <= 0;
               end
               else
                  write_fifo_n <= 1;
            end
```

```verilog
      // Check for EOP or end of EOP
      if (~eop_detected && ~eop_done) begin
        check_crc5 <= 1;
        count <= count + 1; // update count of bits observed
      end
      else if (eop_detected) begin     // first EOP detected
        check_crc5 <= 0;
        count <= count + 1; // update count of bits observed
      end
      else begin                      // if (eop_done == 1)
        if (crc5_data != `CRC5_OK)       // check crc5 output
          status <= `ERROR_CRC5;
        else
          status <= `NORMAL;
        count <= 0;       // reset the count and go to IDLE
        state <= `IDLE;
      end     // of else (if (eop_done == 1))
    end  // of case (state == `IDLE)
//////////////////////////////////////////////////////////////
```

The `DATA phase functions similarly to the `TOKEN phase, except that 16-bit CRC checks are performed instead of 5-bit CRC checks. Another change is due to the fact that a DATA packet can have a variable length depending on the size of the data payload sent along with the DATA packet, which can range from 0 bytes to 1023 bytes. Thus, there needs to be a way to indicate the end of the data payload information in the FIFO. This is accomplished with the special end-of-data indicator, which is written to the FIFO by asserting write_eod at the appropriate time.

```verilog
//////////////////////////////////////////////////////////////
    `DATA: begin  // case (state == `DATA)
      if (store_pid) begin // first, store PID info in FIFO
        fifo[write_ptr] <= {pid, transaction, status};
        write_fifo_n <= 0;
        store_pid <= 0;    // next, store data part in FIFO
      end
      else if (count[2:0] == 3'b000) begin
        fifo[write_ptr] <= data[13:6];  // 2-cycle delayed
        write_fifo_n <= 0;
      end
      else
        write_fifo_n <= 1;
```

```verilog
      // Check for EOP or end of EOP
      if (~eop_detected && ~eop_done) begin
        check_crc16 <= 1;
        count <= count + 1; // update count of bits observed
      end
      else if (eop_detected) begin     // first EOP detected
        check_crc16 <= 0;
        count <= count + 1; // update count of bits observed
      end
      else begin               // eop_done == 1
        if (crc16_data != `CRC16_OK)   // check crc16 output
          status <= `ERROR_CRC16;
        else
          status <= `NORMAL;
        count <= 0;    // then reset the count and go to IDLE
        state <= `IDLE;
        // Write an End-of-Data-Part indicator to the FIFO
        write_eod <= 1;
      end    // of else (if (eop_done == 1))
    end  // of case (state == `DATA)
//////////////////////////////////////////////////////////////////
```

The `HANDSHAKE phase handles Handshake packets observed on the USB cable. In this phase, if store_pid is asserted, the current PID, transaction, and status information is written to the FIFO. Then it waits for the end of the EOP indicator and transitions to the `IDLE phase.

```verilog
//////////////////////////////////////////////////////////////////
      `HANDSHAKE: begin // this covers all cases for "state"
        if (store_pid) begin
          fifo[write_ptr] <= {pid, transaction, status};
          write_fifo_n <= 0;
          store_pid <= 0;
        end
        else
          write_fifo_n <= 1;

        if (eop_done) begin // wait for end of eop sequence
          state <= `IDLE;
        end
      end
    endcase     // of case (state)
  end
end     // of always loop for the state machine
//////////////////////////////////////////////////////////////////
```

The following **always** block is used to detect the end of a USB packet. This is indicated by a single-ended zero on the normally differential pair of data wires D+ and D−. Thus, when Dp and Dn both are detected to be zero, the eop_detected signal is asserted for one clock cycle. Then the eop_done signal is created as a one-clock-cycle delayed version of this eop_detected signal. Both signals are required for synchronization with the rest of the protocol analyzer circuit.

After the end-of-packet **always** block are a set of alias statements and submodule instantiations of the crc5, crc16, and pidr subcircuits. This type of submodule instantiation results in hierarchical circuit descriptions, which are useful for handling circuit complexity. Note that *named association*, in which local signal names are associated with the port names of the submodules, is used in order to clarify the subcircuit connections being used.

```verilog
////////////////////////////////////////////////////////////////
// Subroutine to detect end-of-packet and end of end-of-packet
always @(negedge reset_n or posedge clk) begin
  if (~reset_n) begin
    eop_detected <= 0;
    eop_done <= 0;
  end
  else begin
    if (~Dp && ~Dn) begin
      eop_detected <= 1;  // set when first (0, 0) detected
      eop_done <= 0;
    end
    else if (eop_detected) begin  // end of the EOP sequence
      eop_done <= 1;
      eop_detected <= 0;
    end
    else begin              // normal data
      eop_done <= 0;
      eop_detected <= 0;
    end
  end
end  // of subroutine to detect the end-of-packet

// Submodule to perform a 5-bit CRC check
assign en_check_crc5 = (check_crc5 && (!destuff));
crc5 crc5_1 (.data(crc5_data), .reset_n(reset_n), .clk(clk),
  .en(en_check_crc5), .din(data[`SHIFT_DEPTH-2]),
  .eop_done(eop_done) );

// Submodule to perform a 16-bit CRC check
assign en_check_crc16 = (check_crc16 && (!destuff));
```

```
crc16 crc16_1 (.data(crc16_data), .reset_n(reset_n), .clk(clk),
  .en(en_check_crc16), .din(data[`SHIFT_DEPTH-2]),
  .eop_done(eop_done) );


// Connections to PID recognizer submodule // to calculate PID
pidr pidr_1 (.pid(pid), .reset_n(reset_n), .clk(clk),
  .data(data) );


endmodule
////////////////////////////////////////////////////////////
```

### 7.3.6  Top-Level Module

The top-level module, shown next, for this USB protocol analyzer simply consists of
a set of submodule instantiations. This type of hierarchical code design is useful for
handling circuit complexity, as it hides the details of the implementation of various
parts of the circuit when looking at the "larger picture."  Half-clock-cycle delayed
versions of the Dp and Dn signals are created in order to synchronize with the Dbinary
Then, connections are made to the usb_pll, nrzi_bin, and usb_interface subcircuits
to complete the circuit description.  Note that the usb_interface subcircuit itself is
described using a set of **always** blocks *and* connections to other subcircuits.

```
////////////////////////////////////////////////////////////
// MODULE: USB1.1 Protocol Analyzer
// Author: Sunggu Lee
// Date Created: ...
// Modified: ...
// Description: This module monitors a USB1.1 connection
//   and stores information on all USB packets that it "sees"
//   into a buffer.  The buffer can then be read using any
//   8-bit parallel interface (such as the PC's parallel port
//   or 8255 PPI).  In this implementation, writing to the
//   buffer stops when the buffer is full // so that the
//   monitoring PC can read this protocol data at its leisure.

// COMMON DEFINITIONS
`include "defs.vh"

// MODULE DECLARATION
module usb_pa (parallel_data, Dbinary, usb_clk, eop,
               reset_n, sys_clk, Dp, Dn, read_data_n );

output [`FIFO_WIDTH-1:0] parallel_data;  // protocol data
output Dbinary, usb_clk, eop; // debug signals
input reset_n;                 // active-low system-wide RESET
input sys_clk;                 // 48MHz system clock
```

```verilog
input Dp, Dn;                    // USB connections
input read_data_n;              // low pulse to read FIFO data


// SIGNAL DECLARATIONS
// for the USB signals
reg Dp_del;  // 1/2-clock cycle delayed version of the Dp
reg Dn_del;  // 1/2-clock cycle delayed version of the Dn


// MAIN CODE
// generate the USB clock from the Dp signal
usb_pll usb_pll_L (.usb_clk_out(usb_clk), .reset_n(reset_n),
                 .sample_clk(sys_clk), .Dp(Dp) );


// convert USB NRZI-format signal to a binary-format signal
nrzi_bin nrzi_bin_L (.Dp_bin(Dbinary), .reset_n(reset_n),
                   .clk(usb_clk), .Dp_rec(Dp), .Dn_rec(Dn) );


// Form a one clock cycle delayed form of (Dp, Dn) signals
//   since Dbinary is also delayed by one clock cycle.
always @(posedge usb_clk) begin
  Dp_del <= Dp;
  Dn_del <= Dn;
end


// connect to the main USB interface circuit
usb_interface usb_interface_L (.fifo_data(parallel_data),
    .eop_detected(eop), .eop_done(eop_done), .reset_n(reset_n),
    .clk(usb_clk), .data_in(Dbinary), .Dp(Dp_del), .Dn(Dn_del),
    .read_fifo_n(read_data_n) );


endmodule
///////////////////////////////////////////////////////////
```

### 7.3.7  Test Bench Code for Entire Circuit

The test bench module tb_usb_pa generates and checks test waveforms for the unit
under test (UUT), and the usb_pa protocol analyzer circuit.  The global constant
definitions in defs.vh are included first. Then, the simulation accuracy and precision
are specified using the **`timescale** directive. Then, several local constants are defined.
The constant `PERIOD0 is used to generate the clock signal used to sample the signals
on the USB cable, and `PERIOD1 is used to generate the USB full-speed clock signal.
After the constant definitions is a module declaration for tb_usb_pa consisting of the
module name and an empty parameter list (since a test bench does not need to have
external inputs or outputs).  After this is a set of signal definitions for signals to be
connected to the UUT followed by a set of definitions for internal signals used to
facilitate the generation and checking of test vectors for the UUT.

The test-vector generation and checking portion of the test bench is designed to test proper operation of the protocol analyzer circuit with several different types of USB packets. This circuit is complex mainly because of the *network protocol* that it implements, which involves a sequence of responses to various types of request packets. Thus, generating test vectors for all combinations of input values is insufficient. It also clearly is impossible to generate all possible input-vector sequences. Thus, this type of circuit is tested by exercising its functionality using sequences of test-vectors based on the sequences of inputs expected to occur during actual operation. In this case, the usb_pa circuit is tested by producing test-vector sequences for the SOF, IN, DATA0, and ACK USB packets.

```verilog
/////////////////////////////////////////////////////////////////
// MODULE tb_usb_pa: Test Bench for "usb_pa.v".
// Author: Sunggu Lee
// Date Created: ...
// Last Modified: ...
// Description: Tests the USB1.1 protocol analyzer circuit.

// COMMON DEFINITIONS
`include "defs.vh"

// LOCAL DEFINITIONS
`timescale 1ps/1ps      // 1ps accuracy &  1ps precision
`define PERIOD0 20833   // sampling clock period (1/48MHz)
`define PERIOD1 83333   // USB1.1 full-speed clock (1/12MHz)
`define TEST_CYCLES 4   // number of separate tests used

// MODULE DECLARATION
module tb_usb_pa ();  // empty parameter list for test bench

// SIGNAL DECLARATIONS
wire [`FIFO_WIDTH-1:0] parallel_data; // parallel output data
reg reset_n;                // active-low system-wide RESET
reg sys_clk;                // 48MHz system clock
reg Dp, Dn;                 // USB connections
wire Dbinary, usb_clk, eop; // debugging signals
reg read_data_n;            // used to read the FIFO

// SIGNALS USED TO CONTROL THE TEST BENCH OPERATION
integer cycle1, cycle2; // used to iterate through test vectors
integer error_count;    // num. errors encountered during test
reg done;               // signals that simulation is done

// Instantiate the UUT (unit under test)
usb_pa usb_pa1 (parallel_data, Dbinary, usb_clk, eop,
                reset_n, sys_clk, Dp, Dn, read_data_n );
```

```verilog
// initialize inputs
initial begin
  sys_clk = 0;         // 48MHz (approximately) sampling clock
  read_data_n = 1;

  cycle1 = 0;  cycle2 = 0;
  error_count = 0;
  done = 0;

  reset_n = 1;      // create a reset pulse
  reset_n = #10000 0;   // 10 ns at '1' value
  reset_n = #100000 1;  // 100 ns '0' pulse
end


// generate the clocks
always #(`PERIOD0/2) sys_clk = ~sys_clk;
////////////////////////////////////////////////////////////
```

In the `tb_usb_pa` test bench, we wish to mimic activity on a USB cable and have the
protocol analyzer circuit pick out and decipher the USB packets being produced.
The test-vector generator must also mimic the behavior of a PC used to read the
USB packet information captured by the `usb_pa` module. Since the USB cable will
carry signals using NRZI signalling on the (D+, D−) pair of wires, the test-vector
generator must also perform in the same manner. The rate of signal transmission is
dictated by `PERIOD1: the clock period for the 12 MHz USB clock. In NRZI signalling,
a 0 is transmitted as a change in signal value, and a 1 is transmitted by maintaining
the old signal value. In order to aid in the creation and interpretation of this test
bench code, the binary forms of the data transmitted are shown in the comments.
Note that the *lsb* data is sent first and the *msb* data is sent last. There are also several
FIFO read requests embedded into the test-vector stream. The first test packet is
an SOF packet, consisting of the inverse of the SOF PID code (1010, used for error
checking), followed by the SOF PID code itself (0101), followed by an 11-bit frame
number, and followed by a 5-bit CRC code. Note that the PID and frame numbers
are sent with the *lsb* first while the CRC code is sent with the *msb* first. The packet
is terminated using two EOP markers, which is the norm.

```verilog
////////////////////////////////////////////////////////////
// main body
// test usb to custom bus interface
always @(negedge sys_clk) begin  // to generate test vectors
  case (cycle1)
    0:  begin // SOF packet (based on actual observed data)
      #`PERIOD1; Dp = 1;  Dn = 0; // IDLE mode
```

```
            #`PERIOD1;  Dp = 0;  Dn = 1; // Generate SYNC ({\it lsb} first)
            #`PERIOD1;  Dp = 1;  Dn = 0;
            #`PERIOD1;  Dp = 0;  Dn = 1;
            #`PERIOD1;  Dp = 1;  Dn = 0;  read_data_n = 0;
            #`PERIOD1;  Dp = 0;  Dn = 1;
            #`PERIOD1;  Dp = 1;  Dn = 0;  read_data_n = 1;
            #`PERIOD1;  Dp = 0;  Dn = 1;
            #`PERIOD1;  Dp = 0;  Dn = 1; // SYNC = 8b'1000_0000
            #`PERIOD1;  Dp = 0;  Dn = 1; // Generate SOF PID
            #`PERIOD1;  Dp = 1;  Dn = 0;  read_data_n = 0;
            #`PERIOD1;  Dp = 1;  Dn = 0;  read_data_n = 1;
            #`PERIOD1;  Dp = 0;  Dn = 1;
            #`PERIOD1;  Dp = 1;  Dn = 0;
            #`PERIOD1;  Dp = 1;  Dn = 0;
            #`PERIOD1;  Dp = 0;  Dn = 1;
            #`PERIOD1;  Dp = 0;  Dn = 1; // SOF PID = 8b'1010_0101
            #`PERIOD1;  Dp = 0;  Dn = 1; // 11-bit frame number
            #`PERIOD1;  Dp = 0;  Dn = 1;
            #`PERIOD1;  Dp = 1;  Dn = 0;
            #`PERIOD1;  Dp = 0;  Dn = 1;
            #`PERIOD1;  Dp = 1;  Dn = 0;
            #`PERIOD1;  Dp = 0;  Dn = 1;
            #`PERIOD1;  Dp = 1;  Dn = 0;  read_data_n = 0;
            #`PERIOD1;  Dp = 1;  Dn = 0;
            #`PERIOD1;  Dp = 0;  Dn = 1;  read_data_n = 1;
            #`PERIOD1;  Dp = 1;  Dn = 0;
            #`PERIOD1;  Dp = 0;  Dn = 1; // frame=11b'000_1000_0011
            #`PERIOD1;  Dp = 1;  Dn = 0; // Generate 5-bit CRC code
            #`PERIOD1;  Dp = 0;  Dn = 1;
            #`PERIOD1;  Dp = 0;  Dn = 1;
            #`PERIOD1;  Dp = 0;  Dn = 1;
            #`PERIOD1;  Dp = 0;  Dn = 1; // CRC = 0_0111, msb first
            #`PERIOD1;  Dp = 0;  Dn = 0;  read_data_n = 0;// EOP
            #`PERIOD1;  Dp = 0;  Dn = 0; // EOP
            #`PERIOD1;  Dp = 1;  Dn = 0;  read_data_n = 1;// IDLE
            #(`PERIOD1 - `PERIOD0);   // hold for one more USB clock
        end
//////////////////////////////////////////////////////////////////
```

The second set of test vectors is for a USB IN packet. An 8-bit SYNC is followed by
the IN PID (including the PID check portion), and followed by an 11-bit device and
endpoint address (the place to receive the data from), and followed by a 5-bit CRC
code. In the middle of the transmission of the device and endpoint address is a
stuffed 0-bit, which has been inserted since the address used includes a portion with
six consecutive 1-bits.

```
/////////////////////////////////////////////////////////////////
   1: begin          // second test : IN packet
     #`PERIOD1;  Dp = 0;  Dn = 1; // Generate SYNC (lsb first)
     #`PERIOD1;  Dp = 1;  Dn = 0;
     #`PERIOD1;  Dp = 0;  Dn = 1;
     #`PERIOD1;  Dp = 1;  Dn = 0;
     #`PERIOD1;  Dp = 0;  Dn = 1;
     #`PERIOD1;  Dp = 1;  Dn = 0;
     #`PERIOD1;  Dp = 0;  Dn = 1;
     #`PERIOD1;  Dp = 0;  Dn = 1; // SYNC = 8b'1000_0000
     #`PERIOD1;  Dp = 0;  Dn = 1; // Generate IN PID
     #`PERIOD1;  Dp = 1;  Dn = 0;
     #`PERIOD1;  Dp = 0;  Dn = 1;
     #`PERIOD1;  Dp = 0;  Dn = 1;
     #`PERIOD1;  Dp = 1;  Dn = 0;
     #`PERIOD1;  Dp = 1;  Dn = 0;
     #`PERIOD1;  Dp = 1;  Dn = 0;
     #`PERIOD1;  Dp = 0;  Dn = 1; // IN PID = 8'b0110_1001
     #`PERIOD1;  Dp = 0;  Dn = 1; // Device & endpoint addr.
     #`PERIOD1;  Dp = 0;  Dn = 1;
     #`PERIOD1;  Dp = 0;  Dn = 1;
     #`PERIOD1;  Dp = 0;  Dn = 1;
     #`PERIOD1;  Dp = 0;  Dn = 1;
     #`PERIOD1;  Dp = 0;  Dn = 1;
     #`PERIOD1;  Dp = 1;  Dn = 0; // stuffed bit = 0
     #`PERIOD1;  Dp = 0;  Dn = 1;
     #`PERIOD1;  Dp = 1;  Dn = 0;
     #`PERIOD1;  Dp = 1;  Dn = 0;
     #`PERIOD1;  Dp = 1;  Dn = 0;
     #`PERIOD1;  Dp = 1;  Dn = 0; // (Device, endp) = 11'h73f
     #`PERIOD1;  Dp = 1;  Dn = 0; // Generate 5-bit CRC code
     #`PERIOD1;  Dp = 0;  Dn = 1;
     #`PERIOD1;  Dp = 0;  Dn = 1;
     #`PERIOD1;  Dp = 1;  Dn = 0;
     #`PERIOD1;  Dp = 1;  Dn = 0; // CRC = 1_0101, msb first
     #`PERIOD1;  Dp = 0;  Dn = 0; // Generate the EOP
     #`PERIOD1;  Dp = 0;  Dn = 0; // EOP
     #`PERIOD1;  Dp = 1;  Dn = 0; // Set to IDLE mode
     #(`PERIOD1 - `PERIOD0);    // hold for one more USB clock
   end
/////////////////////////////////////////////////////////////////
```

The third set of test vectors is for the reply from the device to the previous IN packet, which should be a DATA0 packet with the requested data. Since the direction of data flow in the USB cable will change at this point, a few idle states are included at the beginning of this test. This is followed by an 8-bit SYNC, a DATA0 PID, the 16-bit data payload, and a 16-bit CRC code. A stuffed 0-bit again is included during the transmission of the 16-bit data. Also, an incorrect CRC code is used to test the CRC code-testing capabilities of the usb_pa module.

```
//////////////////////////////////////////////////////////////////
  2: begin            // third test: DATA0 packet reply from device
    #(`PERIOD1 * 2);                  // wait for direction change
    #`PERIOD1;  Dp = 1;  Dn = 0; // IDLE mode
    #`PERIOD1;  Dp = 0;  Dn = 1; // Generate SYNC (lsb first)
    #`PERIOD1;  Dp = 1;  Dn = 0;
    #`PERIOD1;  Dp = 0;  Dn = 1;
    #`PERIOD1;  Dp = 1;  Dn = 0;
    #`PERIOD1;  Dp = 0;  Dn = 1;
    #`PERIOD1;  Dp = 1;  Dn = 0;
    #`PERIOD1;  Dp = 0;  Dn = 1;
    #`PERIOD1;  Dp = 0;  Dn = 1; // SYNC = 8'b1000_0000
    #`PERIOD1;  Dp = 0;  Dn = 1; // Generate DATA0 PID
    #`PERIOD1;  Dp = 0;  Dn = 1;
    #`PERIOD1;  Dp = 1;  Dn = 0;
    #`PERIOD1;  Dp = 0;  Dn = 1;
    #`PERIOD1;  Dp = 1;  Dn = 0;
    #`PERIOD1;  Dp = 0;  Dn = 1;
    #`PERIOD1;  Dp = 0;  Dn = 1;
    #`PERIOD1;  Dp = 0;  Dn = 1; // DATA0(~PID, PID) = 8'hc3
    #`PERIOD1;  Dp = 1;  Dn = 0; // Send the data // 16 bits
    #`PERIOD1;  Dp = 0;  Dn = 1;
    #`PERIOD1;  Dp = 1;  Dn = 0;
    #`PERIOD1;  Dp = 0;  Dn = 1;
    #`PERIOD1;  Dp = 0;  Dn = 1;
    #`PERIOD1;  Dp = 0;  Dn = 1;
    #`PERIOD1;  Dp = 1;  Dn = 0;
    #`PERIOD1;  Dp = 0;  Dn = 1;
    #`PERIOD1;  Dp = 0;  Dn = 1;
    #`PERIOD1;  Dp = 0;  Dn = 1;
    #`PERIOD1;  Dp = 0;  Dn = 1;
    #`PERIOD1;  Dp = 0;  Dn = 1;
    #`PERIOD1;  Dp = 0;  Dn = 1;
    #`PERIOD1;  Dp = 1;  Dn = 0; // stuffed bit = 0
    #`PERIOD1;  Dp = 0;  Dn = 1;
    #`PERIOD1;  Dp = 0;  Dn = 1; // data=1011_1111_0011_0000
    #`PERIOD1;  Dp = 1;  Dn = 0; // Generate 16-bit CRC code
    #`PERIOD1;  Dp = 0;  Dn = 1;
    #`PERIOD1;  Dp = 0;  Dn = 1;
    #`PERIOD1;  Dp = 1;  Dn = 0;  read_data_n = 0;
```

```verilog
    #`PERIOD1;  Dp = 1;  Dn = 0;
    #`PERIOD1;  Dp = 0;  Dn = 1;
    #`PERIOD1;  Dp = 0;  Dn = 1;  read_data_n = 1;
    #`PERIOD1;  Dp = 1;  Dn = 0;  read_data_n = 0;
    #`PERIOD1;  Dp = 0;  Dn = 1;  read_data_n = 1;
    #`PERIOD1;  Dp = 1;  Dn = 0;
    #`PERIOD1;  Dp = 0;  Dn = 1;
    #`PERIOD1;  Dp = 1;  Dn = 0;  read_data_n = 0;
    #`PERIOD1;  Dp = 1;  Dn = 0;
    #`PERIOD1;  Dp = 1;  Dn = 0;  read_data_n = 1;
    #`PERIOD1;  Dp = 0;  Dn = 1;
    #`PERIOD1;  Dp = 0;  Dn = 1; // incorrect CRC code
    #`PERIOD1;  Dp = 0;  Dn = 0; // Generate the EOP
    #`PERIOD1;  Dp = 0;  Dn = 0; // EOP
    #`PERIOD1;  Dp = 1;  Dn = 0; // Set to IDLE mode
    #(`PERIOD1 - `PERIOD0);    // hold for one more USB clock
  end
//////////////////////////////////////////////////////////////////
```

In the fourth set of test vectors, an ACK packet is returned by the PC to the device. Again, a direction change is required in the USB cable. A handshake packet does not require a CRC code. Thus, this packet simply consists of an 8-bit ACK PID (4-bit PID check code followed by a 4-bit PID code) followed by two EOP markers. After this last test packet, a small delay is added to permit completion of the test. Then the done signal is asserted (to enable all of the other **always** blocks to terminate), and this process waits for another **always** block to terminate the entire simulation.

```verilog
//////////////////////////////////////////////////////////////////
    3: begin  // fourth test : ACK returned to device
      #(`PERIOD1 * 2);                // wait for direction change
      #`PERIOD1;
      #`PERIOD1;  Dp = 1;  Dn = 0; // IDLE mode
      #`PERIOD1;  Dp = 0;  Dn = 1; // Generate SYNC (lsb first)
      #`PERIOD1;  Dp = 1;  Dn = 0;
      #`PERIOD1;  Dp = 0;  Dn = 1;
      #`PERIOD1;  Dp = 1;  Dn = 0;
      #`PERIOD1;  Dp = 0;  Dn = 1;
      #`PERIOD1;  Dp = 1;  Dn = 0;
      #`PERIOD1;  Dp = 0;  Dn = 1;
```

```verilog
      #`PERIOD1;  Dp = 0;  Dn = 1; // SYNC = 8b'1000_0000
      #`PERIOD1;  Dp = 1;  Dn = 0; // Generate ACK PID
      #`PERIOD1;  Dp = 1;  Dn = 0;
      #`PERIOD1;  Dp = 0;  Dn = 1;
      #`PERIOD1;  Dp = 1;  Dn = 0;
      #`PERIOD1;  Dp = 1;  Dn = 0;
      #`PERIOD1;  Dp = 0;  Dn = 1;
      #`PERIOD1;  Dp = 0;  Dn = 1;
      #`PERIOD1;  Dp = 0;  Dn = 1; // ACK PID = 8b'1101_0010
      #`PERIOD1;  Dp = 0;  Dn = 0; // Generate the EOP
      #`PERIOD1;  Dp = 0;  Dn = 0; // EOP
      #`PERIOD1;  Dp = 1;  Dn = 0; // Set to IDLE mode
      #(`PERIOD1 - `PERIOD0);   // hold for one more USB clock
    end
  `TEST_CYCLES: begin
    #(`PERIOD1*5);
    done = 1;  // done with all of the tests
    wait(0);   // wait forever;
    end        // other always loop terminates simulation
  default: begin
    $display("Warning: unimplemented test cycle in cycle1.");
    $stop;
    end
  endcase
  cycle1 = cycle1 + 1;  // proceed to next test
end  // of always block to generate tests
//////////////////////////////////////////////////////////////
```

The test-checker **always** block must check whether the usb_pa protocol analyzer circuit correctly analyzed all of the USB packets produced in the previous **always** block. These checks are aided by the fact that it is possible to probe internal signals of submodules in Verilog. Thus, for example, usb_pa1.usb_interface_L.state can be used to check the state signal value in the usb_interface_L module, which is itself a submodule of usb_pa1 (note that instance names are used instead of module names). Using this method and appropriate delay statements, it is possible to check whether the main usb_interface submodule is undergoing all of the proper state transitions. The internal state, transaction, status, and pid values are all checked in this manner for all of the USB packets generated in the previous **always** block. Note that, in the third test, an erroneous 16-bit CRC code was generated. Thus, this error should result in status = `ERROR_CRC16 and an increment of error_count to 1. Both of these facts can be checked on the console output. FIFO outputs can be observed on the parallel_data output lines following low read_data_n pulses in the simulation output.

```verilog
///////////////////////////////////////////////////////////
always @(negedge sys_clk) begin  // to check results of test
  case (cycle2)
    0:  begin    // checks for the first test: SOF packet
  $display("Starting check of USB protocol analyzer circuit.");
      #(`PERIOD1 * 3); // state info. delayed by 3 clock cycles
  $display("Starting to check SOF packet at time %0t.", $time);
      #`PERIOD1;  //Dp = 1;  Dn = 0;  // IDLE mode
      if (usb_pa1.usb_interface_L.state !== `IDLE) begin
        $display("ERROR: No IDLE state at time %0t.", $time);
        error_count = error_count + 1;
      end
      #(`PERIOD1 * 15);
      $display("At time %0t, data is %0x.", $time,
              usb_pa1.usb_interface_L.data);
      #`PERIOD1;
      if (usb_pa1.usb_interface_L.pid !== `SOF_PID) begin
        $display("ERROR: No SOF_PID at time %0t.", $time);
        error_count = error_count + 1;
      end
      #`PERIOD1;  //Dp = 0;  Dn = 1;
      if (usb_pa1.usb_interface_L.state !== `TOKEN) begin
        $display("ERROR: No TOKEN state at time %0t.", $time);
        error_count = error_count + 1;
      end
      #(`PERIOD1 * 14);
      $display("At time %0t, data is %0x.", $time,
              usb_pa1.usb_interface_L.data);
      #(`PERIOD1 * 4);
      #(`PERIOD1 - `PERIOD0);
      if (usb_pa1.usb_interface_L.state !== `IDLE) begin
        $display("ERROR: No IDLE state at time %0t.", $time);
        error_count = error_count + 1;
      end
      if (usb_pa1.usb_interface_L.transaction !==
          `IDLE_TRANSACTION) begin
        $display("ERROR: No IDLE_TRANSACTION at %0t.", $time);
        error_count = error_count + 1;
      end
      if (usb_pa1.usb_interface_L.status !== `NORMAL) begin
        $display("ERROR:  Error status %0x at time %0t.",
              usb_pa1.usb_interface_L.status, $time);
        error_count = error_count + 1;
      end
    end
  end
```

```verilog
1:  begin          // checks for the second test: IN packet
  $display("Checking the IN packet at time %0t.", $time);
  #(`PERIOD1 * 15);
  $display("At time %0t, data is %0x.", $time,
          usb_pa1.usb_interface_L.data);
  #`PERIOD1;
  if (usb_pa1.usb_interface_L.pid !== `IN_PID) begin
    $display("ERROR: No IN_PID at time %0t.", $time);
    error_count = error_count + 1;
  end
  #(`PERIOD1 * 16);
  $display("At time %0t, data is %0x.", $time,
          usb_pa1.usb_interface_L.data);
  #(`PERIOD1 * 4);
  #(`PERIOD1 - `PERIOD0);   // hold for one more USB clock
  if (usb_pa1.usb_interface_L.state !== `IDLE) begin
    $display("ERROR: No IDLE state at time %0t.", $time);
    error_count = error_count + 1;
  end
  if (usb_pa1.usb_interface_L.transaction !==
      `IN_TRANSACTION) begin
    $display("ERROR: No IN_TRANSACTION at %0t.", $time);
    error_count = error_count + 1;
  end
  if (usb_pa1.usb_interface_L.status !== `NORMAL) begin
    $display("ERROR: Error status %0x at time %0t.",
        usb_pa1.usb_interface_L.status, $time);
    error_count = error_count + 1;
  end
end
2: begin    // third test: DATA0 packet reply from device
  #(`PERIOD1 * 2);              // wait for direction change
  $display("Checking DATA0 packet at time %0t.", $time);
  #(`PERIOD1 * 16);
  $display("At time %0t, data is %0x.", $time,
          usb_pa1.usb_interface_L.data);
  #`PERIOD1;
  if (usb_pa1.usb_interface_L.pid !== `DATA0_PID) begin
    $display("ERROR: No DATA0_PID at time %0t.", $time);
    error_count = error_count + 1;
  end
```

```verilog
      #(`PERIOD1 * 16);
      $display("At time %0t, data is %0x.", $time,
               usb_pa1.usb_interface_L.data);
      #(`PERIOD1 * 16);
      $display("At time %0t, data is %0x.", $time,
               usb_pa1.usb_interface_L.data);
      #(`PERIOD1 * 4);
      #(`PERIOD1 - `PERIOD0);   // hold for one more USB clock
      if (usb_pa1.usb_interface_L.state !== `IDLE) begin
        $display("ERROR: No IDLE state at time %0t.", $time);
        error_count = error_count + 1;
      end
      if (usb_pa1.usb_interface_L.transaction !==
          `IN_TRANSACTION) begin
        $display("ERROR: No IN_TRANSACTION at %0t.", $time);
        error_count = error_count + 1;
      end
      if (usb_pa1.usb_interface_L.status !== `NORMAL) begin
        $display("ERROR: Error status %0x at time %0t.",
             usb_pa1.usb_interface_L.status, $time);
        error_count = error_count + 1;
      end  // ERROR expected since 16-bit CRC is incorrect
    end
  3: begin  // fourth test : ACK packet returned to device
     #(`PERIOD1 * 2);              // wait for direction change
     #`PERIOD1;
     $display("Checking the ACK packet at time %0t.", $time);
     #(`PERIOD1 * 16);
     $display("At time %0t, data is %0x.", $time,
              usb_pa1.usb_interface_L.data);
     #(`PERIOD1 * 2);
     if (usb_pa1.usb_interface_L.pid !== `ACK_PID) begin
       $display("ERROR: No ACK_PID at time %0t.", $time);
       error_count = error_count + 1;
     end
     #(`PERIOD1 * 2);
     #(`PERIOD1 - `PERIOD0);   // hold for one more USB clock
     if (usb_pa1.usb_interface_L.state !== `IDLE) begin
       $display("ERROR: No IDLE state at time %0t.", $time);
       error_count = error_count + 1;
     end
    end
```

```verilog
`TEST_CYCLES: begin
   wait(done);  #(`PERIOD1 *10);
$display("Test of USB protocol analyzer circuit complete.");
   if (error_count > 0)
     $display("ERROR: There were %0d errors.", error_count);
   else
     $display("No errors.");
   $display("Note: One error (status = 0x02) expected.");
   $finish;
 end
 default: begin
   $display("Warning: unimplemented test in cycle2.\n");
   $stop;
 end
endcase
cycle2 = cycle2 + 1;  // proceed to next test
end // of always block to check test results

endmodule
///////////////////////////////////////////////////////////////
```

## 7.4   Simulation Results

In order to verify the correct operation of our circuit, it is necessary to simulate it using the test bench as the top-level circuit being simulated. There are two types of outputs from this type of simulation: text output and timing waveform output. The text output can be used to quickly ascertain whether the circuit is performing properly at all of the checkpoints where we have inserted **$display** statements. The timing waveform output can be used to examine the expected input-and output-signals waveforms and to check whether those waveforms are as expected. In this case, using the tb_usb_pa test bench, the timing waveform output and console text output shown in Figure 7.8 are produced as the result of our simulation. The Aldec Active-HDL 5.2 Verilog/VHDL simulation tool was used to produce this output. Note that the console output indicates there is one error in the simulation; however, this is to be expected, since an incorrect 16-bit CRC code purposefully is generated for the data packet during "cycle 1 = 2" of the first block (used for test-vector generation).

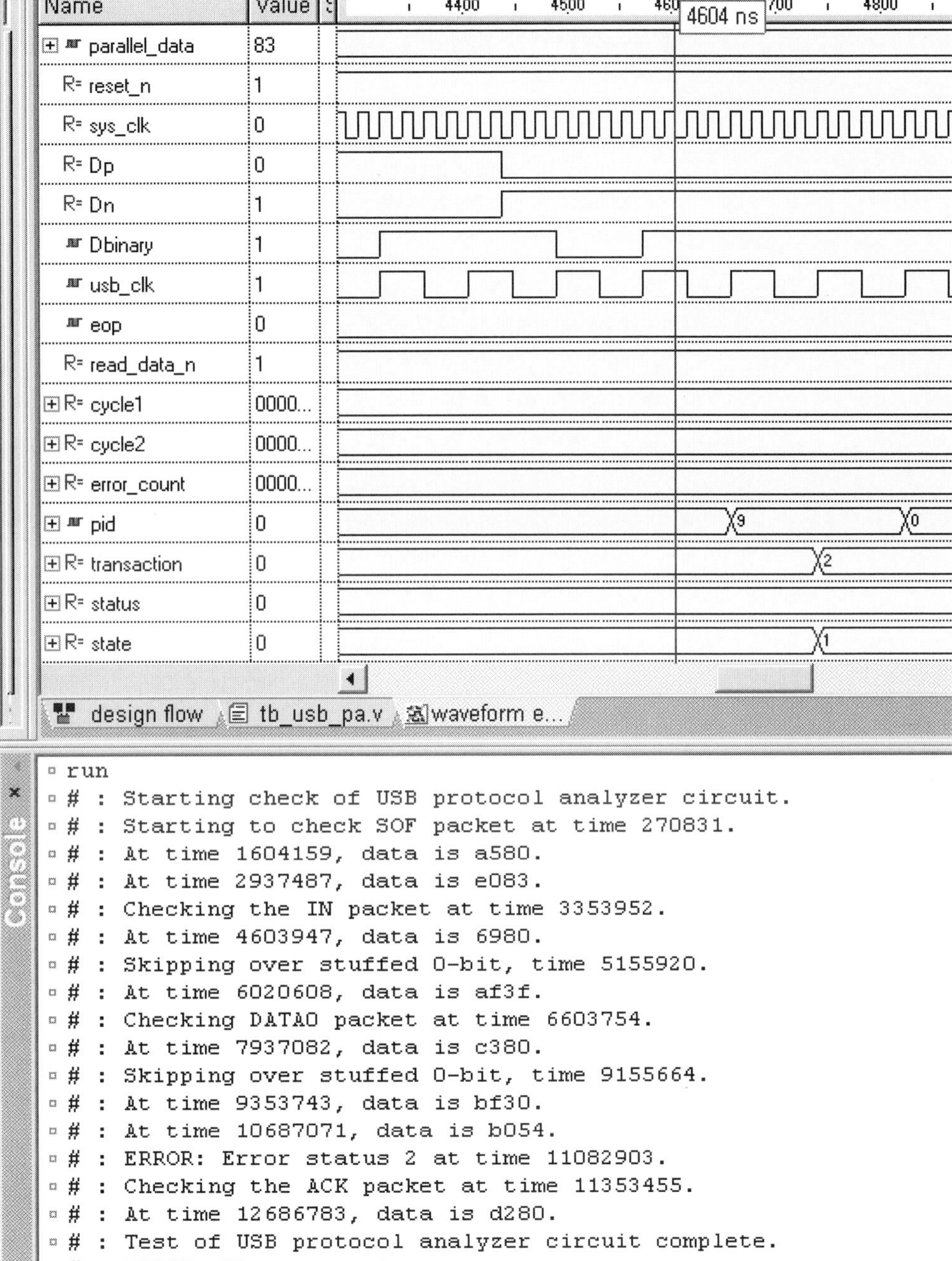



**Figure 7.8**
The simulation output (timing waveforms and console output) for the tb_usb_pa circuit.

## 7.5  Chapter Review

- *Universal Serial Bus (USB)* is a highly flexible, low-cost PC I/O interface that uses *differential signalling*, in which the voltage difference between a pair of wires indicates the logic value being transmitted. In order to permit synchronous data transmission without the use of a separate clock line, USB also uses *NRZI (Non-Return-to-Zero-Inverted)* encoding to embed the clock information in the D+ and D− differential data lines. In this method, logic 0 and logic 1 are indicated by the presence or absence of transitions, a logic 0 is indicated by a transition (positive or negative), and a logic 1 is indicated by no transition.

- Transitions in the D+ and D− differential data lines are used to synchronize the USB receive clock (the clock used to read the data on the receiver's side) with the USB transmit clock (the clock used by the transmitter). Since a long sequence of logic 1's can result in loss of synchronization in the USB receive clock, transitions (representing logic 0's) are needed occasionally to synchronize (i.e., adjust the phase of) the USB receive clock. Thus, after six consecutive 1's are seen by the transmitter, it inserts a logic 0 before sending the next data value (this is referred to as *bit stuffing*). Then, at the receiver side, if six consecutive 1's are observed, the following value (which should be a logic 0) is removed before further processing (this is referred to as *destuffing*).

- Although USB, as its name implies, is a bus, it has the physical topology of a tree with (at most) 127 devices and its root at the root hub port. All requests and responses are broadcast as in a traditional bus, thereby facilitating central control and arbitration from the host (typically a PC). The USB protocol consists of a series of transactions sent during 1 ms frame intervals. All transactions consist of a sequence of packets sent from and to the host. There are ten basic USB packet types, classified into *token* packets (which indicate the transaction being conducted and, thus, the data that follows), *data* packets (which contain the actual data being transmitted), *handshake* packets (which are returned by the device and indicate the presence or absence of data-reception problems), and *special* packets (which are used for control purposes such as changing the data-transmission speed).

- The *cyclic redundancy check (CRC)*, a commonly used method for providing error detection during data transmission, is used in the USB protocol to protect token packets (using a 5-bit code) and data packets (using a 16-bit code). The CRC method basically works by using a $k$-bit *LFSR (linear-feedback shift register)* circuit to divide an $n$-bit "data polynomial" (represented by a binary data stream) by a $k$-bit "CRC generator polynomial" (represented by the structure or, more specifically, the location of exclusive-OR gates in the LFSR circuit) to produce a $k$-bit remainder referred to as the *residue* (represented by the data remaining in the LFSR after all data bits have been shifted through it). The presence of an error in the data is indicated by a residue different from the expected residue. Note, however, that a residue identical to the expected residue does not guarantee the absence of all errors due to the possibility of *aliasing* (identical $k$-bit residues from two different $n$-bit data streams). The CRC codes used in USB nevertheless guarantee the detection of all single-and double-bit errors.

■ This chapter presents the design of a *USB full-speed mode protocol analyzer*, which displays all of the packets being transmitted back and forth in a USB full-speed connection. The protocol-analyzer circuit is designed hierarchically as a connection of eight separate modules. The first interface of this design is a digital phase-locked loop circuit, which is used to produce a USB receive clock for sampling the received NRZI differential data. Also, at the center of this design is a state machine in which the current state is used to encode the current interpretation of the USB packets (and, thus also the transactions), observed thus far. Transaction, USB packet, and error-status information is saved periodically in a FIFO buffer, which can then be read by an external interface to examine the USB packets observed thus far.

# 7.6  Resources

There are numerous books covering the USB protocol. In particular, [Anderson 1997], [Hyde 1999], and [McDowell 1999] were found to be particularly useful references. [Anderson 1997] and [McDowell 1999] provide good general overviews of the USB1.1 protocol. [Hyde 1999] provides a more practical and in-depth treatment of USB1.1 (and a little bit of USB2.0), with an emphasis on the use of a microprocessor-based software approach (including the actual assembly code) for the design of a USB device interface. Of course, the formal specification itself, available from the web pages [USB 1998] (for the USB1.1 spec) and [USB 2000] (for the USB2.0 spec) must also be studied if the reader wishes to examine all of the details of this important protocol. The white paper [USB 2002] describes a state-machine-based approach for the implementation of the digital phase-locked loop circuit required to extract the USB receive clock from the USB NRZI signals sampled. Some of the Verilog code presented in this chapter was based on "library modules" provided in [Zeidman 1999]. Other references on Verilog and synthesis can be obtained from the previous chapters.

## 7.6.1  Bibliography

[7.1] ANDERSON, D., *Universal Serial Bus System Architecture*, Mindshare, Inc., Addison-Wesley, Reading, MA, 1997.

[7.2] HYDE, J., *USB Design by Example*, Intel Corp., John Wiley & Sons, Inc., New York, 1999.

[7.3] McDOWELL, S., and SEYER, M.D., *USB Explained*, Prentice Hall, Upper Saddle River, NJ, 1999.

[7.4] SIERIOREK, D.P., and SWARZ, R. S., *Reliable Computer Systems: Design and Evaluation, 3rd Ed.*, A. K. Peters Ltd., October 1998.

[7.5] USB Implementers Forum, *Universal Serial Bus Specification, Revision 1.1*, September 23, 1998, http://www.usb.org/developers/.

[7.6] USB Implementers Forum, *Universal Serial Bus Specification, Revision 2.0*, April 27, 2000, http://www.usb.org/developers/.

[7.7] USB Implementers Forum, *Designing a Robust USB Serial Interface Engine (SIE)*, 2002, http://www.usb.org/developers/whitepapers/.

[7.8] ZEIDMAN, B., *Verilog Designer's Library*, Prentice Hall, Upper Saddle River, NJ, 1999.

## 7.7 Problems

**P7.1.** Compare and contrast the relative merits and demerits of the USB, FireWire (IEEE 1394), and PCI-X I/O protocols. Construct a table with these three protocols listed across the top and a separate row assigned for each "feature" (i.e., data transfer speed, cost, plug-and-play capability, etc.). Conduct a web search to find the necessary data for these three protocols.

**P7.2.** Suppose that an MP3 song is downloaded from a host PC to an MP3 player using the USB interface. List the sequence of operations and data transfers (both through the software and through the hardware) that occur during this process.

**P7.3.** List the sequence of steps that occur during the initialization phase in order for a new USB device to be recognized and prepared for communication with the host PC.

**P7.4.** If an agreed upon clock frequency and synchronous communication is used, why is a phase-locked loop still necessary when communicating between two independent systems?

**P7.5.** What is the purpose of the SYNC pattern in USB? Why is the SYNC pattern equal to 10000000?

**P7.6.** What is differential signaling? Why is it helpful to use differential signaling to transmit data in a noisy environment?

**P7.7.** What are the pros and cons of using a single-ended mode for the EOP (end-of-packet) indicator? For very high-speed communication, why might it be better to use an alternate method to indicate the end-of-packet? Suggest such an alternative method. Be specific.

**P7.8.** Using a ruler, draw the USB packet waveforms (D+ and D− signals) for a SETUP packet. The relative lengths of all segments of the waveforms must be correct. A ruler can be drawn to aid in viewing the waveforms. Note also that the CRC field must be created correctly and then converted to NRZI format.

**P7.9.** Referring to Problem P7.8, draw the USB packet waveforms for an IN packet.

**P7.10.** Referring to Problem P7.8, draw the USB packet waveforms for an ACK packet.

**P7.11.** Referring to Problem P7.8, draw the USB packet waveforms for a DATA1 packet with the following 16-bit data: 0011110001011010. (Note: A DATA1 packet uses a 16-bit CRC code instead of a 5-bit CRC code.)

**P7.12.** Referring to Problem P7.11, draw the USB packet waveforms for a DATA0 packet with the following 16-bit data: 1110111111110011. (*Hint*: Be sure to include the *stuffed bit*.)

**P7.13.** Give an explanation for the reason that the name *NRZI* is used to describe an encoding method in which a logical '0' is indicated by a transition and a logical '1' is indicated by no transition in the voltage level of the wire carrying the data. Also, compare this with the *NRZ (Non-Return-to-Zero)* encoding method. A description of this latter method can be found through a web search or by referring to books on data communications.

**P7.14.** Suppose that the LFSR structure shown in Figure 7.4 is being used. This structure encodes a 12-bit CRC generator polynomial known as CCITT-12. Assume that the data to be checked is the 16-bit number 0101000011010100011. Use straightforward polynomial division to divide the data polynomial represented by this 16-bit number with the CCITT-12 polynomial. Next, use the LFSR method to determine the remainder computed by shifting through this LFSR structure. (Note: The *lsb* of the data should be entered first.) The two remainder results should be identical. Be sure to show all the steps of this process and of the intermediate results produced.

**P7.15.** The CRC procedure used in the USB protocol is identical with the normal CRC error-checking procedure used, except for the fact that the *inverted* CRC code is appended to the data instead of the noninverted CRC code. The use of this inverted CRC code results in the need to check whether the residue produced with the LFSR at the receiving side is equal to a specific nonzero expected residue value. If the noninverted CRC code were used, the residue at the receiving side simply should be equal to an all-zero pattern. Discuss the pros and cons of using an inverted CRC code versus using a noninverted CRC code.

**P7.16.** When implementing a synchronous sequential digital logic circuit such as a USB device interface, what are the main problems caused by the use of *bit stuffing* and *destuffing* techniques. Be specific.

**P7.17.** Read the [USB 2002] white paper on the use of a state machine to produce a digital phase-locked loop for the USB protocol (refer to the bibliography). Then explain, in detail, just *how* this state machine is used to synchronize the phase of the USB receive clock with the USB transmit clock.

**P7.18.** Referring to Problem P7.17, write a Verilog program to implement the digital phase-locked loop subcircuit required in a USB protocol analyzer or USB device interface circuit. This code should be written using the same I/O ports as the usb_pll subcircuit. Be sure to test your code thoroughly using a test bench and a circuit simulator.

**P7.19.** Write and test a Verilog program to implement a 5-bit CRC generator for USB. Such a program could be used as a subcircuit in a USB device interface circuit. Define (with explanations) and use an appropriate I/O interface for your subcircuit.

**P7.20.** Write and test Verilog programs to implement bit-stuffer and destuffer subcircuits for use with USB. Define (with explanations) and use an appropriate I/O interface for your subcircuit.

# Design of Fast Arithmetic Units

## Important Concepts

- The concept of computer arithmetic and how to compute arithmetic functions (especially addition and multiplication) using digital logic circuits.
- How to design and implement a general pipelined-processing circuit using Verilog.
- How to design a pipelined multiply-accumulate unit using carry-save adders.

*Taylor series expansion* can be used to convert almost any arithmetic operation into a series of addition/subtraction and multiplication/division operations. For example, $sin(z)$—the sine of a number $z$ (in radians)—can be written as

$$\sin(z) = \sum_{n=0}^{\infty} (-1)^n \frac{z^{2n+1}}{(2n+1)!} = z - \frac{z^3}{3!} + \frac{z^5}{5!} - + \cdots \tag{8.1}$$

Note that when this type of expansion is used, an exact result may require an infinite series of addition/subtraction and multiplication/division operations. Since an infinite operation sequence obviously cannot be used in digital-logic circuits, a finite (limited) series of addition, subtraction, multiplication, and division operations (or a look-up table using precomputed values) typically are used to estimate the value returned by an arithmetic operation. Although longer sequences result in more accurate results, a sequence consisting of just the first few terms of the Taylor series expansion typically results in a fairly accurate approximation.

Implementation of arithmetic functions using digital logic circuits or computer programs is referred to as *computer arithmetic*. The most basic operation that needs to be supported by a digital logic circuit used for computer arithmetic is *addition*. As stated above, almost all arithmetic operations can be approximated using a series of addition, subtraction, multiplication, and division operations. Division of a number $x$ by another number $y$ is equivalent to multiplication of $x$ by $1/y$ (the inverse of $y$). The inverse of $y$ in turn can be approximated using a series of multiplication operations.[1] Multiplication of a number $x$ by another number $y$ is equivalent to addition of $x$ with itself, in an accumulative manner, $y - 1$ times. Subtraction of a number $x$ by another number $y$ is equivalent to addition of $x$ with $-y$, where $-y$ can be computed by taking the *2's complement* of the number $y$.[2] Thus, addition (with the use of 2's complement notation) can be used to implement almost *any* type of arithmetic operation. Of course, for faster (and/or more accurate) computation, other methods also may be used to implement some types of arithmetic operations directly.

● ● ● ● ● ● ● ● ● ● ● ● ● ● ● ● ●

## 8.1  Adder Designs

Let us first consider the addition of two 1-bit numbers $a$ and $b$. Since $0 + 0 = 0$, $0 + 1 = 1$, $1 + 0 = 1$, and $1 + 1 = 0$ with an overflow (or carry-out value) of 1, the *sum* and *cout* (carry-out) addition-result bits can be written using the following logic equations.

$$\text{sum} = a'b + ab' = a \oplus b$$

$$\text{cout} = ab$$

---

[1]  The reader interested in the details of this method can refer to the end-of-chapter references.

[2]  Those readers unfamiliar with *2's complement notation* need to review this material from an introductory digital logic or computer organization book. 2's complement notation is also covered briefly in Chapter 1 and Section 8.2.

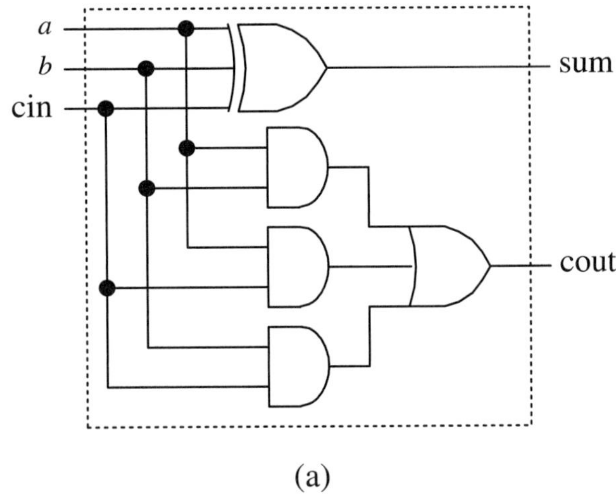

(a)

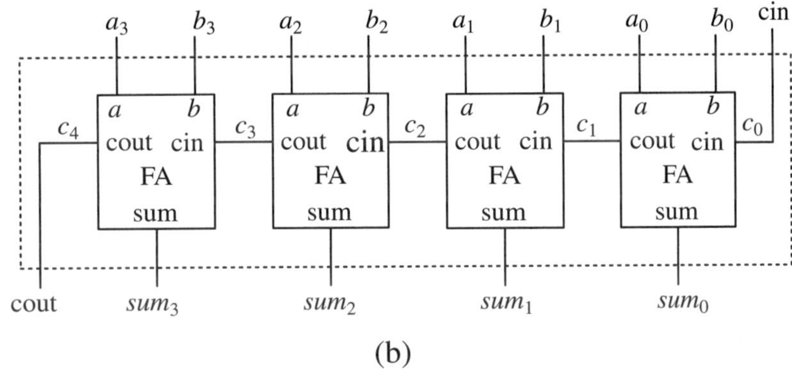

(b)

**Figure 8.1**
A design for (a) a full-adder (FA) and (b) a ripple-carry adder using the full-adder as a subcircuit.

A digital logic circuit implementing these two logic equations is referred to as a *half adder.*

To implement a "more complete" addition function, the 1-bit adder should be able to accept a "carry-in" bit. By using such a carry-in bit, $n$ 1-bit adders could be chained together (with the carry-out of one 1-bit adder connected to the carry-in of the next 1-bit adder) to implement an $n$-bit adder. Thus, instead of a simple half adder, it is necessary to use a 1-bit adder that adds three 1-bit numbers: the carry-in bit *cin* in addition to the inputs $a$ and $b$. This type of 1-bit adder, (referred to as a *full-adder*), can be implemented using the following logic equations (which can be verified by using a truth table).

$$\text{sum} = a'b'cin + a'b\,cin' + a\,b'cin' + a\,b\,cin = a \oplus b \oplus cin$$

$$\text{cout} = a\,b + a\,cin + b\,cin$$

### 8.1.1  Ripple-Carry Adder

The simplest type of adder that can be created for $n$-bit numbers is a *ripple-carry adder*. An $n$-bit ripple-carry adder is created simply by chaining together $n$ 1-bit full adder modules, with the carry-out of one module connected to the carry-in of the next full-adder in the chain. Figure 8.1 shows the design for a 4-bit ripple carry adder.

The main drawback of the ripple-carry adder design is that it is slow. In the worst case (given the most "undesirable" combination of inputs), carry bits can "ripple" through all of the full-adder modules in sequence, thereby changing the carry-out and sum bits of the leftmost full-adder module after a long ripple delay. Let us refer to the delay of a basic logic-gate (AND or OR) as a *logic-gate delay*. For a typical CMOS technology implementation, this delay can range anywhere from about 100 ps to 1 ns. From Figure 8.1, it can be seen that the carry-in input of a full-adder module traverses two logic-gates before reaching the carry-out of that module. Thus, the delay from the carry-in to the carry-out of one full-adder module is two logic-gate delays, and the worst-case delay through an $n$-bit ripple-carry adder is $2n$ logic-gate delays. Given a logic-gate delay of 100 ps (an extremely fast chip) and $n = 32$ bits (the typical size of integers used in computers), this implies an adder delay of 6.4 ns. Then, if we wish to be able to perform at least one addition within a single clock cycle, this implies that our computer must use a system clock with a maximum frequency of $1/(6.4 \times 10^{-9}) = 156$ MHz, which is *much* slower than the clock speeds of current state-of-the-art microprocessors.

### 8.1.2  Carry-Lookahead Adder

A common method used to speed up addition of two $n$-bit numbers is *carry-lookahead*, in which the "carry chain" of a ripple-carry adder is "broken" by computing the carry-in input of a full-adder module *directly* from the primary inputs. Let us consider an $n$-bit ripple-carry adder, which consists of $n$ full-adder modules connected in a "chained-manner. If the $n$ full-adder modules are numbered from $n - 1$ down to 0, then the logic equations for an intermediate full-adder module numbered $i$ can be written as

$$sum_i = a_i \oplus b_i \oplus c_i \tag{8.2}$$

and

$$c_{i+1} = a_i b_i + a_i c_i + b_i c_i \tag{8.3}$$

where $c_{i+1}$ is the carry-out of module $i$, which connects to the carry-in of module $i + 1$ in a ripple-carry adder. The carry chain of a ripple-carry adder can be broken if $c_i$ can be computed directly, instead of waiting for module $i - 1$ to produce $c_i$.

The main reason for the ripple-carry chain is Equation 8.3. Thus, let us rewrite Equation 8.3 as

$$c_{i+1} = a_i b_i + (a_i + b_i)c_i$$

$$= g_i + p_i c_i$$

where $g_i = a_i b_i$ is referred to as the *generate* term (since this is the condition for $c_{i+1}$ to be "generated", regardless of the value of $c_i$) and $p_i = (a_i + b_i)$ is referred to as the *propagate* term (since this is the condition for the $c_i$ term to be "propagated" to the $c_{i+1}$ output[3] Then, denoting the carry-in input to the entire $n$-bit adder as $cin$,

$$c_1 = g_0 + p_0 cin \tag{8.4}$$

$$c_2 = g_1 + p_1 c_1 \tag{8.5}$$

$$= g_1 + p_1(g_0 + p_0 cin) \tag{8.6}$$

$$= g_1 + g_0 p_1 + p_0 p_1 cin \tag{8.7}$$

$$c_3 = g_2 + p_2 c_2 \tag{8.8}$$

$$= g_2 + p_2(g_1 + g_0 p_1 + p_0 p_1 cin) \tag{8.9}$$

$$= g_2 + g_1 p_2 + g_0 p_1 p_2 + p_0 p_1 p_2 cin \tag{8.10}$$

$$c_{i+1} = \sum_{j=0}^{i}\left(g_j \prod_{k=j+1}^{i} p_k\right) + \prod_{k=0}^{i} p_k cin \tag{8.11}$$

The above equations, with the generalized form shown in Equation 8.11, demonstrate how the carry-in inputs for each full-adder module in the $n$-bit adder can be derived directly from the primary inputs (instead of using outputs from previous full-adder modules).

A *carry-lookahead adder* is an adder in which the carry-in input for each full-adder module is computed directly from primary inputs using Equation 8.11. However, for a large value of $i$, Equation 8.11 implies the use of a large number of AND and OR gates, including some with large numbers of inputs. In particular, $i+1$ AND gates, including one with $i+2$ inputs, and an OR gate with $i+2$ inputs are required to produce $c_{i+1}$. Thus, a practical carry-lookahead adder uses "mini-adders", with each mini-adder consisting of (at most) four full-adder modules. Each of the full-adder modules in a mini-adder uses Equation 8.11 to create the carry-in for that module. Thus, since the $c_i$ term necessary for module $i$ relies only on the values of $a_i$, $b_i$, and $cin$, there is no carry chain. Therefore, all outputs of a 4-bit carry-lookahead mini-adder will be ready after $1 + 2 + 2 = 5$ logic-gate delays (one logic-gate delay to form the $g_i$ and $p_i$ terms, followed by two logic-gate delays to form the $c_i$ terms, and followed by two logic-gate delays to form the sum and carry-out terms).

An $n$-bit carry-lookahead adder can be formed in several ways using the 4-bit mini-adder as a subcircuit. In the simplest implementation, the 4-bit mini-adders simply can be cascaded together in a ripple chain, with the carry-out of one mini-adder stage connected to the carry-in of the next mini-adder stage. This implementation, which can be referred to as a *CLA with ripples between stages*, has the disadvantage of being a bit slow, although it is still faster than the ripple-carry adder. For a faster implementation, the carry-lookahead logic can be implemented in a separate module, referred to as a *carry-lookahead generator*. Then the carry-lookahead generators

---

[3]   Note that the propagate term could also be defined as $p_i = (a_i \oplus b_i)$, since a value of $c_i = 0$ will only be propagated to $c_{i+1}$ if exactly one of $a_i$ or $b_i$ (and not both) is equal to 0.).

can be connected in a hierarchical tree-like structure. The equations for the "group" generate and propagate terms used in these carry-lookahead generators are derived in Section 4.2.2. There will be one carry-lookahead generator for each set of four mini-adders (with four full-adders in each mini-adder). Thus, a 16-bit adder will consist of four mini-adders and one carry-lookahead generator, and a 64-bit adder will consist of 16 mini-adders and five carry-lookahead generators (one is used to combine four carry-lookahead generators at a lower level). Figure 4.3(a) shows the design for this type of *hierarchical carry-lookahead adder*.

### 8.1.3 Carry-Save Adder

There are many applications in which a large number of addition operations need to be performed quickly. For such applications, it can be beneficial to use an adder structure referred to as a *carry-save adder (CSA)*.

Consider the basic design for a full-adder. It has three inputs ($a_i$, $b_i$, and $c_i$) and two outputs ($sum_i$ and $c_{i+1}$). Since this is true for all of the $n$ full-adders in an $n$-bit adder, it clearly is possible to add three $n$-bit numbers to produce two $n$-bit results. Thus,

$$a_{n-1}a_{n-2}\ldots a_0 + b_{n-1}b_{n-2}\ldots b_0 + c_{n-1}c_{n-2}\ldots c_0$$

$$= 0sum_{n-1}sum_{n-2}\ldots sum_0 + cout_n cout_{n-1}\ldots cout_1 0$$

$$(8.12)$$

where both the *sum* and *cout* terms have been extended to $n+1$ bits. However, note that a 0-bit is inserted to the left of the *msb* of *sum*, and another 0-bit is inserted to the right of the *lsb* of *cout*. For the $k$th bit inputs, the sum bit is produced in position $k$ while the carry-out bit is produced in position $k+1$.

An adder structure that performs addition in the above manner is referred to as a *carry-save adder (CSA)*. Of course, the two $(n+1)$-bit outputs of a CSA have to be added together to produce one final summation result. However, if there are a large number of $n$-bit numbers to be added, then it is possible to *delay* the addition into one final summation result. Figure 8.2 shows the design for an $n$-bit CSA.

Let us consider the design of an adder used to add a large number of $n$-bit inputs. The two results from the first CSA, which is used to add three $n$-bit inputs,

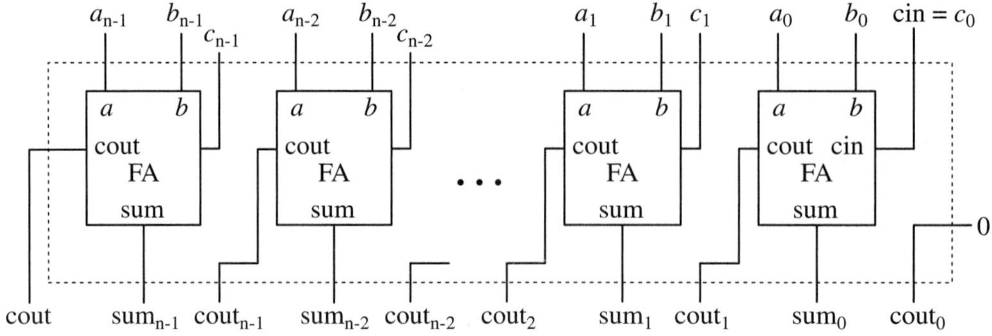


**Figure 8.2** An $n$-bit carry-save adder (CSA).

can be added to a fourth $n$-bit input using another CSA. Then the two results from the second CSA can be added to a fifth $n$-bit input using a third CSA, and so on. Eventually, the two results from the last CSA have to be added together to form one final summation result—this last addition step can be performed using a ripple-carry adder or carry-lookahead adder. Of course, it also is possible to perform additions using several CSAs in parallel (the outputs of which can be added using additional CSAs), provided that sufficient numbers of $n$-bit inputs are available. The fastest method to add a large set of $n$-bit inputs using CSAs involves structuring the CSAs into a 3:2 tree structure (i.e., an inverted tree structure in which each internal node has three inputs and two outputs, and the number of levels in the tree is the minimum possible). Such a CSA tree structure is referred to as a *Wallace-tree adder*. The number of levels in this type of structure is left as a problem for the reader.

## 8.2  Multiplier Designs

In general, *multiplication* can be implemented using a series of addition steps. Multiplication of a number $A$ (the multiplier) with another number $B$ (the multiplicand) corresponds to addition of $A$ with itself $B$ times. For a more efficient implementation, we can multiply the digits of $A$ with $B$ to form partial products, and then add those partial products to produce the final result (the product)—this is the normal method used for the multiplication of decimal numbers taught in elementary school. With binary numbers, the same basic method can be used. However, with binary numbers, the multiplication of a multiplier digit with the multiplicand has one of two results: zero or a left-shifted copy of the multiplicand. In fact, the multiplication of a binary number with a power of two can be implemented as a left- or right-shift operation. For example, multiplication with $2^3 = 8$ corresponds to a 3-bit left shift, and multiplication with $2^{-2} = 0.25$ corresponds to a 2-bit right shift.

For multiplication of *signed* binary numbers, the use of 2's-complement notation permits us to use a method similar to that described above. Negative and positive multiplicands (and partial products) can be accommodated simply by using *sign extension*, in which the sign bit (the *msb* of the original data word) is copied into all extra bit positions to the left of the sign bit when extending the size of the data word. Thus, if an 8-bit negative multiplicand is extended to 16 bits, then the leftmost 8 bits are assigned the sign bit, which should be 1 since the multiplicand is negative. Likewise, if an 8-bit positive multiplicand is extended to 16 bits, then the leftmost 8 bits again are filled with the sign bit, which in this case is 0.

In order to accommodate negative multipliers, we simply need to use a common definition of 2's-complement notation, which states that the 2's complement of an $n$-bit number $A = a_{n-1}a_{n-2}\ldots a_1a_0$ simply is equal to $2^n - A$.[4] If the multiplier $A$ is

---

[4]  Recall from introductory digital logic that the 2's complement of an $n$-bit number $A$ can be obtained by first taking the 1's complement of $A$ (inverting all of the bits of $A$) and then adding 1 to the *lsb* of $A$. However, as the reader can verify, the same result can be obtained by computing $2^n - A$.

a positive number, then $A$ can be represented in binary form as $0\,a_{n-2}a_{n-3}\ldots a_1a_0$. Converting this to a decimal number representation;

$$A = \sum_{i=0}^{n-2} a_i 2^i$$

If $A$ is a negative number, then $A$ can be represented as $1\,a_{n-2}a_{n-3}\ldots a_1a_0$. Thus,

$$-A = 2^n - \left(2^{n-1} + \sum_{i=0}^{n-2} a_i 2^i\right)$$

This, in turn, implies that the original number $A$ can be written as

$$A = \sum_{i=0}^{n-2} a_i 2^i - 2^{n-1}$$

The only difference in this case, from the positive multiplier case, is the subtraction of $2^{n-1}$. Thus, since a negative multiplier has as its *msb* bit $a_{n-1} = 1$ and a positive multiplier has $a_{n-1} = 0$, we can use the following general decimal number-conversion method for a positive or negative multiplier:

$$A \text{ (written as } a_{n-1}a_{n-2}\ldots a_1a_0) = \sum_{i=0}^{n-2} a_i 2^i - a_{n-1}2^{n-1} \tag{8.13}$$

Based on the above observations about positive and negative multipliers and multiplicands, we can perform 2's-complement multiplication of two $n$-bit numbers $A$, (written as $a_{n-1}a_{n-2}\ldots a_1a_0$) and $B$ in a uniform manner. First, the partial products $a_0B$, $a_1 2^1 B$, $\ldots$, $a_{n-2}2^{n-2}B$, must be formed. Then, before adding all of these partial products, all partial products need to be extended to $2n$ bits by "sign extension." Next, all of these sign-extended partial products must be added together. Finally, the term $a_{n-1}2^{n-1}B$ needs to be subtracted from this result (this term also needs to be sign-extended to $2n$ bits). This will have the result of subtracting $2^{n-1}B$ only if $A$ is negative. This final result should be equal to $A \times B$, irrespective of the sign of $A$ or $B$.

<table><tr><td>**Example 8.1**</td><td>**2's Complement Number Conversion and Multiplication**</td></tr></table>

In order to illustrate the validity of the conversion method shown in Equation 8.13, let us consider the conversion of several 4-bit 2's complement numbers. Consider a typical positive number such as 0110. This is the binary notation for the decimal number 6, which can be derived as $0 \times 2^0 + 1 \times 2^1 + 1 \times 2^2$. Next, consider the negative number 1110. According to the conversion method based on Equation 8.13, this number is equal to $1 \times 2^3$ subtracted from the decimal equivalent of 0110. Thus, 1110 is converted to $6 - 8 = -2$, as it should be. The reader can confirm that this conversion method works for any 2's complement positive or negative number of any bit length.

Next, let us illustrate the process of multiplying two 2's complement numbers. Let $A = 0110$ and $B = 1101$. Written in decimal notation, $A = 6$ and $B = -3$. The partial products are $a_0 B = 0 \times 1101 = 0000$, $a_1 2^1 B = 1 \times 2 \times 1101 = 11010$, and $a_2 2^2 B = 1 \times 2^2 \times 1101 = 110100$. Sign extension to 8 bits changes these partial products to 00000000, 11111010, and 11110100. Addition of these three terms results in 11101110, which is the 2's complement notation for $-18$. Since $a_3 = 0$, no additional steps are necessary and the above number is our final result (as it should be), since $6 \times -3 = -18$. Now suppose that $A = 1110$ and $B = 1101$, which are equal to $A = -2$ and $B = -3$ when converted to decimal notation. Then, based on the process described above, multiplication of $A$ and $B$ can be accomplished by following the same procedure as that used for $A = 0110$ and $B = 1101$ prior to the final subtraction of $a_{n-1} 2^{n-1} B$, Thus, since $a_3 2^3 B = 1101000$, we need to subtract 11101000 ($2^3 B$ sign-extended to 8 bits) from 11101110, which results in 00000110 (equal to the decimal number 6, as expected).

### 8.2.1   Combinational Multiplier

The most straightforward method of implementing a multiplier consists of using AND gates to form the partial products (by multiplying the bits of the multiplier with the multiplicand and then left shifting the result by an appropriate amount) and then adding all of the partial products using a set of CSAs. The CSAs can be arranged into a linear structure or a tree structure, depending on speed and hardware complexity constraints. Regardless of the structure used, a standard ripple-carry or carry-lookahead adder is required in order to add the outputs of the CSA in the last stage. Figure 8.3 shows the structure for a simple 2's complement combinational multiplier implemented using an array of CSAs and a carry-lookahead adder in the final stage.

### 8.2.2   Sequential Multiplier

For multiplication of $n$-bit numbers with $n$ large, the use of a combinational multiplier involves a large number of hardware components. For example, if two 64-bit unsigned numbers (double precision) are to be multiplied using a linear CSA structure with a ripple-carry adder following the last CSA, then 64 AND gates (to form the 64 partial products) and 4,032 full-adders ($64 \times 62$ full-adders in 62 CSAs and 64 full-adders in the ripple-carry adder). Since this is an extremely large amount of hardware, an alternative method that is often used (for example, in general purpose multiprocessor chips) is *sequential multiplication*.

Sequential multiplication involves looking at the bits of the multiplier one by one and then adding partial products in an accumulative manner. The addition occurs

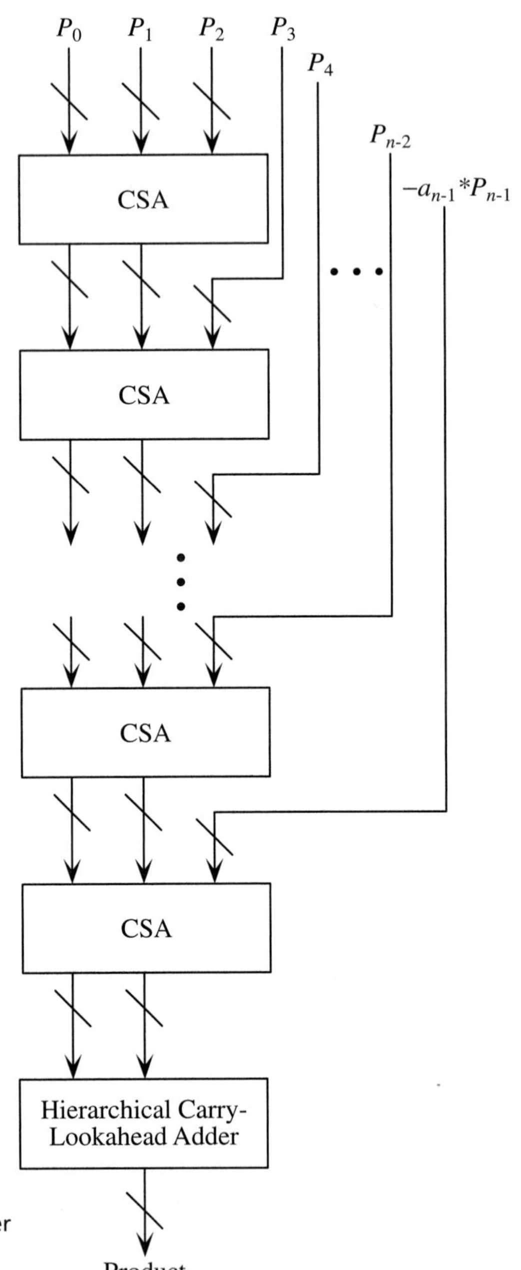



**Figure 8.3**
A design for a
2's complement
combinational multiplier
implemented using
carry-save adders.

in consecutive clock cycles, with a common register used for the accumulation of
partial products. Figure 8.4 shows a block diagram of a design that can be used for
sequential multiplication.

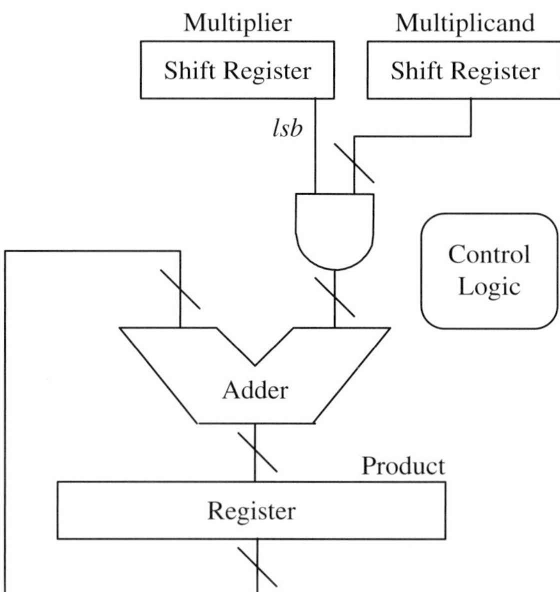


**Figure 8.4**
A sequential multiplier design.

### 8.2.3   Fast Multiplication

There are various methods that can be used for fast multiplication. In general, combinational multiplication is faster than sequential multiplication, since we do not have to wait for intermediate results to be stored into the partial-product register.

For both combinational and sequential multiplication, a common method used to speed up multiplication is *multiplier recoding*. In multiplier recoding, the multiplier is rewritten in a manner such that the multiplication of the multiplier with the multiplicand is accomplished using (preferably) a smaller number of partial-product additions and subtractions than when the original multiplier is used. Two commonly used recoding methods are *Booth recoding* (also known as *differential recoding*) and *modified-Booth recoding*. Given a multiplier $A = a_{n-1}a_{n-2}\ldots a_1a_0$ and a multiplicand $B$, the product is formed from the addition of the partial products $P_i == a_i2^i B$. Since the bits of $A$ are either 1 or 0, this corresponds to the addition of only those partial products $P_i$ with $a_i = 1$. Thus, if the number of 1 bits in the multiplier $A$ can be reduced, the number of additions required is also reduced.

In Booth recoding, the bits of the multiplier are recoded in order to remove long sequences of 1 bits. This method is based on the following observation: when traversing the bits from the *msb* to the *lsb* (left to the right), a long sequence of 1 bits in a binary number is equivalent to an addition at the start of that sequence followed by a subtraction at the end end of that sequence. Thus, for example, a binary number such as 01110 is equivalent to addition of $2^4$ followed by subtraction of $2^1$. The reader can check that $2^4 - 2^1 = 14$ is the decimal equivalent of the binary number 01110.

It can be proven that this type of equivalence holds for any sequence of consecutive 1 bits in a binary number. To denote this type of transformation, an addition can be denoted by a 1 in the corresponding bit position, while a subtraction is denoted by a $\bar{1}$ in the corresponding bit position. Thus, if the multiplier $A = 001110$, its recoded value is $A^* = 010 0\bar{1}0$.

The recoded multiplier $A^*$ tells us that the partial product $P_4 = 2^4 B$ must be added, and the partial product $P_1 = 2^1 B$ must be subtracted in order to compute the product $A \times B$. Booth recoding also is referred to as *differential recoding*, because $A^*$ corresponds to a one-dimensional differentiation of the original number $A$ (the "slope" of the bit sequence in $A$, when viewed from the left to the right, is $+1$ when going from $a_4$ to $a_3$, $-1$ when going from $a_1$ to $a_0$, and 0 at all other bit positions).

Booth recoding works for 2's-complement signed numbers as well as unsigned numbers. To implement this method, the bits of an $n$-bit multiplier $A = a_{n-1}a_{n-2} \cdots a_1 a_0$ are traversed from the *msb* to the *lsb*. Two bits of $A$ are considered at a time and a 0 bit is appended to the right of the *lsb* bit $a_0$. Then, $a_i^* = 1$ when $a_i a_{i-1} = 01$, and $a_i^* = \bar{1}$ when $a_i a_{i-1} = 10$. For all other $a_i a_{i-1}$ sequences, $a_i^*$ is set to 0. Thus, for example, $A = 00110011$ results in $A^* = 010\bar{1}010\bar{1}$, and $A = 11110011$ results in $A^* = 000\bar{1}010\bar{1}$.

Other recoding methods can be used to reduce the number of requisite additions and subtractions beyond those required with Booth recoding. Consider the multiplier $A = 01010101$. The Booth recoding for this number results in $A^* = 1\bar{1}1\bar{1}1\bar{1}1\bar{1}$. Thus, the Booth recoded multiplier $A^*$ requires twice the number of additions and subtractions as in the original multiplier $A$! In order to avoid this type of anomaly, three consecutive bits of the multiplier $A$ can be considered in order to identify isolated 1 and isolated 0 bits. An isolated 1 bit, such as in the sequence 0001000, can be recoded to 0001000 (instead of 001$\bar{1}$000 as in Booth recoding). Equivalently, an isolated 0 bit, such as in the sequence 1110111, can be recoded to 000$\bar{1}$000 (instead of 00$\bar{1}$1000 as in Booth recoding). This conversion is also correct, since $-2^4 + 2^3 = -2^3$.

Modified-Booth recoding involves using the Booth recoding method modified to better handle isolated 1 and isolated 0 bits in the multiplier. Thus, runs of 1's and 0's are handled using the normal Booth recoding, while isolated 1 and isolated 0 bits are recoded as single addition (1) or subtraction ($\bar{1}$) steps. The modified-Booth recoding for each sequence of three bits is shown in Table 8.1.

In order to use this table to produce the recoded multiplier, let us first set $a_n = a_{n-1}$ and $a_{-1} = 0$. Then, all recoded bits of the multiplier can be computed by considering $i = 0, 2, 4, \ldots, n-1$ (or $i$ up to $n-2$ for $n$ even) and using Table 8.1. Note that all recoded bits can be computed in parallel. Thus, for example, $A = 01010011$ (length $n = 8$) results in $A^{**} = 0101010\bar{1}$ ,since using Table 8.1 with $i = 0$ produces ($a_1^{**} = 0, a_0^{**} = \bar{1}$), $i = 2$ produces ($a_3^{**} = 0, a_2^{**} = 1$), $i = 4$ produces ($a_5^{**} = 0, a_4^{**} = 1$), and $i = 6$ produces ($a_7^{**} = 0, a_6^{**} = 1$). Likewise, $A = 111101101$ ($n = 9$) results in $A^{**} = 00001 0\bar{1}01$ since using Table 8.1 with $i = 0$ produces ($a_1^{**} = 0, a_0^{**} = 1$), $i = 2$ produces ($a_3^{**} = 0, a_2^{**} = \bar{1}$), $i = 4$ pro-

| Original Bits $a_{i+1}a_i a_{i-1}$ | Recoded Bits $a_{i+1}^{**} a_i^{**}$ | Comment |
|---|---|---|
| 000 | 00 | Run of 0's |
| 001 | 01 | Start of run of 1's |
| 010 | 01 | Isolated 1 |
| 011 | 10 | End of run of 0's |
| 100 | $\overline{1}0$ | End of run of 1's |
| 101 | $0\overline{1}$ | Isolated 0 |
| 110 | $0\overline{1}$ | Start of run of 0's |
| 111 | 00 | Run of 1's |

duces ($a_5^{**} = 0, a_4^{**} = \overline{1}$), $i = 6$ produces ($a_7^{**} = 0, a_6^{**} = 0$), and $i = 8$ produces $a_8^{**} = 0$ ($a_9 = a_8 = 1$ is used here).

How can Booth or modified-Booth recoding be used to make multiplication faster? Let us consider only the modified-Booth recoding method, as this recoding method results in better performance than Booth recoding. As explained above, modified-Booth recoding can be used to reduce the number of additions and subtractions required to perform the multiplication. This reduction in add/subtract operations can be used to speed up either sequential or combinational multiplication.

Sequential multiplication involves a sequence of shift-and-add or shift-and-subtract operations. If the shift requires one clock cycle and the following add or subtract requires one clock cycle, then reducing the number of add/subtracts required clearly reduces the number of clock cycles required to produce the product. Most of the time, a multiplier recoded using modified-Booth recoding requires the same or fewer add/subtract operations than the corresponding unrecoded multiplier. Given an equal likelihood of 0 and 1 bits, an $n$-bit unrecoded multiplier requires an average of $n/2$ addition operations. Under the same conditions, an $n$-bit recoded multiplier can be shown to require an average of $3n/8$ (for $n$ even) or $(3n + 1)/8$ (for $n$ odd) add/subtract operations (the proof of this fact is left as an exercise).

Combinational multiplication involves the addition of a set of partial-products using a set of CSAs arranged in an array or tree topology. Multiplier recoding can be used to reduce the total number of partial-products that need to be added together. Table 8.1 shows that, for each pair of adjacent bits in the original multiplier (starting from $i = 0$), the recoded multiplier has at most one non-zero bit. Thus, for each pair of adjacent bits in the multiplier, we can choose one partial product term. This results in a 50% reduction in the number of partial-products that have to be added together using a CSA array or tree. Such a reduction clearly reduces the delay of the combinational multiplier, as well as the number of CSAs required.

**Table 8.2:** ▶
Partial-product
terms for each pair
of adjacent recoded
multiplier bits.

| Recoded Bits $a_{i+1}^{**} a_i^{**}$ | Partial-Product Term | Comment |
|---|---|---|
| 00 | 0 | Zero |
| 01 | $P_i$ | $i$ th partial-product |
| 10 | $2P_i$ | $P_i$ shifted left 1 bit position |
| $0\bar{1}$ | $-P_i$ | 2's-complement of $P_i$ |
| $\bar{1}0$ | $-2P_i$ | $-P_i$ shifted left 1 bit position |

One drawback of combinational multiplication using modified-Booth recoding is the fact that producing the partial-product terms required is slightly more involved than with the previous method. Table 8.2 shows the partial-product terms that have to be produced for each pair of adjacent recoded multiplier bits. To form these partial-product terms, a 1-bit left-shift subcircuit, a 2's-complement subcircuit (or 1's complement followed by an addition of 1 at a later stage), and a multiplexer (to choose from among the possible partial-product terms) are necessary.

Although these subcircuits all contribute to the overhead required with this method, this overhead typically is less than the CSA logic that can be saved by having to add 50% fewer partial-products, in addition to which the overall multiplier delay also is reduced.

### 8.2.4  Multiply-Accumulate Units

There are many applications that make extensive use of the following basic operation, commonly referred to as a *multiply-accumulate* operation: $D = A + B \times C$. In particular, one class of applications that uses this as a base operation is *digital signal processing*. Digital signal processors are special-purpose processors that are designed specifically to efficiently perform digital signal processing tasks. Most digital signal processors include a "multiply-and-accumulate" (commonly referred to as MAC) assembly instruction. The efficient execution of a MAC instruction requires a hardware implementation of a MAC instruction, a so-called multiply-and-accumulate (MAC) unit.

From the discussion of combinational and sequential multipliers presented in the previous subsections, it is clear that the MAC operation can be implemented efficiently by simply including the addition step (the accumulate step in the multiply-accumulate operation) at the same level as the additions of the partial-products used in the multiplication step. Thus, for example, in a combinational multiplier implemented using carry-save adders, the addition of $A$ in $D = A + B \times C$ can be performed using one of the CSAs in the multiplier's array of CSAs. Figure 8.5 shows

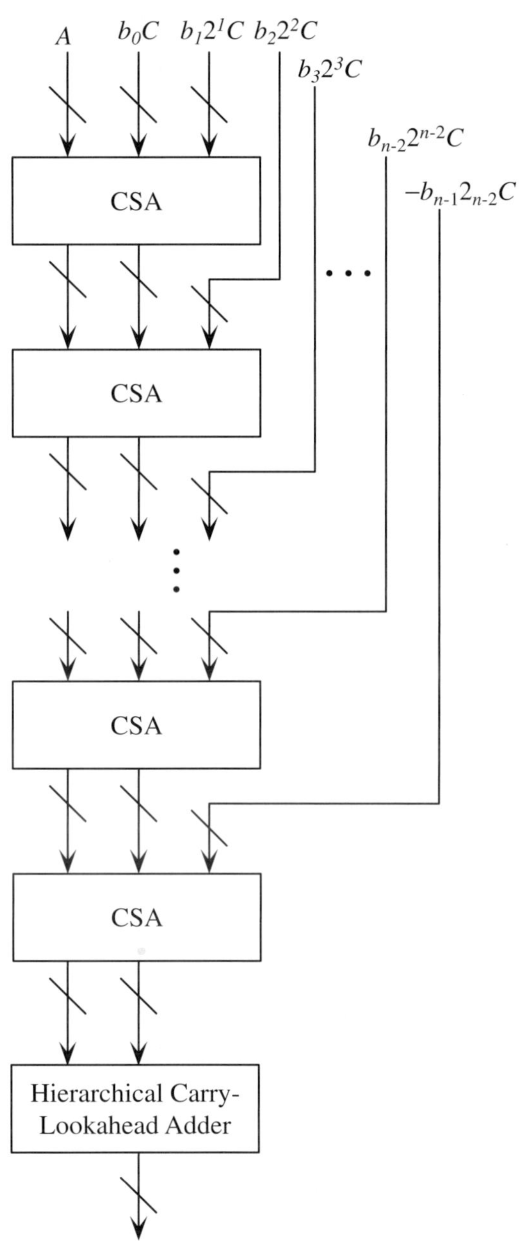


**Figure 8.5**
A combinational-logic implementation of a multiply-and-accumulate (MAC) unit.

the structure of this type MAC unit implementation. Note that a normal ripple carry adder or carry-lookahead adder is required in the last stage in order to add the two data results of the last carry-save adder.

## 8.3  Pipelined Functional Units

### 8.3.1  Introduction to Pipelining

*Pipelining* is a common method used to speed up digital logic hardware. The concept of pipelining can be viewed by analogy with an assembly line used for automobile manufacturing. A partially assembled car moves along a conveyer belt in an automobile manufacturing factory. At fixed points by the side of the moving conveyer belt are special-purpose robots. Each robot performs a specific task related to car assembly. For example, one robot continuously may attach the front windshield of each partially assembled car as it passes by on the conveyer belt. Another robot may attach the tires on partially assembled cars. In this manner, a large number of specialized robots (and humans) can be used to assemble different parts of an automobile as it passes by on the conveyer belt, and fully assembled cars can roll off of the last section of the conveyer belt at a rate of several cars an hour. This concept relies on the use of many specialized robots (and humans) and a large number of cars to be assembled.

How is pipelining, as described above, used in digital logic hardware circuits? Conceptually, the *cars* moving on the conveyer belt in the assembly line are replaced by *data*, and the *robots* working on partially assembled cars as they pass by are replaced by *functional units* that perform operations on the data as it passes by. However, data in a hardware pipeline typically moves at a much faster rate than cars in an assembly line, and some type of coordination is required in order to ensure that all parts of the data arrive and leave functional units in a synchronized manner. This type of coordination is provided by *registers* placed between functional units in a pipeline. The functional units typically (but not always) are combinational-logic units that perform operations on the data (such as addition or subtraction). Thus, in a digital-logic hardware circuit, a pipeline is implemented using several *segments* connected in series, with each segment consisting of a functional unit and a register (referred to as a pipeline register) for intermediate data storage. Assuming that each functional unit can process its data within one clock cycle (of the same clock signal used to clock the pipeline registers), the pipeline can output one data result every clock cycle after the initial time period required to "fill" the pipeline, which will be $m$ clock cycles for an $m$-segment pipeline.

Let us consider an arbitrary circuit such as that shown in Figure 8.6. This circuit consists of several subcircuits, labeled $SC0, SC1, SC2, \ldots$. Data flows through the circuit, as shown in Figure 8.6. However, since we do not know exactly *where* the data is in the circuit at a specific time instant, we must process one data set completely before the next data set can be injected into the circuit.

Pipelining can be used to speed up the processing of the data in the circuit of Figure 8.6. In order to pipeline this circuit, we must partition the circuit into segments. Each segment must have registers (pipeline registers) at all output data paths in order to temporarily hold the data after it has been processed in that segment. Note that,

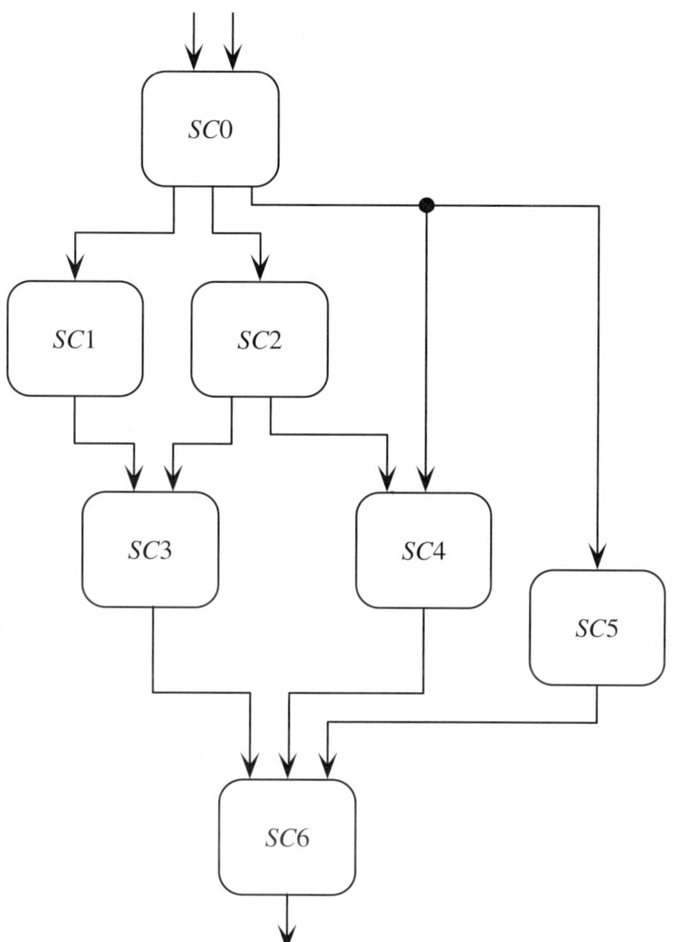


**Figure 8.6**
An arbitrary circuit consisting of several subcircuits.

in order to ensure correct timing, some pipeline registers must even be placed in data paths in which no work has been done. Assuming that a common clock signal is used to clock all pipeline registers, the period of this common clock signal must be not less than the processing delay of the slowest segment in the pipeline. Thus, it clearly is desirable to partition the circuit into segments such that the delays through all segments are approximately equal.

There are many different ways to partition a circuit into pipeline segments. As stated previously, the segments should have approximately equal delays. In addition, since registers (constructed from, for example, D flip-flops) constitute hardware overhead, we should try to minimize the number and sizes of pipeline registers required. The circuit to be pipelined should be partitioned into segments with the above considerations taken into account. The number of segments used should be chosen based on hardware constraints and the desired pipeline clock speed. A larger number of segments will result in a faster clock, while a smaller number of segments will re-

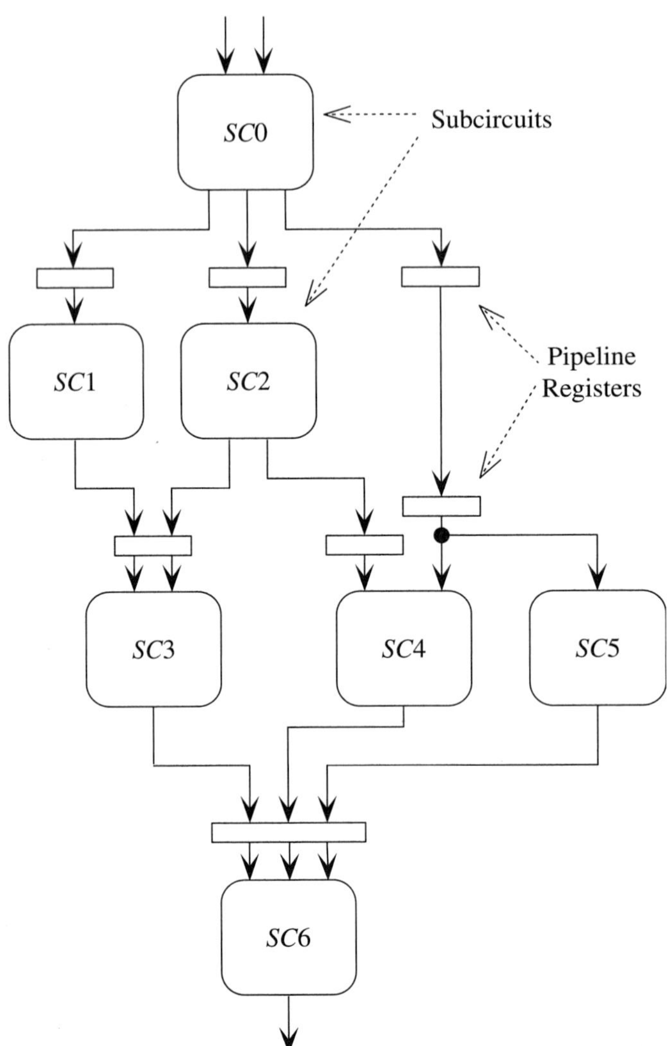


**Figure 8.7**
Partitioning of the
circuit of Figure 8.6
into pipeline segments.

quire a slower clock. However, it also should be noted that, with a larger number of pipeline segments, the delay for the first data item increases because there are delays introduced by the pipeline registers and segments that complete their processing before one full clock period transpires. Figure 8.7 shows one possible method of partitioning the circuit of Figure 8.6 into pipeline segments.

### 8.3.2  Pipelined Multiply-Accumulate Units

Suppose that we wish to design a 16-bit *multiply-accumulate (MAC)* unit such that it accepts data (and outputs results) at a rate of 1/(10 logic-gate delays) (given a logic-gate delay of 100 ps, this would correspond to a 1 GHz data rate, or $10^9$ data sets of

16-bit $A$, $B$, and $C$ numbers per second). This cannot be done using a MAC utilizing a sequential multiplier, since the sequential multiplier would require too many clock cycles (about 16) to produce the result for even one data set. Thus, a combinational logic MAC unit should be used. However, even with a combinational logic MAC unit, pipelining most likely will be necessary in order to achieve the desired data processing rate.

The combinational logic MAC unit design shown in Figure 8.5 is a good candidate for speedup using pipelining. In order to pipeline this type of MAC unit, we must place pipeline registers at appropriate points within the CSA array structure. In deciding where to place these pipeline registers, we should follow the guidelines outlined in the previous subsection: (1) the resulting pipeline segments should have approximately equal delays and (2) pipeline registers should be placed such that pipeline segments are created using the smallest possible pipeline registers.

The delay through a CSA is the same, regardless of the data word size. The delay through an inverter can be ignored, since an inverter can be combined with other basic logic-gates (e.g., $a'b$ can be implemented as one "special" gate instead of an inverter followed by an AND gate). Then the delay through a 1-bit full-adder is two logic-gate delays since 2-level AND-OR circuits can be used to produce both the *cout* and *sum* outputs. It follows that the delay through a CSA is also equal to two logic-gate delays. Thus, five CSAs can be included in a single pipeline segment in order to be able to process data at a rate of 1/(10 logic-gate delays).

Since the multiplication of two 16-bit numbers results in a 32-bit number, the final adder in Figure 8.5 needs to be a 32-bit adder. This final adder can be implemented as a ripple-carry adder or carry-lookahead adder (CLA). If it is a CLA, the CLA can, in turn, take the form of a CLA with ripples between stages or a hierarchical CLA adder.

Let us consider the delays for these various adder options. A 32-bit ripple-carry adder has a delay of $2 \times 32 = 64$ logic-gate delays. Assuming 4-bit CLA mini-adder stages, the CLA with ripples between stages requires seven ripples after the first mini-adder stage, which has a delay of $1 + 2 + 2 = 5$ logic-gate delays (one logic-gate delay to produce the $g_i$ and $p_i$ terms, two logic-gate delays to produce the carry input $c_i$, and two more logic-gate delays to produce the sum and $c_{i+1}$ terms). Since each ripple between stages requires two logic-gate delays (one level of AND-OR logic), a CLA with ripples between stages has an additional delay of $2 \times 7 + 5 = 19$ logic-gate delays. The hierarchical CLA adder, with 4-bit CLA mini-adder stages and 4-bit carry-lookahead generator logic, has an addition delay of $1 + 2 + 2 + 2 + 2 = 9$ logic-gate delays (one logic-gate delay to produce the $g_i$ and $p_i$ terms, two logic-gate delays to produce the $G_i$ and $P_i$ (group generate and group propagate) terms used in the first-level carry-lookahead generator module, two logic-gate delays for the $G_i$ and $P_i$ terms used in the second-level carry-lookahead generator module, two logic-gate delays to produce the carry-in values in the second-level carry-lookahead generator module, and two logic-gate delays to produce the final sum and carry terms from the carry-in value computed by the carry-lookahead generators).

Thus, in order to meet the 1/(10 logic-gate delays) data processing requirement, either the final adder also must be pipelined or the hierarchical CLA adder must be

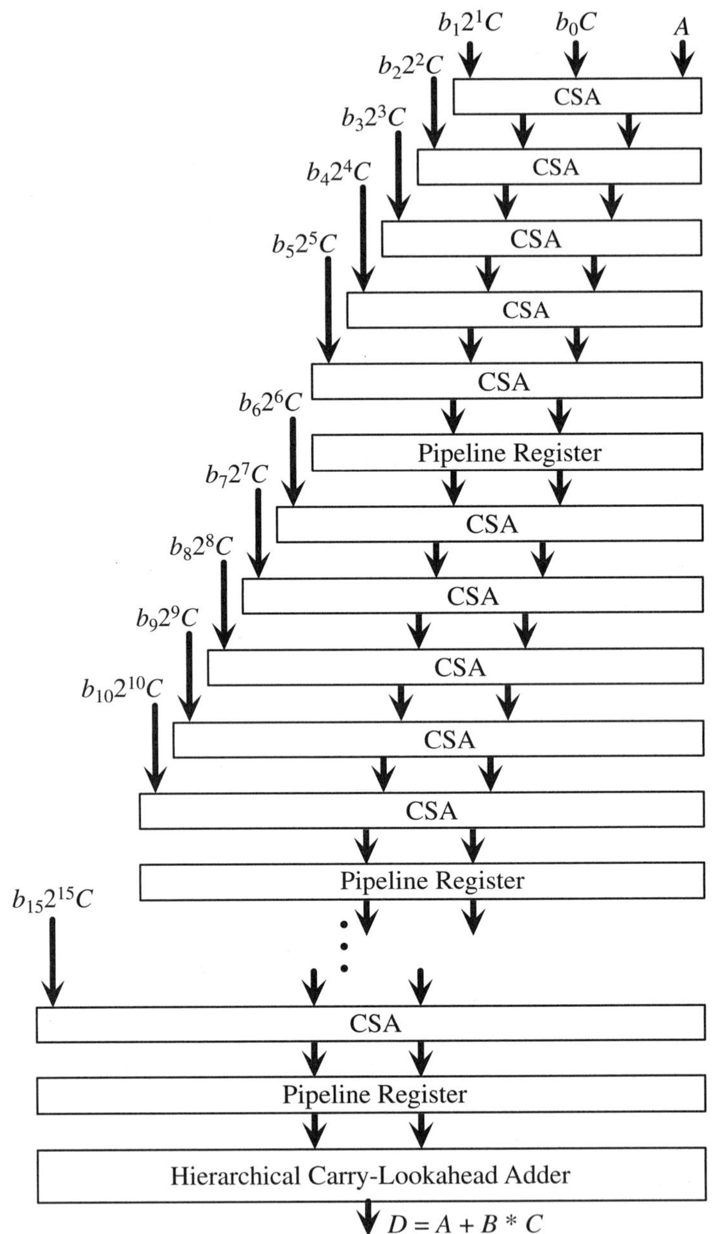



**Figure 8.8**
A multiply-and-accumulate (MAC) unit with pipeline registers inserted at approximately equal delay points.

used. The resulting pipelined MAC unit, which uses a hierarchical CLA adder for the final adder, is shown in Figure 8.8.

There are several interesting points about the design shown in Figure 8.8. First, note that the $A$ term in the $D = A + B \times C$ equation is added in the first CSA block.

Although it is possible to add $A$ *after* the computation of $B \times C$ is complete, $A$ can also be added with the partial-products of $B \times C$ at *any* other time. By adding $A$ in the first pipeline segment, it saves us from having to buffer $A$ in pipeline registers for use in later pipeline segments. Also, there is the fact that the result of $b_0C + b_12^1C$ ranges from bit 0 to bit 17 while the bits of $A$ range from bit 0 to bit 15. On the other hand, the result of $b_{14}2^{14}C + b_{15}2^{15}C$ ranges from bit 14 to bit 31. Thus, adding $A$ with $b_0C + b_12^1C$ results in the maximum bit overlap.

The second important point about the design of Figure 8.8 is the fact that the bit widths of successive CSA blocks increase by one bit each. The carry and sum terms generated from $A + b_0 + b_12^1C$ range from bit 0 to bit 17.[5] Then, since the $b_22^2C$ term ranges from bit 2 to bit 17, the results (new carry and sum terms) generated from the addition with $b_22^2C$ with the outputs of the first CSA block range from bit 0 to bit 18. In a similar manner, the outputs of the third CSA block range from bit 0 to bit 19. Continuing in this manner, the outputs of the last (15th) CSA block range from bit 0 to bit 31. Thus, the hierarchical carry-lookahead adder in Figure 8.8 needs to add a carry term (ranging from bit 1 to bit 31) and a sum term (ranging from bit 0 to bit 30) to produce a final 32-bit summation result and a possible carry-out bit.

Third, as mentioned previously, pipeline registers have been placed after the fifth, tenth, and last (15th) CSA blocks in an attempt to equalize the delays in each pipeline segment. Also important is the fact that the final 2:1 adder (the hierarchical carry-lookahead adder in Figure 8.8) has a delay that is approximately equal to the delays in the other pipeline segments. If the delay of the final 2:1 adder is found to be much longer than the delays of the other pipeline segments, then it would be necessary to insert additional pipeline registers at delay-equalized points *within* the final 2:1 adder.

# 8.4  HDL Implementations

### 8.4.1  HDL Implementation Overview

How can each of the arithmetic unit designs discussed in the previous sections of this chapter be implemented in HDL? Clearly, the most direct method is to use a structural HDL description, which explicitly describes the devices and connections to be used in a hierarchical manner. However, even when a structural HDL description is used, synthesis can produce a circuit structure different from the intended structure due to optimization and the structure of the target architecture.[6]

---

[5] To be more precise, the bits of the carry term range from bit 1 to bit 17, and the bits of the sum term range from bit 0 to bit 16.

[6] Of course, most synthesis tools provide a method (through directives or menu options) to disable optimization and/or hierarchy flattening. By using such options, the structure of the synthesized circuit can be controlled more closely.

In general, it is more desirable to use a behavioral HDL description, since such a description enables the synthesis software to produce a circuit optimized for the target architecture. Thus, for example, if a simple, behavioral addition statement is used in an HDL description, an adder will be created. The specific type of adder created will be dependent on the target architecture and the user-specified synthesis options. Thus, if the circuit is to be optimized for speed, a hierarchical carry-lookahead adder may be produced. On the other hand, if the circuit is to be optimized for hardware area (with no constraint on circuit speed), then a simple ripple-carry adder may be created.

However, there are constraints on the type of circuits that can be produced through synthesis. For example, current synthesis tools cannot produce a sequential multiplier circuit if a simple, behavioral multiplication statement is used. Likewise, current synthesis tools automatically cannot pipeline a combinational logic multiplier circuit that is found to violate the desired delay characteristics for that circuit. Thus, if a sequential circuit is desired, then the corresponding behavioral or structural HDL description must be used. Likewise, if a pipelined circuit is desired, then the specific segments of the pipeline, with their corresponding pipeline registers, must be described in an appropriate manner. After analyzing the delay characteristics of the synthesized pipeline circuit, perhaps using post-synthesis timing analysis and simulation software, the placement and number of pipeline registers used may be modified if necessary.

### 8.4.2 HDL Design for a Pipelined Multiply-Accumulate Unit

Let us now create an HDL design for a pipelined MAC unit, which performs the computation $D = A + B \times C$ using 16-bit data, based on the block diagram shown in Figure 8.8. As shown in this diagram, pipeline registers are inserted into an array of CSAs. Each set of five CSAs and the pipeline registers following them constitute a pipeline segment. There is also a final pipeline segment consisting of the hierarchical carry-lookahead adder following the last CSA stage. Thus, there are a total of four pipeline segments: three segments for the 15 CSAs required to multiply $B$ with $C$ and add $A$ to the result, and a final segment to add the carry and sum outputs out of the last CSA stage.

Note that an initial set of registers is used to buffer the data inputs $A$, $B$, and $C$, since these inputs cannot be assumed to be stable for the full clock cycle required to process them in the first pipeline segment. Also, note that the $B$ and $C$ data values also need to be buffered in pipeline registers, even though these data values are not changed within the CSA array. The reason for this buffering is so that additional sets of $A$, $B$, and $C$ data values can be processed concurrently with the processing of earlier $A$, $B$, and $C$ values in later pipeline segments. ($A$ does not need to be buffered since its value is only required in the first pipeline segment.)

A behavioral description method will be used for the HDL code. However, since we specifically want to create and use CSAs interspersed with pipeline registers (in order to create a pipelined circuit with approximately equal processing delays in each pipeline segment), the bit-level behavior of the CSA structures will be described, instead of simply stating that we want multiplication to be performed. This type of description can be considered as a "lower-level" behavioral description. For the final hierarchical carry-lookahead adder, however, we simply will use a high-level behavioral HDL statement of the form D <= last_carry + last_sum and rely on the use of "optimize for speed" synthesis options to coerce the synthesis tool to produce the desired type of adder for us. Note that even if the final adder produced is not a hierarchical carry-lookahead adder but another adder structure that still meets the required timing specs, then the final design still will be acceptable. More detailed HDL code, such as that shown in Section 4.2.2, also could be used for the hierarchical carry-lookahead adder.

The HDL code will consist of only two modules: one modeling the target circuit and the other modeling a test bench used to test the target circuit. A separate file, named mac_constants, also will be used to store constant values used in these two modules. The target circuit will use several subroutines and functions to perform separately identifiable tasks, some of which may be called more than once.

Once the required subroutines and functions have been defined, the description of the pipelined MAC target circuit (mac_pipe) is fairly simple. There will be five concurrent blocks (always blocks in Verilog) used, one for the initial buffering of the input-data and the others for the four pipeline segments required. Arrays of pipeline registers will be used, with the index of a pipeline register used to indicate the specific pipeline segment it is used in. The first concurrent block, labeled init_buffer, will save the input-data values in pipeline registers on every positive clock edge. The next four concurrent blocks, labeled segment_i (where *i* is the segment number), will perform the tasks for the four segments of the pipeline. Each segment_i block executes concurrently on every positive clock edge—analogous to each worker or robot working on only his task (concurrently with all other workers or robots) , in an automobile assembly line. In order to make this set of concurrent blocks behave properly as a four-segment pipeline, we just have to make certain that pipeline registers are used to store all intermediate data values in addition to the unchanged data values required in subsequent segments.

## Verilog Implementation

The Verilog implementation of the pipelined MAC unit is written based on the design method described earlier in this section. However, since this Verilog code is used to describe hardware (albeit in a behavioral manner), there are some subtle differences with a software program. Most importantly, all always blocks execute concurrently with one another. Also, except for input and output variables used with **functions** and **tasks**, there are no "locally defined" variables—this applies even to index variables used in **for** loops. Thus, if two **always** blocks manipulate the same globally defined signal (such as the index i in a **for** loop), then some strange behavior can result due to the fact that both **always** blocks are manipulating the same signal (or variable) concurrently. Because of this behavior, the safest course of action is to define different variables and signals to be used in different **always** blocks, unless we specifically want those **always** blocks to manipulate and use the same signals in a certain manner. Thus, in the code shown next, there are five **always** blocks with five index variables (i, i0, i1, i2, i3), four copies of the B_S signal (one **always** block does not require it), and several versions of several other signals.

There is another subtle, but important, point regarding the way in which signals are defined and used in order to achieve the desired pipelining behavior. There are two different types of signals: those used to represent the outputs of physical register values (let us call these *registered signals*) and those used simply to aid in the description of the desired behavior of a subcircuit (these latter signals, which will be called *temporary signals*, do not correspond to register-output values). When assigning values to registered signals, *nonblocking assignment* (=) should be used, since all register values should change concurrently on the active edge of the corresponding clock signal. On the other hand, for temporary signals, either *blocking* (=) or nonblocking assignment (=) can be used. In general, it may be preferable to use *blocking assignment* for temporary signals, since such signals are simply "variables" used to aid in behavioral description, and a software program-like sequence of assignment statements is useful for describing desired circuit behavior.

This distinction between the types of assignments used for registered signals and temporary signals is particularly important for pipeline processing, since one pipeline segment (represented by an **always** block) should store values into a pipeline register at the same time that the next pipeline segment is reading values (which should be the values from the previous clock cycle) from the same pipeline register. Thus, for example, in the following code, the segment_1 **always** block stores values into the PP_carry_seg[2] pipeline register using nonblocking assignment, while the segment_2 **always** block starts off by reading the PP_carry_seg[2] value (concurrently with the segment_1 **always** block). For those readers familiar with VHDL, temporary signals behave like VHDL *variable*s, while registered signals behave like VHDL *signal*s.

The Verilog code for this design will start off with a definition of global constants. Since they will be used in several different modules (including the test bench module), these global constants will be defined in a separate file (arbitrarily named mac_constant.vh). The constant `DATA_BITS (defined as the number of bits in the input data words), will be set to 16. Since the multiplication of two $k$-bit numbers will result in a $2k$-bit number, `RESULT_BITS is defined as twice `DATA_BITS. Finally, `PIPE_DEPTH, the number of pipeline segments, is set to four.

```
////////////////////////////////////////////////////////////////////
// Global Constants Used for Pipelined MAC: mac_constants.vh
`define DATA_BITS 16   // number of bits in data
`define RESULT_BITS (DATA_BITS * 2)   // number of bits in result
`define PIPE_DEPTH 4   // number of pipeline segments
////////////////////////////////////////////////////////////////////
```

The main Verilog module for the pipelined MAC is defined next. The ports for this module are one output vector (D), three input vectors (A, B, and C), and a clock input clk. The signal declarations include the **reg** definition for the D output vector and **reg** and **wire** definitions for internal signals. As described above, if a given value, which changes with each new data set, is required in $k$ segments of the pipeline, $k$ separate signals should be defined for that value; these $k$ separate signals may correspond to register outputs or temporary signals. In this case, one signal is defined for A, three signals are defined for B (with an extra signal for the *msb* of B), three signals are defined for C, and four signals are defined for the carry and sum vector results, respectively, of the last CSA blocks in each pipeline segment. Several temporary signals also are defined for use in describing the behavior of the circuitry in the four segments of the pipeline.

```
/////////////////////////////////////////////////////////////////////
// MODULE: Pipelined MAC Unit Example: mac_pipe.v
// Author: Sunggu Lee
// Date: ...
// Description: Implements a pipelined 16-bit multiply-accumulate
//   (MAC) unit.  Performs the computation D <-- A + B * C using
//   a pipelined multiply-accumulate circuit based on inserting
//   pipeline registers in a linear array of carry-save adders.

// Load constant definitions
`include "mac_constants.vh"

// Module Declaration
module mac_pipe (D, A, B, C, clk);
  // INPUT AND OUTPUT PORT DECLARATIONS
  output [`RESULT_BITS-1:0] D;  // want D <-- A + B * C
  input [`DATA_BITS-1:0] A, B, C;
  input clk;                    // pipeline clock signal

  // SIGNAL DECLARATIONS for register variables
  reg [`RESULT_BITS-1:0] D;     // connection to output port
  reg [`DATA_BITS-1:0] A_S0;    // pipeline register for A
  reg [`DATA_BITS-1:0] B_S0;    // pipeline registers for B
  reg [`DATA_BITS-1:0] B_S1;
  reg [`DATA_BITS-1:0] B_S2;
  reg B_S3_msb;                 // need only msb in last segment
  reg [`DATA_BITS-1:0] C_seg[0:`PIPE_DEPTH-2];  // for C
  reg [`RESULT_BITS-1:0] PP_carry_seg[1:`PIPE_DEPTH-1];  // etc.
  reg [`RESULT_BITS-1:0] PP_sum_seg[1:`PIPE_DEPTH-1];

  // SIGNAL DECLARATIONS for temporary intermediate variables
  integer i, i0, i1, i2, i3;       // index variables
  reg [1:0] add_result;            // temporary add result
  reg [`RESULT_BITS-1:0] temp_res;    // temporary result
  reg [`RESULT_BITS-1:0] PP_S0[0:9];  // temporary values
  reg [`RESULT_BITS-1:0] PP_S1[0:11]; // used in pipeline
  reg [`RESULT_BITS-1:0] PP_S2[0:11]; // segments S0-S2
/////////////////////////////////////////////////////////////////////
```

Two Verilog **function**s will be used for simple sign-extension tasks. The **function** convert will take as inputs a data input of size `` `DATA_BITS `` and an offset value (named start). Then it will create an output of size `` `RESULT_BITS ``, in which the data input has been shifted left by the desired offset, with 0 values stored into the rightmost (start) bit positions, and the sign bit (the *msb*—most significant bit—of the original data input) copied into all bit positions to the left of the data input after it has been shifted. The copying of the *msb* value serves to preserve the sign of the original data. The left-shift of the data serves to multiply the data by two with each shift. The **function** comp_convert differs from convert in that it performs a *1's complement* on the result (i.e., it complements all bits before it outputs the result (of size `` `RESULT_BITS ``)). The comp_convert **function** is used for the final subtraction of $b_{n-1}2^{n-1}C$ required for multiplication with 2's-complement numbers. This **function** needs to be used in conjunction with a conditional addition of a carry-in value of 1 in order to perform the required subtraction step (subtraction is equivalent to addition of the 2's-complement value, which in turn is equivalent to addition of the 1's-complement value followed by an addition of 1 in the *lsb* position).

```
///////////////////////////////////////////////////////////////////////
  // Function and Task Definitions
  // Function to shift and sign-extend (B(i) and C) data
  function [`RESULT_BITS-1:0] convert;
    input B_bit;  // b_i value used to compute (b_i C)
    input [`DATA_BITS-1:0] C;   // C data value
    input [31:0] start;        // shift amount (integer)
    begin
      for (i = 0;  i < start;  i = i + 1)
        temp_res[i] = 1'b0;
      for (i = 0;  i < `DATA_BITS;  i = i + 1)
        temp_res[start+i] = B_bit & C[i];
      for (i = `DATA_BITS;  i < `RESULT_BITS-start;  i = i + 1)
        temp_res[start+i] = B_bit & C[`DATA_BITS-1];
      convert = temp_res;  // return result
    end  // of outermost begin for function convert
  endfunction  // of function convert

  // Function to shift, sign-extend, and complement data
  function [`RESULT_BITS-1:0] comp_convert;
    input B_bit;  // b_i value used to compute (b_i C)
    input [`DATA_BITS-1:0] C;   // C data value
    input [31:0] start;        // shift amount
```

```
    begin
      for (i = 0;  i < start;  i = i + 1)
        temp_res[i] = 1'b1;
      for (i = 0;  i < `DATA_BITS;  i = i + 1)
        temp_res[start+i] = ~(B_bit & C[i]);
      for (i = `DATA_BITS;  i < `RESULT_BITS-start;  i = i + 1)
        temp_res[start+i] = ~(B_bit & C[`DATA_BITS-1]);
      comp_convert = temp_res;  // return result
    end  // of outermost begin for function comp_convert
  endfunction  // of function comp_convert
  ///////////////////////////////////////////////////////////////////
```

Next are two Verilog **task** definitions used to define the CSA blocks. The compute_init_CSA **task** describes the first CSA block of the CSA array used in the pipelined MAC unit. The compute_CSA **task** describes the other CSA blocks in this CSA array. The Verilog code for a CSA block corresponds to the design shown in Figure 8.2. The compute_init_CSA **task** computes the inputs into the CSA block as $A$, $b_0 C$, and $b_1 C$. These inputs and their respective carry and sum outputs are described in a bit-by-bit manner in the **task** shown. The compute_CSA subroutine models a general CSA that adds three arbitrary data inputs named data1, data2, and data3. This latter **task** definition is simpler than compute_init_CSA because it is assumed that the three CSA inputs have been converted properly into sign-extended, left-shifted, 32-bit values before this **task** is called.

As explained in Section 8.3.2, the exact size of a CSA block within a multiplier array typically will be less than `RESULT_BITS (and more than `DATA_BITS)—the exact size of the CSA required will depend on its position within the multiplier array. However, in these two **task** definitions, the maximum bit size is used in every CSA block, regardless of its position within the CSA array. Provided that the synthesis software is intelligent enough, it should be able to derive the exact CSA size required and then synthesize it accordingly.

```
  ///////////////////////////////////////////////////////////////////
  // Task to compute CSA values for A, B(0) C, and B(1) C
  task compute_init_CSA;
    output [`RESULT_BITS-1:0] PP_carry; // carry partial products
    output [`RESULT_BITS-1:0] PP_sum;   // sum partial products
    input [`DATA_BITS-1:0] A, B, C;      // A, B, and C data
    begin
      PP_carry[0] = 1'b0;
      for (i = 0;  i <= `DATA_BITS;  i = i + 1) begin
        if (i == `DATA_BITS)
          add_result = A[`DATA_BITS-1] + (B[0] & C[`DATA_BITS-1])
                     + (B[1] & C[i-1]);
```

```verilog
        else if (i == 0)
          add_result = A[i] + (B[0] & C[i]);
        else
          add_result = A[i] + (B[0] & C[i]) + (B[1] & C[i-1]);
        PP_carry[i+1] = add_result[1];
        PP_sum[i] = add_result[0];
      end  // of for loop
    PP_sum[`DATA_BITS+1] = add_result[0];
      for (i = `DATA_BITS+2;  i < `RESULT_BITS;  i = i + 1)
        {PP_carry[i], PP_sum[i]} = add_result;
    end  // of outermost begin for task compute_init_CSA
  endtask  // of task compute_init_CSA


  // Task to compute CSA values from three data inputs
  task compute_CSA;
    output [`RESULT_BITS-1:0] PP_carry; // carry partial products
    output [`RESULT_BITS-1:0] PP_sum;   // sum partial products
    input [`RESULT_BITS-1:0] data1;        // data to be added
    input [`RESULT_BITS-1:0] data2;
    input [`RESULT_BITS-1:0] data3;
    begin
      PP_carry[0] = 1'b0;
      for (i = 0;  i < `RESULT_BITS-1;  i = i + 1) begin
        add_result = data1[i] + data2[i] + data3[i];
        {PP_carry[i+1], PP_sum[i]} = add_result;
      end  // of for loop
      PP_sum[`RESULT_BITS-1] = add_result[0];
    end
  endtask  // of task compute_CSA
//////////////////////////////////////////////////////////////////////
```

The rest of the code for the pipelined MAC circuit consists of five **always** blocks:
one for the initial buffering of the input-data and the others for the four pipeline
segments used. The first **always** block simply saves the input data values in pipeline
registers on every positive clock edge. This block really may not be required if it is
known that the input data will be stable while it is being processed in the first pipeline
segment, The next four **always** blocks perform the tasks for the four segments of the
pipeline. Each **always** block executes concurrently with every other **always** block.
Except for the last **always** block, which models the final 2:1 adder stage, each **always**
block basically consists of (1) code to create or buffer the data inputs for the first CSA
of the corresponding pipeline segment, (2) **for** loops used to create the CSA blocks
for that pipeline segment, (3) code to buffer the data outputs of the last CSA block
in that pipeline segment, and (4) code to buffer data that needs to be relayed to the
next pipeline segment. Calls to compute_init_CSA, compute_CSA, convert, and comp_convert
simply the code for these **always** blocks.

```verilog
///////////////////////////////////////////////////////////////////
// init_buffer: buffers data inputs into initial pipeline reg.
always @(posedge clk) begin
  A_S0 <= A;
  B_S0 <= B;
  C_seg[0] <= C;
end  // of init_buffer


// segment_0: pipeline segment S0
always @(posedge clk) begin
  // create first five CSA levels
  compute_init_CSA(PP_S0[1], PP_S0[0], A_S0, B_S0, C_seg[0]);
  for (i0 = 0;  i0 < 4;  i0 = i0 + 1)
      compute_CSA(PP_S0[2*i0+3], PP_S0[2*i0+2], PP_S0[2*i0+1],
          PP_S0[2*i0], convert(B_S0[i0+2], C_seg[0], i0+2));
  // save intermediate data and results in pipeline registers
  PP_carry_seg[1] <= PP_S0[9];
  PP_sum_seg[1] <= PP_S0[8];
  B_S1 <= B_S0;
  C_seg[1] <= C_seg[0];
end  // of segment_0 always block


// segment_1: pipeline segment S1
always @(posedge clk) begin
  // create next set of five CSA levels
  PP_S1[1] = PP_carry_seg[1];
  PP_S1[0] = PP_sum_seg[1];
  for (i1 = 0;  i1 < 5;  i1 = i1 + 1)
      compute_CSA(PP_S1[2*i1+3], PP_S1[2*i1+2], PP_S1[2*i1+1],
          PP_S1[2*i1], convert(B_S1[i1+6], C_seg[1], i1+6));
  // save intermediate data and results in pipeline registers
  PP_carry_seg[2] <= PP_S1[11];
  PP_sum_seg[2] <= PP_S1[10];
  B_S2 <= B_S1;
  C_seg[2] <= C_seg[1];
end  // of segment_1 always block


// segment_2: pipeline segment S2
always @(posedge clk) begin
  // create next set of five CSA levels
  PP_S2[1] = PP_carry_seg[2];
  PP_S2[0] = PP_sum_seg[2];
  for (i2 = 0;  i2 < 4;  i2 = i2 + 1)
      compute_CSA(PP_S2[2*i2+3], PP_S2[2*i2+2], PP_S2[2*i2+1],
          PP_S2[2*i2], convert(B_S2[i2+11], C_seg[2], i2+11));
  // last CSA level involves conditional subtract of b_{n-1} C
```

```verilog
    compute_CSA(PP_S2[11], PP_S2[10], PP_S2[9], PP_S2[8],
       comp_convert(B_S2[`DATA_BITS-1], C_seg[2], `DATA_BITS-1));
    // save intermediate data and results in pipeline registers
    PP_carry_seg[3] <= PP_S2[11];
    PP_sum_seg[3] <= PP_S2[10];
    B_S3_msb <= B_S2[`DATA_BITS-1];
  end  // of segment_2 always block


  // segment_3: pipeline segment S3 -- performs final addition
  always @(posedge clk) begin
    D <= PP_carry_seg[3] + PP_sum_seg[3] + B_S3_msb;
    // Note 1: 1 added to lsb when (msb(B) = 1), in order to
    //         complete the conditional subtraction of b_{n-1} C.
    // Note 2: Set synthesis options for high speed - in order to
    //         get a delay for this segment similar to other segs
  end  // of segment_3 always block


endmodule
//////////////////////////////////////////////////////////////////
```

### 8.4.3  Test Bench and Simulation Results

The test bench for this circuit will consist of three parts: the connection to the unit under test, a concurrent block (**always** block in Verilog) to generate the test vectors, and a concurrent block to check the application of the test vectors to the unit under test. Two sets of test vectors will be used. The first set, which tests boundary conditions, consists of input values at the boundaries of the ranges for those values. This set consists of two nested **for** loops. The first nested **for** loop generates negative-and positive-input values close to, and including, the zero value. The second nested **for** loop generates all possible combinations of maximum and minimum values for the $A$, $B$, and $C$ inputs. The final set of test vectors will consist of a random set of positive and negative numbers for the $A$, $B$, and $C$ inputs. Test vectors will be generated using simple **for** loops with each set of inputs generated in synch with the falling edge of the system clock signal.

The concurrent block used for test-vector checking will consist of an initial delay, in order to wait for the first pipeline output to be ready, followed by **for** loops almost identical with the concurrent block used for test-vector generation. However, this test-vector checking block will include a "quick and dirty" computation to compute the expected result value followed by a comparison with the output of the unit under test. There will also be numerous "print" statements used to indicate the progress of the simulation and to indicate the presence or absence of errors.

Figure 8.9 shows the simulation output achieved with the application of this test bench to the pipelined MAC unit presented in this chapter. The console output and simulation waveforms can be used to verify the proper operation of this pipelined MAC circuit.

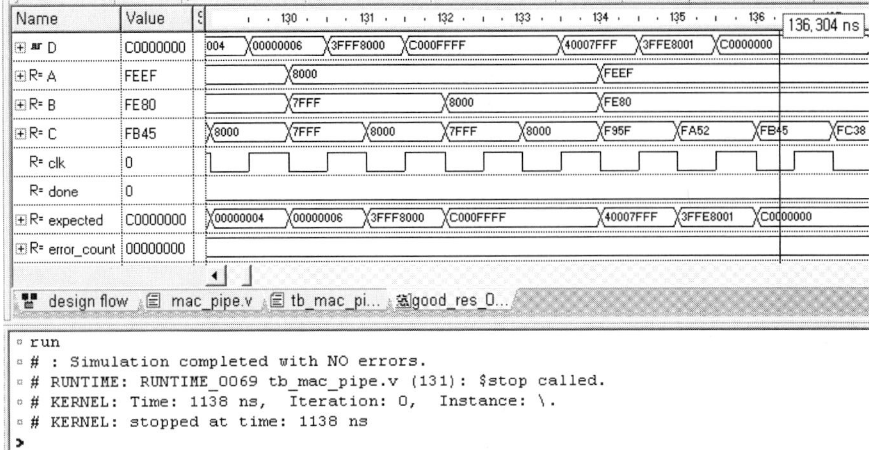

**Figure 8.9**
The simulation output waveforms and console output for the pipelined MAC unit presented in this chapter.

## Verilog Implementation

The Verilog code follows the design method described here. One noteworthy point is the fact that a separate set of index variables are used for the test-vector generation and checking **always** blocks ($i$, $j$, and $k$ versus $ii$, $jj$, and $kk$). This is necessary since even index variables used in different **always** blocks can interfere with each other.

```
////////////////////////////////////////////////////////////////////

// MODULE: Test Bench for Pipelined MAC Unit

// Author: Sunggu Lee

// Date: ...

// Description: Tests the mac_pipe.v module.


// Constant Definitions

`include "mac_constants.vh"  // include common constant defs

`define MAX_DATA 32767        // 2^(`DATA_BITS-1) - 1 = 32767

`define MIN_DATA -32768       // 2^(`DATA_BITS-1) = -32768

`define PERIOD0 1000          // clock period = 1000 ps -> 1 GHz

// Test bench module definition
```

```verilog
module tb_mac_pipe ();
  // SIGNAL DEFINITIONS
  // Signals used to connect to the UUT (unit under test)
  wire [`RESULT_BITS-1:0] D;          // D output out of mac_pipe
  reg [`DATA_BITS-1:0] A, B, C;       // data inputs to mac_pipe
  reg clk = 1'b0;                     // clock signal input
  // Variables used to aid in the definition of the testbench
  reg done = 1'b0;              // signals simulation done
  integer expected;             // expected result value
  integer error_count = 0;      // total number of errors found
  integer i, j, k;              // test vector generation indices
  integer ii, jj, kk;           // test vector checking indices
  integer A_in, B_in, C_in;     // used in test vector checking
  // Connect to the UUT (unit under test)
  mac_pipe uut (D, A, B, C, clk);
  // Generate the clock signal (assumes clk = 1'b0 initially)
  always #(`PERIOD0 / 2) clk = ~clk;

  // Generate the test vectors
  always begin
    // first, generate boundary test vectors
    // generate values close to 0
    for (i = -2;  i <= 2;  i = i + 1)
      for (j = -2;  j <= 2;  j = j + 1)
        for (k = -2;  k <= 2;  k = k + 1) begin
          @(negedge clk);
          A = i;
          B = j;
          C = k;
        end  // of for k loop
    // generate min and max data input values
    for (i = 0;  i <= 1;  i = i + 1)
      for (j = 0;  j <= 1;  j = j + 1)
        for (k = 0;  k <= 1;  k = k + 1) begin
          @(negedge clk);
          if (i == 0) A = `MAX_DATA;
          else        A = `MIN_DATA;
          if (j == 0) B = `MAX_DATA;
          else        B = `MIN_DATA;
          if (k == 0) C = `MAX_DATA;
          else        C = `MIN_DATA;
        end  // of for k loop
```

```verilog
      // next, generate a set of normal test vector values
    for (i = 0;  i <= 9;  i = i + 1)
      for (j = 0;  j <= 9;  j = j + 1)
        for (k = 0;  k <= 9;  k = k + 1) begin
          @(negedge clk);
          A = 41*i-273;    // generate random set of negative
          B = 89*j-384;    // and positive numbers
          C = 243*k-1697;
        end  // of for k loop
    @(posedge done);  // wait for done signal
    $finish;            // terminate the simulation
end  // of always block for test vector generation
// Error checking and reporting task
task error_check;
  input ['RESULT_BITS-1:0] expected;
  input ['RESULT_BITS-1:0] actual;
  begin
    if (expected !== actual) begin
      $display
      ("ERROR: expected (%0d) != actual (%0d) at time %0t ps",
        expected, actual, $time);
      error_count = error_count + 1;
    end  // of if check
  end  // of outermost begin for task error_check
endtask
// Check the test results
always begin
  // first, wait until the pipeline is full
  for (ii = 0;  ii <= 'PIPE_DEPTH;  ii = ii + 1)
    @(negedge clk);
```

```verilog
      // second, check boundary test vectors
      // check values close to 0
      for (ii = -2;  ii <= 2;  ii = ii + 1)
        for (jj = -2;  jj <= 2;  jj = jj + 1)
          for (kk = -2;  kk <= 2;  kk = kk + 1) begin
            @(negedge clk);
            expected = ii + (jj * kk);
            error_check(expected, D);
          end  // of for kk loop
      // check min and max data inputs
      for (ii = 0;  ii <= 1;  ii = ii + 1)
        for (jj = 0;  jj <= 1;  jj = jj + 1)
          for (kk = 0;  kk <= 1;  kk = kk + 1) begin
            @(negedge clk);
            if (ii == 0) A_in = `MAX_DATA;
            else         A_in = `MIN_DATA;
            if (jj == 0) B_in = `MAX_DATA;
            else         B_in = `MIN_DATA;
            if (kk == 0) C_in = `MAX_DATA;
            else         C_in = `MIN_DATA;
            expected = A_in + (B_in * C_in);
            error_check(expected, D);
          end  // of for k loop

      // third, check a set of normal test vector values
      for (ii = 0;  ii <= 9;  ii = ii + 1)
        for (jj = 0;  jj <= 9;  jj = jj + 1)
          for (kk = 0;  kk <= 9;  kk = kk + 1) begin
            @(negedge clk);
            expected = (41*ii-273) + (89*jj-384) * (243*kk-1697);
            error_check(expected, D);
          end  // of for kk loop

      // finally, report results and terminate simulation
      if (error_count > 0)
        $display("ERROR: Simulation completed with %0d errors.",
                 error_count);
      else
        $display("Simulation completed with NO errors.");
      done = 1'b1;
      $stop;
    end  // of always block for test result checking

endmodule
////////////////////////////////////////////////////////////////////
```

## 8.5 Chapter Review

- Most arithmetic functions can be implemented in digital logic hardware using a series of addition and/or subtraction steps. Then, if 2's-complement notation is used for signed binary numbers, subtraction of a number $B$ can be accomplished by addition of $-B$, which can in turn be formed by taking the 2's complement of $B$. Thus, *addition* is the most important basic operation that must be supported in an arithmetic circuit. There are many different ways to implement area-efficient and/or fast adders. A ripple-carry adder is an area-efficient adder in which the carry-out of a full-adder module is connected to the carry-in of the next full-adder module, and so on in an iterative manner. For fast addition, a hierarchical carry-lookahead adder can be used. An $n$-bit CSA is formed using a parallel connection of full adder modules that takes three $n$-bit binary inputs and produces two $n$-bit binary outputs.

- Signed multiplication of an $n$-bit multiplier $A = a_{n-1}a_{n-2}\ldots a_1a_0$ with an $n$-bit multiplicand $B$ can be implemented by adding the partial-products $P_i = a_i 2^i B$, for $i = 0$ to $n - 2$ and then subtracting $P_{n-1} = a_{n-1}2^{n-1}B$. A hardware implementation of such a multiplier can use a linear array of CSAs in which the first CSA adds the first three partial-products, the next CSA adds the two outputs of the first CSA and a fourth partial-product, and so on. After the last CSA, a normal 2:1 adder, such as a ripple-carry adder or hierarchical carry-lookahead adder, must be used to reduce the two outputs of the last CSA into one number, which should be the product of $A$ and $B$. Methods based on multiplier recoding can be used to produce fast sequential or combinational multipliers. Two common multiplier recoding methods are Booth recoding (also known as differential recoding) and modified-Booth recoding.

- Pipelining is a general method that can be used to enhance the *throughput* (the rate at which results are produced) of an arithmetic circuit. In pipelining, there are a set of $m$ functional units (which together accomplish some arithmetic function), separated by registers (referred to as pipeline registers) clocked off of the same clock signal. One functional unit and the pipeline register that follows it constitutes one segment of the pipeline. If each functional unit can process its data in one clock cycle, then the pipeline can output one completely processed data result every clock cycle after the first $m$ clock periods (required to "fill" the pipeline).

- Let us consider the pipelined implementation of a multiply-accumulate (MAC) unit. A pipelined multiply-accumulate unit can be designed using an array of CSAs followed by a hierarchical carry-lookahead adder in the final segment. This circuit can be coded in Verilog using a low-level behavioral description that uses an array of CSAs with pipeline registers interspersed at equal-spaced intervals (in order to have pipeline segments with approximately equal delays).

# 8.6 Resources

[Ciletti 2002] is a useful reference for Verilog coding. [Kreyszig 1999] is a widely used textbook on engineering mathematics. Taylor series expansion and other methods for approximating arithmetic functions are covered in this textbook. [Parhami 2000] is a useful reference for readers that need more material on fast adder/subtracters, multipliers, and other computer arithmetic circuits.

### 8.6.1 Bibliography

[8.1] CILETTI, M.D., *Advanced Digital Design with the Verilog HDL*, Prentice Hall, Upper Saddle River, NJ, 2002.

[8.2] KREYSZIG, E., *Advanced Engineering Mathematics, 8th Ed.*, John Wiley and Sons, Inc., New York, 1999.

[8.3] PARHAMI, B., *Computer Arithmetic: Algorithms and Hardware Designs*, Oxford University Press, New York, 2000.

# 8.7 Problems

**P8.1.** Discuss the pros and cons of implementing a signed-number multiplier using sign-magnitude numbers versus using 2's-complement numbers.

**P8.2.** Discuss the pros and cons of using a look-up table (ROM) to implement a multiplier versus using a combinational multiplier such as that shown in Figure 8.3. For what values of $n$ (the bit-width of the input data) would a ROM-based multiplier be feasible? Suggest approximate viable ranges of input data bit-width values.

*__P8.3.__ At the beginning of this chapter, it is stated that the inverse of a number $y$ can be approximated by using a series of multiplication steps. Show how this can be done. (*Hint*: Consider multiplying both the numerator and denominator with a series of factors that make the denominator approach one.)

**P8.4.** A combinational multiplier designed using an array of carry-save adders (CSAs) shown in Figure 8.3 uses a linear array of CSAs. Given this type of organization and labeling the CSAs from 0 (for the topmost CSA), derive a general formula for the bit-size of the $i$ th CSA (use the smallest bit size possible given the set of inputs that the $i$ th CSA needs to add). What is the bit size and index number of the last CSA in the array (the one whose inputs feed into the hierarchical carry-lookahead adder)?

**P8.5.** A *Wallace-tree adder* uses a tree of CSAs connected in a tree-like manner to compute the addition of a large set of numbers as fast as possible. Considering the fact that a CSA reduces three numbers to two numbers, draw the Wallace-tree structure for the addition of 32 numbers. Given that we wish to design an $n$-bit by $n$-bit multiplier using a Wallace tree structure, how many levels of CSAs are required? Derive a general formula for the number of levels required.

**P8.6.** Figure 8.4 shows a block diagram for a sequential multiplier design. This diagram includes a block labeled "control logic." Propose a set of control signals to be used to control the datapath components in this block diagram, and then design the detailed logic for the "control-logic" block. This control logic design should be based on an algorithm for controlling the datapath components to properly implement a 2's-complement sequential multiplier for $n$-bit numbers.

***P8.7.** In Section 8.2.3, it is stated that, given an equal likelihood of 0 and 1 bits in a multiplier, the average number of add/subtract operations required when using modified-Booth recoding is $3n/8$ for $n$ even and $(3n+1)/8$ for $n$ odd. Prove this assertion. (*Hint*: Use proof by induction.)

**P8.8.** Derive the Booth-encoded and modified-Booth encoded multipliers for the following multiplier values. (Note: the spaces in the numbers shown can be ignored; they are simply there to make the numbers easier to read.)

(a) 0011 1110 1011 1101

(b) 111 1101 1100 0010 0100

(c) 11 0011 0011

(d) 000 0101 0011 0000 1111

**P8.9.** Suppose that we wish to design a fast sequential multiplier using modified-Booth recoding. As with a modified-Booth combinational multiplier, modified-Booth recoding can be used to reduce by half the required number of shift-and-add or shift-and-subtract steps. Draw the block diagram for such a multiplier, and then describe the operation of this multiplier using pseudocode.

***P8.10.** Section 8.2.3 describes, in general terms, how a modified-Booth combinational multiplier can be designed. Based on this description, design the modified-Booth recoding logic. This circuit should have the multiplier and multiplicand as inputs and partial-product terms as outputs, and the number of partial-product terms produced should be half that produced using an unrecoded multiplier.

***P8.11.** Multiplier recoding is a general technique that can be used for fast multiplication. Booth recoding is an improvement over multiplication using unrecoded numbers, while modified-Booth recoding is an improvement over Booth recoding. It is possible to further improve upon the modified-Booth recoding method by using higher-radix recoding methods. Devise recoding methods using radix-8 and radix-16 recoding that improve upon the modified-Booth recoding method. Show how these improved recoding methods can be implemented in hardware.

**P8.12.** In Section 8.2.4, a multiply-accumulate unit is described as an arithmetic unit that performs the general function $\dot{D} = A + B \times C$. However, another type of multiply-accumulate unit can be described as an arithmetic unit that performs the general function $A = A + B \times C$. Then, if $A$ is initialized to 0 and $B$ and $C$ are entered as the inputs to this circuit, $A$ will be the accumulated result of successive $B \times C$ calculations. Modify the pipelined design shown in Figure 8.8 to implement this type of multiply-accumulate circuit. If a small change is made to the design specifications, the circuit could be designed much more simply and efficiently. Suggest what this small change could be, and why such a change would permit us to create a much more efficient design.

**P8.13.** The first three terms of the Taylor series expansion for the $sin(z)$ function is shown at the beginning of this chapter. Using these first three terms to approximate the $sin(z)$ function, design an efficient sequential circuit for implementing the $sin(z)$ function. Assume that 24-bit numbers are used to represent the inputs and constants required in this circuit. The control logic for this circuit simply can be presented as a single labeled block. A pseudocode description should be used to describe the desired operation of this circuit based on the circuit components shown in your block diagram.

**P8.14.** In the combinational multiplier circuit of Figure 8.3, the product requires twice the number of bits in a multiplier input since the multiplication of two $n$-bit numbers results in a $2n$-bit product. However, suppose that we only want to produce an $n$-bit product. This $n$-bit product should consist of the most significant $n$ bits in the original $2n$-bit product. The $n$-bit product should also be rounded up or down, depending on the values of the least significant $n$ bits in the original $2n$-bit product. Assuming that a "round to the nearest even number" method is being used (e.g., 13.5 is rounded to 14 while 26.5 is rounded to 26), show how the combinational-multiplier circuit of Figure 8.3 can be modified to produce an $n$-bit rounded product result. Make your circuit as efficient as possible, and show the bit widths of all CSAs in your design.

**P8.15.** Redesign the circuit of Problem P8.13 as a combinational circuit that uses an array of CSAs to perform the necessary multiplication and addition operations. Using the fact that both multiplications required involve multiplication with constant numbers, make the design as efficient as possible (e.g., a method similar to modified-Booth encoding, except with constant numbers, could be used). Estimate the total delay of your circuit in "logic-gate delays."

**P8.16.** Show how the design of Problem P8.15 can be modified to use a Wallace-tree structure in order to make the resulting circuit as fast as possible. How many levels of CSAs are required in this design? What is the most appropriate type of 2:1 adder to be used in this type of circuit?

**P8.17.** Redesign the circuit of Problem P8.15 as an efficient pipelined circuit with a pipeline clock cycle requirement of ten logic-gate delays.

**P8.18.** A technique known as *wave pipelining* uses the natural delay in the logic gates and wires of a combinational circuit in order to implement pipelining without the aid of pipeline registers. Suggest, in detail, how this could be done. What are the underlying assumptions when using this type of method? What are the main advantages and disadvantages of this type of method?

**P8.19.** Write the Verilog code for the design of Problem P8.17. Then write the test bench and test your design using an HDL simulator.

**P8.20.** How is the test bench for a pipelined arithmetic circuit different from the test bench for the corresponding unpipelined circuit? List and describe as many differences as you can.

# Design of a Pipelined RISC Microprocessor

## Important Concepts

- The basic operation of a microprocessor and how to design a basic microprocessor circuit.

- How to design an instruction pipeline for a RISC microprocessor.

- The instruction set and circuit implementation of a commercial microprocessor, such as the THUMB microprocessor.

This chapter addresses the design of a 16-bit pipelined RISC microprocessor that supports the instruction set of the THUMB[TM] architecture. The THUMB architecture is a standard subset of the 32-bit ARM[TM] architecture,[1] which is the architecture used in the highly successful ARM series of microprocessors.

ARM, which originally stood for *Acorn RISC Microprocessor* and was later modified to *Advanced RISC Microprocessor*, is the brainchild of a company in England that, starting in the early 1980s, developed this architecture as a microprocessor for low-power, embedded applications. This company is unique in that it doesn't produce chips based on the ARM architecture, but simply sells licenses, designs, and associated software for the ARM architecture. As one measure of its success, microprocessor designs based on the ARM architecture are currently the predominant central processing units (CPUs) used in personal digital assistant devices.

The THUMB instruction set is a 16-bit instruction set that is a re-encoded subset of the ARM instruction set. By using the THUMB instruction set, programs can be written in a much more compact manner than when 32-bit ARM instructions are used. Thus, the use of the THUMB instruction set is most suitable for embedded applications with limited on-board memory. However, because it lacks certain interrupt-related instructions, the THUMB instruction set always is used in conjunction with an ARM instruction-set version, typically ARM Version 4 or ARM Version 5. In fact, an application starts in ARM instruction mode, and only switches to THUMB instruction mode when a special ARM instruction is encountered. There are also two THUMB versions. THUMB Version 1 is associated with ARM Version 4. THUMB Version 2, which adds a few instructions to THUMB Version 1, is associated with ARM Version 5.

For the purposes of this textbook, we will ignore the requirement that a THUMB instruction-set needs to be used with an ARM instruction set version, and simply design a pipelined microprocessor that implements most of the instruction set of THUMB Version 1. There are a few instructions, including the interrupt instructions, that are difficult to use and implement in a classroom lab setting. These instructions will be omitted from our implementation. Also, the implementation of many other instructions will be left as lab exercises.

# 9.1 Introduction to Microprocessors

A *microprocessor* is a digital-logic circuit that serves as the *brains* behind hardware circuits designed to control factories, parts of automobiles, computers, internet routers, smart cell phones, and a host of other applications. The reason that a microprocessor can be used for such diverse applications is that it is designed to be *programmed* for use in such applications.

Figure 9.1 shows a block diagram for a basic computer system. As can be seen, it consists of a central processing unit (CPU), memory, and input/output (I/O) devices.

---

[1]   THUMB and ARM are trademarks of ARM Limited.

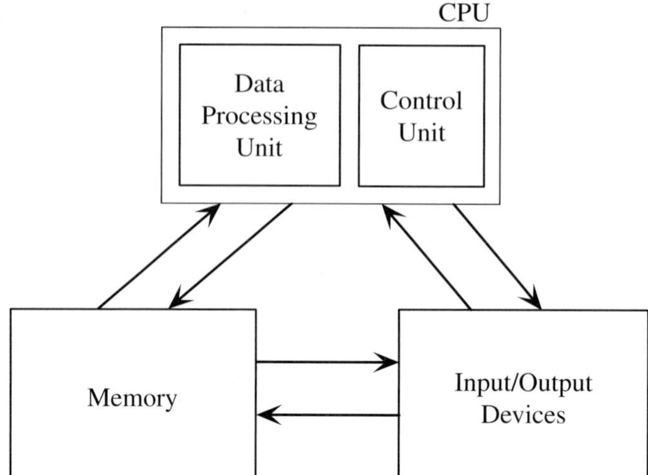


**Figure 9.1**
A block diagram for a basic computer system, consisting of a CPU, memory, and I/O devices.

The CPU in turn consists of a data-processing unit and program control unit. A microprocessor is essentially a CPU on a chip. It also is possible to implement an entire computer (CPU + Memory + I/O) on a single chip, in which case it is referred to as a *microcontroller*. Microcontrollers, which typically use a simple 32-bit or lower CPU architecture and only have a limited amount of memory and I/O ports, are used mostly for embedded applications (such as smart cell phones or intelligent microwaves).

### 9.1.1   Reduced Instruction-Set Computers

In the early days of computer development, computer instruction-sets (and the architectures that they supported) became progressively more and more complex, involving many different addressing modes, multiple-cycle instructions of varying complexity, and rarely used special-purpose instructions. These types of computers, which include the popular Intel Pentium™ series of computers, are referred to as complex instruction-set computers (CISC). However, starting in the early 1980s, a different type of computer architecture, based on relatively simple instructions and referred to as reduced instruction-set computers (RISC), started becoming popular. A RISC architecture has several features that facilitate efficient instruction execution and make it more than simply a computer architecture using a simple instruction-set.

First, a simple and uniform instruction format is used. All instructions are of the same bit length. This enables fast prefetching of instructions since instruction boundaries can be determined without having to decode the previous instructions. Also, a small number of instruction formats are used. By using a small set of common instruction formats, instruction decoding significantly is facilitated. In addition, few addressing modes are supported directly.[2] The addressing modes most commonly supported in an RISC architecture are immediate, direct, and indexed. Most other addressing modes that may be required are formed from these base addressing modes, with perhaps a few predesignated constant-value registers.

Second, a *load-store* architecture typically is used. This is a type of architecture in which all instructions, except for *load* and *store*, read data from registers and store computation results into registers. Since registers are much faster than main memory (and even cache memory), fast instruction execution is possible. This, coupled with the fact that only relatively simple and frequently used instructions are supported, enables the use of a fast instruction-execution cycle, and thus a fast system clock.

Third, a large set of registers, referred to as a *register file*, is used. Since registers are used for temporary data storage during instruction execution, a large register file permits us to only infrequently access main memory or cache. However, a register file also is used typically to permit fast and efficient subroutine execution. Typically, only a subset of the entire register file is visible at any point in time. Then, when a subroutine is called, the visible register set changes. Since the "state" of a program is maintained mostly in the values of the registers that it uses, this method permits us to quickly save the state of the main routine when switching to the subroutine. Also, by creating an overlap between the visible register set for the main routine and the visible register set for the subroutine, data parameters can be passed quickly from the main routine to the subroutine. To permit nested subroutine calls, this system of overlapping visible register sets can be repeated for a limited number of times. Since deeply nested subroutine calls are rare, a small number (about two to eight) of visible register-set regions typically are used, and nested subroutine calls beyond this limit use main memory to save state values.

Fourth, pipelining is used extensively to enable fast instruction execution. The instruction set itself is designed to enable efficient pipelined execution. Most instructions read and write data to registers, thereby resulting in a similar execution delay for most instructions—thus, the pipeline clock cycle can be optimized for this execution delay. There are few, if any, instructions requiring multiple clock cycles for execution. Other architectural and instruction-set features also are designed to facilitate the use of instruction pipelining and the techniques typically used with instruction pipelining.

### 9.1.2   Basic Computer Operation

The operation of a basic computer system can be described using a simple, repetitive fetch-and-execute sequence.[3] This is shown in Figure 9.2. As can be seen from this figure, a special register, typically designated as the Program Counter (PC) or Instruction Pointer (IP), contains the memory address of the instruction to be executed.

---

[2]   An *addressing mode* refers to the way in which a data item is provided with an instruction. Readers unfamiliar with addressing modes and other basic microprocessor concepts should refer to introductory microprocessor or computer architecture textbooks.

[3]   Those readers unfamiliar with microprocessors and computer architecture should refer to textbooks specifically devoted to these topics.

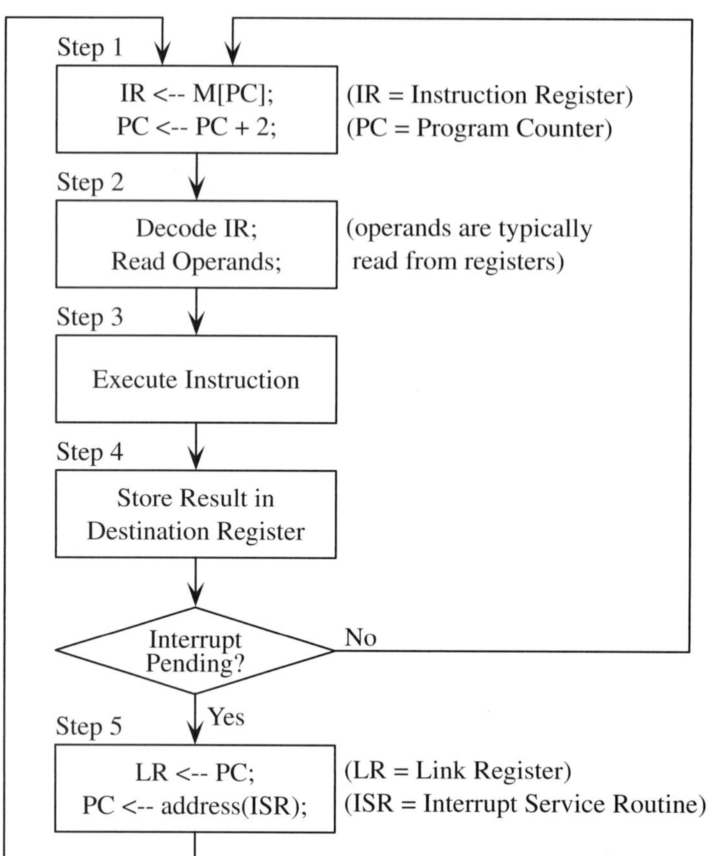



**Figure 9.2**
The fetch-and-execute operation sequence of a basic computer system.

In Step 1, the instruction pointed to by PC (or IP) is fetched from memory (or cache if that instruction has been buffered into the cache) and stored into an Instruction Register (IR). At the same time, the PC is incremented to point to the next instruction, assumed to be at the next sequential-memory address location. In Step 2, the instruction is decoded. In Step 3, the instruction is executed. This step may change the PC value in the case of a branch (or jump) instruction, and also may involve multiple clock cycles if an instruction requiring multiple clock cycles for execution has been decoded. Then, if there is no pending interrupt, the entire process is repeated from Step 1.

The above fetch-and-execute sequence changes slightly if there is an interrupt request (also referred to as an *exception*). If an interrupt-request is received from an external or internal source, the CPU must halt its current instruction sequence, service the interrupt, and then return to the processing of the interrupted instruction sequence. However, the CPU cannot be interrupted in the midst of executing an instruction, since that would make it extremely difficult to return to that instruction once the interrupt has been processed completely. Thus, all interrupt-requests received during the execution of a regular instruction typically are saved in an interrupt-request vector, and an "interrupt pending" flag is asserted. After the execution of the current instruction is completed, the "interrupt pending" flag is checked. If there is a pending interrupt, the current Program Counter (PC) value is stored elsewhere (so that it can be restored upon a return), and the PC value is changed to point to the first instruction of the interrupt-service routine for the highest priority interrupt in the interrupt-request vector. Then the regular fetch-and-execute sequence is rejoined at Step 1 of Figure 9.2. Note that the first instruction of the interrupt-service routine must reset the "interrupt pending" flag in order to avoid entering the interrupt processing path again (the "interrupt pending" flag can be set again later).

●  ●  ●  ●  ●  ●  ●  ●  ●  ●  ●  ●  ●  ●  ●  ●

## 9.2 The THUMB Microprocessor Architecture

The THUMB microprocessor is a modern RISC architecture that has been designed specifically for embedded applications requiring a simple implementation and compact instruction memory usage. As such, it uses a 16-bit instruction format. However, to be compatible with the ARM architecture, of which it is a subset, the THUMB architecture stores and processes data using 32-bit registers and a 32-bit *arithmetic logic unit (ALU)*. Thus, it can be considered as an externally 16-bit and internally 32-bit microprocessor architecture.

### 9.2.1  THUMB Programming Model

The programming model for the THUMB microprocessor is shown in Figure 9.3. There are normally a set of eight visible registers, numbered $R0$ through $R7$. Some instructions also access the Program Counter (PC), which is $R15$; the Link Register,

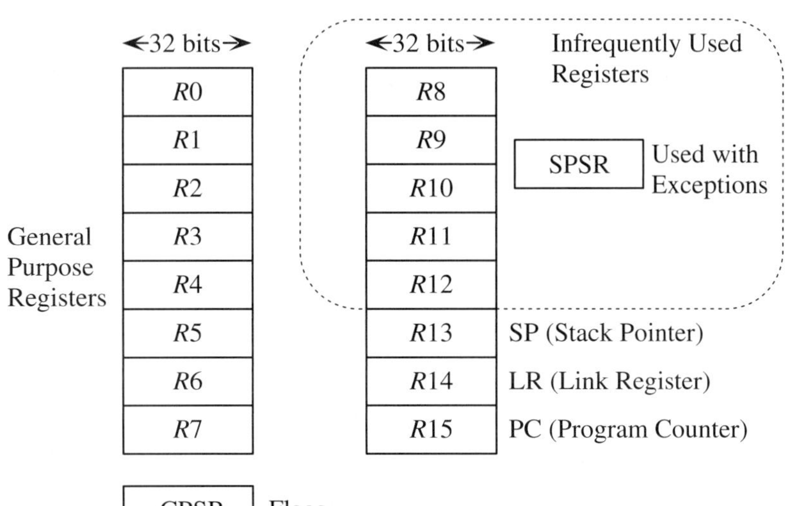


**Figure 9.3**
The programming model for the THUMB microprocessor.

which is $R14$; the Stack Pointer, which is $R13$; and the registers $R8$ through $R12$. As stated above, the Program Counter contains the address of the instruction to be fetched from memory. The Link Register, used with subroutine calls, contains the address of the next instruction after a subroutine call. Thus, when a "return" instruction is executed from within the subroutine, the Link Register points to the instruction to be executed after the "return." The Stack Pointer points to the top of a region of memory referred to as the *stack*, which is used to hold data contents in a last-in-first-out manner. The stack will be discussed in more detail in conjunction with the instructions that use the stack.

There are two *status* registers used to hold information regarding the processor *state*. Of these two status registers, the Saved Program Status Register (SPSR) is used with exceptions, and thus hardly is used with the THUMB architecture, which relies mostly on ARM instructions for exception handling. The Current Program Status Register (CSPR) consists of bits that encode information about the current processor state. Of this information, the most widely used bits are the condition-code flags, which store status information (carry, overflow, etc.) about the results of previous arithmetic or logic instructions. The four condition-code flags are negative (N), zero (Z), carry (C), and overflow (V).

### 9.2.2   Overview of the THUMB Instruction Set

The THUMB instruction set, which is a subset of the ARM instruction set, consists of branch, data processing, status register transfer, load/store, and exception-related instructions.

**Branch Instructions**

There are three main types of branch instructions. There are unconditional branches, conditional-branches, and subroutine calls. Most branch instructions branch to an address at a fixed number of bytes after or before the current instruction (this is referred to as *PC-relative* or *relative addressing*). An unconditional branch allows a forward or backward branch of up to 2 KB. A conditional-branch instruction causes a branch if a certain condition (e.g., the Z flag is set, indicating a zero result from the previous arithmetic or local instruction) is met and sequential instruction execution if the condition is not met. A conditional branch allows a forward or backward branch of up to 256 bytes.

The third type of branch is a subroutine call. This type of branch instruction must allow a "return" to the instruction immediately following the subroutine call instruction. Thus, the Program Counter (PC) value of the instruction after the subroutine call must be saved in a known location before branching to the first instruction of the subroutine. There are two main places where the PC value can be saved: in a register or in memory. In the THUMB architecture, the Link Register ($R14$) is used to store saved PC values. PC values also can be stored in a special section of memory referred to as the *stack*, which is pointed to by the Stack Pointer ($R13$). Note that the first method is more efficient than the second method since a register can be accessed must faster than memory. Since the THUMB architecture uses the Link Register to store saved PC values, a subroutine call is referred to a "Branch-with-Link" instruction in THUMB. Branch-with-Link instructions can branch to a relative address or to an absolute address stored in a register. There are also Branch with Link instructions that change to ARM code execution after the branch; however, these types of instructions will not be considered in this chapter, since we wish to use THUMB instructions exclusively.

The branch instructions, labeled B ⟨target_address⟩ for unconditional branches, B⟨cond⟩ ⟨target_address⟩ for conditional branches, and BL ⟨target_address⟩ for Branch with Link instructions, are shown in Appendix A. In this notation, ⟨cond⟩ refers to a condition code, and ⟨target_address⟩ refers to a relative or absolute branch address. Condition codes used for the THUMB and ARM architectures are shown in Table A.1.

### Data-Processing Instructions

THUMB data-processing instructions, which are a subset of the ARM data-processing instructions, perform operations that work with data values and set the condition codes (negative, zero, carry, or overflow) if those conditions are true after the operation has been performed. There are arithmetic instructions (add, subtract, negate, multiply, divide, etc.), logical instructions (or, and, exclusive-or, etc.), data-movement instructions, shift instructions (logical shift, arithmetic shift, rotate, etc.), and compare and test instructions. Compare instructions perform subtraction of two data values, without saving the subtraction result, and then set the condition codes to indicate whether the data values are equal or which data value is greater than the other. A test instruction performs the logical AND of two data values (without saving the data result) and then sets the condition codes—this is useful for checking the values of certain bits of a data value.

### Load and Store Register Instructions

As mentioned earlier, RISC microprocessors such as THUMB typically use a *load/store* architecture, which means that only load and store instructions can access memory, while normal data-processing instructions must receive data operands from registers and save results in registers. Thus, there are a set of instructions that used to load data values from memory and to store result values into memory. Data can be loaded and stored in increments of 8 bits (1 byte), 16 bits (referred to as a *halfword*), and 32 bits (referred to as a *word*).

Several methods are used to indicate the address in memory where the data should be loaded from or stored into. The address can be the value in a register, the addition of two register values, or the addition of a register value with an immediate offset value that is included in the instruction code. The first method corresponds to a *register addressing* mode while the second and third methods correspond to an *indexed addressing* mode. However, depending on the register used, the third method can also correspond to *relative addressing* (if the register is the PC) or *direct addressing* (if the register value is 0).

### Interrupt and Other Instructions

There are two instructions that are designed to generate exceptions (also referred to as interrupts). The SWI (software interrupt) instruction causes a SWI exception to occur. This instruction typically is used to switch from "User" mode to privileged "Operating System" mode. The breakpoint (BKPT) instruction is used to generate software breakpoints. It can be used in conjunction with debugging hardware or software to temporarily suspend a THUMB program for debugging purposes.

Both types of interrupt instructions involve interrupting the normal fetch-execute operation sequence (as described in Section 9.1.2) and switching to interrupt service routines. In addition, interrupt-service routines must be written to process these types of interrupts correctly. Since this involves knowledge of the specific target application architecture for which the THUMB microprocessor will be used, the SWI and BKPT interrupt instructions will not be implemented in this book.

Two instructions that are not part of the THUMB instruction set, but which are useful for testing and debugging purposes, are HALT and NOP (no operation). As its name implies, the HALT instruction should simply halt the program. A commercial microprocessor typically does not require this instruction because a computer should execute continuously until it is shut down. However, this instruction can be useful for providing a fixed stopping point when the microprocessor is implemented in a test board. Another instruction that is useful particularly with instruction pipelines is the NOP instruction. The NOP instruction can be used to create intervening space between two instructions with data or resource dependencies in an assembly program. As explained earlier, this mechanism, combined with instruction reordering, is useful for resolving data and structural hazards. The HALT instruction is implemented using the opcode for the SWI instruction, and the NOP instruction is implemented using an opcode in the undefined opcode space of the THUMB instruction set.

## 9.3  Instruction Pipeline Design

Any hardware circuit that faithfully implements the entire THUMB instruction set can be considered as a THUMB microprocessor. Such a circuit should be capable of executing any program compiled by a THUMB compiler or written manually in THUMB assembly code. In order to be successful commercially, however, such an implementation should use a fast system clock and be able to execute, on average, one or more instructions per clock cycle. In order to achieve this requirement, pipelining is essential.

An *instruction pipeline* for the THUMB microprocessor can be designed by pipelining the basic fetch-execute operation sequence shown in Figure 9.2. Thus, Figure 9.4 shows a block diagram for a 4-stage pipeline designed in this manner, in which the interrupt path has been omitted to simplify the pipeline design.

When the instructions of an assembly program are executed using an instruction pipeline, we must be careful to ensure that the execution result is the same as when the program is executed using an unpipelined, sequentially executing CPU. Thus, if there are dependencies between two or more instructions present in different stages of the instruction pipeline, those dependencies must be adhered to. Of course the simplest method of solving this problem is to delay the later instruction until the former instruction completes its execution (this is referred to as a

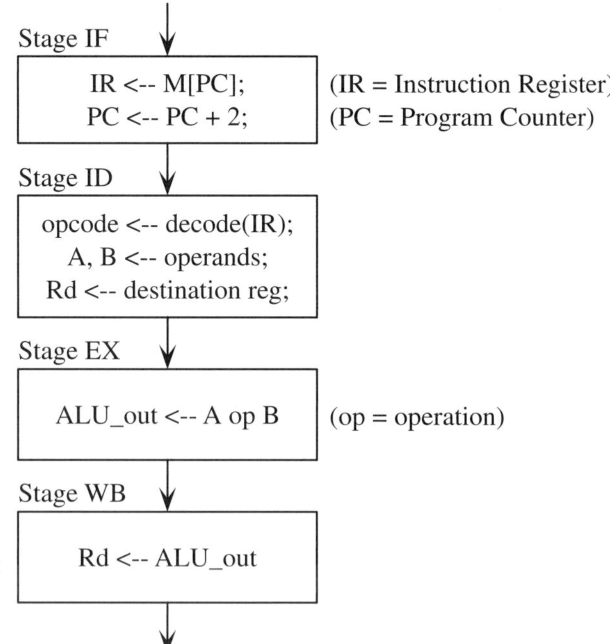



**Figure 9.4**
A four-stage instruction pipeline for the THUMB microprocessor.

*pipeline stall*). However, this will result in an inefficient hardware implementation—in the worst case, the performance will be no better than that of an unpipelined hardware implementation. Thus, the challenge in designing an instruction pipeline is to design it in such a manner that the pipeline is used as fully as possible while adhering to all instruction dependencies.

### 9.3.1  Pipeline Hazards

Conditions that may result in the need to delay the execution of an instruction pipeline are referred to as *pipeline hazards*. There are three main types of pipeline hazards. A *structural hazard* refers to a situation in which two instructions must use a common resource, thereby resulting in a *resource conflict*. For example, a common problem of this form occurs when there is only one memory device and a load or store instruction is encountered. Since instructions must be fetched during each clock cycle, if an instruction in the pipeline requires an access to memory, then a memory-access conflict will result.

The second main type of pipeline hazard is a *data hazard*. In this type of hazard, one instruction changes the value of a register, while a following instruction uses that same register value. If the following instruction reads the register value before the preceding instruction changes the register value, the later instruction may use an incorrect data value, thereby resulting in an incorrect program result. For instance, if an instruction executes $R1 \leftarrow R2 + R3$ and the following instruction executes $R4 \leftarrow R1 + R5$, pipelining of the CPU may result in a second instruction reading the value of register $R1$ before the first instruction has had a chance to update the value of $R1$. This type of data hazard is also referred to as a *read-after-write (RAW)* hazard. Depending upon the manner in which the CPU pipeline is implemented, there also may be other types of data hazards, as detailed in books devoted to computer architecture.

The third main type of pipeline hazard is a *control hazard*. This type of hazard occurs when an unconditional-or conditional-branch instruction is encountered. In the absence of branch instructions, all instructions typically can be fetched in strict sequential order. Then, if a fixed-length instruction format is used, a sequence of instructions can be "prefetched" before the first instruction of that sequence has completed execution. However, if the first instruction of that sequence happens to be a branch instruction, then all subsequent instructions fetched in this manner (assuming a strictly sequential execution order) will be incorrectly fetched instructions.

### 9.3.2 Hazard Prevention Techniques

Many techniques are used for the prevention of pipeline hazards. The use of good hazard prevention techniques will permit the instruction pipeline to be operated at full (or almost full) capacity when executing user programs, thereby increasing the performance of the user programs. New pipeline designs and pipeline hazard prevention techniques are the topics of ongoing research. There are also many well-known techniques for the prevention of all three types of pipeline hazards: structural, data, and control.

**Prevention of Structural Hazards**

Structural hazards commonly are resolved by introducing extra resources or by using pipelining again. A structural hazard can be resolved by replicating the resource for which a conflict may occur. If there is an instruction that accesses memory, then the execution of that instruction may conflict with the fetch of the next instruction from the same memory module. Thus, an instruction pipeline typically uses two separate memories (referred to as *instruction* and *data* memories) for storing instructions and data, respectively. Another example of a commonly replicated resource is the ALU. By replicating the ALU, one ALU can be used for address calculations while another ALU is used for the execution of arithmetic or logic instructions.

For some conflicting resources, it also may be possible to simply pipeline the conflicting resource. An example is a pipelined floating-point execution unit. A floating-point arithmetic instruction takes longer to execute than an integer arithmetic instruction. Thus, if floating-point arithmetic instructions are part of the instruction set, then two or more consecutive floating-point instructions will slow down the instruction pipeline. However, if those floating-point instructions are executed using an arithmetic pipeline, then only the first part of the first floating-point operation needs to be executed before the next floating-point instruction can be accepted into the floating-point arithmetic pipeline unit. Another example of a commonly used pipeline resource is a data buffer. The CPU typically can execute at a much faster rate than memory. Thus, if there are several consecutive instructions that access memory, then the instruction pipeline will be slowed down by having to access slow memory. However, by using a "memory pipeline" (a data buffer consisting of a set of registers connected like a pipeline), only the first part of a memory-access operation needs to be executed before the next instruction can be executed.[4]

## Prevention of Data Hazards

A commonly used method for solving data hazards is to change the order of assembly instructions so that consecutive instructions do not have a "read-after-write" dependency (or any other harmful data dependency). Of course, care must be taken to ensure that the program result does not change when the instructions are reordered. Thus, for those situations in which the compiler can determine that instructions can be reordered without adversely affecting the program result, *instruction reordering* can be used to avoid data hazards.

Another method, which does not require modifications to the assembly program, is based on the use of *data forwarding*. In this method, the CPU hardware must keep track of the instructions currently in the pipeline and the registers used by those instructions. This type of record-keeping typically is done using a data structure, implemented in hardware using a fast cache or registers, referred to as a *scoreboard*. Then, when a "read-after-write" situation is detected, an instruction that obtains its operand from the result of a previous instruction is made to access the output of the ALU directly (instead of from the register file) for that data operand. If an instruction uses the result of an instruction two or more stages in front of it in the pipeline, then the output of the ALU may need to be stored (by using extra register stages) before it is used as a data operand. This type of solution, with the extra hardware required, is shown in Figure 9.5.

## Prevention of Control Hazards

Control hazards can be prevented by reordering the order of assembly instructions or by predicting the location of the next instruction address. The first method, (referred to as a *delayed branch*), makes use of the fact that a branch instruction must make its way through several pipeline stages before the program counter is updated to the target of the branch or the next sequential instruction. For both conditional and unconditional branches, several instructions in the pipeline are processed partially before the branch is completed. If those partially processed instructions are *not* the correct instructions to be executed, then there is a problem. Of course, if those incorrect instructions are prevented from updating any registers or memory locations, then the assembly program still may execute correctly. However, even in this situation, the pipeline resources will have been wasted. In order to prevent this type of waste, a set of $k$ instructions *before* a branch instruction can be moved to positions immediately *after* the branch instruction, where $k$ is the number of the instructions "wasted" by the branch instruction. Then, if the instruction pipeline is executed while disregarding the potential control hazard, the pipeline will be utilized fully, regardless of the result of the branch instruction. Of course, the $k$ instructions

---

[4] A similar type of buffer is an "instruction buffer," which is used to prefetch instructions from memory. Instructions can be prefetched by "guessing" the locations of the next set of instructions to be executed (typically the next set of sequentially located instructions).

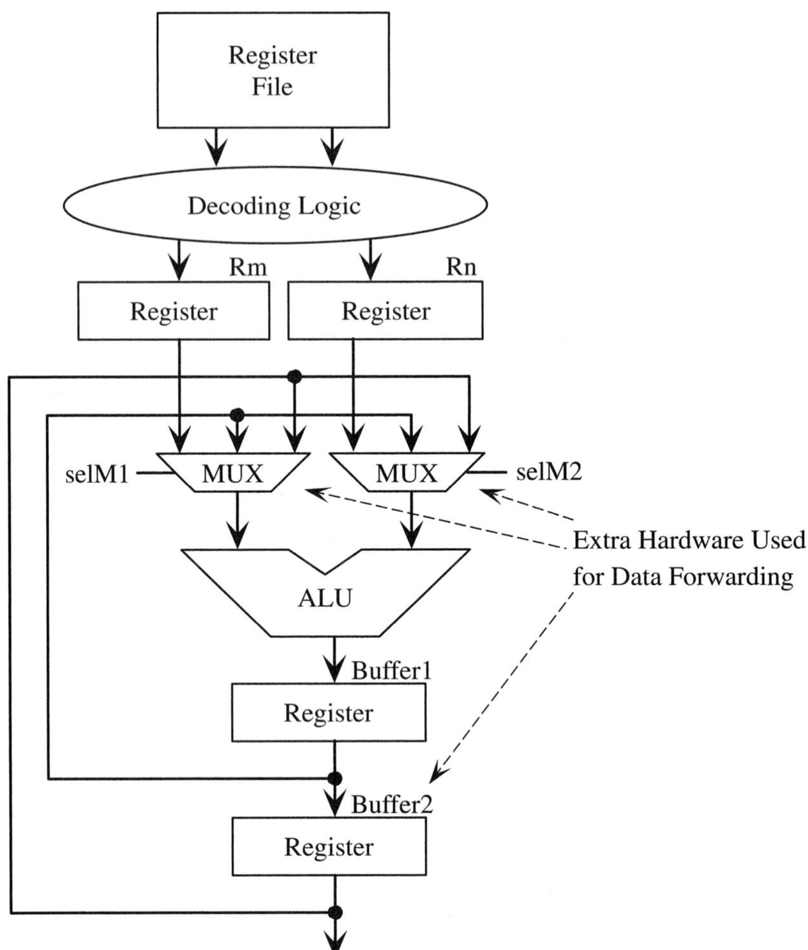



**Figure 9.5**
A block diagram of the hardware used for data forwarding.

chosen to fill the *delay slot* formed by a branch instruction must be independent of the branch instruction. In order to minimize the number $k$, the program counter can be changed at an early pipeline stage when a branch instruction is detected. Figure 9.6 shows an example of the use of this type of delayed-branch method.

The second method, which can be used in combination with the first method described above, involves branch prediction and branch target prediction. In *branch prediction*, a guess is made as to whether a recently fetched instruction is a branch instruction or not. Of course, a branch instruction can be recognized as such after it has been decoded. However, by this time, several other instructions already will have been entered into the instruction pipeline. If a branch instruction could be identified as such *while* it is being fetched, then the next instruction conceivably could be fetched from the target of the branch. How can this be done?

Branch prediction makes use of the fact that a branch instruction typically is part of a loop construct. Thus, a given branch instruction typically will be visited more

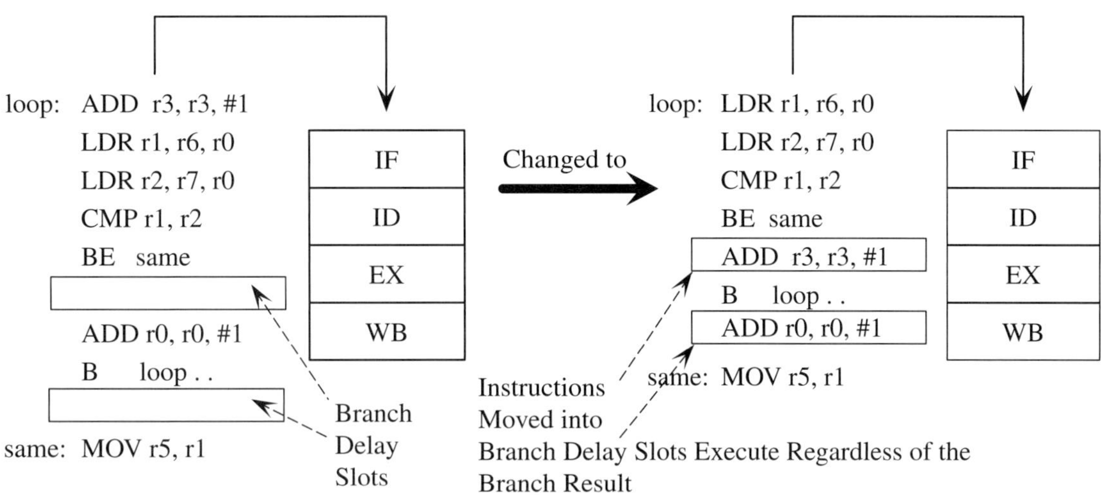



**Figure 9.6**  As shown in this example, with the *delayed-branch* method, *branch-delay slots* caused by "having to delay the pipeline in the case of branch instructions" can be filled by previous unrelated instructions (and the pipeline filled and executed as if there is no control hazard) in order to avoid wasting pipeline resources.

than once during the execution of an assembly program. The first time that a branch instruction is executed, its memory address, the target of the branch (the address to branch to), and whether the branch was taken or not is recorded in a buffer (referred to as a *branch-prediction buffer*). Then, the next time the same branch instruction is encountered, the CPU can determine that the instruction is a branch (before the instruction has even been fetched) by referring to the branch prediction buffer. Furthermore, it can *guess* whether the branch should be taken or not, and then prefetch subsequent instructions from either the branch target or the next sequential address locations, depending on its guess. A program loop typically executes many times before it is exited (otherwise, there would be no point in having a loop in the first case). Thus, an intelligent *guess* of the behavior of a branch instruction is the taken/not-taken behavior of that instruction during the previous loop iteration. In summary, the branch prediction buffer can be used to predict whether an instruction is a branch, whether the branch will be taken or not, and the address of the next instruction in the case of a branch. In order to save buffer space, only part of the address of the branch instruction can be saved in the branch prediction buffer.

### 9.3.3  Pipeline Hazard Solutions Adopted

The pipelined THUMB microprocessor can be designed using the basic block diagram shown in Figure 9.4. In order to handle pipeline hazards, the following solution methods will be used. First, to deal with structural hazards involving the memory, two independently accessible memories (one for instructions and one for data) will be assumed. Thus, there will be two sets of address and data lines. Second, to deal with data hazards and structural hazards involving other resources besides the memory, it will be assumed that the assembly instructions already have been reordered to avoid all such hazards. When creating a test program, this instruction reordering can be performed manually and/or "null" instructions can be inserted as necessary. Third, to deal with control hazards, whenever a branch instruction is encountered in which the branch takes place (i.e., an unconditional branch or a conditional branch in which the condition is met), the following instructions currently in the pipeline are *marked* and subsequent instructions are fetched, starting from the branch-target address. Whenever a *marked* instruction is encountered during the execution stage, the result of that instruction is "nullified" by *not* storing the execution result for that instruction (e.g., for the instruction "$R1 \leftarrow R2 + R3$", the addition result is *not* stored into the $R1$ register).

The above solution approaches are far from optimal. Thus, the reader is encouraged to implement more advanced hazard-prevention techniques as part of a class laboratory. The advanced hazard-prevention techniques adopted can be based on the solution methods described in the previous subsection or by other, novel approaches devised by the reader.

## 9.4  HDL Implementation of the THUMB Pipeline

The pipelined THUMB microprocessor can be implemented in HDL using the overall design method presented in the previous section. The following Verilog design uses an analogous method to implement the "pipeline stages" shown in the block diagram of Figure 9.4. Basically, there must be separate concurrently executing processes for each of the stages of the instruction pipeline for the THUMB microprocessor. Opcodes, instruction arguments, and partially processed instruction results must be passed between these separate processes in a synchronous manner. In addition, if a value computed in an earlier stage is required in a later stage, then that value must be passed through all of the intervening stages using pipeline registers.

The overall design structure, which is used for the Verilog code implementation, is shown in Figure 9.7. In the instruction fetch (IF) stage, the PC register points to a location in the instruction memory. The data at that location is fetched and stored into the IF_IR pipeline register, which also happens to be the instruction register. The PC value also is updated and stored in the IF_PC pipeline register.

The ID (instruction decode) stage decodes the pipeline register values to produce an instruction opcode (stored in ID_opcode) reads the source register values (stored in ID_Rm_Rs and ID_Rn) transfers the binary encoding of the Rd field as ID_Rd_code, and reads and stores the immediate or offset value.

The instruction execute (EX) stage executes the instruction stored in ID_opcode using the source values provided. If the instruction is an arithmetic or logical instruction, the ALU is used, and if the instruction is a fetch or store instruction, the data memory is used.

Finally, in the write back (WB) stage, the result of the executed instruction is written into the register R[EX_Rd_code], which is the register pointed to by the Rd field

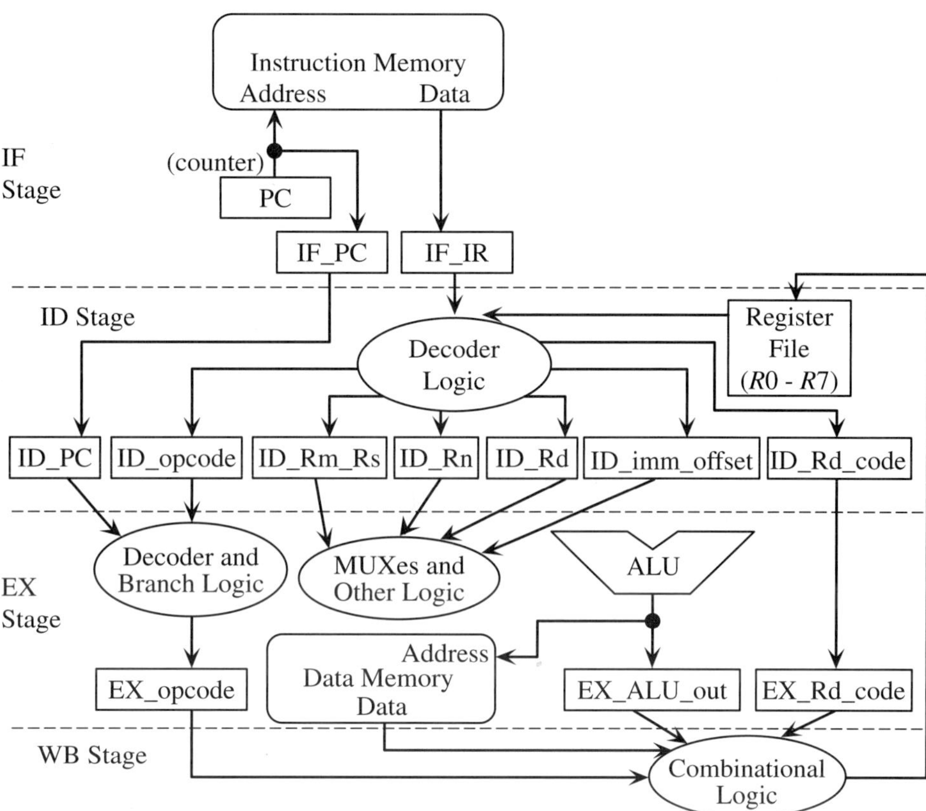


**Figure 9.7**
A block diagram of a design for the pipelined THUMB microprocessor.

of that instruction. Notice how some registers are copied from one stage to the next without even being used in that stage. For example, the IF_PC value is copied to ID_PC in the ID stage. Also, the ID_Rd_code value is copied to EX_Rd_code in the EX stage. As can be seen from the figure, the reason for this seemingly redundant logic is so that the corresponding register value can be used in a later pipeline stage.

### 9.4.1   Verilog THUMB Implementation

The Verilog code for the pipelined THUMB microprocessor implements the four-stage instruction pipeline shown in Figure 9.7. The four pipeline stages are implemented using separate **always** blocks. Pipeline registers are used to transfer data between two adjacent pipeline stages. Pipeline hazards are handled by having independent address and data lines for instructions and data (for structural hazards), by relying on the user (or compiler) to enter instructions in an order that does not cause data dependencies (for data hazards), and by *not* storing the results of instructions (in their respective destination registers) in the pipeline when a branch takes place (in order to handle control hazards).

The Verilog implementation for the pipelined THUMB microprocessor consists of two modules and a file for constant definitions. The thumb_defs.vh file defines values for global constants used by the two modules, including instruction codes for all THUMB instructions supported. The thumb.v file defines a synthesizable circuit (**module** thumb) for the pipelined THUMB microprocessor. Finally, tb_thumb.v is a test bench for the thumb module.

## Constant Definitions

The constants defined in this file consist of global parameter definitions and opcode definitions. Global parameters are bit widths and other "sizes" used in the thumb and tb_thumb modules. Opcodes refer to a set of custom opcodes, and not the original THUMB opcodes, created for all of the supported THUMB instructions. Since there are 59 such instructions, all opcodes can be encoded using six bits. There is also an `UNDEF opcode defined for use with undefined instructions and error situations. There are some instruction encodings that are not used by the THUMB instruction set. Of these unused encodings, the undefined THUMB instruction encoding 0xDE00 is defined in this file so that is can be used as a Null Operation (NOP) instruction.

```
////////////////////////////////////////////////////////////
// GLOBAL PARAMETER DEFINITIONS
`define WORD_SIZE 32     // bits in data registers and address
`define HWORD_SIZE 16    // instruction word size
`define REG_FILE_SIZE 8 // number of general-purpose registers

// OPCODE DEFINITIONS
// Undefined Instruction Thumb Opcode
`define UNDEFINED_INSTRUCTION 16'hDE00  // 16 = `HWORD_SIZE
// Internal Encoding for Thumb Instructions Supported in "thumb.v"
`define ADC        6'b00_0000
`define ADD_1      6'b00_0001
`define ADD_2      6'b00_0010
`define ADD_3      6'b00_0011
`define ADD_5      6'b00_0100
`define ADD_6      6'b00_0101
`define ADD_7      6'b00_0110
`define AND        6'b00_0111
`define ASR_1      6'b00_1000
`define ASR_2      6'b00_1001
`define B_1        6'b00_1010
`define B_2        6'b00_1011
`define BIC        6'b00_1100
`define BL_BLX_H10 6'b00_1101
`define BL_BLX_H11 6'b00_1110
`define CMN        6'b00_1111
`define CMP_1      6'b01_0000
`define CMP_2      6'b01_0001
`define EOR        6'b01_0010
`define LDMIA      6'b01_0011
```

```
`define LDR_1      6'b01_0100
`define LDR_2      6'b01_0101
`define LDR_3      6'b01_0110
`define LDR_4      6'b01_0111
`define LDRB_1     6'b01_1000
`define LDRB_2     6'b01_1001
`define LDRH_1     6'b01_1010
`define LDRH_2     6'b01_1011
`define LDRSB      6'b01_1100
`define LDRSH      6'b01_1101
`define LSL_1      6'b01_1110
`define LSL_2      6'b01_1111
`define LSR_1      6'b10_0000
`define LSR_2      6'b10_0001
`define MOV_1      6'b10_0010
`define MUL        6'b10_0011
`define MVN        6'b10_0100
`define NEG        6'b10_0101
`define ORR        6'b10_0110
`define POP_R0     6'b10_0111
`define POP_R1     6'b10_1000
`define PUSH_R0    6'b10_1001
`define PUSH_R1    6'b10_1010
`define ROR        6'b10_1011
`define SBC        6'b10_1100
`define STMIA      6'b10_1101
`define STR_1      6'b10_1110
`define STR_2      6'b10_1111
`define STR_3      6'b11_0000
`define STRB_1     6'b11_0001
`define STRB_2     6'b11_0010
`define STRH_1     6'b11_0011
`define STRH_2     6'b11_0100
`define SUB_1      6'b11_0101
`define SUB_2      6'b11_0110
`define SUB_3      6'b11_0111
`define SUB_4      6'b11_1000
`define SWI        6'b11_1001
`define TST        6'b11_1010
`define UNDEF      6'b11_1111
////////////////////////////////////////////////////////////////
```

## THUMB Microprocessor Code

The `thumb` Verilog module definition consists of signal declarations, a function definition, continuous assignment statements, and four **always** blocks for the four stages of the instruction pipeline.  The declaration section of the `thumb` Verilog module definition is shown next. A two-memory (one for instructions and one for data) architecture is assumed, since such an architecture prevents structural hazards when a load or store instruction is present in the pipeline. Thus, the outputs and inputs of this CPU module include control signals for both memories, data and address lines to be connected to both memories, an active-low reset input, and a clock input. For the instruction memory, a 32-bit address (instruction_address) and a 16-bit data signal (instruction) are used. For the data memory, both the address and data signals are assumed to be 32-bit signals.

Signal declarations are included for signals connected to this circuit's inputs and outputs, signals connected to the outputs of the registers used in the THUMB programming model shown in Figure 9.3, signals used to aid in the Verilog behavioral description, and signals used for pipeline registers. Pipeline register signals are named with the name of the pipeline stage (IF, ID, etc.) feeding those signals added as prefixes. This pipeline register structure is shown in Figure 9.4. The branch_taken signal is asserted when a branch or branch-with-link instruction causes an abrupt jump to a nonsequential instruction address. When this signal is asserted, all instructions in the pipeline following the branch instruction must be canceled until new instructions are fetched, starting from the branch target address. There are also a few integer variables declared as signals that are used to facilitate behavioral Verilog code description.

```
///////////////////////////////////////////////////////////////////
// MODULE: Pipelined CPU for THUMB microprocessor: thumb.v
// Author: Sunggu Lee
// Created: ...
// Last Modified: ...
// Description: Implements the instruction set for the THUMB
//    microprocessor described in the "ARM Architecture Reference
//    Manual" [ARM 2000].  The SWI (software interrupt) instr.
//    is re-implemented as a HALT.  Currently, a few instructions
//    are not implemented due to a desire to keep the design
//    relatively simple.  Unimplemented instructions are ARM
//    Version 5 Thumb instructions, instructions using the high
//    (R8-R15) registers, LDMIA, STMIA, and PUSH and POP with the
//    PC register or multiple registers.  The specific
//    unimplemented instructions, as listed in [ARM 2000], are
//    ADD(4), BKPT, BLX(H=01), BLX(2), BX, CMP(3), LDMIA, MOV(3),
//    POP(>1 register), POP(PC), PUSH(>1 register), PUSH(PC),
//    STMIA, and SWI (implemented as HALT only).

// INCLUDE files
`include "thumb_defs.vh"     // global constant definitions
```

```
// MODULE DECLARATION
module thumb (read_instruction_n, instruction_address,
                instruction, read_data_n, write_data_n,
                data_address, data, reset_n, clk);
  output read_instruction_n;  // enable read from instruction mem
  output [`WORD_SIZE-1:0] instruction_address; // instr. address
  input [`HWORD_SIZE-1:0] instruction;  // current instruction
  output read_data_n;     // enable read from data memory
  output write_data_n;     // enable write to data memory
  output [`WORD_SIZE-1:0] data_address; // address of data
  inout [`WORD_SIZE-1:0] data;          // current data
  input reset_n;     // active-low RESET signal
  input clk;         // clock signal

  // SIGNAL DECLARATIONS for chip inputs and outputs
  reg read_instruction_n, read_data_n, write_data_n;
  wire [`WORD_SIZE-1:0] instruction_address;
  reg [`WORD_SIZE-1:0] data_address;
  wire [`HWORD_SIZE-1:0] instruction;
  wire [`WORD_SIZE-1:0] data;
  wire reset_n;
  wire clk;

  // SIGNAL DECLARATIONS for internal registers and wires
  reg [`WORD_SIZE-1:0] PC;        // program counter = R15
  reg [`WORD_SIZE-1:0] LR;        // link register = R14
  reg [`WORD_SIZE-1:0] SP;        // stack pointer = R13
  reg [`HWORD_SIZE-1:0] IR;       // instruction register
  reg [`WORD_SIZE-1:0] R [`REG_FILE_SIZE-1:0]; // general regs
  reg N_Flag;             // condition flags (N = negative)
  reg Z_Flag;             // (Z = zero)
  reg C_Flag;             // (C = carry)
  reg V_Flag;             // (V = overflow)
```

```
// SIGNAL DECLARATIONS used to aid in the Verilog description
reg branch_taken;         // indicates a branch has been taken
reg [`WORD_SIZE-1:0] branch_target; // branch target address
reg [`WORD_SIZE:0] ALU_out;      // ALU output
reg [`WORD_SIZE-1:0] DR;         // holds data to be output
integer i, ex_i;                 // index variables
integer found_i, found_ex_i;     // used for searching
// Pipeline Regs (names prefixed with pipeline stage names)
reg [`HWORD_SIZE-1:0] IF_IR;     // IR value stored at end of IF
reg [`WORD_SIZE-1:0] IF_PC, ID_PC;           // saved PC value
reg [`WORD_SIZE-1:0] ID_Rd, ID_Rn, ID_Rm_Rs; // Rd, Rn, Rm/Rs
reg [`WORD_SIZE-1:0] EX_ALU_out;// ALU_out at end of EX stage
reg [5:0] ID_opcode, EX_opcode; // internal Thumb opcode
reg [10:0] ID_imm_offset; // immediate, offset or register list
reg [7:0] EX_imm_offset;  // 8 bits required in WB stage
reg [3:0] ID_cond;        // ARM condition code (same as THUMB)
reg [2:0] ID_Rd_code, EX_Rd_code;  // Rd register number
/////////////////////////////////////////////////////////////////
```

A function named condition_passed is defined for use with the conditional branch instruction. The codes for these condition checks are those shown in Section A3.2 of the [ARM 2000] reference manual. Continuous assignment statements (using the keyword **assign**) are used for aliased signal names (a single wire or bus with two or more names) and tri-state signal outputs.

```
/////////////////////////////////////////////////////////////////
// ASSIGN STATEMENTS
assign instruction_address = PC;  // set instr. address to PC
assign data = (~write_data_n) ? DR : 'bz;  // tri-state data

// Function and Task Definitions
// Function to check condition codes
function condition_passed;
  input [3:0] cond_code;  // 4-bit Thumb/ARM condition code
begin
  case (cond_code)  // codes in Table 3-1 of [ARM 2000]
    4'b0000: condition_passed = Z_Flag;  // EQ (equal)
    4'b0001: condition_passed = ~Z_Flag; // NE (not equal)
    4'b0010: condition_passed = C_Flag;  // CS/HS (carry set)
```

```
    4'b0011: condition_passed = ~C_Flag; // CC (carry clear)
    4'b0100: condition_passed = N_Flag;  // MI (minus)
    4'b0101: condition_passed = ~N_Flag; // PL (plus)
    4'b0110: condition_passed = V_Flag;  // VS (overflow)
    4'b0111: condition_passed = ~V_Flag; // VC (no overflow)
    4'b1000: condition_passed = (C_Flag & (~Z_Flag)); // HI
    4'b1001: condition_passed = ((~C_Flag) & Z_Flag); // LS
    4'b1010: condition_passed = (N_Flag == V_Flag);   // GE
    4'b1011: condition_passed = (N_Flag != V_Flag);   // LT
    4'b1100: condition_passed = ~Z_Flag & (N_Flag == V_Flag);
                                    // GT (greater than)
    4'b1101: condition_passed = Z_Flag & (N_Flag != V_Flag);
                                    // LE (less or eq)
    4'b1110: condition_passed = 1'b1;    // AL (always)
    default: condition_passed = 1'b1; // note: 4'b1111 invalid
  endcase
 end  // of function body
 endfunction  // of function condition_passed
/////////////////////////////////////////////////////////////////
```

The **always** blocks that follow have a one-to-one correspondence with the stages of the instruction pipeline. Intermediate values transmitted between stages are passed through pipeline registers. The following shows the Verilog code for the instruction fetch (IF) stage. In the IF stage, the instruction stored at the address pointed to by the program counter (PC) must be fetched (from the instruction memory) and stored into the IR register. In order to accomplish this, a read-enable signal (read_instruction_n) must be asserted and a valid instruction address stored into PC. Then, during the next clock cycle, the data value returned by the memory device must be latched into the IF_IR pipeline register. The IF stage must perform this operation for all instructions, with a new instruction fetched during each clock cycle. Thus, when the instruction address for the next instruction is being prepared (by storing the address into PC), the current instruction (whose address was output during the previous clock cycle) can be fetched and stored into IF_IR. In the absence of a branch operation, instructions can be fetched in a sequential manner using PC = PC + 2 since each instruction occupies two bytes. Since this PC value will be required in a later pipeline stage, it also must be copied into IF_PC. When a branch operation is detected (from an earlier branch or branch-with-link instruction in the pipeline), the branch target must be loaded into PC. Since the instruction fetched at this time is incorrect (it is the next sequential instruction instead of the instruction after the branch), the IR register simply can be filled with any instruction that can be "nullified" later on in the pipeline (this is referred to as a *pipeline stall* or *bubble*).

```verilog
//////////////////////////////////////////////////////////////////////
  // IF: Instruction Fetch Stage
  always @(negedge reset_n or posedge clk) begin
    if (~reset_n) begin
      PC <= 0;                  // start fetching from location 0
      read_instruction_n <= 0; // and start memory read
    end
    else begin         // on positive clock edge,
      // execute operations and then save values in pipeline reg
      read_instruction_n <= 0; // read next instruction
      if (branch_taken) begin  // determine next instruction
        PC <= branch_target;    // operation for IF stage
        IF_IR <= `UNDEFINED_INSTRUCTION;   // to create a pipeline
        IF_PC <= branch_target;          // stall (bubble)
      end
      else begin
        PC <= PC + 2;           // operation for IF stage
        IF_IR <= instruction; // read instruction and store in IR
        IF_PC <= PC + 2;      // save next instruction address
      end
    end
  end  // of IF stage
//////////////////////////////////////////////////////////////////////
```

In the ID stage, the instruction stored in the IF_IR register is decoded, and data to be used by this instruction in the EX stage is stored in a set of common registers used by the EX stage. The reason for moving data from the register file and portions of the IF_IR register to a set of common registers used by the EX stage is so that the decoders and random logic required to move this data are included as part of this stage and not the EX stage. This **always** block essentially consists of a set of **case** blocks that decode the bits of the IF_IR register to determine the specific instruction being processed. Instruction decoding is made difficult by the fact that different sets of bits in the IF_IR register are used to decode different types of THUMB instructions (i.e., all instructions simply cannot be decoded by looking at a fixed set of IF_IR bits). Whenever possible, the same set of bits are used for the same purpose, especially with similar types of instructions. However, as with the instruction sets of most commercial RISC architectures, many fields of the instruction register must be used for different purposes, depending on the instruction, in order to utilize the fixed set of bits in the instruction register as efficiently as possible.

```verilog
////////////////////////////////////////////////////////////////
// ID: Instruction Decode Stage
always @(posedge clk) begin
  case (IF_IR[`HWORD_SIZE-1:13])
    3'b000: begin // shift by immediate or add/sub
      case (IF_IR[12:11])
        2'b11: begin
          case (IF_IR[10:9])
            2'b10: begin // if imm = 000, same as MOV_2
              ID_opcode <= `ADD_1;
              ID_imm_offset[2:0] <= IF_IR[8:6];
              ID_Rn <= R[IF_IR[5:3]];
              ID_Rd <= R[IF_IR[2:0]];
              ID_Rd_code <= IF_IR[2:0];
            end
            2'b00: begin
              ID_opcode <= `ADD_3;
              ID_Rm_Rs <= R[IF_IR[8:6]];
              ID_Rn <= R[IF_IR[5:3]];
              ID_Rd <= R[IF_IR[2:0]];
              ID_Rd_code <= IF_IR[2:0];
            end
            2'b11: begin
              ID_opcode <= `SUB_1;
              ID_imm_offset[2:0] <= IF_IR[8:6];
              ID_Rn <= R[IF_IR[5:3]];
              ID_Rd <= R[IF_IR[2:0]];
              ID_Rd_code <= IF_IR[2:0];
            end
            default: begin // == case 2'b01
              ID_opcode <= `SUB_3;
              ID_Rm_Rs <= R[IF_IR[8:6]];
              ID_Rn <= R[IF_IR[5:3]];
              ID_Rd <= R[IF_IR[2:0]];
              ID_Rd_code <= IF_IR[2:0];
            end
          endcase
        end // of case IF_IR[12:11] = 2'b11
```

```verilog
        2'b10: begin
          ID_opcode <= `ASR_1;
          ID_imm_offset[4:0] <= IF_IR[10:6];
          ID_Rm_Rs <= R[IF_IR[5:3]];
          ID_Rd <= R[IF_IR[2:0]];
          ID_Rd_code <= IF_IR[2:0];
        end
        2'b00: begin
          ID_opcode <= `LSL_1;
          ID_imm_offset[4:0] <= IF_IR[10:6];
          ID_Rm_Rs <= R[IF_IR[5:3]];
          ID_Rd <= R[IF_IR[2:0]];
          ID_Rd_code <= IF_IR[2:0];
        end
        default: begin // == case 2'b01
          ID_opcode <= `LSR_1;
          ID_imm_offset[4:0] <= IF_IR[10:6];
          ID_Rm_Rs <= R[IF_IR[5:3]];
          ID_Rd <= R[IF_IR[2:0]];
          ID_Rd_code <= IF_IR[2:0];
        end
      endcase
    end // of case IF_IR[15:13] = 3'b000
    3'b001: begin // add/sub/compare/move immediate
      case (IF_IR[12:11])
        2'b10: begin
          ID_opcode <= `ADD_2;
          ID_Rd <= R[IF_IR[10:8]];
          ID_Rd_code <= IF_IR[10:8];
          ID_imm_offset[7:0] <= IF_IR[7:0];
        end
        2'b01: begin
          ID_opcode <= `CMP_1;
          ID_Rn <= R[IF_IR[10:8]];
          ID_imm_offset[7:0] <= IF_IR[7:0];
        end
        2'b00: begin
          ID_opcode <= `MOV_1;
          ID_Rd <= R[IF_IR[10:8]];
          ID_Rd_code <= IF_IR[10:8];
          ID_imm_offset[7:0] <= IF_IR[7:0];
        end
```

```verilog
    default: begin // == case 2'b11
      ID_opcode <= `SUB_2;
      ID_Rd <= R[IF_IR[10:8]];
      ID_Rd_code <= IF_IR[10:8];
      ID_imm_offset[7:0] <= IF_IR[7:0];
    end
  endcase
end // of case IF_IR[15:13] = 3'b001
3'b010: begin // data processing
  casex (IF_IR[12:6])
    7'b000_0101: begin
      ID_opcode <= `ADC;
      ID_Rm_Rs <= R[IF_IR[5:3]];
      ID_Rd <= R[IF_IR[2:0]];
      ID_Rd_code <= IF_IR[2:0];
    end
    7'b000_0000: begin
      ID_opcode <= `AND;
      ID_Rm_Rs <= R[IF_IR[5:3]];
      ID_Rd <= R[IF_IR[2:0]];
      ID_Rd_code <= IF_IR[2:0];
    end
    7'b000_0100: begin
      ID_opcode <= `ASR_2;
      ID_Rm_Rs <= R[IF_IR[5:3]];
      ID_Rd <= R[IF_IR[2:0]];
      ID_Rd_code <= IF_IR[2:0];
    end
    7'b000_1110: begin
      ID_opcode <= `BIC;
      ID_Rm_Rs <= R[IF_IR[5:3]];
      ID_Rd <= R[IF_IR[2:0]];
      ID_Rd_code <= IF_IR[2:0];
    end
    7'b000_1011: begin
      ID_opcode <= `CMN;
      ID_Rm_Rs <= R[IF_IR[5:3]];
      ID_Rn <= R[IF_IR[2:0]];
    end
    7'b000_1010: begin
      ID_opcode <= `CMP_2;
      ID_Rm_Rs <= R[IF_IR[5:3]];
      ID_Rn <= R[IF_IR[2:0]];
    end
```

```verilog
7'b000_0001: begin
  ID_opcode <= `EOR;
  ID_Rm_Rs <= R[IF_IR[5:3]];
  ID_Rd <= R[IF_IR[2:0]];
  ID_Rd_code <= IF_IR[2:0];
end
7'b000_0010: begin
  ID_opcode <= `LSL_2;
  ID_Rm_Rs <= R[IF_IR[5:3]];
  ID_Rd <= R[IF_IR[2:0]];
  ID_Rd_code <= IF_IR[2:0];
end
7'b000_0011: begin
  ID_opcode <= `LSR_2;
  ID_Rm_Rs <= R[IF_IR[5:3]];
  ID_Rd <= R[IF_IR[2:0]];
  ID_Rd_code <= IF_IR[2:0];
end
7'b000_0111: begin
  ID_opcode <= `ROR;
  ID_Rm_Rs <= R[IF_IR[5:3]];
  ID_Rd <= R[IF_IR[2:0]];
  ID_Rd_code <= IF_IR[2:0];
end
7'b000_0110: begin
  ID_opcode <= `SBC;
  ID_Rm_Rs <= R[IF_IR[5:3]];
  ID_Rd <= R[IF_IR[2:0]];
  ID_Rd_code <= IF_IR[2:0];
end
7'b000_1101: begin
  ID_opcode <= `MUL;
  ID_Rm_Rs <= R[IF_IR[5:3]];
  ID_Rd <= R[IF_IR[2:0]];
  ID_Rd_code <= IF_IR[2:0];
end
7'b000_1111: begin
  ID_opcode <= `MVN;
  ID_Rm_Rs <= R[IF_IR[5:3]];
  ID_Rd <= R[IF_IR[2:0]];
  ID_Rd_code <= IF_IR[2:0];
end
```

```
7'b000_1001: begin
  ID_opcode <= `NEG;
  ID_Rm_Rs <= R[IF_IR[5:3]];
  ID_Rd <= R[IF_IR[2:0]];
  ID_Rd_code <= IF_IR[2:0];
end
7'b000_1100: begin
  ID_opcode <= `ORR;
  ID_Rm_Rs <= R[IF_IR[5:3]];
  ID_Rd <= R[IF_IR[2:0]];
  ID_Rd_code <= IF_IR[2:0];
end
7'b000_1000: begin
  ID_opcode <= `TST;
  ID_Rm_Rs <= R[IF_IR[5:3]];
  ID_Rn <= R[IF_IR[2:0]];
end
7'b1100_xxx: begin
  ID_opcode <= `LDR_2;
  ID_Rm_Rs <= R[IF_IR[8:6]];
  ID_Rn <= R[IF_IR[5:3]];
  ID_Rd <= R[IF_IR[2:0]];
  ID_Rd_code <= IF_IR[2:0];
end
7'b01_xxxxx: begin
  ID_opcode <= `LDR_3;
  ID_Rd <= R[IF_IR[10:8]];
  ID_Rd_code <= IF_IR[10:8];
  ID_imm_offset[7:0] <= IF_IR[7:0];
end
7'b1110_xxx: begin
  ID_opcode <= `LDRB_2;
  ID_Rm_Rs <= R[IF_IR[8:6]];
  ID_Rn <= R[IF_IR[5:3]];
  ID_Rd <= R[IF_IR[2:0]];
  ID_Rd_code <= IF_IR[2:0];
end
7'b1101_xxx: begin
  ID_opcode <= `LDRH_2;
  ID_Rm_Rs <= R[IF_IR[8:6]];
  ID_Rn <= R[IF_IR[5:3]];
  ID_Rd <= R[IF_IR[2:0]];
  ID_Rd_code <= IF_IR[2:0];
end
```

```verilog
          7'b1011_xxx: begin
            ID_opcode <= `LDRSB;
            ID_Rm_Rs <= R[IF_IR[8:6]];
            ID_Rn <= R[IF_IR[5:3]];
            ID_Rd <= R[IF_IR[2:0]];
            ID_Rd_code <= IF_IR[2:0];
          end
          7'b1111_xxx: begin
            ID_opcode <= `LDRSH;
            ID_Rm_Rs <= R[IF_IR[8:6]];
            ID_Rn <= R[IF_IR[5:3]];
            ID_Rd <= R[IF_IR[2:0]];
            ID_Rd_code <= IF_IR[2:0];
          end
          7'b1000_xxx: begin
            ID_opcode <= `STR_2;
            ID_Rm_Rs <= R[IF_IR[8:6]];
            ID_Rn <= R[IF_IR[5:3]];
            ID_Rd <= R[IF_IR[2:0]];
            ID_Rd_code <= IF_IR[2:0];
          end
          7'b1010_xxx: begin
            ID_opcode <= `STRB_2;
            ID_Rm_Rs <= R[IF_IR[8:6]];
            ID_Rn <= R[IF_IR[5:3]];
            ID_Rd <= R[IF_IR[2:0]];
            ID_Rd_code <= IF_IR[2:0];
          end
          7'b1001_xxx: begin
            ID_opcode <= `STRH_2;
            ID_Rm_Rs <= R[IF_IR[8:6]];
            ID_Rn <= R[IF_IR[5:3]];
            ID_Rd <= R[IF_IR[2:0]];
            ID_Rd_code <= IF_IR[2:0];
          end
          default: ID_opcode <= `UNDEF;
        endcase
      end // of case IF_IR[15:13] = 3'b010
      3'b011: begin // load/store word/byte immediate offset
        case (IF_IR[12:11])
          2'b01: begin
            ID_opcode <= `LDR_1;
            ID_imm_offset[4:0] <= IF_IR[10:6];
            ID_Rn <= R[IF_IR[5:3]];
            ID_Rd <= R[IF_IR[2:0]];
            ID_Rd_code <= IF_IR[2:0];
          end
```

```verilog
            2'b11: begin
              ID_opcode <= `LDRB_1;
              ID_imm_offset[4:0] <= IF_IR[10:6];
              ID_Rn <= R[IF_IR[5:3]];
              ID_Rd <= R[IF_IR[2:0]];
              ID_Rd_code <= IF_IR[2:0];
            end
            2'b00: begin
              ID_opcode <= `STR_1;
              ID_imm_offset[4:0] <= IF_IR[10:6];
              ID_Rn <= R[IF_IR[5:3]];
              ID_Rd <= R[IF_IR[2:0]];
              ID_Rd_code <= IF_IR[2:0];
            end
            default: begin // == case 2'b10
              ID_opcode <= `STRB_1;
              ID_imm_offset[4:0] <= IF_IR[10:6];
              ID_Rn <= R[IF_IR[5:3]];
              ID_Rd <= R[IF_IR[2:0]];
              ID_Rd_code <= IF_IR[2:0];
            end
          endcase
      end // of case IF_IR[15:13] = 3'b011
      3'b100: begin // other load/store
        case (IF_IR[12:11])
          2'b11: begin
            ID_opcode <= `LDR_4;
            ID_Rd <= R[IF_IR[10:8]];
            ID_Rd_code <= IF_IR[10:8];
            ID_imm_offset[7:0] <= IF_IR[7:0];
          end
          2'b01: begin
            ID_opcode <= `LDRH_1;
            ID_imm_offset[4:0] <= IF_IR[10:6];
            ID_Rn <= R[IF_IR[5:3]];
            ID_Rd <= R[IF_IR[2:0]];
            ID_Rd_code <= IF_IR[2:0];
          end
          2'b10: begin
            ID_opcode <= `STR_3;
            ID_Rd <= R[IF_IR[10:8]];
            ID_Rd_code <= IF_IR[10:8];
            ID_imm_offset[7:0] <= IF_IR[7:0];
          end
```

```verilog
        default: begin // == case 2'b00
          ID_opcode <= `STRH_1;
          ID_imm_offset[4:0] <= IF_IR[10:6];
          ID_Rn <= R[IF_IR[5:3]];
          ID_Rd <= R[IF_IR[2:0]];
          ID_Rd_code <= IF_IR[2:0];
        end
      endcase
    end // of case IF_IR[15:13] = 3'b100
    3'b101: begin // add to SP/PC or Miscellaneous
      case (IF_IR[12:11])
        2'b00: begin // add to PC and store in Rd
          ID_opcode <= `ADD_5;
          ID_Rd <= R[IF_IR[10:8]];
          ID_Rd_code <= IF_IR[10:8];
          ID_imm_offset[7:0] <= IF_IR[7:0];
        end
        2'b01: begin // add to SP and store in Rd
          ID_opcode <= `ADD_6;
          ID_Rd <= R[IF_IR[10:8]];
          ID_Rd_code <= IF_IR[10:8];
          ID_imm_offset[7:0] <= IF_IR[7:0];
        end
        2'b10: begin // add imm_7 to SP
          case (IF_IR[10:8])
            3'b000: begin
              if (IF_IR[7] == 1'b0) begin
                ID_opcode <= `ADD_7;
                ID_imm_offset[6:0] <= IF_IR[6:0];
              end
              else begin // IF_IR[7] == 1'b1
                ID_opcode <= `SUB_4;
                ID_imm_offset[6:0] <= IF_IR[6:0];
              end
            end
            3'b100: begin
              ID_opcode <= `PUSH_R0;
              ID_imm_offset[7:0] <= IF_IR[7:0];
            end
            3'b101: begin
              ID_opcode <= `PUSH_R1;
              ID_imm_offset[7:0] <= IF_IR[7:0];
            end
            default: ID_opcode <= `UNDEF;
          endcase
        end // of case IF_IR[12:11] = 2'b10
```

```verilog
      default: begin // == case 2'b11
        case (IF_IR[10:8])
          3'b100: begin
            ID_opcode <= `POP_R0;
            ID_imm_offset[7:0] <= IF_IR[7:0];
          end
          3'b101: begin
            ID_opcode <= `POP_R1;
            ID_imm_offset[7:0] <= IF_IR[7:0];
          end
          default: ID_opcode <= `UNDEF;
        endcase
      end // of case IF_IR[12:11] = 2'b11 (default)
    endcase
  end // of case IF_IR[15:13] = 3'b101
  3'b110: begin // load/store multiple, branch, interrupt
    case (IF_IR[12:11])
      2'b01: begin
        ID_opcode <= `LDMIA;
        ID_Rn <= R[IF_IR[10:8]];
        ID_imm_offset[7:0] <= IF_IR[7:0];
      end
      2'b00: begin
        ID_opcode <= `STMIA;
        ID_Rn <= R[IF_IR[10:8]];
        ID_imm_offset[7:0] <= IF_IR[7:0];
      end
      default: begin // has (IF_IR[12] == 1'b1)
        if (IF_IR[11:8] == 4'b1110) ID_opcode <= `UNDEF;
        else if (IF_IR[11:8] == 4'b1111) begin
          ID_opcode <= `SWI; // SWI used as HALT instruction
          ID_imm_offset[7:0] <= IF_IR[7:0];
        end
        else begin // others are valid conditional branch
          ID_opcode <= `B_1;
          ID_cond <= IF_IR[11:8];
          ID_imm_offset[7:0] <= IF_IR[7:0];
        end
      end // of case IF_IR[12] = 1'b1 (default)
    endcase
  end // of case IF_IR[15:13] = 3'b110
  3'b111: begin // uncond. branch or subroutine call (BL)
    case (IF_IR[12:11])
      2'b00: begin
        ID_opcode <= `B_2;
        ID_imm_offset <= IF_IR[10:0];
      end
```

```
            2'b10: begin
              ID_opcode <= `BL_BLX_H10;
              ID_imm_offset <= IF_IR[10:0];
            end
            2'b11: begin
              ID_opcode <= `BL_BLX_H11;
              ID_imm_offset <= IF_IR[10:0];
            end  // note: H=01 is not supported in T Variant 4
            default: ID_opcode <= `UNDEF;
          endcase
        end // of case IF_IR[15:13] = 3'b111
        default: ID_opcode <= `UNDEF; // added for completeness
      endcase
      // check for branch, and insert bubble if branch
      if (branch_taken)
        ID_opcode <= `UNDEF;
      // save other pipeline register values
      ID_PC <= IF_PC;
    end  // of ID stage
////////////////////////////////////////////////////////////////////
```

The EX stage performs the actual execution of the instruction. The bulk of the code in this stage again is enclosed within a set of large **case** blocks. However, the **case** blocks used here are simpler than in the ID stage, because the instruction opcode has already been translated to a fixed set of six bits in the ID_opcode pipeline register. In the beginning part of the code for this stage, the read and write control signals for the data memory are initialized. Then, if the branch_taken signal is asserted, the `UNDEF opcode is stored into the EX_opcode pipeline register. A similar action is taken in the ID stage. The idea is that in this stage (if the instruction being processed is a branch instruction that results in a branch), instructions in the IF and ID stages (which will move into the ID and EX stages, respectively) during the next clock cycle must be nullified by asserting branch_taken. If branch_taken is asserted at the beginning of this stage, it means that the previous instruction (which occupied this stage in the previous clock cycle) must have asserted branch_taken.

Some instructions can be executed fairly simply, while other instructions require a more complex behavioral description. Although a few THUMB instructions (such as the MUL multiply instruction) actually may require significantly more execution time (thereby implying multiple clock cycles) than others, all instructions are executed within this single EX stage for simplicity of implementation. Many instructions require the evaluation of condition-flag values (bits in the CSPR register). For those instructions, and for other instructions requiring sequential descriptions, two **case** blocks (one following the other) are used.

```
//////////////////////////////////////////////////////////////////
  // EX: Instruction Execution Stage
  always @(negedge reset_n or posedge clk) begin
    if (~reset_n) begin
      branch_taken <= 1'b0;  // initialize to branch not taken
      read_data_n <= 1'b1;   // disable data memory
      write_data_n <= 1'b1;
    end
    else begin  // on positive clock edge
      read_data_n <= 1'b1;   // set default values for data mem.
      write_data_n <= 1'b1;
      if (branch_taken) begin
        EX_opcode <= `UNDEF;
        branch_taken <= 1'b0;
      end
      else begin
        case (ID_opcode)  // 1st part of EX operations
          `ADC: ALU_out = ID_Rd + ID_Rm_Rs + C_Flag;
          `ADD_1: ALU_out = ID_Rn + ID_imm_offset[2:0];
          `ADD_2: ALU_out = ID_Rd + ID_imm_offset[7:0];
          `ADD_3: ALU_out = ID_Rn + ID_Rm_Rs;
          `ADD_5: EX_ALU_out <= (ID_PC & 33'hfffffffc)
            + {ID_imm_offset[7:0], 2'b00}; // 33 = `WORD_SIZE+1
          `ADD_6: EX_ALU_out <= SP + {ID_imm_offset[7:0], 2'b00};
          `ADD_7: SP <= SP + {ID_imm_offset[7:0], 2'b00};
          `AND: ALU_out = ID_Rd & ID_Rm_Rs;
          `ASR_1: begin
            if (ID_imm_offset[4:0] == 0) begin
              if (ID_Rm_Rs[`WORD_SIZE-1] == 0)
                ALU_out = 0;
              else
                ALU_out = 33'h1ffffffff; //(`WORD_SIZE+1) = 33
            end
            else begin  // imm != 0
              ALU_out = {1'b0, ID_Rm_Rs};
              for (i = 0;  i < `WORD_SIZE;  i = i + 1) begin
                if (i < ID_imm_offset[4:0]) begin
                  {ALU_out[`WORD_SIZE-1:0], ALU_out[`WORD_SIZE]} =
                    {ALU_out[`WORD_SIZE-1:0], ALU_out[`WORD_SIZE]} >> 1;
                  ALU_out[`WORD_SIZE-1] = ID_Rm_Rs[`WORD_SIZE-1];
```

```verilog
             end // of if
           end // of for
         end // of else if imm != 0
      end // of case `ASR_1
      `ASR_2: begin
        if (ID_Rm_Rs[7:0] < `WORD_SIZE) begin
          ALU_out = {1'b0, ID_Rd};
           for (i = 0;  i < `WORD_SIZE;  i = i + 1) begin
             if (i < ID_Rm_Rs[7:0]) begin
               {ALU_out[`WORD_SIZE-1:0], ALU_out[`WORD_SIZE]} =
                 {ALU_out[`WORD_SIZE-1:0], ALU_out[`WORD_SIZE]} >> 1;
               ALU_out[`WORD_SIZE-1] = ID_Rd[`WORD_SIZE-1];
             end // of if
           end // of for
         end // of if (Rs[7:0] < `WORD_SIZE)
         else begin // Rs[7:0] >= `WORD_SIZE
           if (ID_Rd[`WORD_SIZE-1] == 0)
             ALU_out = 0;
           else // Rd[`WORD_SIZE-1] == 1
             ALU_out = 33'h1ffffffff; // (`WORD_SIZE+1) = 33
         end // of else if Rs[7:0] >= `WORD_SIZE
      end // of case `ASR_2
      `B_1: begin  // conditional branch
        if (condition_passed(ID_cond)) begin
          for (i = `WORD_SIZE-1;  i > 8;  i = i - 1)
            ALU_out[i] = ID_imm_offset[7]; // sign-extension
          ALU_out[8:1] = ID_imm_offset[7:0];
          ALU_out[0] = 1'b0;
          branch_taken <= 1'b1;
          branch_target <= ID_PC+2 + ALU_out[`WORD_SIZE-1:0];
        end
      end // of case `B_1
      `B_2: begin  // unconditional branch
        for (i = `WORD_SIZE-1;  i > 11;  i = i - 1)
          ALU_out[i] = ID_imm_offset[10]; // sign-extension
        ALU_out[11:1] = ID_imm_offset[10:0];
        ALU_out[0] = 1'b0;
        branch_taken <= 1'b1;
        branch_target <= (ID_PC+2) + ALU_out[`WORD_SIZE-1:0];
      end // of case `B_2
```

```
`BIC: ALU_out[`WORD_SIZE-1:0] = ID_Rd & (~ID_Rm_Rs);
`BL_BLX_H10: begin
  for (i = `WORD_SIZE-1;  i > 22;  i = i - 1)
    ALU_out[i] = ID_imm_offset[10];
  ALU_out[22:12] = ID_imm_offset[10:0];
  for (i = 11;  i >= 0;  i = i - 1)
    ALU_out[i] = 1'b0;
  LR <= ID_PC + ALU_out[`WORD_SIZE-1:0];
end // of case `BL_BLX_H10
`BL_BLX_H11: begin
  ALU_out = LR + {ID_imm_offset[10:0], 1'b0};
  branch_taken <= 1'b1;
  branch_target <= ALU_out[`WORD_SIZE-1:0];
  LR <= ALU_out[`WORD_SIZE-1:0] | `h00000001;
end // of case `BL_BLX_H11
`CMN: ALU_out = ID_Rn + ID_Rm_Rs;
`CMP_1: ALU_out = ID_Rn - ID_imm_offset[7:0];
`CMP_2: ALU_out = ID_Rn - ID_Rm_Rs;
`EOR: ALU_out = ID_Rd ^ ID_Rm_Rs; // Exclusive-OR
//`LDMIA: null; // currently unimplemented
`LDR_1: ALU_out = ID_Rn + {ID_imm_offset[4:0], 2'b00};
`LDR_2, `LDRB_2, `LDRH_2, `LDRSB, `LDRSH:
        ALU_out = ID_Rn + ID_Rm_Rs;
`LDR_3: ALU_out = {ID_PC[`WORD_SIZE-1:2], 2'b00}
                  + {ID_imm_offset[7:0], 2'b00};
`LDR_4: ALU_out = SP + {ID_imm_offset[7:0], 2'b00};
`LDRB_1: ALU_out = ID_Rn + ID_imm_offset[4:0];
`LDRH_1: ALU_out = ID_Rn + {ID_imm_offset[4:0], 1'b0};
`LSL_1: ALU_out = {1'b0,ID_Rm_Rs} <<ID_imm_offset[4:0];
`LSL_2: if (ID_Rm_Rs[7:0] <= `WORD_SIZE)
          ALU_out = {1'b0, ID_Rd} << ID_Rm_Rs[7:0];
        else
          ALU_out = 0;
`LSR_1: {ALU_out[`WORD_SIZE-1:0], ALU_out[`WORD_SIZE]}
          = {ID_Rm_Rs, 1'b0} >> ID_imm_offset[4:0];
`LSR_2: if (ID_Rm_Rs[7:0] <= `WORD_SIZE)
          {ALU_out[`WORD_SIZE-1:0],ALU_out[`WORD_SIZE]}
            = {ID_Rd, 1'b0} >> ID_Rm_Rs[7:0];
        else
          ALU_out = 0;
`MOV_1: ALU_out[`WORD_SIZE-1:0] =
          {24'b0, ID_imm_offset[7:0]};
```

```verilog
`MUL: ALU_out[`WORD_SIZE-1:0] = ID_Rd * ID_Rm_Rs;
`MVN: ALU_out[`WORD_SIZE-1:0] = ~ID_Rm_Rs;
`NEG: ALU_out = 0 - {ID_Rm_Rs[`WORD_SIZE-1], ID_Rm_Rs};
`ORR: ALU_out[`WORD_SIZE-1:0] = ID_Rd | ID_Rm_Rs;
`POP_R0: begin // only POP of 1 register supported
  data_address <= SP;
  read_data_n <= 1'b0;
  SP <= SP + 4;
  EX_imm_offset[7:0] <= ID_imm_offset[7:0];
end // of case `POP_R0
//`POP_R1: null; // POP(PC) not supported yet
`PUSH_R0: begin // only PUSH of 1 register supported
  ALU_out[`WORD_SIZE-1:0] = SP - 4;
  found_i = 0;
  for (i = 0;  i < `REG_FILE_SIZE; i = i + 1)
    if (ID_imm_offset[i]) // at least 1 bit must = 1
      found_i = i;
  DR <= R[found_i];
  write_data_n <= 1'b0;
  data_address <= ALU_out[`WORD_SIZE-1:0];
  SP <= ALU_out[`WORD_SIZE-1:0];
  EX_imm_offset[7:0] <= ID_imm_offset[7:0];
end // of case `PUSH_R0
//`PUSH_R1: null; // PUSH(PC) not supported yet
`ROR: {ALU_out[`WORD_SIZE-1:0], ALU_out[`WORD_SIZE]} =
        {ID_Rd, 1'b0} >> ID_Rm_Rs[4:0];
`SBC: ALU_out = ID_Rd - ID_Rm_Rs - (~C_Flag);
//`STMIA: null; // not supported yet
`STR_1: ALU_out = ID_Rn + {ID_imm_offset[4:0], 2'b00};
`STR_2, `STRB_2, `STRH_2:
        ALU_out = ID_Rn + ID_Rm_Rs;
`STR_3: ALU_out = SP + {ID_imm_offset[7:0], 2'b00};
`STRB_1: ALU_out = ID_Rn + ID_imm_offset[4:0];
`STRH_1: ALU_out = ID_Rn + {ID_imm_offset[4:0], 1'b0};
`SUB_1: ALU_out = ID_Rn - ID_imm_offset[2:0];
`SUB_2: ALU_out = ID_Rn - ID_imm_offset[7:0];
`SUB_3: ALU_out = ID_Rn - ID_Rm_Rs;
`SUB_4: SP <= SP - {ID_imm_offset[6:0], 2'b00};
`SWI: begin // currently implemented as a HALT instr.
  branch_taken <= 1'b1;
  branch_target <= ID_PC - 2; // repeat SWI forever
end
```

```
    `TST: begin
      ALU_out = ID_Rn & ID_Rm_Rs;
      N_Flag <= ALU_out[`WORD_SIZE-1];
      Z_Flag <= (ALU_out[`WORD_SIZE-1:0] == 0);
    end
    default: ALU_out = 'bx;  // added for completeness
endcase
case (ID_opcode)  // 2nd part of EX operations
  `ADD_1: begin // note: (imm == 000) implies MOV_2
      EX_ALU_out[`WORD_SIZE-1:0] <=ALU_out[`WORD_SIZE-1:0];
      N_Flag <= ALU_out[`WORD_SIZE-1];
      Z_Flag <= (ALU_out[`WORD_SIZE-1:0] == 0);
      if (ID_imm_offset[2:0] != 3'b000) begin  // not MOV_2
        C_Flag <= ALU_out[`WORD_SIZE];
        V_Flag <=
          (ALU_out[`WORD_SIZE] != ALU_out[`WORD_SIZE-1]);
      end
  end
  `ADC, `ADD_2, `ADD_3, `CMN, `CMP_1, `CMP_2,
  `NEG, `SBC, `SUB_1, `SUB_2, `SUB_3, `SUB_4: begin
      EX_ALU_out[`WORD_SIZE-1:0] <= ALU_out[`WORD_SIZE-1:0];
      N_Flag <= ALU_out[`WORD_SIZE-1];
      Z_Flag <= (ALU_out[`WORD_SIZE-1:0] == 0);
      C_Flag <= ALU_out[`WORD_SIZE];
      V_Flag <=
        (ALU_out[`WORD_SIZE] != ALU_out[`WORD_SIZE-1]);
  end
  `ASR_1, `ASR_2, `LSL_1, `LSL_2, `LSR_2: begin
      EX_ALU_out[`WORD_SIZE-1:0] <=ALU_out[`WORD_SIZE-1:0];
      N_Flag <= ALU_out[`WORD_SIZE-1];
      Z_Flag <= (ALU_out[`WORD_SIZE-1:0] == 0);
      C_Flag <= ALU_out[`WORD_SIZE];
  end
  `LDR_1, `LDR_2, `LDR_3, `LDR_4, `LDRB_1, `LDRB_2,
  `LDRH_1, `LDRH_2, `LDRSB, `LDRSH: begin
      data_address <= ALU_out[`WORD_SIZE-1:0];
      read_data_n <= 1'b0;
  end
  `LSR_1: begin
      if (ID_imm_offset[4:0] == 0) begin
        C_Flag <= ALU_out[`WORD_SIZE-1];
        EX_ALU_out[`WORD_SIZE-1:0] <= 'b0;
      end
```

```verilog
                  else begin
                    C_Flag <= ALU_out[`WORD_SIZE];
                    EX_ALU_out[`WORD_SIZE-1:0] <=
                      ALU_out[`WORD_SIZE-1:0];
                  end
                  N_Flag <= ALU_out[`WORD_SIZE-1];
                  Z_Flag <= (ALU_out[`WORD_SIZE-1:0] == 0);
                end // of case `LSR_1
              `AND, `BIC, `EOR, `MOV_1, `MUL, `MVN, `ORR: begin
                EX_ALU_out[`WORD_SIZE-1:0] <=
                  ALU_out[`WORD_SIZE-1:0];
                N_Flag <= ALU_out[`WORD_SIZE-1];
                Z_Flag <= (ALU_out[`WORD_SIZE-1:0] == 0);
              end
              `ROR: begin
                if (ID_Rm_Rs[7:0] == 0)
                  EX_ALU_out[`WORD_SIZE-1:0] <= ID_Rd;
                else if (ID_Rm_Rs[4:0] == 0) begin
                  C_Flag <= ALU_out[`WORD_SIZE-1];
                  EX_ALU_out[`WORD_SIZE-1:0] <= ID_Rd;
                end
                else begin // Rs[4:0] > 0
                  C_Flag <= ALU_out[`WORD_SIZE];
                  EX_ALU_out[`WORD_SIZE-1:0] <=
                    ALU_out[`WORD_SIZE-1:0];
                end
                N_Flag <= ALU_out[`WORD_SIZE-1];
                Z_Flag <= (ALU_out[`WORD_SIZE-1:0] == 0);
              end // of case `ROR
              `STR_1, `STR_2, `STR_3, `STRB_1, `STRB_2, `STRH_1,
              `STRH_2: begin
                data_address <= ALU_out[`WORD_SIZE-1:0];
                DR <= ID_Rd;
                write_data_n <= 1'b0;
              end
              default: EX_ALU_out <= 'bx;
            endcase // of second case
            // then save values into pipeline registers
            EX_opcode <= ID_opcode;
            EX_Rd_code <= ID_Rd_code;
          end // of else (branch not taken)
        end // of else (posedge clk)
      end // of EX stage
      ///////////////////////////////////////////////////////////
```

In the final WB stage, execution results are written into their corresponding destination registers. Some instructions do not require the write back of results into destination registers. For those instructions and the `UNDEF instruction, no action needs to be taken. For other instructions, the execution result (typically the EX_ALU_out register value), possibly padded with 0's, is stored into the register file at the address indicated by EX_Rd_code. An exception to this rule is the `POP_R0 instruction, for which the destination-register value must be computed separately by examining the bits of its offset field.

```
/////////////////////////////////////////////////////////////////////
  // WB: Write Back Stage (final stage)
  always @(negedge reset_n or posedge clk) begin
    if (~reset_n)
      for (i = 0;  i < `REG_FILE_SIZE;  i = i + 1)
        R[i] <= 'hffffffff;  // initialize general registers
    else case (EX_opcode)
      `ADC, `ADD_1, `ADD_2, `ADD_3, `ADD_5, `ADD_6, `AND,
      `ASR_1, `ASR_2, `BIC, `EOR, `LSL_1, `LSL_2, `LSR_1,
      `LSR_2, `MOV_1, `ROR,
      `SBC, `SUB_1, `SUB_2, `SUB_3, `SUB_4:
        R[EX_Rd_code] <= EX_ALU_out;  // write back to Rd
      `LDR_1, `LDR_2, `LDR_3, `LDR_4:
        R[EX_Rd_code] <= data;  // read data from data memory
      `LDRB_1, `LDRB_2: R[EX_Rd_code] <= {24'b0, data[7:0]};
      `LDRH_1, `LDRH_2: R[EX_Rd_code] <=
                  {16'b0, data[`HWORD_SIZE-1:0]};
      `LDRSB: if (data[7])
                R[EX_Rd_code] <= {24'hffffff, data[7:0]};
              else
                R[EX_Rd_code] <= {24'h000000, data[7:0]};
      `LDRSH: if (data[`HWORD_SIZE-1])
                R[EX_Rd_code] <=                // 16 = `HWORD_SIZE
                  {16'hffff, data[`HWORD_SIZE-1:0]};
              else
                R[EX_Rd_code] <=                // 16 = `HWORD_SIZE
                  {16'h0000, data[`HWORD_SIZE-1:0]};
      `POP_R0: begin
        found_ex_i = 0;
        for (ex_i = 0;  ex_i < `REG_FILE_SIZE;  ex_i = ex_i + 1)
          if (EX_imm_offset[ex_i]) // at least 1 bit must = 1
            found_ex_i = ex_i;
        R[found_ex_i] <= data;
      end // of case `POP_R0
```

```
      `UNDEF: found_ex_i = 32'bx;
      default: found_ex_i = 32'bx;
    endcase
  end // of WB stage
endmodule
/////////////////////////////////////////////////////////////////////
```

## Test Bench Code

The test bench for the `thumb` module is different from previous test benches in that it needs to model memory devices in order to simulate and test the target circuit. Since the target circuit is a microprocessor, it functions by repeatedly fetching instructions from memory and then executing those instructions. Thus, in order to test a microprocessor, we must have a memory interface feeding the microprocessor the instructions and data that it asks for.

The description of an accurate memory model can be extremely complex, since typical memory devices such as EPROMs, SRAMs, and DRAMs have complicated behavioral characteristics. However, since this memory device only needs to be used as part of a test bench, a simplified memory model that supplies the requested data after a suitable read delay is sufficient. In addition, since the modeling of a large memory device can consume inordinate amounts of simulation memory and time, reduced-size 256-byte memories will be used for both the instruction memory and the data memory.

As stated in the previous chapters, a test bench consists of connections to the unit under test (UUT), an **always** block used to generate the test vectors, and an **always** block used to check the results of the application of those test vectors. In this case, test-vector generation simply requires us to create reset and clock input signals—the actual test itself is controlled by the contents of the instruction memory. Test vector checking simply is handled by waiting an appropriate number of time units (which can be determined by performing preliminary simulations) and then checking whether the expected outputs appear on the address and data busses. By including "store" instructions in the program stored in the instruction memory, it is possible to observe the values of specific registers through the output data lines. Note that the same mechanism can be used during actual hardware prototype testing in order to observe internal register values.

```
/////////////////////////////////////////////////////////////////////
// MODULE: Test bench for pipelined Thumb microproc: tb_thumb.v
// Author: Sunggu Lee
// Created: ...
// Last Modified: ...
// Description: Tests the module "thumb.v".

// DEFINITIONS
`timescale 1ns/1ns
`define PERIOD1 100      // assume system clock cycle of 10MHz
`define READ_DELAY 80    // delay before memory data is ready
```

```verilog
`define WRITE_DELAY 80    // delay in writing to memory
`define STABLE_TIME 10    // time data is stable after end-of-read
`define MEMORY_SIZE 256   // size of reduced memory is 2^8 words
                          // - use only 8 lowest bits of address
`include "thumb_defs.vh" // include common defs

// MODULE DEFINITION
module tb_thumb();

  // SIGNAL DECLARATIONS for chip inputs and outputs
  wire read_instruction_n;  // control read from instruction mem
  wire [`WORD_SIZE-1:0] instruction_address; // address of instr
  wire [`HWORD_SIZE-1:0] instruction; // current instruction
  wire read_data_n;             // control read from data memory
  wire write_data_n;            // control write to data memory
  wire [`WORD_SIZE-1:0] data_address; // address of data
  wire [`WORD_SIZE-1:0] data;         // current data
  reg reset_n;    // active-low RESET signal
  reg clk;        // clock signal

  // SIGNAL DECLARATIONS for signals being used internally
  reg [`WORD_SIZE-1:0] output_instruction; // instr. memory outp.
  reg [`WORD_SIZE-1:0] output_data;        // data memory output
  reg [`WORD_SIZE-1:0] write_data;         // data to be written

  // instantiate the unit under test
  thumb UUT (read_instruction_n, instruction_address, instruction,
          read_data_n, write_data_n, data_address, data,
          reset_n, clk);

  // initialize inputs
  initial begin
    clk = 0;              // set initial clock value

    reset_n = 1;          // generate a LOW pulse for reset_n
    #(`PERIOD1/4) reset_n = 0;
    #(`PERIOD1 * 2) reset_n = 1;
  end

  // generate the clock
  always #(`PERIOD1/2)clk = ~clk; // period = `PERIOD1

  // model the instruction and data memory devices
  reg [`HWORD_SIZE-1:0] instruction_memory [0:`MEMORY_SIZE-1];
  reg [`WORD_SIZE-1:0] data_memory [0:`MEMORY_SIZE-1];
```

```verilog
// model the read process for the instruction memory device
assign instruction = read_instruction_n ? 'bz
                                            : output_instruction;
always begin

  if (read_instruction_n == 0) begin

    #`READ_DELAY;  // assume no spurious address changes

    output_instruction =
      instruction_memory[instruction_address[7:0]];

    wait ((read_instruction_n == 1) ||

          (output_instruction !=

            instruction_memory[instruction_address[7:0]]));

  end

  else begin  // end of read

    #`STABLE_TIME;

    output_instruction = `WORD_SIZE'bz;

    wait (read_instruction_n == 0);

  end

end  // of always block for instruction memory read


// model the read process for the data memory device
assign data = read_data_n ? `WORD_SIZE'bz : output_data;
always begin

  if (read_data_n == 0) begin

    #`READ_DELAY;  // assume no spurious address changes

    output_data = data_memory[data_address[7:0]];

    wait ((read_data_n == 1) ||

          (output_data != data_memory[data_address[7:0]]));

  end

  else begin  // end of read

    #`STABLE_TIME;

    output_data = `WORD_SIZE'bz;

    wait (read_data_n == 0);

  end

end  // of always block for data memory read
```

```verilog
// model the write process for the data memory device
always begin
  wait (write_data_n == 0);
  write_data = data;
  wait ((write_data_n == 1) || (data != write_data));
  if (write_data_n == 1) begin  // wait for write enable = '1'
    #`WRITE_DELAY;
    data_memory[data_address[7:0]] = write_data;
  end
  else  // data != write_data (data has changed)
    write_data = data;
end  // of always block for data memory write


// store programs and data in the instruction and memories
initial begin
  instruction_memory[0] = 16'h2100; // MOV r1, #0 (R[1] <- 0)
  instruction_memory[2] = 16'h2200; // MOV r2, #0 (R[2] <- 0)
  instruction_memory[4] = 16'h20fc; // MOV r0, #fc (R[0] <- fc)
  instruction_memory[6] = 16'h2909; // CMP r1, #9 (R[1] == 9)?
  instruction_memory[8] = 16'hda04; // BGE 0x14(if >=, goto 20)
  instruction_memory[10] = 16'he001;// B    0x10(goto 0x10 = 16)
  instruction_memory[12] = 16'h3101;// ADD r1, #1 (R[1] += 1)
  instruction_memory[14] = 16'he7fa;// B    0x6  (goto 0x4 = 6)
  instruction_memory[16] = 16'h1852;// ADD r2, r2, r1
  instruction_memory[18] = 16'he7fb;// B    0xc  (goto 0xc = 12)
  instruction_memory[20] = 16'h6042;// STR r2,r0,1*4(data:0x24)
  instruction_memory[22] = 16'hdf00;// SWI(to halt the program)
  // instruction_memory[22] = 16'h4770;// BX  r14  (goto R[14])
  // last instruction in original assembly is a return to main
end


// test output of program to verify proper operation of circuit
initial begin
  #15700; // this is the time when the program output is avail.
  if ( (data_address === 32'h00000100) &&
       (data === 32'h00000024) )
    $display("Data address (0x100) and data (0x24) correct.");
  else
    $display("ERROR: data address = %0x and data = %0x.",
             data_address, data);
  #500;  $display("Simulation completed at time %0t.", $time);
  $finish;  // terminate simulation after 16.2 microseconds
end


endmodule
////////////////////////////////////////////////////////////////
```

### 9.4.2 Test Bench Based Verification

The simulation of the HDL implementation of the pipeline THUMB microprocessor results in the simulation output shown in Figure 9.8. The console output indicates that the simulation completed without any errors, producing the expected data result ($R2 = 0x24$) on the data output lines. The simulation waveforms also can be checked to verify that circuit is behaving as expected.

One interesting aspect of this pipelined circuit, which also applies to the pipelined multiply-and-accumulate unit of Chapter 8, is that the result of an instruction (such as a branch) is only visible a few clock cycles (with the exact number depending on the pipeline structure) *after* the instruction has been fetched from instruction memory. Also, because control hazards are handled by nullifying incorrect instructions *after* they already have been fetched, addresses and instructions near branch points change in a delayed manner. For example, in the part of the simulation waveform shown in Figure 9.8, the instruction 0xEF7A at time 1750 ns, which corresponds to the B 0x6 (branch to location 0x6) instruction, causes a branch (as indicated by branch_taken and branch_address) to location 0x6 at time 2050 ns, and instructions fetched between 1750 ns and 2050 ns are nullified (by not storing the results of those instructions).

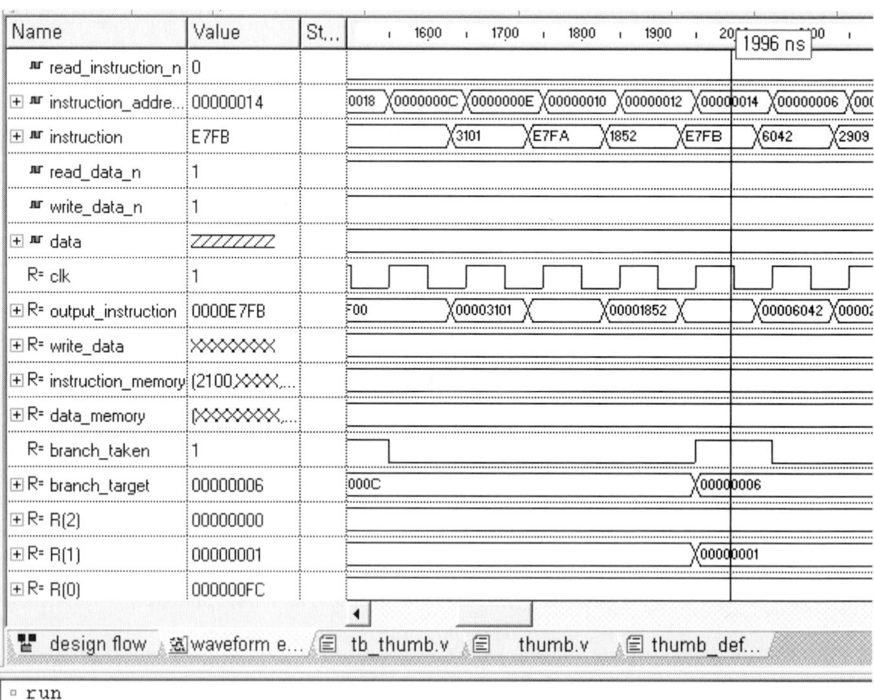

**Figure 9.8**
The simulation output for the pipelined THUMB microprocessor.

## 9.5  Chapter Review

- Microprocessors (which are essentially single-chip CPUs) and microcontrollers (which are essentially single-chip computers (including a CPU, memory, and I/O)) currently are used in all aspects of modern life, including factories, automobiles, personal digital assistants, and microwave ovens. The operation of a CPU can be described as a repetitive fetch-and-execute sequence in which an instruction is fetched from memory, decoded, executed, the next instruction fetched from memory, and so on. Interrupts change this operation sequence slightly by forcing a subroutine jump to an interrupt service routine immediately before the instruction fetch step.

- A RISC (reduced instruction-set computer) is a type of computer architecture that uses a simple and uniform instruction format, a load/store architecture (all instructions except "load" and "store" read and write data from/to registers instead of memory), a register file (large set of registers) for fast temporary data access and fast subroutine calls, and pipelining for fast instruction execution. ARM and THUMB architectures are commercial RISC architectures specifically designed for use with embedded applications such as cell phones and personal digital assistants. In particular, the THUMB architecture uses a 16-bit instruction format and a relatively simple instruction set of approximately sixty instructions. The THUMB instruction set, which is typical of modern microprocessors, consists of branch instructions (unconditional branch, conditional branch, and subroutine call), data-processing instructions (arithmetic, logical, shift, compare, and test), load and store instructions using various addressing modes and data sizes, and software-interrupt and breakpoint instructions.

- The design of an instruction pipeline requires attention to several issues beyond those dealt with for the design of arithmetic pipelines such as the multiply-accumulate unit of Chapter 8. An instruction pipeline must efficiently resolve pipeline hazards: situations in which two or more instructions in different stages of the pipeline interfere with each other. Pipeline hazards consist of structural hazards (contention for common resources), data hazards (contention for common data items) and control hazards (branches that cause sequentially prefetched instructions to become unnecessary). Structural hazards can be dealt with by replicating the contended resource (i.e., separate instruction and data memories can be used in order to handle load and store instructions, which must be executed while subsequent instructions are being fetched). Data hazards can be dealt with by using instruction reordering (reordering the order of assembly instructions in an assembly program without changing the results of the program). Control hazards can be dealt with by *nullifying* all instructions in the pipeline following a branch instruction as soon as the branch target address is determined.

■ This chapter presents the design of a pipelined microprocessor that implements almost all of the THUMB instruction set. The design techniques described are used to create a Verilog implementation of the pipelined THUMB microprocessor that uses a four-stage instruction pipeline implemented by concurrent **always** blocks. The test bench again is found to be a valuable tool for the verification of the HDL implementation. When written properly, the same method used in the test bench can be used to test the actual prototype after the target circuit is implemented in hardware.

## 9.6  Resources

There are numerous books on computer architecture, including [Hayes 2002]. A computer architecture textbook with a focus on pipelined RISC architectures is [Hennessy *et. al.*, 2002]. Information on the ARM and THUMB architectures are available via the web at http://www.arm.com, as well as other sites. [ARM 2000] is a publication detailing the instruction set of the ARM and THUMB architectures.

### 9.6.1  Bibliography

[9.1]  *ARM Architecture Reference Manual*, ARM Limited, 2000.

[9.2]  HAYES, J.P., *Computer Architecture and Organization, 3rd Ed.*, McGraw-Hill, New York, 2002.

[9.3]  HENNESSY, J.L., PATTERSON, D.A., and GOLDBERG, D. *Computer Architecture: A Quantitative Approach, 3rd Ed.*, Morgan Kaufmann, San Francisco, CA, 2002.

[9.4]  http://www.arm.com, home page for ARM Limited, the company that developed the ARM and THUMB architectures.

## 9.7  Problems

*P9.1. Using the complete state-machine design method as described in Chapter 5, design the datapath and control logic for a special-purpose microprocessor that implements only the ADD (one mode), BEQ (branch if Z flag $= 1$), and EOR instructions of the THUMB architecture. Optimize your design for the fact that *only* these three instructions are to be implemented.

P9.2. Using Verilog, implement the complete design for a special-purpose microprocessor that implements only the ADD (one mode), BEQ (branch if Z flag $= 1$), and EOR instructions of the THUMB architecture. Optimize your design for the fact that *only* these three instructions are to be implemented. Implement your design as an unpipelined design, test it using a test bench, and synthesize the circuit using a commercial synthesis tool. Show and explain a block diagram of the synthesized design.

**P9.3.** Referring to the THUMB instruction set listing in [ARM 2000] or Appendix A, derive the 16-bit instruction codes for the SBC (subtract with carry), TST (test), BL (subroutine call), and BCC (branch on carry clear) instructions. In constructing these instruction codes, it can be assumed that $H = 11$, $Rm = 2$, $Rn = 3$, $Rd = 4$, and all immediate operands are 0.

**P9.4.** Referring to the THUMB instruction set listing in [ARM 2000] or Appendix A, derive the 16-bit instruction codes for the BX (branch with exchange), ASR (arithmetic shift right, all modes), LSL (logical shift left, all modes), and MUL (multiply) instructions.

**P9.5.** Explain how the LR (Link Register) is used with subroutine calls and returns. Suppose that the stack was to be used, instead of LR, for this purpose. Explain how this could be done.

**P9.6.** It is mentioned that the compare instruction TST is implemented by performing subtraction and then setting the flag bits without saving the result of the subtraction itself. Discuss the pros and cons of this type of method versus simply implementing TST using a combinational logic comparator circuit.

**P9.7.** Consider the clock rate of a modern, commercial microprocessor. This type of microprocessor typically uses a system clock with a frequency of several hundred megahertz or even several gigahertz. Since a frequency of 1 GHz corresponds to a clock period of 1 ns, a DRAM module (which typically has an access delay of several tens of nanoseconds) clearly cannot be used to feed the instructions required in the IF stage of an instruction pipeline. How then can an instruction pipeline be operated at the speed of the system clock, assuming a 1 GHz system clock? What kinds of devices and mechanisms are required?

***P9.8.** Given an interrupt instruction such as the SWI instruction or even a hardware-generated interrupt, suggest at least two methods for providing the starting address of the corresponding interrupt-service routine.

**P9.9.** Hand assemble a program to compute the average of a set of $n$ values stored in data memory. Show the assembly instructions used, their 16-bit codes, and the addresses where those codes (and the data) are stored. Assume that the value $n$ and the $n$ data values are stored in data memory starting at address 0xa0. Refer to [ARM 2000] or Appendix A for the instruction codes for all assembly instructions used.

***P9.10.** Suggest a method for avoiding data hazards that does not require instruction reordering or delaying instruction execution. Show how this method could be implemented in hardware. (*Hint*: Try to use the execution result of an instruction in the EX stage as a direct input during the next clock cycle.)

***P9.11.** Besides the RAW (read-after-write) type of data hazard described in this chapter, what other types of data hazards are possible? For this question, assume that different instructions can require different numbers of clock cycles in the EX (execution) stage; thereby, possibly resulting in later instructions finishing earlier than the earlier instructions.

**P9.12.** The NOP (no operation) instruction can be used as a *general* solution method for resolving all types of hazards: structural, data, and control. Explain how this can be done. What are the drawbacks of this approach?

**P9.13.** PUSHes and POPs of more than one register were not implemented in the Verilog code shown in this chapter. What complications are introduced by the implementation of multiple-register PUSHes and POPs? Could a similar problem result with the MUL instruction? How? Suggest possible methods for dealing with this type of problem.

**P9.14.** Identify (using labeled arrows) all hazards in the assembly program shown in the tb_thumb test bench. How are these hazards handled in the thumb architecture implemented? What effect do these hazards have on the simulation output waveform?

**P9.15.** The method used to resolve control hazards in this chapter is inefficient in that it wastes time and pipeline resources. Suggest a more efficient method to resolve control hazards. How would this solution affect the assembly program and the simulation output for the tb_thumb test bench?

**P9.16.** Referring to the instruction pipeline design diagram of Figure 9.7, explain the need for each of the pipeline registers shown.

**P9.17.** Consider the pipeline design shown in Figure 9.7. It clearly can be seen that some of the stages will require more time than other stages. For example, the WB stage, which simply writes the computed result to the destination register, probably will require much less time than the IF stage (which requires a new instruction to be read from instruction memory) or the EX stage (which may require a fairly long execution time for some instructions). As stated in Chapter 8, a pipeline should be designed so that the delays in each stage are approximately equal. Let us assume that memory access (for both the instruction and data memories) requires $3D$ delay and instruction execution for all instructions except load and store require $2D$ delay, where $D$ is the delay required in the ID stage and in the WB stage. Then, how could the pipeline design of Figure 9.7 be modified to equalize the delays in each stage of the pipeline? What effects would such a design modification have?

**P9.18.** Modify the Verilog thumb module to accommodate the LDMIA (Load Multiple Increment After) and STMIA (Store Multiple Increment After) THUMB instructions. What implementation problems are presented by these two instructions? Suggest several possible methods to solve these problems.

**P9.19.** Extend the tb_thumb test bench so that it includes tests to verify the read and write capabilities of the data memory. Draw the simulation waveform for this part of the test bench.

***P9.20.** Extend the test bench shown in the Verilog module tb_thumb so that it includes tests for *all* of the instructions implemented. Be sure to include a method of verifying the correctness of your results (at the output pins) for all instructions tested.

# THUMB Instruction Set Listing

This appendix lists all of the instructions of the THUMB instruction set. This material is a summary of the information presented in the *ARM Architecture Reference Manual*, which can be obtained by contacting ARM Limited (www.arm.com). The instructions are listed in alphabetical order, with the instruction code and a short description given for each instruction. Instruction parameters are denoted using the ⟨*param*⟩ notation, where ⟨*param*⟩ is to be replaced with an actual operand or data value. Immediate operands are denoted using #immedX, where X is the number of bits occupied by the immediate operand.

## A.1 ADC ⟨Rd⟩, ⟨Rm⟩

```
Bits:   15 14 13 12 11 10  9  8  7  6  5  4  3  2  1  0
Code:    0  1  0  0  0  0  0  1  0  1|   Rm   |   Rd
```

This is an "Add with Carry" instruction. It adds the data in the register ⟨*Rd*⟩, with the data in the register ⟨*Rm*⟩, and the carry (C) flag bit, and stores the result in ⟨*Rd*⟩. This instruction affects the N, Z, C, and V condition-code flags.

Let us consider an example of the usage of this instruction. Suppose that the 64-bit data in {*R2*, *R3*} (high 32 bits in register *R2* and low 32 bits in register *R3*) is to be added with the 64-bit data in {*R6*, *R7*} (high 32 bits in register *R6* and low 32 bits in register *R7*). This can be done by first adding *R3* with *R7*, which will set the C flag bit if there is a carry-out, and then executing "ADC R2, R6".

## A.2 ADD ⟨Rd⟩, ⟨Rn⟩, #immed3

```
Bits:   15 14 13 12 11 10  9  8  7  6  5  4  3  2  1  0
Code:    0  0  0  1  1  1  0| immed3 |   Rn   |   Rd
```

This is an "Add" instruction. It simply adds the data in the register ⟨*Rn*⟩, with the 3-bit immediate data *immed3*, and then stores the result in ⟨*Rd*⟩. Note that since it is a simple ADD and not an ADC, it does not add the carry (C) flag bit. It affects the N, Z, C, and V condition-code flags.

## A.3  ADD $\langle Rd \rangle$, immed8

```
Bits:  15 14 13 12 11 10  9  8  7  6  5  4  3  2  1  0
Code:   0  0  1  1  0|  Rd   |         immed8
```

This "Add" instruction adds the 8-bit immediate data *immed8* with the data in register $\langle Rd \rangle$ and stores the result in $\langle Rd \rangle$. It affects the N, Z, C, and V condition-code flags.

## A.4  ADD $\langle Rd \rangle$, $\langle Rn \rangle$, $\langle Rm \rangle$

```
Bits:  15 14 13 12 11 10  9  8  7  6  5  4  3  2  1  0
Code:   0  0  0  1  1  0  0|  Rm   |  Rn   |  Rd
```

This "Add" instruction adds the data in register $\langle Rm \rangle$ with the data in register $\langle Rn \rangle$ and stores the result in register $\langle Rd \rangle$. It affects the N, Z, C, and V condition-code flags.

## A.5  ADD $\langle Rd \rangle$, $\langle Rm \rangle$

```
Bits:  15 14 13 12 11 10  9  8  7  6  5  4  3  2  1  0
Code:   0  1  0  0  0  1  0  0|H1|H2|  Rm   |  Rd
```

This "Add" instruction adds the data in register $\langle Rd \rangle$, with the data in register $\langle Rm \rangle$, and stores the result in register $\langle Rd \rangle$. The condition codes (flags) are not affected. A special feature of this instruction is that the registers $\langle Rd \rangle$ and $\langle Rm \rangle$ can range from $R0$ to $R15$. In order to accommodate 16 possible registers, the $H1$ bit is used as the most significant bit for the binary encoding of the $Rd$ register, while the $H2$ bit is used as the most significant bit for the binary encoding of the $Rm$ register.

Note that this instruction is not implemented in our HDL code, since the instruction uses the high registers $R8$ through $R15$, which we did not wish to support.

## A.6  ADD $\langle Rd \rangle$, PC, #immed8 $\times$ 4

```
Bits:  15 14 13 12 11 10  9  8  7  6  5  4  3  2  1  0
Code:   1  0  1  0  0|  Rd   |         immed8
```

This "Add" instruction adds the data in the Program Counter, (PC), with the immediate value *immed8* $\times$ 4, and stores the result in the register $\langle Rd \rangle$. Also, the two least significant bits in the PC are ANDed with 0 before storing the result in $\langle Rd \rangle$. This instruction is used to create and store a 32-bit word-aligned PC-relative address (the address is a multiple of 4). Note also that the condition codes (flags) are not affected.

## A.7   ADD ⟨Rd⟩, SP, #immed8 × 4

```
Bits:  15 14 13 12 11 10  9  8  7  6  5  4  3  2  1  0
Code:   1  0  1  0  1|  Rd   |        immed8
```

This "Add" instruction adds the data in the Stack Pointer (SP), with the immediate value *immed*8 × 4, and stores the result in the register ⟨*Rd*⟩.  This instruction is used to create and store a 32-bit word-aligned SP-relative address (the address is a multiple of 4). Note also that the condition code flags are not affected.

## A.8   ADD SP, #immed7 × 4

```
Bits:  15 14 13 12 11 10  9  8  7  6  5  4  3  2  1  0
Code:   1  0  1  1  0  0  0  0  0|        immed7
```

This "Add" instruction adds the data in the Stack Pointer (SP), with the immediate value *immed*7 × 4, and stores the result back into SP. Note that the condition code flags are not affected.

## A.9   AND ⟨Rd⟩, ⟨Rm⟩

```
Bits:  15 14 13 12 11 10  9  8  7  6  5  4  3  2  1  0
Code:   0  1  0  0  0  0  0  0  0  0|  Rm   |  Rd
```

This is a logical instruction that ANDs the data in register ⟨*Rd*⟩, with the data in register ⟨*Rm*⟩, and stores the result back into ⟨*Rd*⟩. This instruction updates the N and Z flags based on its final result.

## A.10   ASR ⟨Rd⟩, ⟨Rm⟩, #immed5

```
Bits:  15 14 13 12 11 10  9  8  7  6  5  4  3  2  1  0
Code:   0  0  0  1  0|   immed5    |  Rm   |  Rd
```

This is an arithmetic-shift instruction. An arithmetic shift differs from a logical shift in that an arithmetic shift of a data word retains the *sign* of the data word when performing the shift (by copying the sign bit into the *msb* position when shifting right), while a logical shift simply shifts in 0-bits into the empty slots left by the shift. The data in the register $\langle Rm \rangle$ is arithmetic-shifted right by the amount *immed5*, and then stored into the register $\langle Rd \rangle$. The shift amount ranges from 1 to 32, where a shift of 32 bits is denoted by *immed5* $= 00000$, and shifts from 1 to 31 bits are denoted by the binary value of *immed5*. This instruction updates the N, Z, and C flags based on the result of the shift. The most recent shifted-out bit is stored as the C flag.

## A.11 ASR $\langle$Rd$\rangle$, $\langle$Rs$\rangle$

```
Bits:  15 14 13 12 11 10  9  8  7  6  5  4  3  2  1  0
Code:   0  1  0  0  0  0  0  1  0  0|   Rs   |    Rd
```

This is an arithmetic-shift-right instruction that shifts register $\langle Rd \rangle$ by the amount indicated in register $\langle Rs \rangle$, where the value of the latter is interpreted as an unsigned binary number. An arithmetic-shift-right operation is a shift-right operation in which the sign bit is copied into the *msb* bit position as the register is shifted right. Thus, a positive value remains positive after the shift, and a negative value remains negative after the shift. This instruction updates the N, Z, and C flags based on the result of the shift. The C flag attains the value of the most recent shifted-out bit.

## A.12 B$\langle$cond$\rangle$, $\langle$target_address$\rangle$

```
Bits:  15 14 13 12 11 10  9  8  7  6  5  4  3  2  1  0
Code:   1  1  0  1|    cond    |     signed_immed8
```

This is a conditional-branch instruction that causes a branch to a PC-relative address if the condition being checked is true. The address to branch to is computed as $PC \leftarrow PC+$ the 32-bit sign-extended form of *signed_immed8* $\times 2$. The $n$-bit sign-extended form of a $k$-bit number is that number extended to $n$ bits by copying the sign bit (*msb*) of the original $k$-bit number into all bit positions to the left of the original sign bit. By using sign extension, a $k$-bit number can be extended to a larger number of bits while retaining its original 2's complement signed value. The condition to be checked for this conditional-branch instruction uses the condition code (flag) checks shown in Table A.1.

| Code | Mnemonic | Meaning | Flags Checked |
|------|----------|---------|---------------|
| 0000 | EQ | Equal | $Z = 1$ |
| 0001 | NE | Not equal | $Z = 0$ |
| 0010 | CS/HS | Carry set | $C = 1$ |
| 0011 | CC/LO | Carry clear | $C = 0$ |
| 0100 | MI | Minus | $N = 1$ |
| 0101 | PL | Plus | $N = 0$ |
| 0110 | VS | Overflow | $V = 1$ |
| 0111 | VC | No overflow | $V = 0$ |
| 1000 | HI | Unsigned higher | $(C = 1)$ and $(Z = 0)$ |
| 1001 | LS | Unsigned less or equal | $(C = 0)$ or $(Z = 1)$ |
| 1010 | GE | Greater or equal | $(N = V)$ |
| 1011 | LT | Less | $(N \neq V)$ |
| 1100 | GT | Greater | $(Z = 0)$ and $(N = V)$ |
| 1101 | LE | Less or equal | $(Z = 1)$ or $(N \neq V)$ |
| 1110 | AL | Always | None (unconditional) |

● ● ● ● ● ● ● ● ● ● ● ● ● ● ● ● · · ·

# A.13  B ⟨target_address⟩

```
Bits:  15 14 13 12 11 10  9  8  7  6  5  4  3  2  1  0
Code:   1  1  1  0  0|          signed_immed11
```

This is an unconditional-branch instruction that causes a branch to a PC-relative address. The address to branch to is computed as $PC \leftarrow PC+$ the 32-bit sign-extended form of *signed_immed*11 $\times$ 2. The *n*-bit sign-extended form of a *k*-bit number is that number extended to *n* bits by copying the sign bit (*msb*) of the original *k*-bit number into all bit positions to the left of original sign bit. By using sign extension, a *k*-bit number can be extended to a larger number of bits while retaining its original 2's complement signed value.

## A.14  BIC ⟨Rd⟩, ⟨Rm⟩

This is a "bit clear" instruction that performs the bit-wise AND of the value in register ⟨*Rd*⟩, with the complement (all bits inverted) of the value in register ⟨*Rm*⟩, and stores the result back into ⟨*Rd*⟩. The N and Z condition flags are affected by this instruction according to the final value of the ⟨*Rd*⟩ register.

## A.15  BKPT ⟨immed8⟩

```
Bits:  15 14 13 12 11 10  9  8  7  6  5  4  3  2  1  0
Code:   1  0  1  1  1  1  1  0|        immed8
```

This is a software-breakpoint instruction. The execution of this instruction results in a subroutine call to a breakpoint exception handler. The ⟨*immed8*⟩ value is an immediate value that is ignored by the hardware but can be used by a debugger to store additional information about the breakpoint.

Note that this instruction is not implemented in our HDL code, since it requires additional software and hardware support.

## A.16  BL ⟨target_address⟩ or BLX ⟨target_address⟩

```
Bits:  15 14 13 12 11 10  9  8  7  6  5  4  3  2  1  0
Code:   1  1  1|  H  |           offset11
```

These two instructions are subroutine-call instructions. BL is a "Branch with Link" instruction, used to call a THUMB subroutine, while BLX is a "Branch with Link and Exchange" instruction, used to call an ARM subroutine. BL or BLX is used depending on whether the subroutine being called uses THUMB or ARM assembly code. The "H" field is used to select variants of these instructions. To allow for a large offset for the target address, this instruction is translated by the assembler into a set of two consecutive THUMB instructions. The first THUMB instruction uses $H = 10$ (this mode causes *offset*11 to be used as the high part of the branch offset), and the second THUMB instruction uses $H = 01$ or $H = 11$ (this mode causes *offset*11 to be used as the low part of the branch offset).

The operation of this set of instructions varies depending on the $H$ value. If $H = 10$, then the Link Register (LR) is updated as $LR \leftarrow PC + sign\_extend$ (*offset*11 $<< 12$). If $H = 11$, then the Program Counter (PC) is updated as

$PC \leftarrow LR + sign\_extend(offset11 << 1)$ and the Link Register is updated as $LR \leftarrow$ (*address of next instruction*) ORed with 1 (bitwise-OR). If $H = 00$, then the instruction is an unconditional-branch instruction (the encoding is equivalent to B ⟨target_address⟩). The $H = 01$ mode is only supported in T variants of 5 and above and results in the same sequence of operations as the $H = 11$ mode, except that the T (trap) flag is also cleared.

Note that the BLX form of this instruction is not implemented in our HDL code, since only THUMB code, and not ARM code, is supported.

## A.17  BLX ⟨Rm⟩

```
Bits:  15 14 13 12 11 10  9  8  7  6  5  4  3  2  1  0
Code:   0  1  0  0  0  1  1  1  1|H2|   Rm   |   SBZ
```

This is a "Branch with Link and Exchange" (subroutine call) instruction that is used to branch to a THUMB or ARM subroutine at the address stored in the register ⟨Rm⟩ (with bit $H2$ used as the high bit of this register number). The T flag is also set to bit 0 of ⟨Rm⟩. Thus, the sequence of operations for this instruction arc $LR \leftarrow$ (*address of next instruction*) ORed with 1, $T\ flag \leftarrow Rm(0)$, and $PC \leftarrow Rm(31:1) << 1$.

Note that this instruction is not implemented in our HDL code since this instruction accesses ARM code and uses registers $R8$ through $R15$, which are two features that are not supported in our HDL code.

## A.18  BX ⟨Rm⟩

```
Bits:  15 14 13 12 11 10  9  8  7  6  5  4  3  2  1  0
Code:   0  1  0  0  0  1  1  1  0|H2|   Rm   |   SBZ
```

This is a "Branch with Exchange" instruction that is used to branch between ARM code and THUMB code. The instruction BX R414 commonly is used as a subroutine return instruction since $R14$ is the Link Register (LR), which typically stores the subroutine return address. As before, bit $H2$ contains the high bit and ⟨Rm⟩ contains the low three bits of the number of the register containing the branch address. This instruction first performs $T\ flag \leftarrow Rm(0)$ and then $PC \leftarrow Rm(31:1) << 1$.

Note that this instruction is not implemented in our HDL code, since this instruction accesses ARM code and uses registers $R8$ through $R15$, which are two features that are not supported in our HDL code.

## A.19  CMN ⟨Rn⟩, ⟨Rm⟩

```
Bits:  15 14 13 12 11 10  9  8  7  6  5  4  3  2  1  0
Code:   0  1  0  0  0  0  1  0  1  1|   Rm   |   Rd
```

This is a "Compare Negative" instruction that compares the value of register ⟨Rn⟩ with the negation of the value of register ⟨Rm⟩. Compare instructions typically are implemented as subtract instructions that do not store the subtraction result but only set the condition flags based on the subtraction result. Thus, this instruction first performs $Rn + Rm$ (subtraction of the negative of $Rm$ is the same as addition of $Rm$) and then sets the Z, N, C, and V condition-code flags based on the result of the addition.

## A.20  CMP ⟨Rn⟩, #immed8

```
Bits:  15 14 13 12 11 10  9  8  7  6  5  4  3  2  1  0
Code:   0  0  1  0  1|   Rn   |           immed8
```

This "Compare" instruction subtracts the immediate value $immed8$ from the value of the register ⟨Rn⟩ and then sets the Z, N, C, and V condition-code flags based on the result of the subtraction.

## A.21  CMP ⟨Rn⟩, ⟨Rm⟩

```
Bits:  15 14 13 12 11 10  9  8  7  6  5  4  3  2  1  0
Code:   0  1  0  0  0  0  1  0  1  0|   Rm   |   Rn
```

This "Compare" instruction subtracts the value in register ⟨Rm⟩ from the value in register ⟨Rn⟩ and then sets the Z, N, C, and V condition-code flags based on the result of the subtraction.

## A.22  CMP ⟨Rn⟩, ⟨Rm⟩

```
Bits:  15 14 13 12 11 10  9  8  7  6  5  4  3  2  1  0
Code:   0  1  0  0  0  1  0  1|H1|H2|    Rm    |    Rn
```

This "Compare" instruction subtracts the value in register ⟨Rm⟩ from the value in register ⟨Rn⟩ and then sets the Z, N, C, and V condition-code flags based on the result of the subtraction. The difference with the previous instruction is that registers $R0$ through $R15$ can be used in this instruction. The fields $H1$ and $H2$ store the high bits for registers ⟨Rn⟩ and ⟨Rm⟩, respectively. At least one of the two registers must be a register in the range $R8$ through $R15$.

Note that this instruction is not supported in our HDL code, since the registers $R8$ through $R15$ are not supported.

## A.23  EOR ⟨Rd⟩, ⟨Rm⟩

```
Bits:  15 14 13 12 11 10  9  8  7  6  5  4  3  2  1  0
Code:   0  1  0  0  0  0  0  0  0  1|    Rm    |    Rd
```

This is an "Exclusive-OR" instruction that forms the exclusive-OR of the contents of ⟨Rd⟩ with the contents of ⟨Rm⟩ and stores the result back into register ⟨Rd⟩. The operation performed is a bit-wise (bit by bit) logical operation. This instruction affects the Z and N condition-code flags.

## A.24  LDMIA ⟨Rn⟩!, ⟨registers⟩

```
Bits:  15 14 13 12 11 10  9  8  7  6  5  4  3  2  1  0
Code:   1  1  0  0  1|    Rn    |      register_list
```

This instruction, which stands for "Load Multiple Increment After," is a multiple load instruction that loads a non-empty set of registers from memory starting at the address specified in ⟨Rn⟩. LDMIA, when used with STMIA (store multiple), permits efficient block copies. The parameter ⟨registers⟩ is a list of the registers to be loaded, separated by commas and surrounded by braces (e.g., $R0$, $R5$, $R6$). The parameter ⟨Rn⟩ contains the address of the memory location where the data is to be loaded from. Consecutive 32-bit data words are stored into the registers listed in ⟨registers⟩, starting from the lowest numbered register. ⟨Rn⟩ then is updated to point to the next memory location after the last data item loaded. In the instruction encoding, *register_list* is an 8-bit vector, with bit $i$ $(0 \leq i \leq 7)$ set to 1 if register $i$ is included in $\langle registers \rangle$.

Note that this instruction is not supported in our HDL code, since this instruction requires multiple accesses to memory, which implies variable instruction execution times that result in complications in the design of the instruction pipeline.

## A.25  LDR $\langle Rd \rangle$, $\langle Rn \rangle$, #immed5 × 4

```
Bits:  15 14 13 12 11 10  9  8  7  6  5  4  3  2  1  0
Code:   0  1  1  0  1|    immed5    |   Rn   |   Rd
```

This "Load Register" instruction loads the register $\langle Rd \rangle$ with the 32-bit data contained in the memory location addressed by $\langle Rn \rangle + (immed5 \times 4)$.  For aligned memory word access, it is required that the two least significant bits of $\langle Rn \rangle$ be 0.

## A.26  LDR $\langle Rd \rangle$, $\langle Rn \rangle$, $\langle Rm \rangle$

```
Bits:  15 14 13 12 11 10  9  8  7  6  5  4  3  2  1  0
Code:   0  1  0  1  1  0  0|   Rm   |   Rn   |   Rd
```

This "Load Register" instruction loads the register $\langle Rd \rangle$ with the 32-bit data contained in the memory location addressed by $\langle Rn \rangle + \langle Rm \rangle$. For aligned memory word access, it is required that the two least significant bits of the address thus formed be 0.

## A.27  LDR $\langle Rd \rangle$, PC, #immed8 × 4

```
Bits:  15 14 13 12 11 10  9  8  7  6  5  4  3  2  1  0
Code:   0  1  0  0  1|   Rd   |         immed8
```

This "Load Register" instruction loads the register $\langle Rd \rangle$ with the 32-bit data contained in the memory location addressed by $(PC(31:2) << 2) + (immed8 \times 4)$.

## A.28  LDR ⟨Rd⟩, SP, #immed8 × 4

```
Bits:   15 14 13 12 11 10  9  8  7  6  5  4  3  2  1  0
Code:    1  0  0  1  1|  Rd   |         immed8
```

This form of the "Load Register" instruction is useful for accessing stack data. The register ⟨Rd⟩ is loaded with the 32-bit data stored at the address $SP + (immed8 \times 4)$. For aligned memory word access, it is required that the two least significant bits of the address thus formed be 0.

## A.29  LDRB ⟨Rd⟩, ⟨Rn⟩, #immed5

```
Bits:   15 14 13 12 11 10  9  8  7  6  5  4  3  2  1  0
Code:    0  1  1  1  1|    immed5   |  Rn   |  Rd
```

This "Load Register Byte" instruction loads the register ⟨Rd⟩ with the data byte at the memory location addressed by $⟨Rn⟩ + immed5$. The data byte retrieved is zero-extended to a 32-bit word (zeroes are inserted into the 24 most significant bit positions) before it is stored into the destination register.

## A.30  LDRB ⟨Rd⟩, ⟨Rn⟩, ⟨Rm⟩

```
Bits:   15 14 13 12 11 10  9  8  7  6  5  4  3  2  1  0
Code:    0  1  0  1  1  1  0|  Rm   |  Rn   |  Rd
```

This "Load Register Byte" instruction loads the register ⟨Rd⟩ with the data byte at the memory location addressed by $⟨Rn⟩+⟨Rm⟩$. The data byte retrieved is zero-extended to a 32-bit word (zeroes are inserted into the 24 most significant bit positions) before it is stored into the destination register.

## A.31  LDRH ⟨Rd⟩, ⟨Rn⟩, #immed5 × 2

```
Bits:   15 14 13 12 11 10  9  8  7  6  5  4  3  2  1  0
Code:    1  0  0  0  1|    immed5   |  Rn   |  Rd
```

This "Load Register Halfword" instruction loads the register ⟨Rd⟩ with the 16-bit data at the memory location addressed by $⟨Rn⟩+(immed5 \times 2)$. For aligned memory access, the address thus formed must be an even number. The 16-bit data retrieved is zero-extended to a 32-bit word (the most significant 16 bits are filled with zeroes) before it is stored into the destination register.

## A.32 LDRH ⟨Rd⟩, ⟨Rn⟩, ⟨Rm⟩

```
Bits:  15 14 13 12 11 10  9  8  7  6  5  4  3  2  1  0
Code:   0  1  0  1  1  0  1|   Rm   |   Rn   |   Rd
```

This "Load Register Halfword" instruction loads the register ⟨*Rd*⟩ with the 16-bit data at the memory location addressed by ⟨*Rn*⟩ + ⟨*Rm*⟩. For aligned memory access, the address thus formed must be an even number. The 16-bit data retrieved is zero-extended to a 32-bit word (zeroes are inserted into the 16 most significant bit positions) before it is stored into the destination register.

## A.33 LDRSB ⟨Rd⟩, ⟨Rn⟩, ⟨Rm⟩

```
Bits:  15 14 13 12 11 10  9  8  7  6  5  4  3  2  1  0
Code:   0  1  0  1  0  1  1|   Rm   |   Rn   |   Rd
```

This "Load Register Signed Byte" instruction loads the register ⟨*Rd*⟩ with the data byte at the memory location addressed by ⟨*Rn*⟩+⟨*Rm*⟩. The data byte retrieved is sign extended to a 32-bit word (the sign bit, bit 7 of the original data byte, is copied into the 24 most significant bit positions) before it is stored into the destination register.

## A.34 LDRSH ⟨Rd⟩, ⟨Rn⟩, ⟨Rm⟩

```
Bits:  15 14 13 12 11 10  9  8  7  6  5  4  3  2  1  0
Code:   0  1  0  1  1  1  1|   Rm   |   Rn   |   Rd
```

This "Load Register Signed Halfword" instruction loads the register ⟨*Rd*⟩ with the 16-bit data at the memory location addressed by ⟨*Rn*⟩ + ⟨*Rm*⟩. For aligned memory access, the address thus formed must be an even number. The 16-bit data retrieved is sign extended to a 32-bit word (the sign bit, bit 15 of the original data halfword, is copied into the 16 most significant bit positions) before it is stored into the destination register.

## A.35 LSL ⟨Rd⟩, ⟨Rm⟩, #immed5

```
Bits:  15 14 13 12 11 10  9  8  7  6  5  4  3  2  1  0
Code:   0  0  0  0  0|    immed5      |   Rm   |   Rd
```

This "Logical Shift Left" instruction shifts the value in the register ⟨*Rm*⟩ left by the amount *immed5* and then stores the result in register ⟨*Rd*⟩. Zeroes are shifted into the bit positions vacated by the shift, and the Z, N, and C condition-code flags are updated based on the result of the shift. In particular, the C flag contains the most recent shifted-out bit.

## A.36 LSL ⟨Rd⟩, ⟨Rs⟩

```
Bits:  15 14 13 12 11 10  9  8  7  6  5  4  3  2  1  0
Code:   0  1  0  0  0  0  0  0  1  0|   Rs   |   Rd
```

This "Logical Shift Left" instruction shifts the value in the register ⟨*Rd*⟩ left by the amount indicated in the least significant byte of ⟨*Rs*⟩. Zeroes are shifted into the bit positions vacated by the shift, and the Z, N, and C condition-code flags are updated based on the result of the shift. In particular, the C flag contains the most recent shifted-out bit.

## A.37 LSR ⟨Rd⟩, ⟨Rm⟩, #immed5

```
Bits:  15 14 13 12 11 10  9  8  7  6  5  4  3  2  1  0
Code:   0  0  0  0  1|    immed5      |   Rm   |   Rd
```

This "Logical Shift Right" instruction shifts the value in the register ⟨*Rm*⟩ right by the amount indicated in *immed5* and stores the result in register ⟨*Rd*⟩. The value $immed5 = 0$ is used to indicate a shift of 32 bits, while all other shift values are equal to the binary value of *immed5*. Zeroes are shifted into the bit positions vacated by the shift, and the Z, N, and C condition-code flags are updated based on the result of the shift. In particular, the C flag contains the most recent shifted-out bit.

## A.38  LSR ⟨Rd⟩, ⟨Rs⟩

```
Bits:  15 14 13 12 11 10  9  8  7  6  5  4  3  2  1  0
Code:   0  1  0  0  0  0  0  0  1  1|   Rs   |   Rd
```

This "Logical Shift Right" instruction shifts the value in the register ⟨Rd⟩ right by the amount indicated in the least significant byte of ⟨Rs⟩. Zeroes are shifted into the bit positions vacated by the shift, and the Z, N, and C condition-code flags are updated based on the result of the shift. In particular, the C flag contains the most recent shifted-out bit.

## A.39  MOV ⟨Rd⟩, #immed8

```
Bits:  15 14 13 12 11 10  9  8  7  6  5  4  3  2  1  0
Code:   0  0  1  0  0|   Rd   |       immed8
```

This "Move" instruction moves the 8-bit immediate value *immed8* into the register ⟨Rd⟩. The Z and N condition-code flags are updated based on the final ⟨Rd⟩ value.

## A.40  MOV ⟨Rd⟩, ⟨Rn⟩

```
Bits:  15 14 13 12 11 10  9  8  7  6  5  4  3  2  1  0
Code:   0  0  0  1  1  1  0  0  0  0|   Rn   |   Rd
```

This "Move" instruction copies the value in register ⟨Rn⟩ into the register ⟨Rd⟩. The Z and N condition-code flags are updated based on the final ⟨Rd⟩ value. Note that the encoding for this instruction is the same as the ADD Rd, Rn, #0 instruction.

## A.41  MOV ⟨Rd⟩, ⟨Rm⟩

```
Bits:  15 14 13 12 11 10  9  8  7  6  5  4  3  2  1  0
Code:   0  1  0  0  0  1  1  0|H1|H2|   Rm   |   Rd
```

This form of the "Move" instruction is used for moves from or to high registers (one the registers in the range $R8$ through $R15$). The destination register is indicated by ⟨Rd⟩ (least significant three bits) and $H1$ (most significant bit). The source register is indicated by ⟨Rm⟩ (least significant three bits) and $H2$ (most significant bit). For proper operation, at least one of $H1$ or $H2$ must be a 1. The condition-code flags are not affected by this instruction.

Note that this instruction is not supported in our HDL code, since the registers $R8$ through $R15$ are not supported.

## A.42  MUL ⟨Rd⟩, ⟨Rm⟩

```
Bits:  15 14 13 12 11 10  9  8  7  6  5  4  3  2  1  0
Code:   0  1  0  0  0  0  1  1  0  1|   Rm    |   Rd
```

This "Multiply" instruction multiplies the contents of register ⟨Rm⟩, with the contents of register ⟨Rd⟩, and stores the result in ⟨Rd⟩. Although the multiplication of two 32-bit numbers results in a 64-bit number in general, only the least significant 32 bits are stored into ⟨Rd⟩ by this instruction. The Z and N condition-code flags are updated according to the final result stored in ⟨Rd⟩.

## A.43  MVN ⟨Rd⟩, ⟨Rm⟩

```
Bits:  15 14 13 12 11 10  9  8  7  6  5  4  3  2  1  0
Code:   0  1  0  0  0  0  1  1  1  1|   Rm    |   Rd
```

This "Move NOT" instruction moves the complement (all bits inverted) of the value of ⟨Rm⟩ into the register ⟨Rd⟩. The Z and N condition-code flags are updated according to the final result stored in ⟨Rd⟩.

## A.44  NEG ⟨Rd⟩, ⟨Rm⟩

```
Bits:  15 14 13 12 11 10  9  8  7  6  5  4  3  2  1  0
Code:   0  1  0  0  0  0  1  0  0  1|   Rm    |   Rd
```

This "Negate" instruction stores $0 - ⟨Rm⟩$ into the register ⟨Rd⟩. The Z, N, C, and V condition-code flags are updated based on the result of the subtraction.

## A.45  ORR ⟨Rd⟩, ⟨Rm⟩

```
Bits:  15 14 13 12 11 10  9  8  7  6  5  4  3  2  1  0
Code:   0  1  0  0  0  0  1  1  0  0|   Rm    |   Rd
```

This "Logical OR" instruction computes the bit-wise OR of the value in ⟨Rd⟩, with the value in ⟨Rm⟩, and stores the result in register ⟨Rd⟩. The Z and N condition-code flags are updated based on the result of this logical operation.

## A.46  POP ⟨registers⟩

```
Bits:  15 14 13 12 11 10  9  8  7  6  5  4  3  2  1  0
Code:   1  0  1  1  1  1  0| R|     register_list
```

This "Pop Multiple Registers" instruction pops a sequence of 32-bit data items from the stack and stores them into a non-empty set of registers, which may include the registers $R0$ through $R7$ and $PC$. The registers into which the data items are to be popped are indicated by *register_list*, where a 1 in bit $i$ indicates that register $i$ is to loaded. The $R$ bit is set to 1 if the PC is to be loaded and to 0 otherwise. At least one register must be included in ⟨*registers*⟩. Depending on the bits that are set, the registers are loaded in order starting from $R0$ to $R7$ to $PC$.

Note that this instruction only is supported in our HDL code as a single POP instruction. The register corresponding to the first bit found, when scanning *register_list* from bit 0 to bit 7, is loaded from the stack and all other registers are left unaffected. This change is made since pops of a variable number of data items require a variable amount of execution time, which complicates the design of the instruction pipeline.

## A.47  PUSH ⟨registers⟩

```
Bits:  15 14 13 12 11 10  9  8  7  6  5  4  3  2  1  0
Code:   1  0  1  1  0  1  0| R|     register_list
```

This "Push Multiple Registers" instruction pushes a non-empty set of registers, which may include the registers $R0$ through $R7$ and $LR$, onto the stack. The registers to be pushed onto the stack are indicated by *register_list*, where a 1 in bit $i$ indicates that register $i$ is to be pushed. The $R$ bit is set to 1 if $LR$ is to be pushed onto the stack, and 0 otherwise. At least one register must be included in ⟨*registers*⟩. Depending on the bits that are set, the registers are pushed in order starting from $LR$ to $R7$ to $R0$.

Note that this instruction only is supported in our HDL code as a single PUSH instruction. The register corresponding to the first bit found, when scanning *register_list* from bit 0 to bit 7, is pushed onto the stack and all other registers are ignored. This change is made since pushes of a variable number of data items requires a variable amount of execution time, which complicates the design of the instruction pipeline.

# A.48  ROR ⟨Rd⟩, ⟨Rs⟩

```
Bits:  15 14 13 12 11 10  9  8  7  6  5  4  3  2  1  0
Code:   0  1  0  0  0  0  0  1  1  1|   Rs    |   Rd
```

This "Rotate Right" instruction rotates the number in the register ⟨Rd⟩ right by the amount indicated in the least significant byte of ⟨Rs⟩, and then stores the result in ⟨Rd⟩. Since this is a *rotate* instruction, each bit shifted out of the *lsb* position of ⟨Rd⟩ is stored back into the *msb* position (which is vacated by the shift operation) of ⟨Rd⟩. The Z, N, and C condition-code flags are updated based on the final value stored in ⟨Rd⟩. In particular, the C flag attains the value of the most recently shifted-out bit.

# A.49  SBC ⟨Rd⟩, ⟨Rm⟩

```
Bits:  15 14 13 12 11 10  9  8  7  6  5  4  3  2  1  0
Code:   0  1  0  0  0  0  0  1  1  0|   Rm    |   Rd
```

This "Subtract with Carry" instruction stores the value of ⟨Rd⟩ − ⟨Rm⟩ − (*not C*) into the ⟨Rd⟩ register, where $C$ refers to the value of the C condition-code flag. This instruction is useful for multiple word subtractions. The carry bit (C) behaves as the complement of the "borrow" after the right halves of two 64-bit numbers are subtracted from one another. The Z, N, C, and V condition-code flags are updated based on the result of the subtraction operation performed for this instruction. Note that the C flag will again be the complement of the "borrow" resulting from this subtraction.

# A.50  STMIA ⟨Rn⟩!, ⟨registers⟩

```
Bits:  15 14 13 12 11 10  9  8  7  6  5  4  3  2  1  0
Code:   1  1  0  0  0|   Rn    |     register_list
```

This "Store Multiple Increment After" instruction stores the contents of a non-empty set of registers into memory starting at the address specified in ⟨Rn⟩. Bit $i$ of *register_list* is set to 1 if register $i$ is to be stored into memory. The registers are stored in sequence from $R0$ to $R7$ (provided that the corresponding bit $i$ in *register_list* is 1), and ⟨Rn⟩ is then updated to point to the memory address after the location where the contents of the last register are stored.

Note that this instruction is not supported in our HDL code, since this instruction requires multiple accesses to memory, which implies variable instruction execution times that result in complications in the design of the instruction pipeline.

## A.51 STR ⟨Rd⟩, ⟨Rn⟩, #immed5 × 4

```
Bits:  15 14 13 12 11 10  9  8  7  6  5  4  3  2  1  0
Code:   0  1  1  0  0|    immed5    |   Rn   |   Rd
```

This "Store" instruction stores the 32-bit data in the register ⟨Rd⟩ into memory at the address specified by ⟨Rn⟩ + (immed5 × 4). For proper aligned memory access, the least significant two bits of the address must be equal to 0.

## A.52 STR ⟨Rd⟩, ⟨Rn⟩, ⟨Rm⟩

```
Bits:  15 14 13 12 11 10  9  8  7  6  5  4  3  2  1  0
Code:   0  1  0  1  0  0  0|   Rm   |   Rn   |   Rd
```

This "Store" instruction stores the 32-bit data in the register ⟨Rd⟩ into memory at the address specified by ⟨Rn⟩ + ⟨Rm⟩. For proper aligned memory access, the least significant two bits of the address must be equal to 0.

## A.53 STR ⟨Rd⟩, SP, #immed8 × 4

```
Bits:  15 14 13 12 11 10  9  8  7  6  5  4  3  2  1  0
Code:   1  0  0  1  0|   Rd   |        immed8
```

This "Store" instruction stores the 32-bit data in the register ⟨Rd⟩ into memory at the address specified by $SP + (immed8 × 4)$. For proper aligned memory access, the least significant two bits of the address must be equal to 0. This instruction is useful for storing items onto the stack.

## A.54 STRB ⟨Rd⟩, ⟨Rn⟩, #immed5

```
Bits:  15 14 13 12 11 10  9  8  7  6  5  4  3  2  1  0
Code:   0  1  1  1  0|    immed5    |   Rn   |   Rd
```

This "Store" instruction stores the least significant byte (8 bits) of the data in the register ⟨Rd⟩ into memory at the address specified by ⟨Rn⟩ + immed5.

## A.55   STRB ⟨Rd⟩, ⟨Rn⟩, ⟨Rm⟩

```
Bits:  15 14 13 12 11 10  9  8  7  6  5  4  3  2  1  0
Code:   0  1  0  1  0  1  0|   Rm    |   Rn   |   Rd
```

This "Store" instruction stores the least significant byte (8 bits) of the data in the register ⟨Rd⟩ into memory at the address specified by ⟨Rn⟩ + ⟨Rm⟩.

## A.56   STRH ⟨Rd⟩, ⟨Rn⟩, #immed5 × 2

```
Bits:  15 14 13 12 11 10  9  8  7  6  5  4  3  2  1  0
Code:   1  0  0  0  0|    immed5    |   Rn   |   Rd
```

This "Store" instruction stores the least significant 16 bits of the data in the register ⟨Rd⟩ into memory at the address specified by ⟨Rn⟩ + (immed5 × 2). For proper aligned memory access, the least significant bit of the address must be equal to 0.

## A.57   STRH ⟨Rd⟩, ⟨Rn⟩, ⟨Rm⟩

```
Bits:  15 14 13 12 11 10  9  8  7  6  5  4  3  2  1  0
Code:   0  1  0  1  0  0  1|   Rm    |   Rn   |   Rd
```

This "Store" instruction stores the least significant 16 bits of the data in the register ⟨Rd⟩ into memory at the address specified by ⟨Rn⟩ + ⟨Rm⟩. For proper aligned memory access, the least significant bit of the address must be equal to 0.

## A.58   SUB ⟨Rd⟩, ⟨Rn⟩, #immed3

```
Bits:  15 14 13 12 11 10  9  8  7  6  5  4  3  2  1  0
Code:   0  0  0  1  1  1  1| immed3 |   Rn   |   Rd
```

This "Subtract" instruction computes ⟨Rn⟩ − immed3 and stores the result in register ⟨Rd⟩. The Z, N, C, and V flags are updated based on the results of this subtraction operation.

## A.59   SUB ⟨Rd⟩, #immed8

```
Bits:  15 14 13 12 11 10  9  8  7  6  5  4  3  2  1  0
Code:   0  0  1  1  1|   Rd   |         immed8
```

This "Subtract" instruction computes $\langle Rd \rangle - immed8$ and stores the result in register $\langle Rd \rangle$. The Z, N, C, and V flags are updated based on the results of this subtraction operation.

## A.60   SUB ⟨Rd⟩, ⟨Rn⟩, ⟨Rm⟩

```
Bits:  15 14 13 12 11 10  9  8  7  6  5  4  3  2  1  0
Code:   0  0  0  1  1  0  1|   Rm   |   Rn   |   Rd
```

This "Subtract" instruction computes $\langle Rn \rangle - \langle Rm \rangle$ and stores the result in register $\langle Rd \rangle$. The Z, N, C, and V flags are updated based on the results of this subtraction operation.

## A.61   SUB SP, #immed7 × 4

```
Bits:  15 14 13 12 11 10  9  8  7  6  5  4  3  2  1  0
Code:   1  0  1  1  0  0  0  0  1|        immed7
```

This "Subtract" instruction computes $SP - (immed7 \times 4)$ and stores the result back into the SP register. This instruction is used to allocate memory on the top of the stack. It does not affect the condition-code flags.

## A.62   SWI ⟨immed8⟩

```
Bits:  15 14 13 12 11 10  9  8  7  6  5  4  3  2  1  0
Code:   1  1  0  1  1  1  1  1|         immed8
```

This is a "Software Interrupt" instruction that is used to call operating system service routines. The parameter $\langle immed8 \rangle$ is an 8-bit immediate value that is used by the operating system to determine the service being requested. This parameter simply is ignored by the CPU.

Note that this instruction is not supported, as described previously by our HDL code. The above implementation would require a close interface with an operating system, which is outside the scope of the CPU design being attempted. Instead, this "Software Interrupt" instruction simply is used as a "Halt" instruction, which suspends the operation of the CPU in our HDL code.

## A.63  TST ⟨Rn⟩, ⟨Rm⟩

```
Bits:  15 14 13 12 11 10  9  8  7  6  5  4  3  2  1  0
Code:   0  1  0  0  0  0  1  0  0  0|    Rm    |    Rn
```

This "Test" instruction forms the bit-wise AND of the contents of register ⟨*Rn*⟩ with the contents of register ⟨*Rm*⟩. The N and Z condition-code flags are then updated based on the result of this bit-wise AND operation. This instruction is used to check if a particular set of bits in a register includes 1 bits. Similar to a "Compare" (CMP) instruction, it does not store its operation result in a destination register; it simply updates the N and Z flags to indicate the result of its test.

## B.1 Condensed Digital Logic Review

**P1.1**  Presence or absence of light (optical communication); a sine-wave voltage waveform with different phases for logic 0 and logic 1.

**P1.3**  In CMOS technology (and in a few other technologies used to implement digital logic circuits), "natural" implementation of digital logic gates and devices result in inverted output values. Such devices can also be considered as having negative-logic inverted inputs and noninverted outputs. This latter viewpoint (negative logic) can help facilitate the analysis of circuits with several such devices chained together in series.

**P1.5**  Solution shown in the latter half of Section 3.4.1.

**P1.7**  A 2-input NOR gate with its two inputs tied together implements an inverter. A 2-input NOR gate with an inverter (created from a 2-input NOR gate) tied to the output of the first NOR gate implements a 2-input OR function. Lastly, based on DeMorgan's Theorem, a 2-input NOR gate with inverters tied to both inputs forms a 2-input AND gate. Thus, since inverters, OR gates, and AND gates can be implemented using 2-input NOR gates exclusively, the 2-input NOR is a universal gate.

**P1.9**  We can either connect the first four or the last four variables to the four select inputs of the 16:1 MUX. Let us choose the former ($a$ connected to $sel_3$, $b$ connected $sel_2$, etc.). Then, input $I0$ will be routed to the MUX output when $(a, b, c, d) = (0, 0, 0, 0)$. The minterms 0 ($=$ binary 00000) and 1 ($=$ binary 00001) correspond to this setting of the $a, b, c,$ and $d$ variables. Of these two possible minterms, only minterm 1 is included in the function $f$. Thus, $e$ is connected to the MUX input $I0$. Likewise, the $I1$ input will be routed to the MUX output with minterms 2 and 3. Since neither of these minterms are included in $f$, logic 0 (or ground) should be connected to MUX input $I1$. Similarly, since minterm 4 is not included in $f$ but minterm 5 is, $e$ should be connected to MUX input $I2$. Next, since minterm 6 is included in $f$ but minterm 7 is not, $\bar{e}$ should be connected $I3$. Continuing in this manner, logic 1 is connected to $I4$, logic 0 is connected to $I5$ through $I7$, $\bar{e}$ is connected to $I8$, logic 0 is connected to $I9$ and $I10$, $\bar{e}$ is connected to $I11$, logic 0 is connected to $I12$, $e$ is connected to $I13$, logic 0 is connected to $I14$, and logic 1 is connected to $I15$.

**P1.11**  For the SOP expression, we can use a four-variable K-map and a "circle-the-1's" approach. For the POS expression, the same K-map can be used with a "circle-the-0's" approach. Following these methods, we can obtain the following results.

> SOP: g = A'C' + C'D + BD + ABC
> POS: g = (B + C') (A + C' + D) (A' + C + D)

Since the SOP expression has 9 literals and the POS expression has 8 literals, the POS expression is simpler.

**P1.13**  From the solution to Problem P1.11, we can observe that the POS expression should be used since it is simpler. A straightforward 2-level OR-AND logic implementation of the $g$ POS expression corresponds exactly to the form shown in the P1.11 solution. To convert this to a NAND-only implementation, we note that an OR gate is equivalent to a NAND gate with inverted inputs. Also, an AND gate is equivalent to a NAND gate with an inverter connected to the output. Thus, a NAND-only implementation of $g$ consists of a NAND gate $n1$ with inputs B' and C, a NAND gate $n2$ with inputs A,' C, and D,' and a NAND gate $n3$ with inputs A, C,' and D' on the first level. Then, on the second level, there should be a 3-input NAND gate its inputs connected to the outputs of $n1$, $n2$, and $n3$. Finally, the output of the second-level 3-input NAND gate should be connected to an inverter formed using a 2-input NAND gate with its two inputs tied together (a "NAND inverter"). The NAND-only implementation described above can be converted into a 2-input NAND-only implementation by replacing the 3-input NAND gates used with their equivalent 2-input NAND implementations. A 3-input NAND gate with inputs $x$, $y$, and $z$ is equivalent to a 2-input NAND gate $n6$ with inputs $x$ and $y$ followed by a NAND inverter $n7$ followed by another 2-input NAND gate with its inputs tied to the output of $n7$ and $z$.

**P1.15**  In a Moore machine, the outputs are dependent on the current state only. Thus, this figure can be modified by creating another combinational logic block named $C2$. Referring to the original combinational logic block as $C1$, the current state values should connect to both $C1$ and $C2$; the outputs should come from $C2$ only.

**P1.17**  The five-step FSM method requires the construction of a state diagram, a state assignment, a state transition table, logic equations, and a logic diagram. For the state assignment, let us use a simple binary assignment of the state variables. Then Figure B.1 shows the state diagram, state transition table, logic equations, and logic diagram for this problem.

**P1.19**  The [Wakerly 2002] textbook includes a more detailed discussion of metastability. In particular, it includes another "synchronizer" circuit consisting of a chain of three D flip-flops. The original input enters the first D flip-flop and the synchronized input exits from the third D flip-flop. The first two D flip-flops use a clock input named $fastclk$, while the third D flip-flop uses the system clock input $clk$. A clock divider circuit is used to produce $fastclk$ by dividing the system clock signal $clk$ by a constant factor $N$.

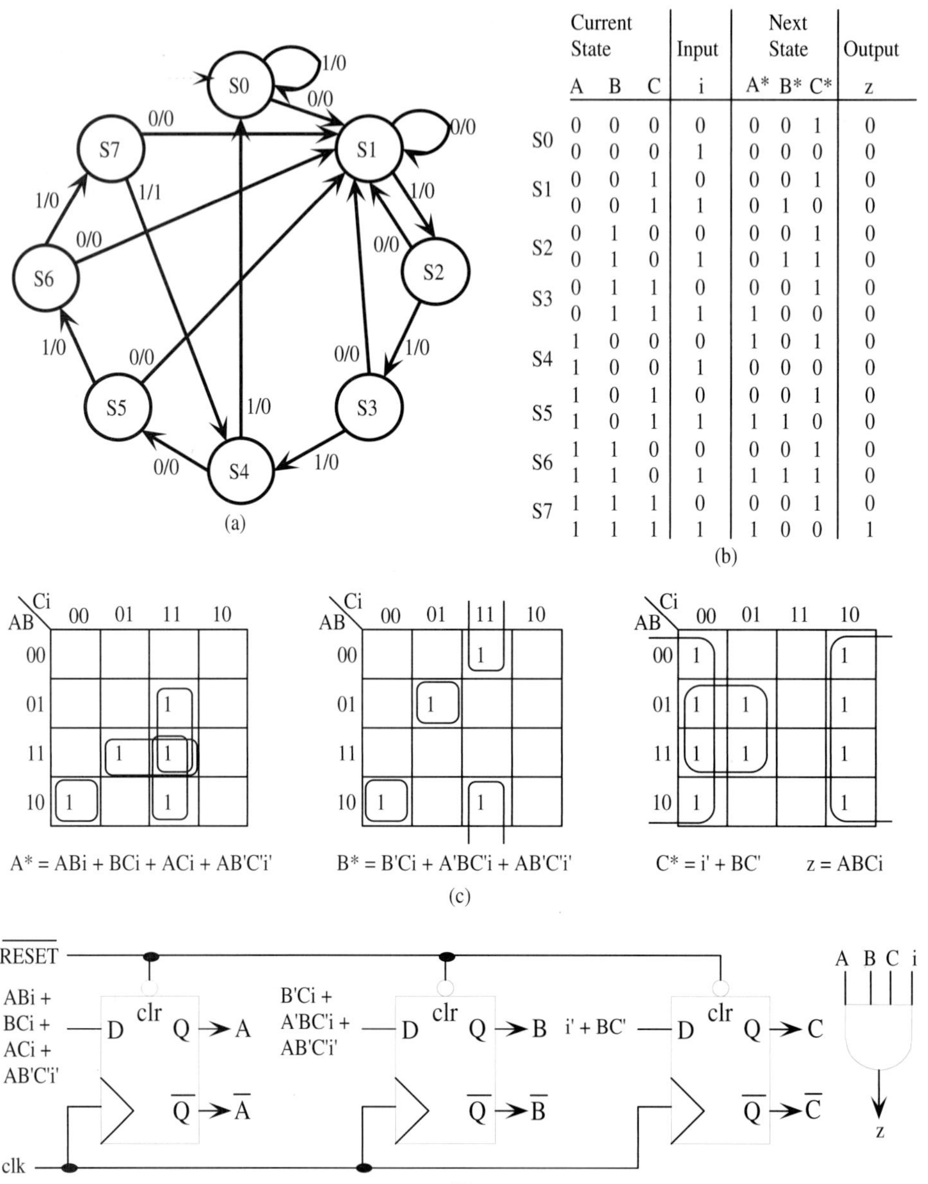

| Current State | | | Input | Next State | | | Output |
|---|---|---|---|---|---|---|---|
| A | B | C | i | A* | B* | C* | z |
| 0 | 0 | 0 | 0 | 0 | 0 | 1 | 0 |
| 0 | 0 | 0 | 1 | 0 | 0 | 0 | 0 |
| 0 | 0 | 1 | 0 | 0 | 0 | 1 | 0 |
| 0 | 0 | 1 | 1 | 0 | 1 | 0 | 0 |
| 0 | 1 | 0 | 0 | 0 | 0 | 1 | 0 |
| 0 | 1 | 0 | 1 | 0 | 1 | 1 | 0 |
| 0 | 1 | 1 | 0 | 0 | 0 | 1 | 0 |
| 0 | 1 | 1 | 1 | 1 | 0 | 0 | 0 |
| 1 | 0 | 0 | 0 | 1 | 0 | 1 | 0 |
| 1 | 0 | 0 | 1 | 0 | 0 | 0 | 0 |
| 1 | 0 | 1 | 0 | 0 | 0 | 1 | 0 |
| 1 | 0 | 1 | 1 | 1 | 1 | 0 | 0 |
| 1 | 1 | 0 | 0 | 0 | 0 | 1 | 0 |
| 1 | 1 | 0 | 1 | 1 | 1 | 1 | 0 |
| 1 | 1 | 1 | 0 | 0 | 0 | 1 | 0 |
| 1 | 1 | 1 | 1 | 1 | 0 | 0 | 1 |



**Figure B.1**   The solution to Problem P1.17: (a) state diagram, (b) state transition table, (c) logic equations, and (d) logic diagram.

# B.2 Digital Logic Design Using Hardware Description Languages

**P2.1**  The pros and cons of schematic- versus HDL-based design entry are listed in Section 2.1.

**P2.3**  The following is a possible sequence of RTL statements for an 8:1 MUX.

```
sel0 := (not S2) & (not S1) & (not S0);
sel1 := (not S2) & (not S1) & S0;
...
sel7 := S2 & S1 & S0;
out := (sel0 & I0) | (sel1 & I1) | (sel2 & I2) | (sel3 & I3) |
       (sel4 & I4) | (sel5 & I5) | (sel6 & I6) | (sel7 & I7);
```

The following is a sequence of RTL statements for a 3:8 decoder.

```
O0 := (not I2) & (not I1) & (not I0);
O1 := (not I2) & (not I1) & I0;
...
O7 := I2 & I1 & I0;
```

The following is a sequence of RTL statements for an 8-bit even parity checker.

```
f := not (I7 ^ I6 ^ I5 ^ I4 ^ I3 ^ I2 ^ I1 ^ I0);
```

**P2.5**  Let us use the following NAND-gate circuit. NAND-gate $n1$ has inputs $a$ and $b$ and output $c$. NAND-gate $n2$ has inputs $a$ and $c$ and output $d$. NAND-gate $n3$ has inputs $c$ and $b$ and output $e$. Finally, NAND-gate $n3$ has inputs $d$ and $e$ and output $f$. $f$ is assumed to be output of the exclusive-OR circuit that we are implementing. Then, given delays of 2 ns for each NAND-gate, the resulting timing diagram is shown in Figure B.2. The $f$ signal waveform is a 6 ns delayed form of the exclusive OR of $a$ and $b$, but the 1-pulse starts after only a 4 ns delay. This is due to the fact that there are signal paths with both 4 ns and 6 ns delays to the output $f$ from the two inputs $a$ and $b$.

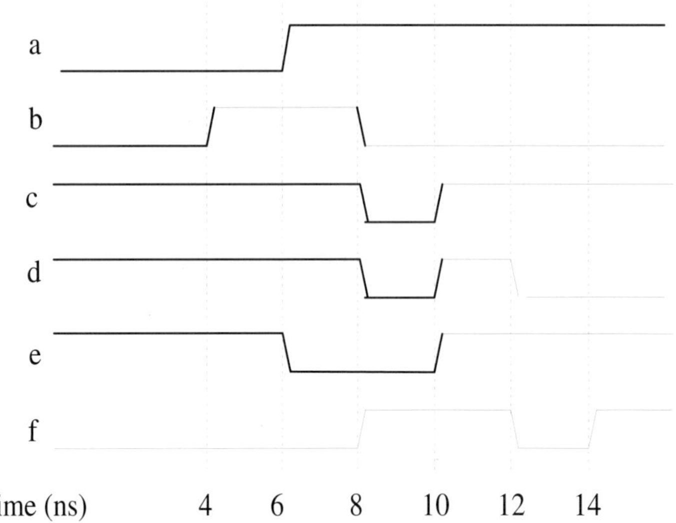

**Figure B.2**   The timing diagram for Problem P2.5.

**P2.7**   Although two path from the inputs to the output $f$ go through both the NAND-gate and the OR-gate, there is also a more direct path from the signal $b$ to the output $f$ that goes through only the OR-gate. Since $b$ is initially logic 1, the OR of any value with this $b$ value will be a logic 1. Thus, since the OR-gate has a 2 ns delay, the $f$ value will be a logic 1 after a 2 ns delay.

**P2.9**   Many different methods can be used to send a logic 1 followed by a logic 0 from one end of a wire to the other. Such methods include the following: (1) raising the voltage to +5 V and then dropping the voltage to 0 V, (2) injecting a current of 10 mA and then halting the injection of current, and (3) applying +100 mV followed by $-100$ mV on one wire and $-100$ mV followed by +100mV on another wire. These methods have the following pros and cons: (1) is simple to implement but may be a bit slow and require a significant amount of power, (2) may require less power than method (1) but requires a circuit capable of controlling the amount of current produced even with large variations in the type of load presented at the end of the signal wire, and (3) is fast, immune to noise on the wire, and low-power but requires two wires to send a single bit value. Other possible methods include voltage transitions, sinusoidal voltage waves with two different frequencies, sinusoidal voltage waves with two different amplitudes, monochromatic light passed through an optical fiber, etc.

**P2.11**   For this problem, the student should write a C or C++ program first. Then, once the program has been debugged and found to work satisfactorily, the algorithmic steps used in the program can be converted into RTL format.

**P2.13**  Let us use $TB$ to refer to the toothbrush and $TP$ to refer to the toothpaste. Then, an RTL algorithm for a toothbrush/toothpaste vending machine circuit can be written as follows.

1.  $MONEY \leftarrow 0;$
    $drop\_TB \leftarrow 0;$
    $drop\_TP \leftarrow 0;$
2.  while (TRUE) do begin
2a. if ($select\_TB$ and ($MONEY \geq 1$)) then
    $MONEY \leftarrow MONEY - 1;$
    $drop\_TB \leftarrow 1;$
2b. else if ($select\_TP$ and ($MONEY \geq 2$)) then
    $MONEY \leftarrow MONEY - 2;$
    $drop\_TP \leftarrow 1;$
    end if;
2c. wait for 1 clock cycle;
2d. $drop\_TB \leftarrow 0;$
    $drop\_TP \leftarrow 0;$
2e. if ($dollar\_detected$) then
    $MONEY \leftarrow MONEY + 1;$
    end if;
    end while;

# B.3  Introduction to Verilog and Test Benches

**P3.1**  Behavioral code follows a C-like syntax to describe how the target circuit should operate or behave. A structural description describes a module as an interconnection of submodules. Behavioral Verilog code has the advantages of simpler usage for complex circuits, a higher level of abstraction leading to a lower probability of design errors due to the use of automated design tools, and more possibilities for logic optimization using *synthesis* tools. The disadvantages of behavioral Verilog are less control over the exact logic circuit created, possibly incorrect circuits resulting from slightly erroneous behavioral descriptions, and less control over circuit delays in different parts of the circuit. The advantages and disadvantages of structural Verilog coding are the opposite of those for behavioral Verilog.

**P3.3**  The "other" major hardware description language, VHDL, uses a nine-values logic system. Several of these logic values are used to describe different "strengths" of logic signals. Such strength characteristics are useful for more precise modeling of transistor-based logic circuits, in which conflicting assignment of two or more logic values from two or more sources can result in valid logic outputs. Besides the four logic values used in Verilog, the five other logic values used in VHDL are U (uninitialized), W (weak unknown), L (weak zero), H (weak one), and - (don't care).

**P3.5**    The following is the behavioral Verilog code for a 16-bit BCD adder.

```verilog
///////////////////////////////////////////////////////////////////
// MODULE: 16-bit BCD adder
// Name: Sunggu Lee
// Created: ...
// Last Modified: ...
// Description: Implements a 16-bit binary-coded decimal (BCD)
//    adder. Module written assuming only valid 16-bit BCD inputs.
module bcd_adder_16 (sum, cout, a, b, cin);
  output [15:0] sum;   // the summation result
  output cout;         // the carry-out bit
  input [15:0] a;      // one of the BCD addition inputs
  input [15:0] b;      // the other BCD addition input
  input cin;           // the carry-in bit

  reg [15:0] sum;
  reg cout;

  always @(a or b or cin) begin
    sum[4:0] = a[3:0] + b[3:0] + cin;
    if (sum[4:0] > 9)
      sum[4:0] = sum[4:0] + 6;
    sum[8:4] = a[7:4] + b[7:4] + sum[4];
    if (sum[8:4] > 9)
      sum[8:4] = sum[8:4] + 6;
    sum[12:8] = a[11:8] + b[11:8] + sum[8];
    if (sum[12:8] > 9)
      sum[12:8] = sum[12:8] + 6;
    {cout, sum[15:12]} = a[15:12] + b[15:12] + sum[12];
    if ({cout, sum[15:12]} > 9)
      {cout, sum[15:12]} = {cout, sum[15:12]} + 6;
  end
endmodule
///////////////////////////////////////////////////////////////////
```

The following is the structural Verilog code for a 16-bit BCD adder, written in a hierarchical manner.

```verilog
///////////////////////////////////////////////////////////////////
// MODULES: 1-bit full adder, 4-bit BCD adder, and 16-bit BCD adder
// Name: Sunggu Lee
// Created: ...
// Last Modified: ...
// Description: This set of modules implements a 16-bit BCD adder
//    in a hierarchical manner.
```

```verilog
module full_adder (sum, cout, a, b, cin);
  output sum, cout;
  input a, b, cin;

  assign sum = a ^ b ^ cin;
  assign cout = ( (a && b) || (a && cin) || (b && cin) );
endmodule
///////////////////////////////////////////////////////////////////
// 4-bit BCD adder subcircuit definition
module bcd_adder_4 (sum, cout, a, b, cin);
  output [3:0] sum;    // the summation result
  output cout;         // the carry-out bit
  input [3:0] a;       // one of the BCD addition inputs
  input [3:0] b;       // the other BCD addition input
  input cin;           // the carry-in bit

  wire [4:1] c;        // for intermediate carry signals
  wire [3:0] bin_sum;  // binary summation result

  full_adder g1 (bin_sum[0], c[1], a[0], b[0], cin);
  full_adder g2 (bin_sum[1], c[2], a[1], b[1], c[1]);
  full_adder g3 (bin_sum[2], c[3], a[2], b[2], c[2]);
  full_adder g4 (bin_sum[3], c[4], a[3], b[3], c[3]);
  assign sum[0] = bin_sum[0];
  assign sum[1] = ( (~bin_sum[3] && bin_sum[1] && ~c[4]) ||
                    (bin_sum[3] && bin_sum[2] && ~bin_sum[1]) ||
                    (bin_sum[0] && c[4]) );
  assign sum[2] = ( (~bin_sum[3] && bin_sum[2]) ||
                    (bin_sum[2] && bin_sum[1]) );
  assign sum[3] = ( (bin_sum[3] && ~bin_sum[2] && ~bin_sum[1]) ||
                    (bin_sum[1] && c[4]) );
  assign cout = ( (bin_sum[3] && bin_sum[1]) ||
                  (bin_sum[3] && bin_sum[2]) || c[4]);
endmodule
///////////////////////////////////////////////////////////////////
// 16-bit BCD adder defined using 4-bit BCD adder subcircuit
module bcd_adder_16 (sum, cout, a, b, cin);
  output [15:0] sum;   // the summation result
  output cout;         // the carry-out bit
  input [15:0] a;      // one of the BCD addition inputs
  input [15:0] b;      // the other BCD addition input
  input cin;           // the carry-in bit

  wire [2:0] ic;       // intermediate carry results
```

```
   bcd_adder_4 g1 (sum[3:0], ic[0], a[3:0], b[3:0], cin);
   bcd_adder_4 g2 (sum[7:4], ic[1], a[7:4], b[7:4], ic[0]);
   bcd_adder_4 g3 (sum[11:8], ic[2], a[11:8], b[11:8], ic[1]);
   bcd_adder_4 g4 (sum[15:12], cout, a[15:12], b[15:12], ic[2]);
endmodule
////////////////////////////////////////////////////////////////////
```

**P3.7** A textual description of the resulting timing waveforms will be used. The reset_n signal starts out at 0, changes to 1 at 25 ns, and then changes to 0 at 125 ns. The clk signal starts out at 0, changes to 1 at 50 ns, changes to 0 at 100 ns, changes to 1 at 150 ns, and so on in a periodic manner (with a 100 ns period). The x signal starts out at 0, changes to 1 at 200 ns, and then changes to 0 at 500 ns. The state 2-bit signal starts out at 2'bXX, changes to 0 at 25 ns due to the reset_n signal, changes to 2 at 150 ns (this is when the clk signal changes from 0 to 1), changes to 0 at 250 ns, changes to 1 at 350 ns, changes to 2 at 550 ns, changes to 0 at 650 ns, and then changes to 1 at 750 ns. The z output signal starts out at 1'bX, changes to 0 at 25 ns, changes to 1 at 150 ns, changes to 0 at 250 ns, changes to 1 at 550 ns, changes to 0 at 650 ns, and then changes to 1 at 750 ns. Finally, all waveforms stop at 800 ns due to the **\$finish** statement executed at that time.

**P3.9** The following is the structural Verilog code for a 32-bit ripple-carry subtracter that uses the 1-bit full-subtracter of Example 3.1 as a submodule.

```
////////////////////////////////////////////////////////////////////
// MODULE: 32-bit ripple-carry (borrow) subtracter
// Name: Sunggu Lee
// Created: ...
// Last Modified: ...
// Description: Implements a 32-bit subtracter as a "ripple"
//    device in a structural 2-level hierarchical manner.
module subtracter_32 (res, borrow_out, a, b, borrow_in);
   output [31:0] res;   // subtraction result
   output borrow_out;   // the borrow-out bit
   input [31:0] a;      // the number to subtract from
   input [31:0] b;      // the number to subtract
   input borrow_in;     // the borrow-in bit

   wire [31:1] ib;      // for intermediate borrow signals

   full_subtract_struct g0 (res[0], ib[1], a[0], b[0], borrow_in);
   full_subtract_struct g1 (res[1], ib[2], a[1], b[1], ib[1]);
   full_subtract_struct g2 (res[2], ib[3], a[2], b[2], ib[2]);
   // ... continue on in this manner
   full_subtract_struct g30 (res[30], ib[31], a[30], b[30], ib[30]);
   full_subtract_struct g31 (res[31], borrow_out, a[31], b[31], ib[31]);
endmodule
////////////////////////////////////////////////////////////////////
```

**P3.11**   An 8-bit Gray-code counter can be defined in a hierarchical manner as shown below.

```
//////////////////////////////////////////////////////////////////
// MODULES: Two types of 2-bit Gray-code counters and an 8-bit
//              Gray-code counter
// Name: Sunggu Lee
// Created: ...
// Last Modified: ...
// Description: Implements an 8-bit Gray-code counter in a
//   hierarchical manner.  The first module defined is a 2-bit
//   Gray-code counter that counts up and then down.
module gcc_2_updown (count, next_clk, reset_n, clk);
  output [1:0] count;  // 2-bit Gray code output
  output next_clk;       // used to enable next counter
  input reset_n;         // active-low reset
  input clk;             // counter clock

  reg [1:0] count;
  reg next_clk;          // used for clock input to next block
  reg high_bit;          // used to increase count to 3 bits

  always @(negedge reset_n or posedge clk) begin
    if (~reset_n) begin
      count <= 0;
      next_clk <= 0;
      high_bit <= 0;
    end
    else begin
      case ({high_bit, count})
        3'b000: count <= 2'b01;
        3'b001: begin
                count <= 2'b11;
                next_clk <= 0;
              end
```

```verilog
                3'b011: count <= 2'b10;
                3'b010: begin
                          high_bit <= 1;
                          next_clk <= 1;
                        end
                3'b110: count <= 2'b11;
                3'b111: begin
                          count <= 2'b01;
                          next_clk <= 0;
                        end
                3'b101: count <= 2'b00;
                default: begin  // case 3'b100
                          high_bit <= 0;
                          next_clk <= 1;
                        end
            endcase
        end // of else
    end // of always
endmodule
//////////////////////////////////////////////////////////////////////
// A 2-bit Gray-code counter that counts up
module gcc_2_up (count, reset_n, clk);
  output [1:0] count;   // 2-bit Gray code output
  input reset_n;        // active-low reset
  input clk;            // counter clock

  reg [1:0] count;
  reg prev_toggle_en;

  always @(negedge reset_n or posedge clk) begin
    if (~reset_n) begin
      count <= 0;
      prev_toggle_en <= 0;
    end
    else begin
      case (count)
        2'b00: count <= 2'b01;
        2'b01: count <= 2'b11;
        2'b11: count <= 2'b10;
        default: count <= 2'b00;
      endcase
    end // of else
  end // of always
endmodule
//////////////////////////////////////////////////////////////////////
// An 8-bit Gray-code counter, defined in a hierarchical manner
```

```
module gcc_8 (gray_code, reset_n, clk);
  output [7:0] gray_code; // Gray-code counter output
  input reset_n;          // active-low reset
  input clk;              // system clock

  wire [2:0] n; // intermediate enable signals (toggle type)
  reg clk2;     // a clock signal with twice the period of clk

  initial clk2 <= 0;
  //always @(posedge clk) clk2 <= ~clk2;

  gcc_2_updown g1 (gray_code[1:0], n[0], reset_n, clk);
  gcc_2_updown g2 (gray_code[3:2], n[1], reset_n, n[0]);
  gcc_2_updown g3 (gray_code[5:4], n[2], reset_n, n[1]);
  gcc_2_up     g4 (gray_code[7:6],       reset_n, n[2]);
endmodule
////////////////////////////////////////////////////////////////////
```

**P3.13**   Verilog code that must be synthesized is written in a significantly different manner from test bench Verilog code. A few of the main differences between synthesizable Verilog and test bench Verilog are listed below.

- A test bench module typically does not include any external inputs or outputs while a synthesizable module normally includes external inputs and outputs.

- Test bench Verilog code includes delay statements in order to control the timing waveforms generated. Synthesizable Verilog code does not use delay statements.

- Test bench Verilog code normally use exact equality (===) and exact inequality (!==) for error checking while synthesizable Verilog code uses logical equality (==) and logical inequality (!=).

- A test bench can make liberal use of integers and other data types while minimal-sized bit vectors are used almost exclusively in synthesizable Verilog code in order to save hardware.

- In a test bench, a sequence of actions can be written as one long sequence of actions separated by appropriate delay statements. In synthesizable Verilog, sequences of actions are typically written as separate **case** statements within an **always** block activated by the positive (or negative) edge of a clock signal; consecutive actions are executed in consecutive clock periods.

**P3.15**   The Verilog keyword **reg** refers to a registered signal. This type of signal can be manipulated like a C language variable from within an **always** or **initial** block. The keyword **wire** refers to a "passive" signal that can only change value due to an external stimulus (much like a physical electrical wire). An **integer** is equivalent to a 32-bit **reg** signal. The keywords **input**, **output**, and **inout** refer to the external ports of a circuit. Once a signal enters a circuit through an external port, it must be carried as a **wire** or **reg** signal. The **input**, **output**, and **inout** terms refer to external input, output, and bidirectional ports, respectively.

**P3.17**   Unexpected behavior can occur when using nonblocking statements (<=) involving the same signal from within two (or more) **always** blocks. Consider the following example, involving an integer signal i used in two concurrent **initial** blocks.

```
/////////////////////////////////////////////////////////////////
module test_conc();
   integer i;
   reg [7:0] a, b;

   initial begin
     a = 0;
     b = 0;
   end

   initial begin
     for (i = 0;  i < 10;  i = i + 1) begin
       a = a + 1;
       #100;
     end
   end

   initial begin
     for (i = 0;  i < 20;  i = i + 1) begin
       b = b + 1;
       #200;
     end
   end
endmodule
/////////////////////////////////////////////////////////////////
```

Since the first **for** loop iterates over i from 0 to 9 and the second **for** loop iterates over i from 0 to 19, it might be expected that the final value for a would be 10 and b would be 20. However, simulation of this code using Aldec Active-HDL reveals that the final value for a is 7 and b is 13. In addition, the i signal is found to change in the following sequence: 0, 1, 3, 4, 6, 7, 9, 10, 11, 12, 13, etc. This is clearly very strange behavior, brought on by the fact that a common signal is read and written in two separate concurrent blocks.

# B.4 High-Level Verilog Coding for Synthesis

**P4.1**   The follows the Verilog code for a simple thermostat device and its test bench.

```verilog
////////////////////////////////////////////////////////////////////////
// MODULE: A simple thermostat circuit (digital logic portion)
// Name: Sunggu Lee
// Created: ...
// Last Modified: ...
// Description: Implements a simple thermostat.

// Constant Definitions
`define TOO_LOW 3  // heater turn-on threshold
`define TOO_HIGH 4 // air conditioner turn-on threshold

module thermostat (heater_on, ac_on, temp, desired, reset_n);
  output heater_on;    // signal to turn on the heater
  output ac_on;        // signal to turn on the air conditioner
  input [7:0] temp;    // the measured room temperature
  input [7:0] desired; // the desired room temperature setting
  input reset_n;       // active-low system reset

  reg heater_on, ac_on;

  always @(reset_n or temp or desired) begin
    if (~reset_n) begin
      heater_on <= 0;
      ac_on <= 0;
    end
    else if (temp < desired - `TOO_LOW) begin
      heater_on <= 1;
      ac_on <= 0;
    end
    else if (heater_on && (temp > desired))
      heater_on <= 0;
    else if (temp > desired + `TOO_HIGH) begin
      ac_on <= 1;
      heater_on <= 0;
    end
    else if (ac_on && (temp < desired))
      ac_on <= 0;
  end // of always
endmodule
////////////////////////////////////////////////////////////////////////
```

```verilog
// MODULE: Test bench for thermostat circuit.
// Name: Sunggu Lee
// Created: ...
// Last Modified: ...
// Description: Tests the "thermostat" mopdule.

`timescale 1ns/100ps // 1ns simulation accuracy & 1ns precision
`define TOO_LOW 3    // heater turn-on threshold
`define TOO_HIGH 4   // air conditioner turn-on threshold
// Test bench module definition
module tb_thermostat ();
  // Signals used to connect to the UUT (unit under test)
  reg [7:0] temp, desired; // measured and desired temperatures
  reg reset_n;             // active-low system reset
  wire heater_on, ac_on;   // heater and a/c turn-on signals
  // Internal signals
  integer error_count = 0; // total number of errors found

  // Connect to the UUT (unit under test)
  thermostat uut (heater_on, ac_on, temp, desired, reset_n);

  // Generate the test vectors
  initial begin // for reset_n and desired inputs
    $display("Start simulation.");
    reset_n = 1;
    temp = 50;
    #20 reset_n = 0;
    #100 reset_n = 1;
    #50 desired = 72;
    #200 desired = 78;
    #200 desired = 68;
  end // of initial
  always begin // for temp input
    if ($time < 300)
      temp = temp + 1;
    else if ($time < 600)
      temp = temp - 1;
    else if ($time > 1000) begin
      $display("Terminating simulation at time %0d.", $time);
      $finish;
    end
    else
      temp = temp + 1;
    #10;
  end // of always
```

```verilog
// Check the test results
always begin
  #50;
  if ($time < 200)
    error_check(1'b0, heater_on, 1'b0, ac_on);
  else if ($time < 250)
    error_check(1'b1, heater_on, 1'b0, ac_on);
  else if ($time < 300)
    error_check(1'b0, heater_on, 1'b0, ac_on);
  else if ($time < 400)
    error_check(1'b0, heater_on, 1'b1, ac_on);
  else if ($time < 800)
    error_check(1'b1, heater_on, 1'b0, ac_on);
  else if ($time < 850)
    error_check(1'b0, heater_on, 1'b0, ac_on);
  else if ($time < 1000)
    error_check(1'b0, heater_on, 1'b1, ac_on);
  else if (error_count > 0)
    $display("ERROR: There were %0d errors!", error_count);
  else
    $display("Simulation complete with NO errors.");
end  // of always block for test vector checking

task error_check; // error check and display task
  input expected_heater_on, heater_on, expected_ac_on, ac_on;
  begin
    if ({expected_heater_on, expected_ac_on} !==
          {heater_on, ac_on}) begin
      $display
      ("ERROR! Expected (%0d, %0d), got (%0d, %0d) at time %0t ns.",
          expected_heater_on, expected_ac_on, heater_on, ac_on,
          $time/10);
      error_count = error_count + 1; // update global value
    end  // of if
  end  // of outermost begin for function error_check
endtask
endmodule
/////////////////////////////////////////////////////////////////////
```

**P4.3**   The following is the Verilog code for the digital logic circuit portion of a telephone answering machine.

```verilog
/////////////////////////////////////////////////////////////////////
// MODULE: Telephone answering machine (digital logic portion)
// Author: Sunggu Lee
```

```verilog
// Created: ...

// Last Modified: ...

// Description: Implements a telephone answering machine.


// DEFINITIONS

`define TIMER_BITS      5

`define STEP_BITS       3

`define PICKUP_RINGS    4

`define MESSAGE_LENGTH 10

`define RECORD_LENGTH  30


// MODULE DECLARATION

module telephone (pickup, play_message, play_beep, record_on,

                  messages, reset_n, clk, hangup, rings);

  output pickup;        // signal to pick up the phone

  output play_message;  // signal to play the outgoing message

  output play_beep;     // signal to play a "beep" sound

  output record_on;     // signal to record the incoming message

  output [3:0] messages; // number of recorded messages

  input reset_n;        // active-low RESET signal

  input clk;            // clock signal

  input hangup;         // signal observed when phone is off-hook

  input [3:0] rings;    // number of rings on the phone


  // SIGNAL DECLARATIONS

  // Signals connected to module outputs

  reg pickup, play_message, play_beep, record_on;

  reg [3:0] messages;

  // Internal signals

  reg [`TIMER_BITS-1:0] timer; // internal timer signal

  reg [`STEP_BITS-1:0] step;   // used with state machine
```

```verilog
always @(posedge clk) begin
  if (~reset_n || hangup) begin // synchronous reset
    pickup <= 0;
    play_message <= 0;
    play_beep <= 0;
    record_on <= 0;
    if (~reset_n)
      messages <= 0;
    step <= 0;
  end
  else begin  // each "step" is one step of RTL algorithm
    case (step)
      0: if (rings >= `PICKUP_RINGS) begin
             pickup <= 1;
             step <= 1;
           end
      1: begin
           play_message <= 1;
           timer <= 0;
           step <= 2;
         end
      2: if (timer >= `MESSAGE_LENGTH) begin
             play_message <= 0;
             play_beep <= 1;
             step <= 3;
           end
           else
             timer <= timer + 1;
      3: begin
           play_beep <= 0;
           record_on <= 1;
           messages <= messages + 1;
           timer <= 0;
           step <= 4;
         end
      4: if (timer >= `RECORD_LENGTH) begin
             record_on <= 0;
             step <= 0;
           end
           else
             timer <= timer + 1;
      default: step <= 0;
    endcase // of case (state)
  end // of else
end // of always
```

```
endmodule
//////////////////////////////////////////////////////////////////////
```

**P4.5**    The following shows a Verilog **always** block for the required combinational logic devices.

```verilog
//////////////////////////////////////////////////////////////////////
// MODULES: An 8:1 MUX, 3:8 decoder, and an even parity checker
// Name: Sunggu Lee
// Created: ...
// Last Modified: ...
// Description: Implements 3 simple combinational logic circuits.
module test_comb (mux_out, dec_out, even_parity, I, sel)
  output mux_out;        // output of 8:1 multiplexer (MUX)
  output [7:0] dec_out;  // outputs of 3:8 decoder
  output even_parity;    // output of 8-bit even parity checker
  input [7:0] I;         // inputs to MUX and parity checker
  input [2:0] sel;       // inputs to MUX and decoder

  reg mux_out;
  reg [7:0] dec_out;
  reg even_parity;
  reg [3:0] sum;         // used in even parity calculation
  integer k;             // index variable

  always @(I or sel) begin  // sensitive to any input changes
    case (sel)   // for the 8:1 MUX
      3'b000: mux_out <= I[0];
      3'b001: mux_out <= I[1];
      3'b010: mux_out <= I[2];
      3'b011: mux_out <= I[3];
      3'b100: mux_out <= I[4];
      3'b101: mux_out <= I[5];
      3'b110: mux_out <= I[6];
      default: mux_out <= I[7];
    endcase

    for (k = 0;  k < 8;  k = k + 1)  // for the 3:8 decoder
      if (k == sel)
        dec_out[k] <= 1;
      else
        dec_out[k] <= 0;
```

```
     for (k = 0;  k < 8;  k = k + 1)  // for the parity checker
        sum = sum + I[k];
     if ((sum % 2) == 0)  // modulus operation
        even_parity = 1;
     else
        even_parity = 0;
  end // of always block
endmodule
////////////////////////////////////////////////////////////////////
```

**P4.7**   Although the same code as that shown for the priority 4:2 encoder can be used, this would result in wasted hardware logic when synthesized due to the extra logic used to enforce priority order. Thus, the **always** block shown in the code for the priority 4:2 encoder should be changed to the following.

```
////////////////////////////////////////////////////////////////////
  always @(i3 or i2 or i1 or i0) begin
    valid <= 1;  // initially, assume valid input
    case ({i3, i2, i1, i0}) // case block with don't care values
       4'b1000: encoded <= 2'b11;  // i3 asserted
       4'b0100: encoded <= 2'b10;  // i2 asserted
       4'b0010: encoded <= 2'b01;  // i1 asserted
       4'b0001: encoded <= 2'b00;  // i0 asserted
       default: begin  // invalid input combination
                    encoded <= 2'bXX;
                    valid <= 0;  // input found to be invalid
                end
    endcase
  end // of always
////////////////////////////////////////////////////////////////////
```

**P4.9**   Since the logic for all $P_{i*m+m-1:i*m}$ and $G_{i*m+m-1:i*m}$ show the exact same pattern for all values of $i$ in both the logic equations and Figure 4.2(b), let us show the required equivalence for only the $i = 0$ case. Figure 4.2(b) uses slightly different notation from the logic equations. Thus, if we convert to the notation used in the logic equations and use temporary variables named $K_j$, we can write the equations for the group propagate terms as follows.

$$K_1 = I0 \cdot I2 + I0 \cdot I3 + I1 \cdot I2 + I1 \cdot I3$$
$$= a_3 a_2 + a_3 b_2 + b_3 a_2 + b_3 b_2$$
$$= a_3(a_2 + b_2) + b_3(a_2 + b_2)$$
$$= (a_3 + b_3)(a_2 + b_2)$$
$$= p_3 p_2$$
$$K_2 = I0 + I1$$
$$= a_1 + b_1$$
$$= p_1$$
$$P_{3:0} = I0 \cdot I2 \cdot I3 + I1 \cdot I2 \cdot I3$$
$$= a_0 K_1 K_2 + b_0 K_1 K_2$$
$$= a_0(p_1)(p_3 p_2) + b_0(p_1)(p_3 p_2)$$
$$= p_3 p_2 p_1(a_0 + b_0)$$
$$= p_3 p_2 p_1 p_0$$

Likewise, we can derive the equations for the group generate terms as follows.

$$K_3 = I0 \cdot I1 + I0 \cdot I2 \cdot I3 + I1 \cdot I2 \cdot I3$$
$$= a_1 b_1 + a_1 a_0 b_0 + b_1 a_0 b_0$$
$$= a_1 b_1 + a_0 b_0(a_1 + b_1)$$
$$= g_1 + g_0 p_1$$
$$K_4 = I_0 + I1 \cdot I2 + I1 \cdot K_3 + I2 \cdot K_3$$
$$= b_3 + a_2 b_2 + a_2 K_3 + b_2 K_3$$
$$= b_3 + a_2 b_2 + K_3(a_2 + b_2)$$
$$= b_3 + g_2 + (g_1 + g_0 p_1)p_2$$
$$= b_3 + g_2 + g_1 p_2 + g_0 p_1 p_2$$
$$K_5 = I_0 \cdot I1 \cdot I2 + I_0 \cdot I1 \cdot I3 + I_0 \cdot I2 \cdot I3$$
$$= b_3 a_2 b_2 + b_3 a_2 K_3 + b_3 b_2 K_3$$
$$= b_3 g_2 + b_3(a_2 + b_2)K_3$$
$$= b_3 g_2 + b_3 p_2(g_1 + g_0 p_1)$$
$$= b_3(g_2 + g_1 p_2 + g_0 p_1 p_2)$$

$$G_{3:0} = I_2 + I_0 \cdot I_1$$

$$= K_5 + a_3 K_4$$

$$= b_3(g_2 + g_1 p_2 + g_0 p_1 p_2) + a_3(b_3 + g_2 + g_1 p_2 + g_0 p_1 p_2)$$

$$= a_3 b_3 + (b_3 + a_3)(g_2 + g_1 p_2 + g_0 p_1 p_2)$$

$$= g_3 + g_2 p_3 + g_1 p_2 p_3 + g_0 p_1 p_2 p_3$$

**P4.11**  The following is the Verilog code for a comprehensive test bench that can be used to test a 16-bit BCD adder.

```verilog
////////////////////////////////////////////////////////////////////////////
// MODULE: Test Bench for BCD Adder: tb_bcd_adder_16.v
// Author: Sunggu Lee
// Date: ...
// Description: Tests the bcd_adder_16 BCD adder circuit.

`timescale 1ns/100ps // 1ns simulation accuracy & 100ps precision
// Constant Definitions
`define LENGTH 16     // length of input BCD number
`define MAX_BCD 9999 // maximum BCD input value possible
`define DELAY  10     // delay value used in testing (10 ns)

// Test bench module definition
module tb_bcd_adder_16 ();
  // SIGNAL DEFINITIONS
  // Signals used to connect to the UUT (unit under test)
  reg [`LENGTH-1:0] bcd1, bcd2;    // BCD inputs to the UUT
  //reg cin;                       // carry-in input to UUT
  wire [`LENGTH:0] sum;            // {cout, sum} output of UUT
  // Internal signals
  integer error_count = 0;    // total number of errors found
  integer i, j, k, cin;       // test vector generation indices
  reg [`LENGTH-1:0] temp;     // used as a temporary data holder
  reg [3:0] digit;            // used to hold single decimal digit
  reg [`LENGTH:0] expected;   // used to check test results

  // Connect to the UUT (unit under test)
  bcd_adder_16 uut (sum[`LENGTH-1:0], sum[`LENGTH], bcd1, bcd2, cin[0]);
```

```verilog
// Generate the test vectors and check the results
initial begin
  $display("Start simulation.");
  for (cin = 0;  cin <= 1;  cin = cin + 1) begin // repeat twice
  // generate all possible input values, since max = 9999
  for (i = 0;  i <= `MAX_BCD;  i = i + 1) begin
    for (k = 0;  k <= `MAX_BCD;  k = k + 1) begin
    temp = i;
    for (j = 0;  j < (`LENGTH/4);  j = j + 1) begin
      digit = temp % 10; // look at least sig. decimal digit
      bcd1[j*4 + 3] = digit[3];  // assign test vector input
      bcd1[j*4 + 2] = digit[2];
      bcd1[j*4 + 1] = digit[1];
      bcd1[j*4] = digit[0];
      temp = temp / 10;          // move to next decimal digit
    end // of for (j ... )

    temp = k;
    for (j = 0;  j < (`LENGTH/4);  j = j + 1) begin
      digit = temp % 10; // look at least sig. decimal digit
      bcd2[j*4 + 3] = digit[3];  // assign test vector input
      bcd2[j*4 + 2] = digit[2];
      bcd2[j*4 + 1] = digit[1];
      bcd2[j*4] = digit[0];
      temp = temp / 10;          // move to next decimal digit
    end // of for (j ... )

    expected = i + k + cin;      // computation of expected value
    #`DELAY;          // wait for result to be computed
    if (expected !== expected) begin  // test result check
      $display
      ("ERROR: expected (%0b) != actual (%0b) at time %0t ps",
        expected, sum, $time);
      error_count = error_count + 1;
    end  // of if
  end // of for (k ... )
  end // of for (i ... )
  end // of for (cin ... )
```

```verilog
      #`DELAY;
      if (error_count > 0)
        $display("ERROR: There were %0d errors!", error_count);
      else
        $display("Simulation complete with NO errors.");
      $finish;            // terminate the simulation
    end  // of always block for test vector generation and checking

endmodule
////////////////////////////////////////////////////////////////////////
```

**P4.13**   Since all top-level Verilog statements execute concurrently with each other, two **always** blocks will also execute concurrently. Thus, if a common signal is checked and updated in two separate **always** blocks, then there will be a "race" to check and update the common signal. This will result in unpredictable results. For example, in the following code, the common **reg** signal state will be an even number at certain times and an odd number at others. However, both the simulation and actual implementation of this circuit will result in unpredictable results regarding the times when state is even or odd.

```verilog
////////////////////////////////////////////////////////////////////////
// Example demonstrating a problem with two always blocks
  always @(negedge reset_n or posedge clk)
    if (~reset_n)
      state <= 0;
    else
      state <= state + 1;

  always @(posedge clk)
    state <= state + 2;
////////////////////////////////////////////////////////////////////////
```

**P4.15**   The following shows a set of continuous assignment statements that can be used to implement a D latch. This set of statements is derived directly from the logic diagram of Figure 1.16 (or Figure 1.15 with an inverter connected between the S and R inputs).

```verilog
////////////////////////////////////////////////////////////////////////
// Example of a D latch implemented with continuous assignments
  assign Q = ~(~(D && En) && Q_bar);
  assign Q_bar = ~(Q && ~(En && ~D));
////////////////////////////////////////////////////////////////////////
```

**P4.17**   To insert a specific delay into a circuit that must be synthesized, we can use an enable signal that is enabled after the desired delay or a set of D flip-flops connected as a shift register. These two methods are demonstrated by the following example code.

```verilog
//////////////////////////////////////////////////////////////////
// Examples of delay insertion
  // Method 1: use a counter output value as an enable signal
  assign delayed_data = (count >= `DELAY1) ? data : 'bZ;
  always @(negedge reset_n or posedge clk) // implements a counter
    if (~reset_n)
      count <= 0;
    else
      count <= count + 1;

  // Method 2: use a shift register connection
  always @(posedge clk) begin
    q1 <= data;
    q2 <= q1;
    q3 <= q2;
    q4 <= q3;  // q4 is a 3-clock-cycle delayed version of data
  end
//////////////////////////////////////////////////////////////////
```

**P4.19**    The following shows the required Verilog code.

```verilog
//////////////////////////////////////////////////////////////////
// An 8-bit encoder, defined using "// synopsys parallel case"
  always @(data) begin
    casex (data) // synopsys parallel case
      8'b1xxxxxxx: encoded <= 3'b111;
      8'bx1xxxxxx: encoded <= 3'b110;
      8'bxx1xxxxx: encoded <= 3'b101;
      8'bxxx1xxxx: encoded <= 3'b100;
      8'bxxxx1xxx: encoded <= 3'b011;
      8'bxxxxx1xx: encoded <= 3'b010;
      8'bxxxxxx1x: encoded <= 3'b001;
      8'bxxxxxxx1: encoded <= 3'b000;
      8'b00000000: encoded <= 3'bXXX;
    endcase
  end // of always
//////////////////////////////////////////////////////////////////
```

**P4.21**    With the indicated change, the "previous" (during the last clock cycle) `prev_data` signal value will be checked for equality with `8'b0111_1110`. This will have the effect of *delaying* the output of the circuit by one clock cycle. Thus, in the corresponding timing diagram, the `detected` signal will become a logic 1 one clock cycle *after* the last 0 of the `8'b0111_1110` sequence is entered into the circuit.

# B.5 State Machine Design

**P5.3** (a) Let us assume that there is an active-low reset signal named reset_n, an 8-bit register temp storing the current measured room temperature, and an 8-bit register desired storing the desired temperature setting. The desired register value is changed when the user adjusts the desired temperature setting on the thermostat. The outputs of this circuit are a heater_on signal to turn on the heater and a ac_on signal to turn on the air conditioner. For simplicity, it will also be assumed that the room temperature will be allowed to vary by a fixed amount from the desired temperature before the heater or air conditioner is turned on or off. Then the following shows one possible simple pseudocode solution for this problem.

1. $heater\_on \leftarrow 0$;
   $ac\_on \leftarrow 0$;
2. while (TRUE) do begin
2a. if $(temp < (desired - THRESHOLD))$ then
   $heater\_on \leftarrow 1$;
   $ac\_on \leftarrow 0$;  //to prevent conflicts
2b. else if $(heater\_on$ and $(temp > desired))$ then
   $heater\_on \leftarrow 0$;
2c. else if $(temp > (desired + THRESHOLD))$ then
   $ac\_on \leftarrow 1$;
   $heater\_on \leftarrow 0$;  //to prevent conflicts
2d. else if $(ac\_on$ and $(temp < desired))$ then
   $ac\_on \leftarrow 0$;
   end if;
   end while;

(b) The above pseudocode solution can also serve as the RTL program solution for this problem.

**P5.5** The main difference between a *finite state machine (FSM)* and a *state machine* is that the former model is only suitable for small synchronous sequential circuits with small numbers of inputs and outputs while the latter model can be used for large synchronous sequential circuits with large numbers of inputs and outputs. The design of a circuit based on the FSM model begins with a *state diagram* while the state machine method begins with a *pseudocode* description. Also, the state diagram used with the FSM model is different from the state diagram used with the state machine model. In an FSM state diagram, transitions between states are labeled as $value_1, value_2 \ldots /output_1, output_2, \ldots$, where $value_i$ is the value of the $i$'th primary input and $output_j$ is the value to assign to the $j$'th primary output signal. In a state machine state diagram, transitions are labeled as $condition_i/action_j$, where $condition_i$ is a general combinational logic condition check and $action_j$ is an action to be performed (typically an RTL assignment).

Thus, for example, in a circuit with primary inputs $a$ and $b$ and a primary output $f$, an FSM state diagram may have a transition from a state $S5$ to another state $S7$ labeled as $0, 1/1$ while a state machine state diagram would label the corresponding transition as $(a = 0 \; and \; b = 1)/(f \leftarrow 1)$.

**P5.7** Based on Table 5.1, the following logic equations can be derived.

$$NE_1 = E_1 + E_0 \cdot \overline{less\_or\_equal}$$

$$NE_0 = \overline{E_1} \cdot \overline{E_0} + E_0 \cdot less\_or\_equal$$

$$sel = \overline{E_1} \cdot \overline{E_0}$$

$$loadF = \overline{E_1} \cdot \overline{E_0} + E_0 \cdot less\_or\_equal$$

$$loadC = \overline{E_1} \cdot E_0$$

$$resetLatch = \overline{E_1} \cdot \overline{E_0}$$

$$incC = E_0 \cdot less\_or\_equal$$

$$setLatch = E_1$$

**P5.9** The five-step state machine method requires the creation of a pseudocode solution, RTL program, datapath, state diagram, and control logic. The following is a possible pseudocode solution for this problem.

```
1.  A ← multiplier;
    B ← multiplicand;
    C ← 0;  //variable used to hold final product
2.  for i = 1 to 8 do
2a.  C[15 : 8] ← C[15 : 8] + (A[0] and B);
2b.  C ← C >> 1;  //shift 0 into C[0]
2c.  A ← A >> 1;
    end for;
```

The following RTL program can be derived from this pseudocode.

```
1.  A ← multiplier;
    B ← multiplicand;
    P ← 0;  //variable used to hold final product
    count ← 8;  //used for the ∩∩for" loop
2.  if (count > 0) then
2a.  P[15 : 8] ← P[15 : 8] + (A[0] and B);
2b.  P ← P >> 1;  //shift 0 into P[15]
    A ← A >> 1;
    count ← count − 1;
    Go to Step 2
    end if;
```

The datapath that can be created for this circuit, based on the RTL statements used in the above RTL program, can be described as follows. There is an 8-bit shift register named $A$, an 8-bit register named $B$, a 16-bit shift register named $P$, and a 4-bit down-counter named $count$. The two 8-bit numbers to be multiplied are connected to the inputs of the $A$ and $B$ registers and the constant value 8 is connected to the input of the $count$ counter. There is also an 8-bit adder with its left input coming from the most significant 8 bits of $P$ and its right input coming from the output of 8 AND gates; in these AND gates, the $i$'th AND gate is the AND of the lsb of $A$ and $B[i]$. The control signals used to control the operation of these datapath components are $loadA$ (for loading the $A$ register with external data), $rshiftA$ (to enable right-shifting of register $A$), $loadB$ (for loading $B$), $clearP$ (for clearing $P$ to 0), $loadP$ (for loading $P$ with the result of the addition of $P[15:8]$ with $(A[0]\ and\ B)$), $rshiftP$ (to enable right-shifting of $P$), $loadC$ (for loading the $count$ counter with the constant value 8), and $decC$ (to enable down-counting of the $count$ counter). A 4-input NOR gate can also be connected to the four output bits of the $count$ counter; the output of this NOR gate, labeled $zero$, becomes a 1 when the $count$ value becomes 0.

From the above datapath and RTL programs, a state diagram can be derived. This state diagram can consist of four states, labeled $S0$, $S1$, $S2$, and $S3$. The initial state is state $S0$. There is an unlabeled transition (indicating an unconditional transition) from state $S0$ to state $S1$. There is another unlabeled transition from $S1$ to $S2$. There is a transition from state $S2$ to $S1$ labeled with $\overline{zero}$ and a transition from state $S2$ to $S3$ labeled with $zero$. In state $S0$, $loadA$, $loadB$, $clearP$, and $loadC$ are all set to 1. In state $S1$, $loadP$ is set to 1. In state $S2$, $rshiftA$, $rshiftP$, and $decC$ are set to 1. State $S3$ is an empty state with no outgoing transitions.

The control logic can then be designed from the state diagram. In this case, since there are four states, we can use two D flip-flops to encode the state values. Given a straightforward binary encoding with the state variables $f$ and $g$, the next-state inputs (the D inputs to the two D flip-flops) can be derived as

$$nf = \overline{f} \cdot g + f\overline{g} \cdot zero$$

$$ng = \overline{f} \cdot \overline{g} + f \cdot \overline{g} \cdot \overline{zero} + f \cdot \overline{g} \cdot zero$$

$$= \overline{f} \cdot \overline{g} + f \cdot \overline{g}$$

$$= \overline{g}$$

Next, the control signals can be derived as

$$loadA = loadB = clearP = loadC = \overline{f} \cdot \overline{g}$$

$$loadP = \overline{f} \cdot g$$

$$rshiftA = rshiftP = decC = f \cdot \overline{g}$$

A system clock signal $clk$ is connected to the clock inputs of both current-state D flip-flops and all registers and counters in the datapath. Finally, an active-low system reset signal *reset_n* can be connected to the clear inputs of both current-state D flip-flops and all registers and counters in the datapath.

**P5.11**    This type of solution requires pseudocode, an RTL program, a datapath, a state diagram, and a control logic design. The pseudocode and RTL program for this problem have already been presented in the solution to Problem P5.3.

Based on the RTL program, the following type of datapath, written in textual form, can be designed. There are two R-S latches named heater_on and ac_on. The R and S inputs to the heater_on latch are labeled R_ho and S_ho (R_ao and S_ao are the corresponding inputs to the ac_on register). There are two seven-bit registers named temp and desired. There is a seven-bit adder/subtracter unit with a carry-in input named cin and an sub control input (addition is performed when sub is 0 and subtraction is performed when this value is 1). The inputs to this adder/subtracter are the output of the desired register and the output of a seven-bit-wide 2:1 MUX. The data inputs to this MUX are the constants THRESHOLD and 0, and its select input is labeled sel. Then there is another seven-bit subtracter unit. The msb of the output of this subtracter, which indicates whether the result is negative or not, is labeled less. The temp register value is automatically updated every few minutes using an external temperature sensor and an analog-to-digital converter. The desired register value is changed when the user changes the temperature setting of the thermostat.

The state diagram, based on the datapath and RTL program described above, consists of nine states labeled S0 through S8. The control signals R_ho and R_ao are asserted in state S0, which is the initial state of this state machine. There is an unconditional transition from S0 to S1. State S1, which corresponds to the condition check of Step 2a in the RTL program, includes just the control signal sub. Then, since all other control signals will be made logic 0 by the control logic circuit (by default), this will result in the desired action of desired - THRESHOLD. There is a transition from S1 to S2 labeled with less. There is a transition from S1 to S3 labeled with less'. The control signals listed in S2 are S_ho and R_ao. State S3, which corresponds to the condition check of Step 2b, includes the control signals sub and cin (for the RTL operation desired - THRESHOLD + 1). There is a transition from S3 to S4 labeled with (heater_on *and* less'). There is another transition from S3 to S5 labeled with (heater_on' *or* less). The only control signal listed in S4 is R_ho (used to turn off the heater). State S5, which corresponds to the condition check of Step 2c, includes the control signal cin (for the RTL operation desired + THRESHOLD + 1). There is a transition from S5 to S6 labeled with less'. There is a second transition from S5 to S7 labeled with less. State S6 includes the control signals S_ao and R_ho. State S7, which corresponds to the condition check of Step 2d, includes the control signal sel (for the RTL operation desired + 0). There is a transition from S7 to S8 labeled with (ac_on *and* less). There is a second transition from S7 to S1 labeled with (ac_on' *or* less'). State S8 includes the control signal R_ao. Finally there are unconditional transitions to S1 from states S2, S4, S6 and S8.

If the one-hot state encoding method is used, then the following control logic design can be derived from the datapath and the state diagram. There are nine D flip-flops with outputs labeled S0 through S8 (corresponding to the nine states in the state diagram). The preset input of the S0 D flip-flop is connected to an active-low reset signal named reset_n. The reset inputs of all other D flip-flops are also connected to this reset_n signal. The system clock signal clk is connected to the clock inputs of all D flip-flops. The constant 0 is connected to the D input of the S0 D flip-flop. The logical combination (S0 *or* S2 *or* S4 *or* S6 *or* S8 *or* (S7 *and* (ac_on' *or* less'))) is connected to the D input of the S1 D flip-flop. (S1 *and* less) is connected to the D input of the S2 D flip-flop. (S1 *and* less') is connected to the D input of the S3 D flip-flop. (S3 *and* heater_on *and* less') is connected to the D input of the S4 D flip-flop. (S3 *and* (heater_on' *or* less)) is connected to the D input of the S5 D flip-flop. (S5 *and* less') is connected to the D input of the S6 D flip-flop. (S5 *and* less) is connected to the D input of the S7 D flip-flop. (S7 *and* ac_on *and* less) is connected to the D input of the S8 D flip-flop. Finally, the following logic equations can be derived for the generation of the control signals.

$$R\_ho = S0 + S4 + S6$$

$$R\_ao = S0 + S2 + S8$$

$$sub = S1 + S3$$

$$S\_ho = S2$$

$$cin = S3 + S5$$

$$S\_ao = S6$$

$$sel = S7$$

**P5.15**   Let us assume a 2 Hz clock signal. The pedestrian crosswalk signal light will be designed such that the red and green lights are turned on for 32 seconds each. The green light will start blinking at a 0.25 Hz rate when there are 16 seconds left, at a 0.5 Hz rate when there are 8 seconds left, and at a 1 Hz rate when there are 4 seconds left. Then the synthesizable Verilog code for this circuit can be written as follows.

```
///////////////////////////////////////////////////////////////////
// MODULE: Pedestrian Crosswalk Signal Control Circuit
// Author: Sunggu Lee
// Date: ...
// Description: Implements a pedestrian crosswalk system that
//    blinks at a faster rate as the remaining crossing time
//    decreases.

module crosswalk (green, red, reset_n, clk);
   output green;   // the green "walk" light
   output red;     // the red "don't walk" light
   input reset_n; // active-low system reset
   input clk;      // a 2 Hz system clock
```

```verilog
reg green, red;
reg [6:0] count;  // counter output

always @(negedge reset_n or posedge clk) begin
  if (~reset_n) begin
    green <= 0;
    red <= 1;
    count <= 0;
  end
  else begin
    casex (count)
      7'b00xxxxx: begin  // stay green for 16 seconds
                    green <= 1;
                    red <= 0;
                  end
      7'b01000xx: green <= 0;  // 0.25 Hz rate for 8 seconds
      7'b01001xx: green <= 1;
      7'b01010xx: green <= 0;
      7'b01011xx: green <= 1;
      7'b011000x: green <= 0;  // 0.5 Hz rate for 4 seconds
      7'b011001x: green <= 1;
      7'b011010x: green <= 0;
      7'b011011x: green <= 1;
      7'b0111000: green <= 0;  // 1 Hz rate for 4 seconds
      7'b0111001: green <= 1;
      7'b0111010: green <= 0;
      7'b0111011: green <= 1;
      7'b0111100: green <= 0;
      7'b0111101: green <= 1;
      7'b0111110: green <= 0;
      7'b0111111: green <= 1;
      default: begin  // stay red for 32 seconds
                 green <= 0;
                 red <= 1;
               end
    endcase
    count <= count + 1;  // increment the counter
  end // of else
end // of always
endmodule
////////////////////////////////////////////////////////////////
```

Since the target circuit is a simple circuit that is correct if it outputs the green and red signals with the proper timing, the test bench for this circuit can be constructed in a simple manner and the output verified by examining the simulation output waveform directly. The logical equivalence of the synthesized circuit with a manually designed circuit can be verified using a method similar to that used for Problem P4.9. The following shows a simple test bench for the crosswalk circuit.

```verilog
///////////////////////////////////////////////////////////////////////
// MODULE: Test Bench for Pedestrian Crosswalk Control Circuit
// Author: Sunggu Lee
// Date: ...
// Description: Tests the "crosswalk" circuit.  To verify the
//    output, simulate for about 2 minutes and examine the
//    simulation output waveforms.

// Timescale and constant definitions
`timescale 10ms/10ms  // 10 millisecond simulation accuracy
`define PERIOD 50      // used to create a 2 Hz clock signal

// Test bench module definition
module tb_crosswalk();
  // SIGNAL DEFINITIONS
  // Signals used to connect to the UUT (unit under test)
  reg reset_n, clk;  // inputs to the UUT
  wire green, red;   // outputs from the UUT

  // Connect to the UUT
  crosswalk uut (green, red, reset_n, clk);

  // generate the reset_n and clk waveforms
  initial begin
    reset_n <= 1;
    clk <= 0;
    #(`PERIOD/4) reset_n <= 0;
    #`PERIOD reset_n <= 1;
  end
  always #(`PERIOD/2) clk <= ~clk;
endmodule
///////////////////////////////////////////////////////////////////////
```

**P5.25**  A test bench for the LCD interface circuit is shown below. The timing diagram for this circuit will show a sequence of commands and data being sent out, with the proper timing, to the LCD interface circuit. For the answer to the timing diagram question, the student should test out his/her code using a Verilog simulation tool and then draw the portions of the simulation waveform corresponding to a few (perhaps 3-4) of these LCD command/data combinations.

```verilog
//////////////////////////////////////////////////////////////////
// MODULE: Test Bench for LCD Interface Circuit
// Author: Sunggu Lee
// Date: ...
// Description: Tests the "lcd3" circuit.

// Timescale and constant definitions
`timescale 1ns/1ns
`define DB_BITS 8    // width of DB data
`define PERIOD 100   // clock period

// Begin module definition
module tb_lcd3 ();
  wire A0, CS1_n, CS2_n, RD_n, WR_n;      // UUT outputs
  wire [`DB_BITS-1:0] DB;                 // UUT data output
  reg RES_n, clk;                         // UUT inputs

  // Connections to the UUT
  lcd3 UUT (A0, CS1_n, CS2_n, RD_n, WR_n, DB, RES_n, clk);

  // Generate input waveforms
  initial begin
    clk = 0;
    RES_n = 0;
    #(`PERIOD * 2) RES_n = 1;
  end

  always #(`PERIOD / 2) clk = ~clk;
endmodule
//////////////////////////////////////////////////////////////////
```

$\bullet \quad \bullet \quad \bullet \quad \bullet \quad \bullet \quad \bullet \quad \bullet \quad \bullet \quad \bullet \quad \bullet \quad \bullet \quad \bullet \quad \bullet$

# B.6 FPGAs and Other Programmable Logic Devices

**P6.1**   Yes. A RAM can also be used to implement arbitrary combinational logic circuits. A RAM *can* be used as a programmable logic device (PLD). The main difficulty with using a RAM in this manner is the fact that it is, in general, a *volatile* memory device, which means that the stored data (corresponding to the circuit implemented) disappears once power is removed from the device. However, in an application where power will be supplied all of the time, a RAM may still be used in this manner. There are also special types of RAMs, referred to as *nonvolatile RAMs*, that can be used for this purpose. A flash memory device is an example of a nonvolatile RAM.

**P6.3**   To design a 4-bit binary counter, the finite state machine method can be used. The state diagram will simply consist of a set of 16 states, with each state corresponding to a count value. A binary assignment of the state variables can be used for the state encoding. The state transition table consists of 16 table entries, with the next state for a given state corresponding to the next count value. To derive the logic equations, K-maps can be used. Finally, the derivation of the control logic can use the PAL18L4 device. Given $A$, $B$, $C$, and $D$ for the current state variables and $NA$, $NB$, $NC$, and $ND$ for the next state variables, the following logic equations can be derived using K-maps based on the state transition table.

$$ND' = D$$

$$NC' = C'D' + CD$$

$$NB' = B'C' + B'D' + BCD$$

$$NA' = A'B' + A'C' + A'D' + ABCD$$

Note that inverted logic equations have been derived (using a circle-the-0's method) in order to use a PAL device with inverted outputs. For the control logic design, there will be four D flip-flops whose outputs will be $A$, $B$, $C$, and $D$. The $\overline{clr}$ inputs of all D flip-flops will be connected to an active-low system reset signal named reset_n. The $clk$ inputs of all D flip-flops will be connected to the system clock signal clk. The D inputs of these four flip-flops, named $NA$, $NB$, $NC$, and $ND$, will be connected to the four outputs of a PAL18L4 device. The $A$, $B$, $C$, and $D$ outputs of these four flip-flops will be connected to the $I_1$, $I_2$, $I_3$, and $I_4$ inputs, respectively, of the PAL18L4 device. Referring to Figure 6.7, instead of the connections shown, the $I_1'$ and $I_2'$ crosspoints will be connected in the first row, the $I_1'$ and $I_3'$ crosspoints will be connected in the second row, the $I_1'$ and $I_4'$ crosspoints will be connected in the third row, and the $I_1$, $I_2$, $I_3$, and $I_4$ crosspoints will be connected in the fourth row. This will result in the $NA$ signal being output as $\overline{Y_1}$. In a similar manner, the seventh, eighth, and ninth rows will have connections corresponding to the logic equation for $NB'$. Likewise, the eleventh and twelfth rows will have connections corresponding to the logic equation for $NC'$. Finally, to produce $ND$, the $\overline{Y_4}$ output will be used and a connection will be made at the $I_4$ crosspoint of the fifteenth row.

**P6.7**   A large EPROM chip can be used to implement an 8-bit multiplier by simply storing the multiplication results as data in the EPROM and connecting two 8-bit input numbers as a 16-bit address to the EPROM. The result of multiplying two 8-bit numbers is a 16-bit product. If the EPROM chip is a 16-bit-wide device with at least 16 address inputs, then one chip will be sufficient. However, memory devices are typically only available up to 8 bits. In such a case, two EPROM chips will have to be used. Thus, for example, if a 1 Mbit (with a 128K $\times$ 8 configuration) EPROM chip is used, then there will be 8 data bits and 17 address bits (since $2^{17} = 128\text{K}$). Two such chips will be required. If the address bits are labeled A16 through A0, then the 8-bit multiplicand can be connected to the A15 through A8 pins of both EPROM chips. The 8-bit multiplier can be connected to the A7 through A0 pins of both EPROM chips. The A16 pin of both chips can simply be tied to ground. Then the 8 data bits of the first and second EPROM chips will be $P_{15} \ldots P_8$ and $P_7 \ldots P_0$, respectively, where the bits $P_{15} P_{14} \ldots P_0$ correspond to the 16 bits of the product. The following shows part of the data stored in the two EPROM chips. All addresses and data are shown in hexadecimal format. The recommended method for generating this type of data is to write a simple high-level program (e.g., a C language program).

| Address(both chips) | EPROM1 data | EPROM2 data |
| --- | --- | --- |
| 0000 | 00 | 00 |
| 0001 | 00 | 00 |
| 0002 | 00 | 00 |
| . . . | | |
| 0101 | 00 | 01 |
| 0102 | 00 | 02 |
| . . . | | |
| 0502 | 00 | 0A |
| 0503 | 00 | 0F |
| 0504 | 00 | 14 |
| . . . | | |
| 0A03 | 01 | 2C |
| 0A04 | 01 | 90 |
| 0A05 | 01 | F4 |
| 0A06 | 02 | 58 |
| . . . | | |

**P6.9** For a combinational multiplier, it is simpler to use a PROM instead of a PLA. With a PROM, we simply have to generate multiplication table data and store it as data in the PROM. With a PLA, we must derive separate logic equations, based on the multiplicand and multiplier bits, for all product bits. However, if a synthesis tool is available to convert a high-level description into these logic equations, then a PLA can be still be used to implement circuits with large numbers of inputs such as multipliers. The main advantage of using a PLA instead of PROM to implement this type of circuit is that the resulting PLA circuit typically is smaller than the corresponding PROM circuit. With a PROM, all *minterms* of the sum-of-products implementation of a circuit must be encoded as data in the PROM. However, with a PLA, the product and sum terms of the sum-of-products expression must be encoded into the PLA. The latter method will clearly require fewer terms, thereby resulting in a smaller overall circuit. Another advantage of PLAs over PROMs is that a PLA can accept a large number of external inputs, while the number of external inputs that a PROM can use is limited by the number of address pins that it uses, which in turn is determined by its data capacity.

**P6.11** An LUT is a small PROM device. It can be used to implement an arbitrary combinational logic circuit in the same manner as a PROM or EPROM device. Its use for the implementation of such circuits is limited by the number of address inputs that it has. Thus, if a 4-input LUT is used, this corresponds to a PROM with 4 address lines, and any combinational logic function of up to 4 variables can be implemented.

**P6.13**  The answer, of course, is a cache. To use a Xilinx Block RAM as a cache, we simply have to connect the appropriate control, address, and data lines to the Block RAM. Referring to the Block RAM shown in Figure 6.12, the system reset signal can be connected to RSTA, an appropriate clock signal (with a period corresponding to the access time of the Block RAM) can be connected to CLKA, a cache enable signal can be connected to ENA, and a cache write enable signal can be connected to WEA. The number of available address lines and data lines depends on the data configuration used. Assuming 16 data lines and the 4096-bit Block RAM shown, there are $log_2 256 = 8$ address lines. Then, if nine such devices are connected in parallel and the cache tag occupies 2 bytes, a 4K-byte cache with a cache line size of 16 bytes can be implemented. Given a 20-bit physical address, a direct cache can be implemented by connecting the high order 8 bits of the address to the ADDRA inputs of the nine Block RAMs in parallel (refer to Chapter 9 for information on cache operation and organization). The next 8 bits of the address, corresponding to the cache tag, are entered into the Block RAM as data entries within the tag portion of the set of nine Block RAMs (the first Block RAM). The cache data is entered into the latter set of eight Block RAMs. The cache data is accessed through the DOA outputs of the Block RAMs. To write new data into the cache, input data is connected to the DIA inputs of the Block RAMs. If bidirectional data lines are to be connected to the Block RAMs, then the bidirectional data lines need to be converted into unidirectional data lines by using a set of tri-state buffers.

**P6.15**  The main difference between *synchronous* and *asynchronous* FPGA configuration is that the former enters configuration data timed to a clock signal, while the latter enters configuration data independently of any clock signal. For asynchronous FPGA configuration, a *handshaking* protocol is required, with a *REQ* signal sent by the programming device every time a new data item is entered, and an *ACK* signal sent by the FPGA device every time it safely receives a new data item. If the data connection is unreliable, a *NACK* signal may be used to indicate unsuccessful data reception, which may necessitate retransmission of the previous data item.

# B.7  Design of a USB Protocol Analyzer

**P7.3**  The sequence of steps that occur during the initialization phase are listed in Section 7.1.2.

**P7.5**  The SYNC pattern in USB is used to permit facilitate phase locking of the receive clock with the transmit clock. The reason that the SYNC pattern is 10000000 is because non-return-to-zero (NRZI) encoding is being used, in which a logic 0 is indicated by a 1-to-0 or 0-to-1 transition and a logic 1 is indicated by the absence of a transition in the data signal. Then, since the lsb is sent first, a SYNC pattern of 10000000 will result in a sequence of seven transitions in the D+ and D- wires followed by a non-transition. This sequence of seven consecutive signals is sufficient to enable the receive clock to become phase-locked with the transmit clock.

**P7.7** USB full-speed mode uses a sequence of two single-ended 0's to indicate an end-of-packet (EOP). The main advantages of this method, over other methods, are that it is simple and short. The main disadvantage is that the (D+, D-) pair of wires no longer constitute a purely differential data signal. Another disadvantage is that this complicates the design of the receive circuitry, since it must now handle single-ended as well as differential signals. With very high-speed communication, single-ended signals cannot be reliably transmitted over potentially noisy channels. Thus, a special packet pattern is typically used for the EOP indicator. For instance, 01111110 is a commonly used EOP pattern.

**P7.9** The timing waveform for an IN packet will be similar to the waveform shown in Figure 7.3. However, since the process ID (PID) for an IN packet is 0001, 0001 will be present in the PID type field and 1110 will be present in the PID check field. The timing waveform for an IN packet corresponds to the simulation waveform created by the Verilog test bench code shown in page 285, which is the second test set in the tb_usb_pa.v Verilog code. This code shows an 8-bit SYNC, an IN PID type field, an IN PID check field, a device and endpoint address of 0x73f, and a CRC code of 10101.

**P7.11** The timing waveform for a DATA1 packet will be similar to the simulation waveform created by the Verilog test bench code shown in page 286, which corresponds to the third test set in the tb_usb_pa.v Verilog code. However, the DATA1 PID code of 10110100 will be used instead of the DATA0 PID code of 00111100. Also, the data sent will be 0x3c5a instead of 0xbf30. The procedure described in Section 7.1.5 can be used to create the 16-bit CRC code for this data field. Since this procedure is tedious to follow in a manual manner, the CRC code can be generated automatically using a simple C language program.

**P7.13** There are many different ways to encode 0 and 1 binary values into a voltage signal sent on a wire. In the return-to-zero (RZ) method, a logic 0 is represented by 0 V and a logic 1 is represented by a high voltage value followed by 0 V. Thus, the line voltage always "returns" to the 0 V level. In the non-return-to-zero (NRZ) method, a logic 0 is represented by 0 V and a logic 1 is represented by a high voltage value (it does *not* return back to the 0 V level). The non-return-to-zero-inverted (NRZI) method is named as such because it adopts a modification of the NRZ method in which a logic 0 is represented by an *inversion* of the previous voltage level while a logic 1 is represented by maintaining the previous voltage level. Note, however, that the opposite convention (a voltage transition for a logic 1 and a non-transition for a logic 0) is also commonly ascribed to the NRZI term as used in computing networking.

**P7.15**   When checking whether a hardware circuit is faulty or not, it is important to check the circuit in a manner such that the likelihood of a faulty circuit passing the check is minimized.  An all-zero result can easily be produced by a faulty circuit.  Thus, one disadvantage of using the noninverted CRC code (as opposed to the inverted CRC code) is that the effective fault coverage is lowered by the fact that there is a higher probability of noise or a faulty circuit producing the correct all-zero CRC residue pattern.

**P7.17**   Figure 2 of the [USB 2002] white paper shows a state machine that can be used to implement a digital phase-locked loop for use with an interface circuit for USB devices.  This state machine consists of a set of 12 states.  The sequence of states visited determines the shape of the USB receive clock.  When it is perfectly synchronized to the USB transmit clock, it sequences through a set of four states.  When the USB receive clock is discovered to be oscillating at a faster or slower rate than the USB transmit clock, the state machine transitions through more or fewer states in order to adjust the phase of the USB receive clock.  The exact states used are explained in the text describing the operation of the state machine in Figure 2 of the [USB 2002] white paper.

# B.8  Design of Fast Arithmetic Units

**P8.1**   A signed-number multiplier for sign-magnitude numbers has the following pros and cons when compared to a 2's complement multiplier.  One advantage is the fact that, for a similar range of positive and negative numbers, the sign-magnitude multiplier effectively multiplies two numbers with fewer bits than a 2's complement multiplier.  Consider numbers in the range $-127$ to $+127$.  For this range of numbers, both the sign-magnitude notation and 2's complement notation require at least 8 bits (although with 8 bits, the 2's complement notation can also include the number $-128$).  Then a 2's complement multiplier needs to multiply two 8-bit numbers.  However, since the sign calculation can be treated separately, a sign-magnitude multiplier only needs to multiply the magnitude portions, which correspond to 7-bit unsigned numbers, of the two 8-bit numbers.  Also, since this is an unsigned multiplication, partial products do not need to be extended to $2n$ bits ($= 16$ if $n = 8$) before being added together to form the final product.  On the other hand, a disadvantage of the sign-magnitude multiplier is the fact that a separate computation is required to handle the signs.  Another disadvantage is the fact that, since signed numbers are typically represented in 2's complement notation in digital logic and computer systems, a conversion process may be necessary in order to convert 2's complement numbers into sign-magnitude numbers before the sign-magnitude multiplication takes place (the product also has to be converted back into a 2's complement format).

**P8.3**   The inverse of a number $y$ can be approximated in the following manner. Let us assume that $y$ is a positive number (if it is not, we can simply take the 2's complement of $y$ before proceeding further). First, $y$ is scaled into another number $A$ between 0.5 and 1.0. This can be done by multiplying $y$ by a power of 2, corresponding to a right or left shift, such that the most significant 1-bit of $y$ appears immediately after the decimal point. More precisely stated, the range of $A$ can include 0.5 but not 1.0. Let $B = 1 - A$. Then, if we compute $1/(1 - B)$, the result can be multiplied by an appropriate power of 2 to produce the desired result $1/y$. Since $0.5 \leq A < 1.0$, it follows that $0.0 < B \leq 0.5$. Also, $0.0 < B^2 \leq 0.25$, $0.0 < B^4 \leq 0.0625$, etc. If we multiply the numerator and denominator of $1/(1 - B)$ by appropriate factors, we can make the denominator $(1 - B^k)$, where $k$ is itself a positive power of 2. If $k$ is even modestly large, this will make the denominator very close to 1. The resulting numerator will then be very close to the desired result $1/A = 1/(1 - B)$. The following sequence of equations shows the factors that have to be multiplied to the numerator and denominator of $1/(1 - B)$ to accomplish this result.

$$\frac{1}{1 - B^2} = \frac{1}{1 - B}\,\frac{1 + B}{1 + B}$$

$$\frac{1}{1 - B^4} = \frac{1}{1 - B^2}\,\frac{1 + B^2}{1 + B^2}$$

$$\frac{1}{1 - B^8} = \frac{1}{1 - B^4}\,\frac{1 + B^4}{1 + B^4}$$

$$\cdots$$

**P8.5**   Figure B.3 shows a Wallace tree, used to add 32 numbers, constructed using several carry-save adders (CSAs) and one carry lookahead adder (CLA).

   The design shown here uses 8 levels of CSAs and 1 CLA level. Let us compute the number of CSA levels required for general numbers of inputs. Let us first consider a binary adder tree, constructed using normal adders that reduce 2 inputs to one output. Given $n$ inputs, such a tree clearly requires $\lceil log_2 n \rceil$ levels of adders (e.g., 8 numbers can be added using $log_2 8 = 3$ adder levels, with 4 adders used in the first level, two adders used in the second level, and 1 adder used in the third level). In an analogous manner, since a CSA reduces 3 inputs to 2 outputs, the number of CSA levels required to reduce $n$ inputs to 1 final output (the sum) is $\lceil log_{3/2} n \rceil = \lceil log_{1.5} n \rceil$. However, the last level has to be a 2-to-1 adder such as a CLA or ripple-carry adder since a CSA cannot produce one final output. Applying this equation to $n = 32$ results in 9, which is equal to the number of levels of adders (including the final CLA) used in Figure B.3.

**P8.7**   Given an $n$-bit modified-Booth-recoded multiplier $X^{**}$, the non-0 bits in $X^{**}$ correspond to the add/subtract operations required to produce the product of the original multiplier $X$ with the multiplicand $Y$. Given $n = 2$, the possible multiplier values are 00, 01, 10, and 11. Each of these multiplier values are equally likely. The corresponding $X^{**}$ values are $00, 01, 1\bar{0}$, and $0\bar{1}$. Since each of these values are equally likely, the average number of add/subtract operations are

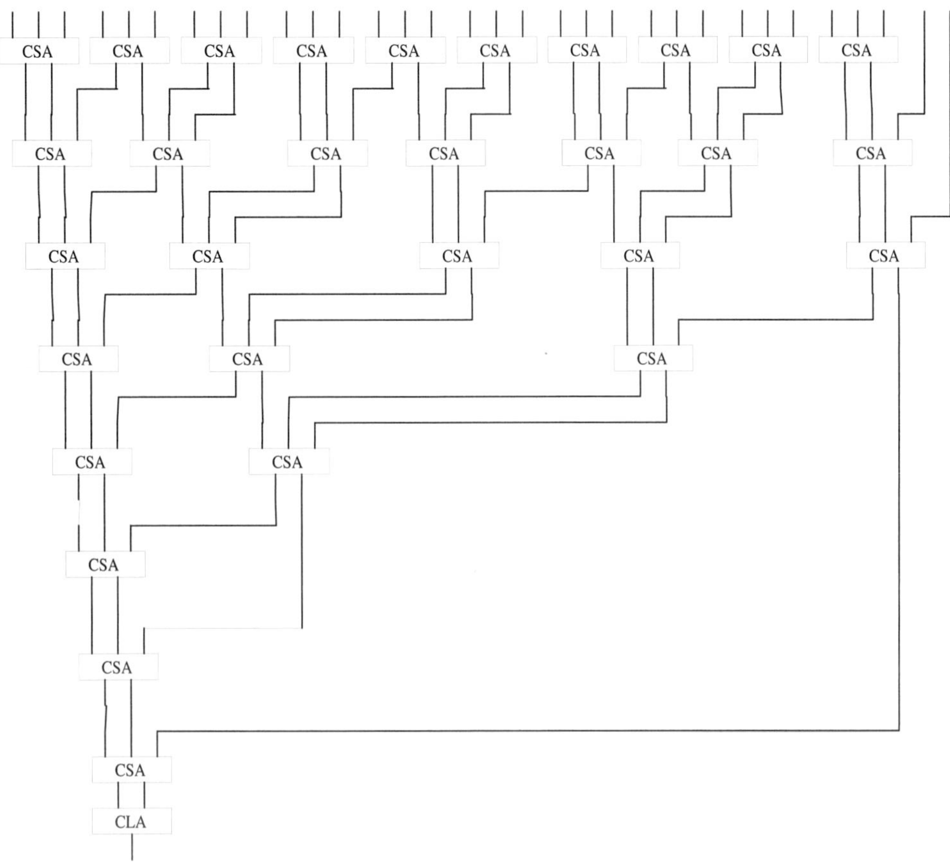

**Figure B.3**   The Wallace tree diagram used in the solution of Problem P8.5.

$(0+1+1+1)/4 = (3\times2)/8$. Then, given $n = 3$, the possible multiplier values are $000, 001, 010, 011, 100, 101, 110,$ and $111$. The corresponding $X^{**}$ values $000, 001,$ $1\bar{1}0, 10\bar{1}, \bar{1}00, \bar{1}01, 0\bar{1}0,$ and $00\bar{1}$. Since each of these values are equally likely, the average number of add/subtract operations are $(0+1+2+2+1+2+1+1)/8$ $= (3 \times 3 + 1)/8$. Thus, the statement given in this problem is shown to hold for the base cases of $n = 2$ and $n = 3$. It also holds for the simpler base case of $n = 1$.

Next, let us make an induction hypothesis for $n = k$ and show that the assertion to be proven hold to for $n = k + 1$. We must consider two cases: $k$ even and $k$ odd. Let us first consider the case when $k$ is even. It is hypothesized that the number of non-0 recoded bits are $3k/8$ when $n = k$. Then, we must show that there are an average of $(3 \times (k + 1) + 1)/8$ add/subtract operations when $n = k+1$. Extending the number of bits from $n = k$ to $n = k+1$ increases the number of possible bit patterns by a factor of 2. If we disregard the "new" msb, then the total number of non-0 bits in all of the possible modified-Booth-recoded is also increased by a factor of 2. This increases both the numerator and denominator of the $3k/8$ count by a factor of 2. Next, referring to the modified-Booth recoding procedure described in this chapter, it is noted that recoded bits are computed for only *even* values of $i$ using Table 8.1, where the $(k+1)$-bit multiplier $X = x_k x_{k-1} \dots x_i \dots x_1 x_0$. Thus, since $k$ is even, a $(k+1)$-bit multiplier requires one additional computation (beyond those used for a $k$-bit multiplier) using Table 8.1. To use Table 8.1, we need an $x_{k+1}$ bit, which can be formed by copying the $x_k$ bit (corresponding to sign extension). This results in the following set of bits to be recoded and the recoded results.

Multiplier Bits     Recoded Bits

| $x_{k+1}$ | $x_k$ | $x_{k-1}$ | $x_{k+1}^{**}$ | $x_k^{**}$ |
|---|---|---|---|---|
| 0 | 0 | 0 | 0 | 0 |
| 0 | 0 | 1 | 0 | 1 |
| 1 | 1 | 0 | 0 | $\bar{1}$ |
| 1 | 1 | 1 | 0 | 0 |

Only these four entries are possible because bit $x_{k+1}$ is simply a sign-extension of bit $k_k$. From this table, it can be seen that there are only two non-0 recoded bits in this set of four new entries. All other patterns have already been counted in the term $(2 \times 3k)/(2 \times 8)$. Thus, the average number of non-0 recoded bits for $n = k + 1$ is given as follows.

$$\frac{2 \times 3k}{2 \times 8} + \frac{2}{4} = \frac{3k + 4}{8} = \frac{3(k + 1) + 1}{8}$$

Next, let us consider the case of $n = k + 1$ with $k$ odd. The induction hypothesis states that, with $n = k$, the average number of non-0 recoded bits is $(3k + 1)/8$. We must show that, with $n = k + 1$, the average number of non-0 recoded bits is $3(k+1)/8$. Extending the number of bits from $n = k$ to $n = k+1$ increases the number of possible bit patterns of the multiplier by a factor of 2. If the msb is disregarded, then there are two sets of identical patterns to the right of the msb. Thus, in the computation of the average number of non-0 recoded bits, both the numerator and denominator of the $(3k+1)/8$ term need be increased by a factor of 2. Recoded bits are computed for only *even* values of $i$ using Table 8.1, where the $(k+1)$-bit multiplier $X = x_k x_{k-1} \dots x_i \dots x_1 x_0$. Then, since $k$ is odd, no further computations are required beyond the computation with $i = k - 1$. However, when $n = k$, the computation with $i = k - 1$ included only values for $x_k$ that were copies of $x_{k-1}$. With $n = k$, four additional $x_k x_{k-1} x_{k-2}$ sequences need to be considered in the computation for $i = k - 1$. The following shows all eight sequences possible with $i = k - 1$.

Multiplier Bits Recoded Bits

| $x_{k+1}$ | $x_k$ | $x_{k-1}$ | $x_{k+1}^{**}$ | $x_k^{**}$ |
|---|---|---|---|---|
| 0 | 0 | 0 | 0 | 0 |
| 0 | 0 | 1 | 0 | 1 |
| 0 | 1 | 0 | 0 | 1 |
| 0 | 1 | 1 | 1 | 0 |
| 1 | 0 | 0 | $\bar{1}$ | 0 |
| 1 | 0 | 1 | 0 | $\bar{1}$ |
| 1 | 1 | 0 | 0 | $\bar{1}$ |
| 1 | 1 | 1 | 0 | 0 |

Of these eight sequences, the middle four rows show the new sequences. The first two and last two rows, contributing a total of two non-0 recoded bits, have already been included into the $((3k+1) \times 2)/(8 \times 2)$ term. Although the middle four rows contribute a total of four non-0 recoded bits, two of these bits have already been included into the $((3k+1) \times 2)/(8 \times 2)$ term since the bits to the right of the msb are identical with the $n = k$ situation. This leaves two new non-0 recoded bits out of the eight sequences for $i = k - 1$. Thus, the average number of non-0 recoded bits with $n = k + 1$ can be computed as follows.

$$\frac{2 \times (3k+1)}{2 \times 8} + \frac{2}{8} = \frac{3k+1+2}{8} = \frac{3(k+1)}{8}$$

**P8.9**   Figure B.4 shows a block diagram for a fast sequential multiplier that uses modified-Booth recoding.

Given two $n$-bit numbers $X$ (the multiplier) and $Y$ (the multiplicand), the sequential multiplier operates as shown below. Let $x_{-1}$ (the bit to the right of the lsb of $X$) be initialized to 0 and $x_n$ (the bit to the left of the msb of $X$) be initialized to $x_{n-1}$. This algorithm is based on the modified-Booth procedure described in this chapter.

```
1.  P ← 0;
2.  for (j = 0; j ≤ (n − 1)/2; j = j + 1)
2a. i = j * 2;
2b. use Table ~8.1 to compute recoded bits;
2c. switch (recoded bits)
        00: A = 0;
        01: A = Y;
        10: A = 2Y;
        0̄1: A = −Y;
        1̄0: A = −2Y;
2d. P = P + A;  (store into most significant half)
2e. X = X >> 2;
2f. P = P >> 2;
    end for;
```

**P8.13** The block diagram for a circuit to compute $sin(z)$ based on the approximation $sin(z) = z - z^3/6 + z^5/120$ can be described as follows. There needs to be a 48-bit by 48-bit multiplier block, a 48-bit adder/subtracter block, and three 48-bit wide 2:1 multiplexers. Even though the input value $z$ is assumed to be a 24-bit value, 48-bit blocks are used because multiplication of two 24-bit values results in a 48-bit product. Although multiplication of this product by another 24-bit (or 48-bit) number results in a 72-bit (or 96-bit) product, this latter product can be rounded off to 48-bits (for simplicity). Then the following shows the pseudocode for a control circuit that can be used to implement the desired circuit.

1. $z2 = z \times z;$
2. $z3 = z2 \times z;$
3. $z5 = z2 \times z3;$
4. $term2 = z3 \times (1/6);$ (use precomputed value for $1/6$)
5. $term3 = z5 \times (1/120);$ (use precomputed value for $1/120$)
6. $sin\_z = z - term2;$
7. $sin\_z = sin\_z + term3;$

**P8.15** The circuit designed for Problem P8.13 must now be converted into an efficient combinational logic circuit. To accomplish this transformation, let us examine the steps of the algorithm for this circuit one by one. Referring to the pseudocode algorithm shown for Problem P8.13, we note that Step 1 involves the multiplication of two 24-bit values. By using modified-Booth encoding, this

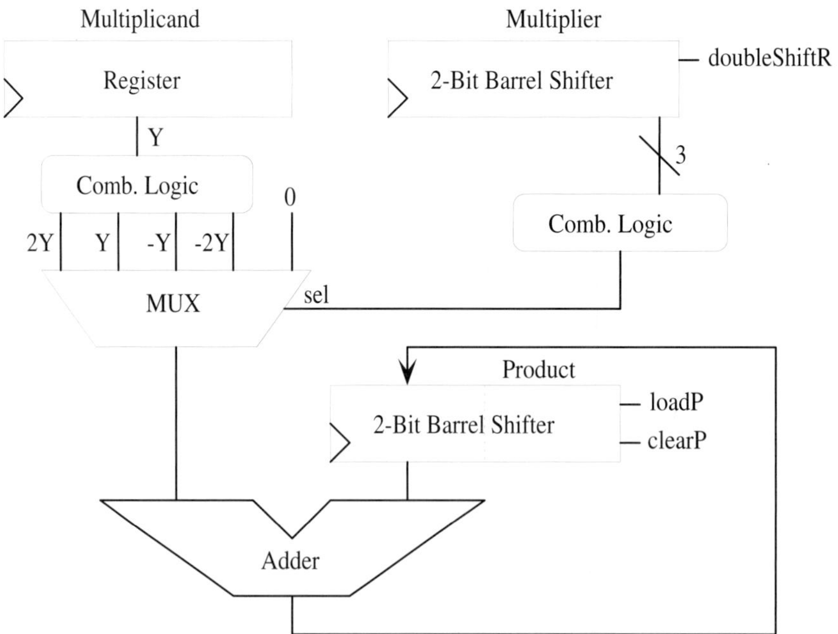


**Figure B.4**  A modified-Booth sequential logic multiplier.

multiplication can be accomplished with the addition of 12 terms. Using the solution presented for Problem P8.5, this can be accomplished with a Wallace tree with $\lceil log_{1.5} 12 \rceil - 1 = 5$ CSA levels and 1 CLA. Step 2 involves the multiplication of a 48-bit number by a 24-bit number. This can again be accomplished by the addition of 12 terms. Step 3 involves the multiplication of two 48-bit numbers, which can be accomplished using 24 additions. This requires a Wallace tree with $\lceil log_{1.5} 24 \rceil - 1 = 7$ CSA levels and 1 CLA. Step 4, which can be executed concurrently with Step 3, can be accomplished efficiently by using the fact that it is multiplying a 48-bit number by a *constant* value. Thus, by examining the bits of the constant value and using modified-Booth recoding, it is possible to accomplish this multiplication using adding or subtracting only those $2^i Y$ terms corresponding to non-0 bits in the modified-Booth recoded version of the constant value. Likewise, Step 5, which needs to follow Step 3, can be accomplished efficiently (with only a few additions) by using modified-Booth recoding and the fact that multiplication is being performed with a constant value. Finally, Steps 6 and 7 can be combined into a single operation involving one CSA block followed by a CLA block. This results in a total of $5 + 5 + 7 + a + 1 = 18 + a \leq 18 + 7 = 25$ CSA levels, where $a$ is the number of CSA levels required for the constant multiplier involving the constant $(1/120)$, and $1 + 1 + 1 + 1 + 1 = 5$ CLA levels. Of course, there is also a little bit of additional logic used for the computation of the modified-Booth-recoded terms at several levels. Since a CSA block has a delay of approximately 2 logic-gate delays and a CLA block (for 48-bit addition) has a delay of approximately 10 logic-gate delays, the total delay through this combinational logic circuit is approximately $25 \times 2 + 10 \times 5 = 100$ logic-gate delays (using an upper bound value for the delays of the constant multipliers and ignoring the delays required for modified-Booth recoding).

**P8.17**   To pipeline the circuit of Problem P8.15, we simply have to insert pipeline registers after every $k$ levels of CSA blocks, where $k$ is the largest number of cascaded CSA blocks that results in a delay of 10 logic-gate delays or less. Since a CSA block has a 2 logic-gate delay, there needs to be a pipeline register after every set of 5 cascaded CSA blocks (or 5 CSA levels). There also needs to be a pipeline register after every CLA block, since a CLA block requires approximately 10 logic-gate delays. Note that if a multiplier has 7 levels of CSA blocks and 1 CLA block, then 3 pipeline registers are required. Considering the sequence of operations required, the total number of pipeline registers required are $1 + 1 + 2 + 2 + 1 = 7$ for the CSA levels and 5 for the CLA levels. By inserting pipeline registers after every 5 or 2 CSA levels and each CLA level, the circuit of Problem P8.15 can be converted into a pipelined circuit with a clock period of 10 logic-gate delays. The delay for the first output is $12 * 10$ logic-gate delays (since there are 12 pipeline stages), after which consecutive outputs can be produced after every 10 logic-gate delays.

# B.9 Design of a Pipelined RISC CPU

**P9.1**    The state machine method requires a pseudocode solution, an RTL program, a datapath design, a state diagram, and a control logic design. Using the flow diagram of Figure 9.2, the following pseudocode solution, which can also serve as an RTL program, can be devised for this problem.

1.  $IR \leftarrow M[PC]$;
    $PC \leftarrow PC + 2$;
2.  case (IR) begin ~//refer to Appendix A for bit fields
2a.  ∩∩ADD'': $Rd \leftarrow Rm + Rn$;
2b.  ∩∩BEQ'': if (EQ) then $PC \leftarrow target$;
2c.  ∩∩EOR'': $Rd \leftarrow Rm$ exclusive-OR $Rd$;
    end case;
3.  Repeat from Step 1;

Based on the RTL program shown above and the bit fields shown in Appendix A, the following datapath can be devised. There is a 16-bit register named $IR$ that is loaded from the data output of an external memory module. There is a 32-bit counter named $PC$ that can be loaded from the output of the $IR$ register (the sign-extended form of twice the value in bits 7 through 0 of the $IR$ register). This counter counts up in increments of 2 when the incPC control signal is asserted. The output of the $PC$ counter is used as the address input to the external memory module. The read control signal for the memory module can be active all of the time. There is a simple combinational logic circuit (a type of decoder) that is used to decode the instruction opcode stored in the $IR$ register. There is a set of 8 32-registers, one of which can be selected for writing using 3 select lines, and another of which can be selected for reading using another set of 3 select lines. Since both the ADD and EOR instructions require reading from two registers and writing to one register at the same time, a 32-bit register named $Rm$ will be used to store a data operand. For the BEQ instruction, there will be a 1-bit register named $Z$ (set by a previous instruction if the ALU result is 0). The ADD and EOR instructions will use a 32-bit ALU that can perform addition or an exclusive-OR operation. Finally, there will be loadIR, loadPC, loadZ, loadRm, loadRd, selRd (select Rd as one input to the ALU - the other is Rm) and selAdd (select addition as the ALU function) control signals attached to their respective datapath components. The loadZ signal is used to load the $Z$ register with 1 if the output of the ALU is 0 (i.e., an ADD instruction results in a zero result, indicating it has added a number with its negative value) and 0 otherwise.

The state diagram for this problem can then be designed in the following manner, where state transitions are from one state to its next consecutively numbered state unless otherwise indicated.

S0:  loadIR, incPC
S1:  loadRm
S2:  if (IR = ∩∩ADD″) selAdd, loadRd, loadZ;
     if ((IR = ∩∩BEQ″) and (Z = 1)) loadPC;
     if (IR = ∩∩EOR″) selRd, loadRd;
     (go to state S0)

Based on the above state diagram and datapath design, the control logic can be designed as a state machine consisting of two D flip-flops and a little bit of combinational logic. Assuming binary state encoding with the state variables A and B (and next state inputs NA and NB), the following logic equations can be derived for the control logic.

$$NB = A'B'$$

$$NA = B$$

$$loadIR = incPC = A'B'$$

$$loadRm = B$$

$$selAdd = loadZ = A(IR = \text{``}ADD\text{''})$$

$$loadPC = A(IR = \text{``}BEQ\text{''})Z$$

$$selRd = A(IR = \text{``}EOR\text{''})$$

$$loadRd = A(IR = \text{``}ADD\text{''}) + A(IR = \text{``}EOR\text{''})$$

**P9.3**  Using the given assumptions and the descriptions in Appendix A, SBC = 0x4194, TST = 0x4213, BL = 0x7800, and BCC = 0xD300.

**P9.5**  In the THUMB architecture, the LR register is used with subroutine calls and returns in the following manner. When a subroutine call is made, the next consecutive address (after the subroutine call instruction) is stored in the LR register before the jump to the subroutine address is made. Then, when a return is required, the PC is loaded with the value in the LR register.

    Most other CPU architectures use the stack for storing the return address required for a subroutine call. Upon encountering a subroutine call instruction, the next consecutive address (after the subroutine call instruction) is pushed onto the stack before the jump to the subroutine address is made. Then, when a return is required, the PC is loaded with the value popped off of the top of the stack. The main advantage of this method, as opposed to the LR register method, is that nested subroutine calls can be implemented using this same mechanism. The main disadvantage is the fact that a memory access operation is required in order to store and load the return address (as opposed to simply writing to and reading from a register).

**P9.7**  The instructions used by a microprocessor with a 1 GHz system clock must be fed by a cache. For an access time of 1 ns, the cache will most likely have to be "on chip"; i.e., on the same chip as the microprocessor. If a single cache module cannot achieve this access time requirement, then there must be several cache modules accessed in an interleaved manner. Also, since data accesses by previous instructions in the instruction pipeline will interfere with instruction accesses during the first instruction pipeline stage, it is preferable to have separate instruction and data caches. For most instructions, the microprocessor should be able to get load its instruction pipeline from the instruction cache. However, since the instruction cache will be limited in size, some instruction fetches will result in cache misses, which will require access to the DRAM main memory. During a cache miss, the instruction pipeline will have to stalled or other work will have to be performed (such as the fetch of the next instruction, provided that it is available in the cache). In order to minimize cold-start cache misses (cache misses during initial system startup), the cache can be filled with a set of consecutive instructions from the default starting address during system startup. Other methods, described in this chapter and in other books on computer architecture, are also available for minimizing cache misses.

**P9.11**    Other types of data hazards include write-after-write (WAW) and write-after-read (WAR). These types of data hazards can occur if instructions can be executed and completed out-of-order in the instruction pipeline. In a WAW hazard, an instruction A writes a data item, which is subsequently overwritten by another instruction that has entered the pipeline earlier than instruction A. In a WAR hazard, an instruction B writes a data item, which is subsequently read by another instruction that has entered the pipeline earlier than instruction A. WAW and WAR hazards occur when earlier instructions are overtaken by later instructions in the pipeline due to pipeline stalls or long execution times. Note, however, that read-after-read (RAR) is *not* a data hazard since exchanging the order of reads between two instructions in the pipeline does not change the semantics of program execution.

**P9.13**    Multiple-register PUSHes and POPs require several accesses to the stack segment portion of main memory. Such operations require long execution times and multiple clock cycles to complete. If a fast parallel multiplier (multiplier implemented in combinational logic) is not available, then a multiply (MUL) will also require multiple clock cycles to execute. Since all other instructions supported in this text's THUMB implementation can be executed in one clock cycle, the addition of multiple-register PUSHes and POPs complicates the design of the EX stage. In addition, when such a multiple-clock-cycle instruction enters the EX stage, all later instructions, including those instructions in the IF and ID stages, must be stalled until the EX stage instruction completes.

By modifying the thumb.v Verilog code, it is nevertheless possible to include support for instructions such as multiple-register PUSHes and POPs. Upon encountering such an instruction in the EX stage, a "stall" signal must be sent to the IF and ID stages. When the IF stage receives this signal, it must temporarily suspend the fetching of new instructions from main memory. In addition, both the IF and ID stages must temporarily suspend the transfer of instructions to later stages. Finally, the code for the EX stage must be modified so that an instruction can be processed over several clock cycles (by using the multiple-clock-cycle algorithm execution method described in Section 3.4.3).

**P9.15**   Section 9.3.2 describes several techniques commonly used to deal with control hazards in commercial microprocessors. The adoption of these techniques, such as delayed branch, branch prediction, and branch target prediction, will result in faster program execution than the method adopted in `thumb.v` ("nullifying" the instructions in the IF and ID stages upon encountering a branch). For the `tb_thumb` test bench, the adoption of these techniques will result in the execution of the test program using a fewer total number of clock cycles. In the original simulation waveform output, the address changes in the sequence 0x000c, 0x000e, 0x0010, 0x0012, 0x0014, 0x0006, 0x0008 ... even though the instruction at address 0x000e is a branch to address 0x0006. This is shown in Figure 9.8. This occurs because the effect of the branch only takes place three clock cycles *after* the branch instruction is fetched. The instructions at addresses 0x0010, 0x0012, and 0x0014 are nullified when the branch instruction at address 0x000e exits the EX stage. If a more efficient control hazard prevention method is used, then the the the addresses (in the simulation of the improved code) will change in the sequence 0x000c, 0x000e, 0x0010, 0x0006, 0x0008, ... (only one wasted instruction at address 0x0010) or in the sequence 0x000c, 0x000e, 0x0006, 0x0008, ... (no wasted instructions).

**P9.17**   In order to equalize the delays in each pipeline stage given the delay assumptions stated in this problem, an 8-stage instruction pipeline (instead of the original 4-stage pipeline) could be used. The stages of this modified pipeline, listed in order, would be IF1, IF2, IF3, ID, EX1, EX2, EX3, and WB. The reason that the IF stage has been expanded into 3 stages is because the IF delay is three times the ID delay. Likewise, the EX stage has been expanded into 3 stages because the LOAD and STORE instructions each require 3D delay. Although the other instructions only require 2D delay, the pipeline must be designed to accommodate *all* instructions. Since data fetches from the instruction and data memories are now spread out over three stages each, memory interleaving (or cache interleaving) must be used. This implies that there must be three instruction memory modules and three data memory modules. Consecutive sequences of three instructions must be stored in separate instruction memories in order to permit concurrent IF operations from three pipeline stages. Likewise for the data memories. Finally, since execution of non-load-store instructions require 2D delay, there must be two copies (or 2-stage pipeline implementations) of all instruction execution modules.

**P9.19**   In order to extend the `tb_thumb` test bench to include data memory read and write tests, we simply have to extend the "test program" currently stored in memory addresses 0x0000 to 0x0017 of the instruction memory modeled in `tb_thumb`. Thus, the instruction memory can be modified starting from memory address 0x0014 (decimal 20) as follows.

```
instruction_memory[20] = 16'h6042;// STR r2,r0,1*4(data:0x24)
instruction_memory[22] = 16'h60c1;// STR r1,r0,3*4(data:0x9)
instruction_memory[24] = 16'h6843;// LDR r3,r0,1*4(load 0x24)
instruction_memory[26] = 16'h68c4;// LDR r4,r0,3*4(load 0x9)
instruction_memory[28] = 16'hdf00;// SWI(to halt the program)
```

The simulation waveform resulting from this modified test bench can be described as follows. Most of the waveform remains unchanged. The simulation waveform is simply extended by three clock cycles during the last part of the simulation waveform. During these three extra clock cycles, the results of the new instructions stored in memory address 22, 24, and 26 are displayed. The instruction at memory address 22 (STR r1,r0,3*4) results in $data = 0x9$. Next, the instruction at memory address 24 (LDR r3,r0,1*4) results in $data = 0x24$. Finally, the instruction at memory address 26 (LDR r4,r0,3*4) results in $data = 0x9$.

# Index